Automating Building Energy Management for Accelerated Building Decarbonization: System Architecture and the Network Layer

IEEE Press
445 Hoes Lane
Piscataway, NJ 08854

About the Author

In 1984, Dr. James Kempf graduated from University of Arizona with a Ph.D. in Systems Engineering and Computer Science and immediately went to work in Silicon Valley. He has worked for a variety of large tech companies, including Hewlett-Packard, Sun Microsystems, NTT Docomo, Ericsson, and Equinix, primarily in research on programming languages, operating systems, networking, blockchain, and cloud computing, and is the author of 30 patents and 3 books on information technology topics. From 2000 through 2006, Dr. Kempf was active in the Internet Engineering Task Force, co-chairing two working groups and serving on the Internet Architecture Board for 3 years. In 2019, Dr. Kempf started consulting to apply his tech background to renewable energy and building decarbonization, and since 2023 Dr. Kempf has served as CTO for Whygrene, a virtual powerplant startup based in Seattle. Dr. Kempf volunteered with the IEEE Blockchain Technology Task Force from 2019 through 2022, and is a Senior Member of IEEE.

Preface

Decarbonization of society's energy use is the central technical and political challenge of the early twenty-first century. Buildings represent approximately 33% of direct and indirect carbon emissions worldwide. The DOE report "Decarbonizing the U.S. Economy by 2050: A National Blueprint for the Building Sector" published in April 2024 states that the goal is to "reduce U.S. building emissions 65% by 2035 and 90% by 2050 vs. 2005 while enabling net-zero emissions economy wide and centering equity and benefits to communities." Since the electrical gird is slowly moving toward full decarbonization through the deployment of wind, water, and solar power, strategies for building decarbonization primarily involve electrifying all building energy uses, including space and hot water heating which today are often fueled by fossil fuels. Strategies for building decarbonization include reducing energy use overall through energy efficiency measures and controlling it to avoid times of grid peak load when grid power generates more carbon emissions. Another strategy involves utilizing carbon-free electricity generated locally by building mounted solar PV panels or small wind turbines, storing the energy in a battery, and switching to the stored renewable electricity during times of higher grid carbons emission intensity. In addition to these measures, the successful widespread acceptance and deployment of building energy management requires a building energy management system to maintain the lifestyle and levels of comfort, health, and safety expected by the occupants.

While there are many books discussing building decarbonization at a high level, the hardware for building electrification (HVAC systems, hot water heaters, etc.), energy efficiency measure, and designing buildings for energy efficiency, very few deal with the details of the information and communication technology needed to manage building energy use. This book focuses on the information and communication technologies for designing and implementing a building energy management system, and specifically, on the network architecture and protocols for communicating between a collection of building sites with controllable appliances and a building energy management system hosted in a public cloud or in an on-premises data center. Chapters 1 and 2 define the problem of building energy management and present a blueprint for its solution, with Chapter 1 presenting a detailed analysis of the building decarbonization problem including energy management strategies and Chapter 2 describing the architecture of a building energy management system. Chapters 3 through 5 deal with standard link layer technologies such as Ethernet and Wi-Fi, the IP stack, and the cybersecurity measures needed to protect the building energy management system from a major security breach. While much of this material can be found in other places, such as standards documents, gathering it together in one place and putting it into the context of a building energy management system should help in understanding how these protocols can be useful in designing, implementing, and managing the communication system to control building energy use.

Chapters 6 through 9 contain material specific to networking protocols and APIs for building energy management. Chapter 6 deals with the BACnet protocol from the American Society of Heating, Refrigerating and Air-Conditioning Engineers (ASHRE), an over 30-year-old network protocol for building automation, and thus energy management. BACnet is primarily deployed in large commercial and government buildings and holds a key place today in any concerted effort to decarbonize larger buildings through controlling energy use. Chapter 7 delves into protocols for residential and small commercial buildings, namely the newer "smart home" Internet of Things (IoT) protocols developed in the early 2000s and deployed by the large smart home platform vendors such as Amazon, Google, Apple, and Samsung. The systems built around these protocols have yet to achieve the deployment volume and impact envisioned when the protocols were introduced. However, their promise remains undiminished and newer, more recently developed protocols such as Matter, which is also discussed in the chapter, may help increase the appeal and deployment momentum of IoT technology in residential and small commercial buildings. Chapter 8 discusses application-level protocols, APIs, and the architectures which they enable, namely the client-server and event-based application architectures, while Chapter 9 presents their application to energy management through manufacturer-supported APIs between the building energy management system and devices such as solar PV inverters. Chapter 9 also discusses the APIs between the utility and the building energy management system and local protocols for controlling solar PV systems and batteries. These topics complete the description of how to design the end-to-end communication links allowing a building energy management system to obtain data from sensors in the remote building site, remain informed about utility conditions, and exercise supervisory control over equipment at the building site computed based on the information obtained from the devices and the utility.

This book is designed to serve as a text for a two-semester upper level undergraduate or graduate engineering course in designing building energy management systems, or as a reference for an engineer or technical building manager designing or overseeing such a system. Originally, my intent in writing it was to cover the full space of information and communication technology for building energy management and decarbonization, but as writing proceeded, the need for maintaining technical depth precluded covering everything in one volume. Consequently, it focuses on defining the problem, presenting an architecture to solve it, and then presenting technical details on a key part of the architecture – the network layer –toward successfully implementing a solution. I hope the result inspires teachers to incorporate technology for building energy management into their curricula, students to learn about it, and practitioners to build the systems necessary to fully decarbonize the built environment by 2050.

Acknowledgments

Special thanks go to Tanya Barham, CEO and founder of Community Energy Labs, and John Powers, CSO (formerly CEO) and co-founder of Elexity (formerly Extensible Energy) for introducing me to the topic building energy control technology and its importance to decarbonizing the economy. I had the good fortune to consult for CEL and Extensible Energy while Tanya and John were in the early stages of building their building energy control startups. Building a startup is hard work, but the success of Tanya's and John's companies is absolutely critical if society is to reach the goal of carbon neutrality by 2050, and I hope they succeed.

I also thank James Dice, founder and CEO of Nexus Labs, a company involved in helping building managers and others understand the technical details of smart building technology, how the technologies play out in products from vendors, and how those products can help automate building management. As with so many relationships during the pandemic, James and I have never met, but I was a regular on his webinars and am a newsletter subscriber. They have been a major inspiration for the work in this book, particularly the generational model of smart building technology architecture in Chapter 2.

Acronyms

** Numbers **

10BASE-T	10 Mbit/sec Ethernet
100BASE-TX	100 Mbit/sec Ethernet
1000BASE-T	1 Gbit/sec Ethernet
3GPP	Third-Generation Partnership Project
4G	Fourth-Generation Cellular Protocol, i.e., LTE
5G	Fifth-Generation Cellular Protocol
6LoWPAN	IPv6 over Low-Power Wireless Personal Area Network

** A **

AAA	Authentication, Authorization, and Accounting
ACK	Acknowledgement
ACL	Access Control List
ADU	Application Data Unit
AES	Advanced Encryption Standard
AIFS	Arbitrary Distributed Inter-Frame Space
AMI	Advanced Metering Infrastructure
AMQP	Advanced Message Queuing Protocol
ANSI	American National Standards Institute
AODV	Ad-hoc On-demand Distance Vector
APDU	Application Protocol Data Units
API	Application Programming Interface
ARP	Address Resolution Protocol

ARPANET	Advanced Research Projects Agency Network
ASCII	American Standard Code for Information Interchange
ASHRAE	American Society of Heating, Refrigeration, and Air-conditioning Engineers
ASN	Absolute Slot Number
ASN.1	Abstract Syntax Notation 1
ATT	Attribute Profile
AH	Authentication Header
AXFR	Absolute Zone Transfer Protocol

**** B ****

BACnet	Building Automation Control network
BACnet/SC	BACnet Secure Connect
BAS	Building Automation System
BBMD	BACnet Broadcast Management Device
BDB	Base Device Behavior
BESS	Battery Energy Storage System
BEMS	Building Energy Management System
BESS	Battery Energy Storage System
BGP	Border Gateway Protocol
BIBB	BACnet Interoperability Building Block
BLE	Bluetooth Low Energy
BOOTP	Bootstrap Protocol
BPDU	Bridge Protocol Data Units
Bps	Bits per second
BSS	Basic Service Set
BSSID	Basic Service Set ID
BTP	Bluetooth Transport Protocol
BTU	British Thermal Units
BVLC	BACnet Virtual Link Control
BVLL	BACnet Virtual Link Layer

**** C ****

CA	Certificate Authority
CAISO	California Independent System Operator
CAT M1	Category M1 LTE wireless protocol

CCMP or CCM*	Counter Mode Cipher Block Chaining Message Authentication Code Protocol
CEC	California Energy Commission
CID	Company ID
CIDR	Classless Interdomain Routing
CIM	Common Information Model
CLI	Command Line Interface
COBS	Consistent Overhead Byte Stuffing
COP	Coefficient of Performance
CORS	Cross-Origin Resource Sharing
COV	Change of Value
CP	PAE Controlled Port state machine
CRL	Certificate Revocation List
CRUD	Create, Read, Update, and Delete
CSA	Connectivity Standards Alliance
CSIP	Common Smart Inverter Profile
CSMA/CA	Carrier Sense Multiple Access with Collision Avoidance
CSMA/CD	Carrier Sense Multiple Access with Collision Detection
CSML	Control System Markup Language
CSV	Comma Separated Value
CTA	Consumer Technology Association
CTE	Constant Tone Extension
CW	contention window

** D **

DAG	Directed Acyclic Graph
DALI	Digital Addressable Lighting Interface
DAO	Destination Advertisement Object
DAQ	Device Automated Qualification
DCF	Distributed Co-ordination Function
DER	Distributed Energy Resource
DERMS	Distributed Energy Resource Management System
DHCP	Dynamic Host Configuration Protocol
DIFS	Distributed Inter-Frame Space
DIO	DODAG Information Object
DODAG	Destination Oriented Acyclic Graph
DOE	Department of Energy
DNP3	Distributed Network Protocol 3

DNS	Domain Name System
DR	Demand Response
DRAS	Demand Response Automation Server
DRLC	Demand Response and Load Control
DSO	Distribution System Operator
DTLS	Datagram Transport Layer Security

** E **

EAP	Extensible Authentication Protocol
EAPOL	Extensible Authentication Protocol over LAN
ECC	Elliptic Curve Cryptosystem
ECDSA	Elliptic Curve Digital Signature Algorithm
ECN	Explicit Congestion Notification
EDCF	Enhanced Distributed Co-ordination Function
EIA	Energy Information Administration
EPA	Environmental Protection Agency
EPRI	Electric Power Research Institute
ESS	Extended Service Set
ESP	Encapsulating Security Header
ETS	Engineering Tool Software
EUI	Extended Unique Identifier
EV	Electric Vehicle
EVSE	Electric Vehicle Supply Equipment

** F **

FDO	Firmware Device Onboard
FIFO	First-In-First-Out
FQDN	Fully Qualified Domain Name
FTP	File Transfer Protocol

** G **

GAP	Generic Access Profile
GATT	Generic Attribute Profile
Gbps	Gigabits Per Second

GEB	Grid-Interactive Efficient Buildings
GHz	Gigahertz
GSS-API	Generic Security Service Application Program Interface
GUI	Graphical User Interface

**** H ****

HAN	Home Area Network
HCI	Host Controller Interface
HLDE	High Level Dialog on Energy
HMAC	Hashed Message Authentication Code
HTML	Hypertext Markup Language
HTTP	Hypertext Transfer Protocol
HTTPs	Hypertext Transfer Protocol secure
HVAC	Heating, Ventilation, and Air Conditioning

**** I ****

IA	Interoperability Area
IAM	Identity and Access Management
IANA	Internet Assigned Numbers Authority
ICANN	Internet Corporation for Assigned Names and Numbers
ICT	Information and Communication Technologies
ID	Identifier
IEC	International Electrotechnical Commission
IETF	Internet Engineering Task Force
I/G	Individual/Group
IKE	Internet Key Exchange
IoT	Internet of Things
IP	Internet Protocol
IPSec	Internet Protocol Security
IPv4	Internet Protocol, version 4
IPv6	Internet Protocol, version 6
ISM	Industrial, Scientific, and Medical
ISO	Independent System Operator
ISP	Internet Service Provider
IT	Information Technology
ITU	International Telecommunication Union

ITU-T	International Telecommunication Union Telecommunication Standardization Committee
IXFR	Incremental Zone Transfer Protocol

** J **

JSON	JavaScript Object Notation
JWK	JSON Web Key
JWKS	JSON Web Key Set
JWT	JSON Web Token

** K **

Kb	Kilobyte
Kbps	Kilobits per second
Kw	Kilowatt
Kwh	Kilowatt-hour

** L **

L2CAP	Logical Link Control and Adaptation Protocol
LAN	Local Area Network
LEED	Leadership in Energy and Environmental Design
LoRa	Long-Range Wireless Protocol
LoRaWAN	Long-Range Wide Area Network
LTE	Long Term Evolution

** M **

MAC	Medium Access Control
MACsec	MAC Layer Security, IEEE 802.1AE
Mb	Megabyte
MBAP	Modbus Application Protocol
Mbps	Megabits per second
MHz	Megahertz
MIDAS	Market Informed Demand Automation Server
MIME	Multimedia Internet Mail Extension

MKA	MACsec Key Agreement
MQTT	Message Queuing Telemetry Transport
MRP	Message Reliability Protocol
mTLS	Mutual Transport Layer Security
MTU	Maximum Transmission Unit

** N **

NAK	Negative Acknowledgement
NAS	Network Access Server
NAT	Network Address Translation
NB-IoT	Narrow Band Internet of Things LTE wireless protocol
NFC	Near Field Connectivity
NIST	National Institute of Standards in Technology
NLC	Network Lighting Control
NOC	Node Operational Credentials
NRDC	Natural Resources Defense Council
NSF	National Science Foundation

** O **

OCSP	Online Certificate Status Protocol
OFDMA	Orthogonal Frequency Division Multiple Access
OIDC	OpenID Connect
OpenADR	Open Automated Demand Response
OSI	Open Systems Interconnect
OSPF	Open Shortest Path First
OT	Operation Technology
OTA	Over the Air
OUI	Organizationally Unique Identifier

** P **

PAC	Port Access Controller
PACP	Port Access Control Protocol
PACH	Periodic Advertising Channel
PAKE	Password Authenticated Key Exchange
PAN	Personal Area Network

PDP	Policy Decision Point
PDU	Protocol Data Unit
PEP	Policy Enforcement Point
PICS	Protocol Implementation Conformance Statement
PIFS	Point Inter-Frame Space
PKCE	Proof Key for Code Exchange
PKI	Public Key Infrastructure
PLC	Power Line Communication
PoE	Power Over Ethernet
PSK	Pre-shared Key
PTP	Point to point
Pub-Sub	Publish-Subscribe
PV	Photovoltaic

** Q **

QoS	Quality of Service
QUIC	Quick UDP Internet Connections

** R **

RADIUS	Remote Authentication Dial-In User Service
RAM	Random Access Memory
RBAC	Role-Based Access Control
REED	Router-Eligible End Device
REST	Representational State Transfer
RFC	Request for Comments
RIB	Routing Information Base
RIR	Regional Internet Registry
RPL	IPv6 Routing Protocol for Low-Power and Lossy Networks
RSA	Rivist-Shamir-Adleman
RSNE	Robust Security Network Element
RSO	Regional System Operator
RTO	Regional Transmission Operator
RTS/CTS	Request-to-Send/Clear-to-Send

** S **

SaaS	Software as a Service
SA	Security Association

SAD	Security Association Database
SAE	Simultaneous Authentication of Equals
SAML	Security Assertion Markup Language
SAN	Subject Alternate Name
SASL	Simple Authentication and Security Layer
SCADA	Supervisory Control and Data Acquisition
SDK	Software Development Kit
SHA1, SHA3, SHA256	Secure Hashing Algorithm 1, 3 and 256
SID	Set Identifier
SIFS	Short Inter-Frame Space
SIM	Subscriber Identity Module
SM	Security Manager
SMCU	Smart Inverter Control Unit
SOAP	Simple Object Access Protocol
SPD	Security Policy Database
SPI	Security Parameters Index
Spiffe	Secure Production Identity Framework for Everyone
Spire	Spiffe Runtime Environment
SSH	Secure Shell
SSID	Service Set ID
SSL	Secure Socket Layer
SSO	Single Sign-On

** T**

TLS	Transport Layer Security
TCP	Transmission Control Protocol
TKIP	Temporal Key Integrity Protocol
TLV	Type-Length-Value
TOU	Time of Use
TSO	Transmission System Operator

** U **

UCM	Universal Communications Module
UDMI	Universal Device Management Interface
UDP	User Datagram Protocol
U/L	Unique/Local
URI	Uniform Resource Identifier
URL	Uniform Resource Locator
URN	Uniform Resource Name

USB	Universal Serial Bus
UTC	Universal Time Coordinated
USGBC	US Green Building Council
UUID	Universally Unique Identifier

** V **

V2B	Vehicle to Building
V2G	Vehicle to Grid
VAV	Variable Air Volume
VEN	Virtual End Node
VLAN	Virtual Local Area Network
VM	Virtual Machine
VOC	Volatile Organic Compound
VTN	Virtual Top Node

** W **

WAN	Wide Area Network
WEP	Wired Equivalent Privacy
WLAN	Wireless Local Area Network
WPA	WiFi Protected Access
WPAN	Wireless Personal Area Network

** X **

xdd	Extended Data Definition
XML	eXtended Markup Language

** Y **

YAML	YAML Ain't a Markup Language

** Z **

ZDD	Zigbee Direct Device
ZVD	Zigbee Virtual Device

1

Introducing Grid-interactive Efficient Buildings (GEBs)

Since the beginning of the twenty-first century, the scope of the global climate crisis has become increasingly clear. Wildfires due to extreme drought, gigantic hurricanes and typhoons, and other weather emergencies have become more common. The alteration in habitat for plants and animals adapted to a cooler climate has resulted in accelerated extinction rates. Every year seems to set a new record for high temperatures due to increasing greenhouse effect heating from the buildup of greenhouse gases, such as carbon dioxide and methane, released by fossil fuel use.

In fits and starts, humanity has slowly come around to recognizing that climate change is a serious problem, as the symptoms become more visible, and that the cause of the problem is the buildup of greenhouse gases in the atmosphere since the Industrial Revolution. Policymakers and politicians in many countries have instituted varying degrees of support for policies to decarbonize the energy industry. The United Nations has developed a global roadmap for decarbonizing the energy industry by 2050 through the high-level dialog on energy (HLDE), a series of meetings between heads of state and government and energy ministers [1]. The dialog establishes a societal goal of having net carbon neutrality for the energy industry by 2050 toward mitigating the climate emergency. The built environment is a major consumer of fossil fuel energy so decarbonizing the built environment is a critical goal for the transition to a decarbonized society. Since residential and commercial buildings make up the bulk of the building stock in industrialized countries, this book is focused on information and communication technology (ICT) solutions for controlling energy use in residential and commercial buildings.[1]

1 In US government surveys, commercial buildings include schools, churches, and government buildings. Industrial buildings account for around 2% of energy use, not counting the energy used for industrial processes themselves. Many of the technologies discussed in this book are applicable to nonindustrial process energy consumption functions of industrial buildings.

Automating Building Energy Management for Accelerated Building Decarbonization: System Architecture and the Network Layer, First Edition. James Kempf.

Published 2025 by John Wiley & Sons, Inc.

In this chapter, we start with an introduction to the scope of the building decarbonization problem and characterize the energy use functions that the operation of buildings represents. We then discuss strategies for mitigating carbon emissions in buildings and how, at a high level, the BEMS contributes to building decarbonization through measurement, reporting, and control of energy use. Next, we present an introduction to how buildings that are highly efficient (i.e. use very little energy to fulfill the same functions compared to peers) and are connected to and can be managed by the grid can serve as a resource for managing the electrical grid more efficiently. In the last section, we discuss such energy efficiency rating systems as the Leadership in Energy and Environmental Design (LEED) and Energy Star and the role they play in building decarbonization.

1.1 Scope of the Building Decarbonization Problem

Buildings in aggregate are a major consumer of energy and therefore a major source of carbon dioxide and other greenhouse gases associated with fossil fuel consumption. Buildings account for one-third of combined direct (from burning of fossil fuels) and indirect (from electricity) energy consumption [2] globally and up to 40% for residential and commercial buildings in the United States [3]. Considering indirect energy alone, buildings consume more than 70% of the United States electricity output [4]. Since two-thirds of the buildings in existence in 2022 will still be around in 2050 [5], policy changes encouraging the construction of new buildings with decarbonized energy supplies will not be enough to fully decarbonize the built environment by 2050. Therefore, decarbonizing existing buildings represents a critical step in the overall societal goal of having a net carbon zero society by 2050.

We can get an idea of the scope of the problem by considering the direct greenhouse gas emissions from the operation of residential and commercial buildings, that is, energy not originating from electrical consumption. As shown in Figure 1.1, residential and commercial buildings in the United States account for 13% of the direct greenhouse gas emissions in 2019 [6]. This is the fourth largest emissions sector, after transportation, electricity (for all uses), and industry. Residential and commercial buildings use fossil energy primarily for space conditioning, hot water production, cooking, and clothes drying. Focusing on space conditioning, in the most recent United States government surveys, 47% of the residential buildings and 50% of the commercial buildings utilized fossil gas for space conditioning. Other fossil fuels (oil and propane) are also used in various parts of the country. The United States has around 100 million single-family homes, 5.2 million multifamily residential buildings, and 5.5 million commercial buildings [7]. If as noted above, two-thirds of these buildings are still standing by 2050, the United States alone will need to somehow

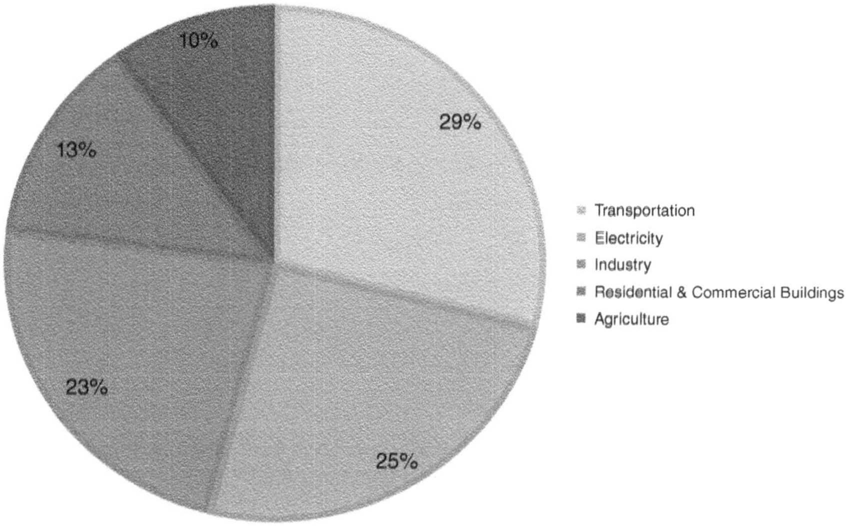

Figure 1.1 Total United States Greenhouse Gas Emissions by Economic Sector 2019. *Source:* Adapted from EPA [6].

retrofit 34.8 million buildings with energy that has no carbon emissions to meet the societal goal of net zero carbon emissions from the built environment. Buildings with fossil-fueled hot water, cooking, and clothes drying appliances will similarly need to be supplied with nonfossil energy.

In contrast with the building sector, decarbonization of the electrical grid has been underway since the early part of the century. In 2023, 21.3% of the United States grid was powered by renewables (including large hydro) [8]. Solar and wind are now the cheapest of any electricity generation technology to deploy and operate, and in 2023, 86% of the newly deployed electrical generation capacity was wind and solar [9]. Figure 1.2 compares the levelized cost of wind and solar vs. nonrenewable generation from 2008 to 2020. Given the cost advantages of renewables coupled with policy incentives from local, regional, and national governments to deploy them, a path toward decarbonization of the electrical grid by 2050 seems clear.

Grid decarbonization will clearly reduce building carbon emissions from indirect energy consumption, but as discussed above, the built environment's direct energy consumption currently comes from fossil fuels. While nonelectrical replacement technologies with no carbon emissions are still under investigation, electrical appliances with the same functionality as fossil-fueled appliances are available today as consumer products. Given the rapid and accelerating pace of grid decarbonization, the most straightforward and least

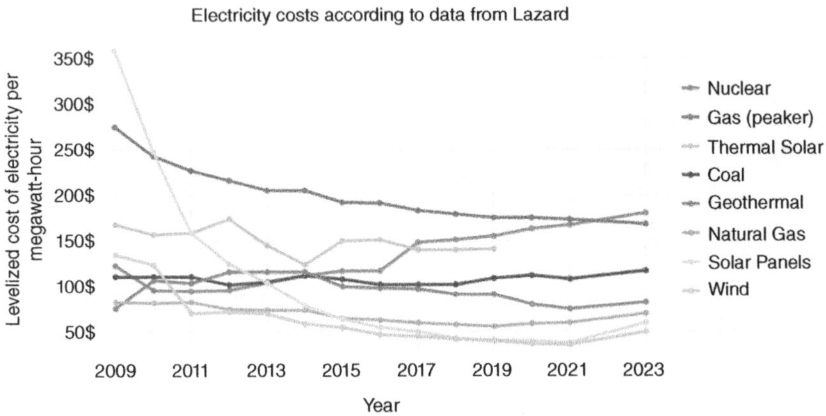

Figure 1.2 Comparative Levelized Cost of Energy per MWh from Different Sources. *Source:* Mir-445511 [10]/Mir-445511/Wikimedia Commons /CC-BY-SA-4.0.

expensive path for building decarbonization is therefore replacing fossil-fueled appliances with electrical appliances, a process sometimes called "beneficial electrification" [11].

1.2 What Are Grid-Interactive Efficient Buildings (GEBs)?

Once full building electrification is in place, a variety of ICTs can be brought to bear on the problem of controlling building energy use and operational carbon emissions. Improving how electricity is consumed and reducing the amount of electricity used through these ICT control measures can significantly reduce energy costs, increase the use of renewably sourced electricity, and contribute to grid stability, thereby improving a building's carbon emissions profile. Coupling these measures with physical changes to reduce overall building energy use like improved insulation and more energy-efficient windows can in some cases make the operating cost

LBNL on Grid-Interactive Efficient Buildings

Lawrence Berkeley National Laboratory has been the leading research institution in the United States working on GEBs and has been a thought leader for many years in the area of building load flexibility. This longer webinar walks through their 2021 report cited in [4].

of powering a fully electrified building lower than powering a building with fossil fuel. Buildings that feature full electrification with physical efficiency improvements, ICT infrastructure for electricity consumption measurement and control, and integrated renewable power sources such as rooftop solar and batteries are called GEBs by the United States Department of Energy (DOE) [12]. Such buildings are grid interactive in the sense that they have the ICT control technologies allowing them to adapt to grid conditions signaled by the utility or distribution system operator (DSO) and are efficient due to the physical efficiency improvements. The ICT technologies involved, often grouped under the general term the "Internet of Things" (IoT), consist of sensors for measuring building and external conditions, control devices for changing the energy consumption patterns in the building in response to measured data from the sensors, networking protocols and hardware for communicating sensor data and control signaling to and from centralized control points and for communicating with the DSO, and software with algorithms for projecting energy use and carbon emissions over a particular time period and calculating the optimal control signaling to reduce them.

1.3 How Do GEBs Advance the Goal of Energy Decarbonization?

The United States DOE has estimated that over the next 20 years, national adoption of GEBs in the currently electrified building stock could be worth $100–$200 billion in savings to the United States electrical power system. The savings are accompanied by reductions in carbon emissions of 80 million tons per year by 2030 or 6% of the total electricity sector carbon emissions [4]. The cost savings come from utilizing energy generated onsite by building integrated distributed energy resources (DERs) such as solar, reducing electricity consumption overall from efficiency measures, and shifting consumption to times when the cost of electricity is lower through load flexibility. Carbon emission reductions come from utilizing onsite generated renewable electricity and shifting consumption away from times when demand for power is high and more fossil-fueled "peaker" plants are generating to times when more renewably generated electricity is available on the grid. These estimates were based on conservative assumptions about grid decarbonization and building electrification, for example, that building owners don't replace their fossil fuel appliances until their end of life. Policy changes and incentives that could accelerate the pace of building electrification were not modeled.

An additional benefit of GEBs is their indirect contribution to societal decarbonization. Energy efficiency and load flexibility free up additional electricity for uses which today are powered by fossil fuels. As space conditioning, hot water production, and other uses are electrified, the load on the grid increases.

Widespread adoption of electric vehicles (EVs) will further increase electricity demand, resulting in the need for more supporting grid infrastructure. By making buildings more efficient, allowing their load to be flexibly scheduled, and using more onsite generated renewable energy, GEBs reduce the need for adding new generation and allow existing generation to be utilized more efficiently. GEBs can thereby help to reduce the need for additional investment in grid infrastructure, facilitating a transition to a more reliable and affordable decarbonized grid.

Finally, GEBs can play a role in grid stabilization as more renewables are integrated into the grid. Renewable generation is inherently more variable than fossil generation due to the variability of the underlying wind and solar energy sources, changing the grid from a system with dispatchable generation to a system with variable generation. On the load side, loads in the past have been mostly fixed. Through load flexibility, GEBs can match load to supply even with non-dispatchable generation. In addition, the incorporation of building-mounted solar and batteries allows GEBs to offer grid services such as resource adequacy. Aggregations of GEBs with onsite solar and batteries can act as "virtual power plants" [13], storing power during the day when solar-generated electricity is plentiful and loads are light, and releasing it in the evening when the sun goes down and loads ramp up as people come home from work and switch on their space conditioning appliances and ovens.

1.4 Characterizing Building Loads in Commercial and Residential Buildings

At a high level, people expect certain services from their appliances. As Amory Lovins, the founder of Rocky Mountain Institute, said: "People don't want raw kilowatt-hours or lumps of coal or barrels of sticky black goo. They want hot showers, cold beer, comfort, mobility, illumination" [14]. In some cases, the services need to be delivered at a particular time, and in others, delivery can be accelerated or deferred to a more favorable time for the grid. People may also have a particular range around which their expectation of satisfactory delivery is structured. For example, people's perception of room comfort is not set at a fixed temperature. It varies throughout the day and between people, typically within a certain range. Yet, most control systems for appliances assume that delivery of their service involves a fixed setting that then consumes a fixed amount of power for a certain amount of time until the need for the service has been satisfied. As a result, the grid was designed in the twentieth century for fixed loads, where the load requires a constant amount of power for the duration over which the device is running.

1.4.1 The Three-Dimensional Load Flexibility Criteria Space

Today many smart appliances' load can be modulated, ramping their power draw up or down depending on control signaling, or the appliance's operation can be prescheduled or deferred. The amount of the service output varies depending on the amount of power provided to the device and the time the device is scheduled to operate. If the service delivered by the device is satisfactory at a lower power draw or a different time, building control systems have an opportunity to institute power and energy control regimes that can substantially reduce a building's energy and power usage as well as the building's indirect carbon emissions.

Building loads can therefore be roughly characterized along two independent flexibility dimensions with respect to the control flexibility of the appliances generating them [15]:

- Fixed vs. Schedulable in Time – Some appliances such as medical equipment need to run at a particular time and cannot be rescheduled, while others, such as heating, ventilation, and air conditioning (HVAC) equipment, can be scheduled to run when power or energy is cheaper or has lower carbon content.
- Constant vs. Modulated Power Draw – Some appliances such as TV sets may need to run at constant power to provide their service, while others such as LED lighting can be varied within a certain range, reducing overall energy usage.

An additional flexibility constraint derives from the level of satisfaction of the building's occupants with the service provided by the appliance [4]:

- Satisfied vs. Dissatisfied – Within a range around an appliance setpoint (temperature, lusmens, etc.), the occupants are satisfied with the service (thermal comfort, brightness, etc.), but outside that range they are dissatisfied.

These three flexibility dimensions – schedulability, level of power modulation, and occupant satisfaction – provide a constraint space within which building energy and power consumption can be optimized.

1.4.2 Types of Residential and Commercial Building Loads and How to Make Them Flexible

Every five years, the United States Energy Information Administration (EIA) performs a half-decadal survey of energy use in residential and commercial buildings. The last such survey for which results have been published was conducted in 2020 for residential buildings (results published in 2023) [16] and 2018 for commercial buildings (results published in 2022) [17]. The pie charts of these surveys for electricity are shown in Figure 1.3 for residential buildings and Figure 1.4 for commercial buildings, respectively.

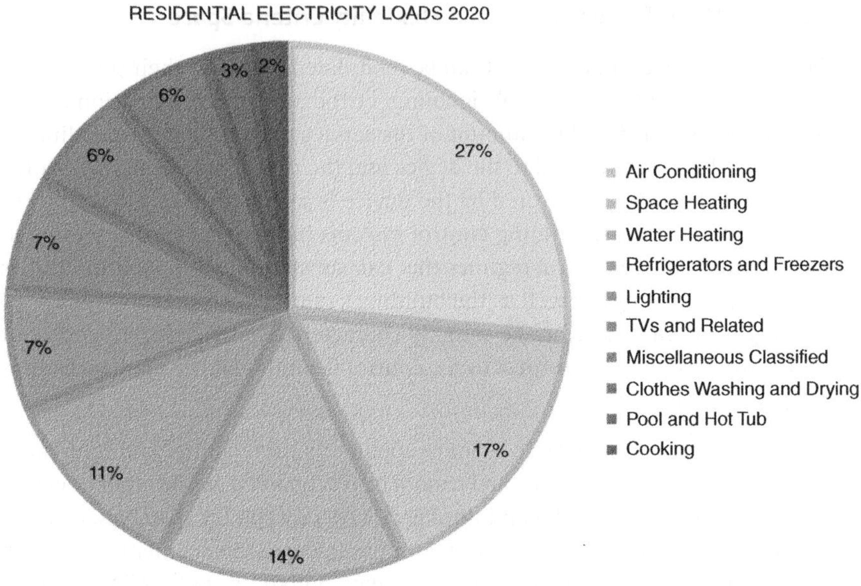

Figure 1.3 Types of United States Residential Electrical Loads 2018. *Source:* EIA [16]/U.S. Energy Information Administration/Public domain.

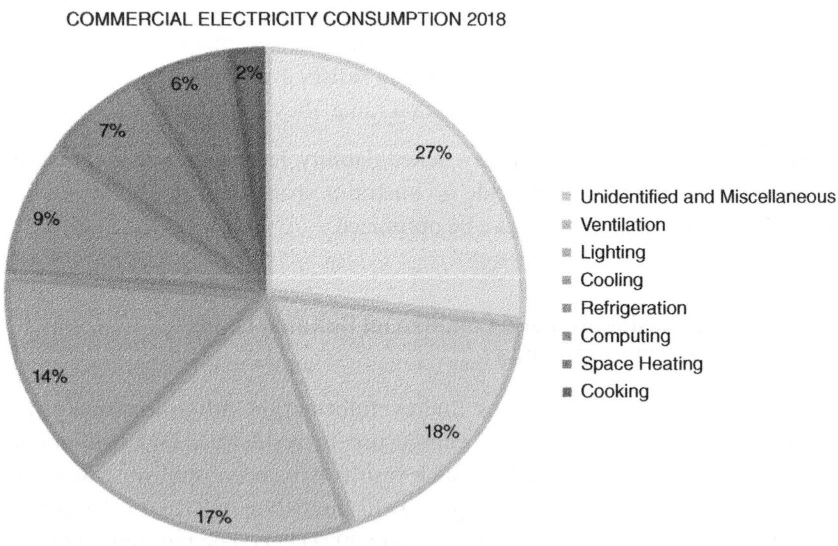

Figure 1.4 Types of United States Commercial Loads 2018. *Source:* EIA [17]/U.S. Energy Information Administration/Public domain.

In the residential survey, HVAC loads (combined heating, cooling, humidity control, and ventilation) consume the most amount of electricity, a combined 44%. The next largest load type for residential buildings is water heating at 14%, followed by refrigerators and freezers at 11%, lighting at 7%, TV and consumer electronics at 7%, clothes washing and drying at 6%, miscellaneous classified loads at 6%, pool and hot tub equipment at 3%, and cooking at 2%. The miscellaneous classified loads cover ceiling fans, dehumidifiers, humidifiers, dishwashers, and EV charging which, individually, are less than 2%.

Table 1.1 classifies these load types within the three-dimensional load flexibility constraint space presented in Section 1.4.1. In some cases, whether the load can be controlled depends on the type of technology used to implement the appliance. For example, HVAC appliances that use variable speed DC motors on heat pumps or resistive heating are modulated but those with fixed speed AC motors for heat pumps are not. HVAC systems are schedulable, modulated (if the technology allows it), and have a service satisfaction range. Building occupants will typically have preferences for thermal comfort over certain ranges, although for multi-occupant spaces, the thermal comfort range needs to be constructed as the intersection of the ranges for all the occupants to ensure that no one is dissatisfied.

Table 1.1 Residential Load Types Classified According to the Three-Dimensional Flexibility Criteria

Load Type	Load Percent (%)	Schedulable	Modulated	Service Satisfaction Range
HVAC	44	Yes	Yes if variable speed motors or resistive	Yes
Hot water	14	Yes	Yes if variable speed motors or resistive	No
Refrigeration and freezing	11	Yes	No	No
Lighting	7	No	Yes	Yes
TVs and consumer electronics	7	No	No	No
Clothes washing and drying	6	Yes	No	No
Pool and hot tub equipment	3	Yes	Yes	No
Cooking	2	No	No	No

Hot water heating is also schedulable and modulated with the same conditions on the technology as for HVAC; however, the service satisfaction range is usually fairly narrow to nonexistent. If the building occupants want a hot shower, the temperature must be high enough, typically 120 °F, in order to provide a comfortable temperature when mixed with cold water to prevent scalding. Since these two load types constitute more than half the residential load and have good controllability characteristics, prospects are excellent for reducing energy consumption and carbon pollution from residential buildings by allowing control systems to schedule and modulate the appliances within the occupants' satisfaction range.

Refrigeration and freezing loads are schedulable as long as the temperature remains below 40 °F for refrigeration and 0 °F for freezing, the temperatures recommended by the United States Food and Drug Administration (FDA) to prevent bacterial growth [18], but they are not modulated and have no service satisfaction range. Lighting systems are not schedulable since light is needed when the indoor space is too dark for the occupants' activity and not at some other time, but they are modulated, in that the brightness can be changed depending on the activity and the occupant's vision. Occupant's expectation of service will vary, with, for example, people having poorer vision needing brighter lights for reading than those with better vision. Clothes washing and drying are schedulable, since the appliances can be put on timers to run when electricity is cheap or has high renewable content, but they have no service satisfaction range since the clothing must be either clean and dry or not. Pool and hot tub equipment are both schedulable and modulated (if the appliances allow it), whereas TVs, other consumer electronics, and cooking appliances are not, since the occupant's expectation is that the appliances will perform their function when turned on and not at some other time. Additional savings can therefore be achieved by including lighting, refrigeration and freezing, clothes drying, and pool and hot tub equipment within the scope of control systems.

In the commercial survey, HVAC loads (heating, cooling, and ventilation combined) were the largest load type at 41%, followed by unidentified and miscellaneous loads at 27%, lighting at 17%, refrigeration at 9%, computing at 7%, and cooking at 2%. In addition, the unidentified and miscellaneous loads contain water heating and office equipment, which were, individually, less than 2%. With the largest category of loads listed as unidentified, classifying the loads within the three-dimensional load flexibility constraint space naturally won't yield a complete picture. However, Table 1.2 presents a classification of the identified load types. As in the residential case, HVAC is the largest load type and has good controllability characteristics, so prospects for achieving savings through control systems are good. Similarly, refrigeration and lighting have modest control characteristics. Lighting in particular can be switched off automatically if no one is occupying the space.

Table 1.2 Commercial Load Types Classified According to the Three-Dimensional Flexibility Criteria

Load Type	Load Percent (%)	Schedulable	Modulated	Service Satisfaction Range
HVAC	41	Yes	Yes if variable speed motors or resistive	Yes
Lighting	17	No	Yes	Yes
Refrigeration and freezing	9	Yes	No	No
Computing	7	No	No	No
Cooking	2	No	No	No

1.5 The Role of the BEMS

A key component in implementing a GEB is the BEMS. The BEMS acts as the control center for collecting data from sensors and appliances and implementing control actions. The BEMS collects information from appliances and sensors throughout the building utilizing network protocols designed for remote monitoring of building energy data. The BEMS may also collect data from external sources on the Internet, for example, from weather websites. It then uses data from local and external sources to project control setpoints for the various energy-generating, storing, and using appliances forward over a time horizon, based on internal algorithms and strategies programmed into it by the building manager or owner. Control actions are then scheduled for execution at various points during the time horizon and conveyed to the appliances using the building monitoring network. The strategies are designed to reduce energy consumption and carbon emissions.

BEMS varies in complexity depending on its implementation architecture. On the low end, a BEMS may run on a small, single-board computer like a Raspberry Pi and only control a few appliances. Or the BEMS may consist of an on-premises gateway that communicates building data to and supervisory control actions from a more sophisticated control platform running as a commercial service in a cloud data center. Many commercial BEMS consist of a complete platform running on a server in the building, monitoring, and controlling all the load-generating devices in the building. In the next chapter, the architectures of these common BEMS deployment patterns will be explored in more detail.

An important distinction needs to be drawn between a BEMS and a "smart home" platform. An example of a smart home platform is Apple HomeKit [19].

Smart home devices often have capabilities that match well with energy management, like smart lighting and smart plugs that can be scheduled through a smartphone app to switch on or off automatically at particular times. However, the goal of a smart home platform is primarily to provide comfort and convenience rather than energy efficiency and decarbonization. For example, the building owner may program the lights and home entertainment center in their home to switch on just prior to their arrival back from work in the afternoon, to give the home a more welcoming feel. Platforms for smart home control typically lack any sophisticated control algorithms for scheduling energy consumption to reduce carbon emissions or to increase efficiency. Some smart home platforms like the Google Nest API [20] may provide remote access to a limited class of energy control devices, such as thermostats enabled with Wi-Fi connectivity. In commercial buildings, automated security platforms that control some appliances like elevators, key-carded doors, and security cameras are similar; in that, their fundamental purpose is to secure the building from unauthorized access and not to enable reduced energy use and decarbonization.

1.6 Strategies for Decarbonization and Cost Reduction

With a BEM in place, loads can be managed to optimize consumption for reduced carbon emissions and cost within the three-dimensional load flexibility constraint space outlined in Section 1.4.1 Depending on the type of load and its flexibility characteristics, different strategies can be implemented. If the building has onsite DERs such as solar panels, batteries, or EV chargers in place, additional strategies are possible.

1.6.1 Strategies Based on Load Flexibility

There are three basic strategies for utilizing load flexibility [21]:

- Deferral of consumption during periods when the grid is under stress like late summer afternoons in warmer climates. For a number of years, utilities have had demand response (DR) programs that compensate customers for reducing demand during grid stress periods, though in some jurisdictions such programs have been limited to commercial and industrial customers that consume large amounts of power.
- Scheduling consumption strategically to achieve particular cost or decarbonization goals. For cost reduction, consumption can be scheduled during a different time period than when energy costs are high or the building is likely to incur a large demand charge. Alternatively, for carbon emission reduction,

consumption can be scheduled for times when the grid has lots of renewably generated energy available and reduced during times when the carbon content of the electricity is higher.

- If an appliance's service can't be deferred or rescheduled, modulating the power draw of the appliance within the occupants' expectation of service to reduce energy consumption during times when energy costs or carbon content of the electricity is higher, and ramping up power draw when energy costs are reduced or more renewably generated power is available.

The second strategy is particularly useful for thermal control appliances such as HVAC and hot water heaters. These devices can utilize the thermal mass of the medium they are heating or cooling – the building for an HVAC system and the water in the tank for a hot water heater – as a "thermal battery." The thermal battery is "charged" by heating or cooling the mass to above or below the desired setpoint when electricity is cheaper or has less carbon content and allowed to discharge – by turning the appliance off – when the opposite is the case. As long as the temperature during the preconditioning period and the subsequent thermal battery "discharge" period remains within the range of the occupants' perception of satisfactory service, such a strategy can be used to reduce a building's indirect carbon emissions and to reduce costs.

Scheduling energy use for times when electricity is cheaper is often easier than scheduling for times when carbon content is lower. While retail electricity tariffs are known by the building manager or owner and can be programmed directly into the BEMS or obtained from the DSO's website, real-time information about the carbon content of electricity may be more difficult to obtain. In some jurisdictions, transmission system operators (TSOs) or independent system operators (ISOs) publish a real-time feed of the carbon content. For example, the California Independent System Operator (CAISO), which serves as the TSO for California, publishes a real-time feed of the carbon content for the electricity in the CAISO wholesale market on its website [22]. If such information is available, the BEMS can combine data from the TSO or DSO on the electricity's carbon content together with data from onsite DERs installed in the building, if any, to minimize a building's indirect carbon emissions. In addition, the BEMS can utilize historical and day-ahead estimates to set up provisional schedules in advance for load flexibility situations like HVAC precooling or preheating or hot water tank preheating.

1.6.2 Strategies Based on Utilizing Generated and Stored Renewable Energy from Onsite DERs

Onsite solar panels, batteries, and electric vehicle supply equipment (EVSE, car chargers) are different types of building-mounted DERs that can be selectively controlled to serve building loads while achieving decarbonization or cost goals.

Strategies for controlling them can be implemented by the BEMS, just as for flexible loads. Carbon emissions reduction can be achieved by strategically scheduling electricity generated from solar panels or stored in batteries or EVs to service different loads for times when electricity from the grid has higher carbon content. The solar-generated electricity can either directly power appliances or charge batteries or EVs which then later discharge to power the building at peak grid stress times when utilities typically bring fossil gas "peaker" plants online.

DERs can also be scheduled to reduce energy costs using the following strategies [21]:

- If the building is on a variable rate, "time of use" tariff, scheduling consumption of electricity generated from solar panels directly or stored in batteries or EVs during times when the DSO energy tariff is in the peak period can reduce energy costs.
- For commercial buildings that have an additional demand charge tariff, scheduling DERs can be used to reduce the building's power draw when the load would result in a demand charge exceeding the cost target, according to projections from the BEMS.

EV charging comes into play as a GEB resource specifically for EVs and EVSE enabled for vehicle to grid (V2G) or vehicle to building (V2B). V2G-capable infrastructure allows EVs to both charge from and discharge to the building's electrical network. If the EVs are charged when building-mounted solar is abundant and discharged to the building when the grid has more carbon-based energy on it, the EV batteries can serve as an adjunct or replacement for an onsite stationary battery. The two cost reduction strategies outlined in the above list can similarly be used to reduce costs. EVs connected to an EVSE without V2G capability can be additionally treated like the appliances described in Section 1.4.1: as a schedulable, modulated load with specific user preferences. User preferences around EV charging are typically that the EV is always charged sufficiently for commuting to work in the morning and commuting home in the evening. The load flexibility strategies described in Section 1.6.1 can then be utilized to optimize EV charging respecting the user preferences.

1.7 Building Energy Impact of Flexibility Strategies

In Figure 1.5, the impact of these flexibility strategies on a building's energy, carbon emissions, and load curves is summarized. The first three strategies – Deferral, Rescheduling, and Modulation – involve controlling load. Deferral moves the operation of smart appliances such as smart dishwashers until after the peak period. Rescheduling moves the operation of appliances that heat or cool such as

Flexibility Strategy	Energy/Carbon Impact	Example	Benefit
Deferral	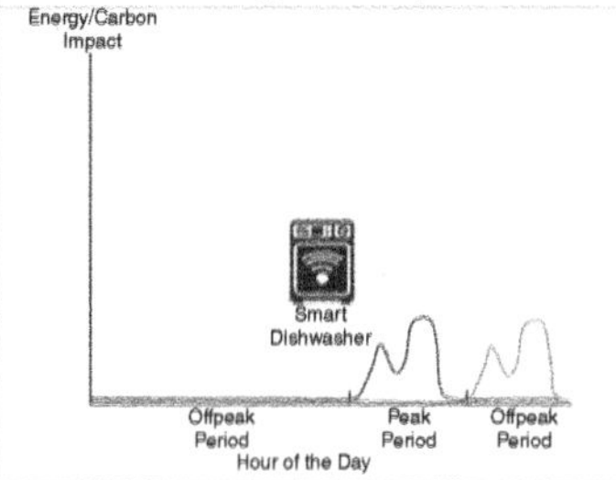	Defer operation of smart appliances such as dishwashers, cloths washers and dryers, etc. to run during offpeak hours.	Reduces energy consumption during peak tariff periods when the price and carbon content of electricity tends to be higher.
Rescheduling	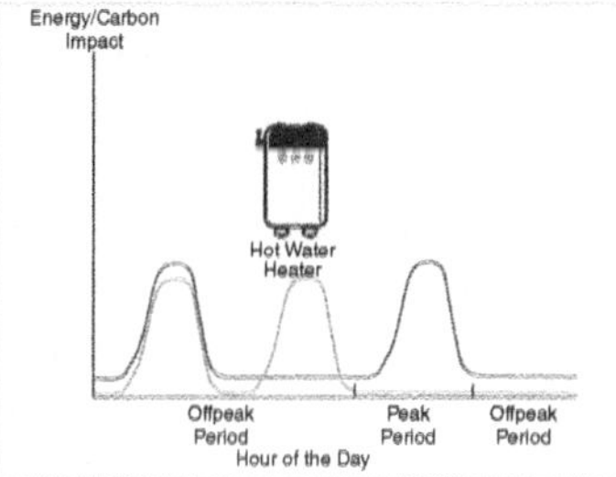	Reschedule operation of smart appliances such as HVAC systems and water heaters that can preheat or precool the a thermal mass to an offpeak period, so that an acceptable temperature is maintained during the peak period.	Reduces energy consumption during peak tariff periods when the price and carbon content of electricity tends to be higher.
Modulation	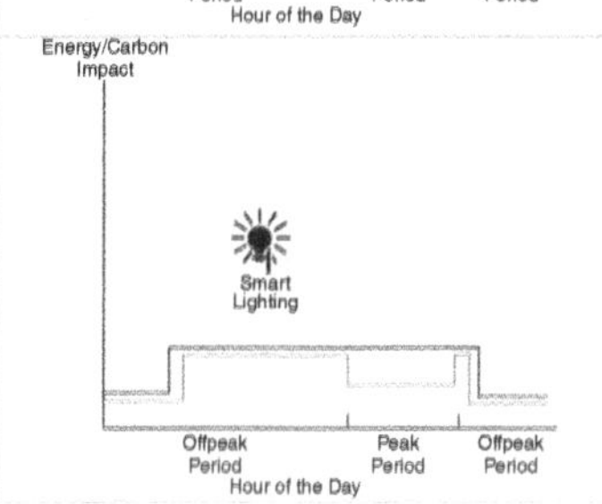	Modulate the energy usage of smart appliances such as lighting that cannot be deferred or rescheduled to use less energy during the peak period.	Reduces energy consumption during peak tariff periods when the price and carbon content of electricity tends to be higher.
Discharge Scheduling for Peak Tariff/Carbon Emission Period	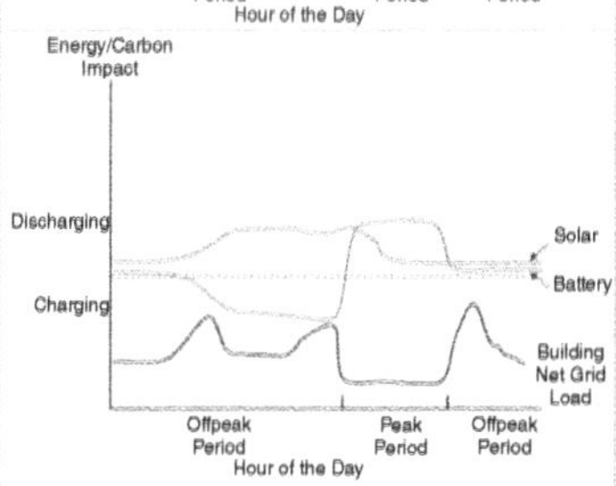	Schedule battery/EV charging for offpeak hours when building installed solar is generating and discharge during peak period when solar generates less and the cost and carbon content of electricity is higher.	Reduces energy consumption during peak tariff periods when the price and carbon content of electricity tends to be higher, utilizes more locally generated energy.
Discharge Scheduling to Reduce Monthly Demand Charge	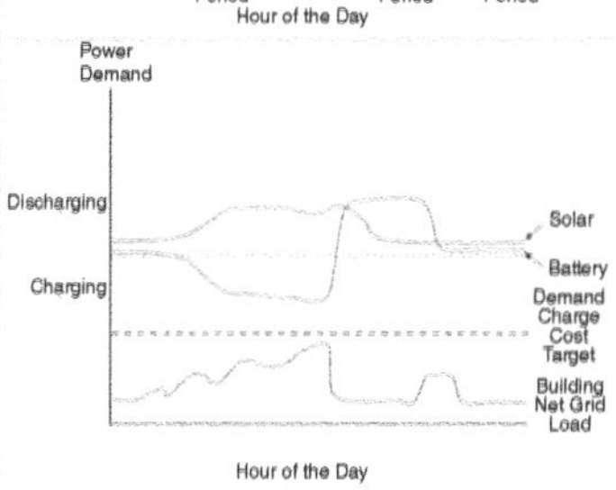	Schedule battery/EV charging for times when the building load is minimal and discharge as the load nears a monthly demand charge cost target load, in order to minimize demand charges.	Reduces building demand charges by using more locally generated power when demand cost nears a preset monthly cost target.

Figure 1.5 GEB Flexibility Strategies and Their Impact on the Building Energy, Carbon, and Load Curves

HVAC systems or hot water heaters, so that they preheat or precool a thermal mass during an off-peak period, then "discharge" during the peak period, using the accumulated heating or cooling capacity to condition the building rather than operate the appliance. Modulation reduces the power draw of smart appliances such as lighting during the peak period without violating the building occupants' perception of acceptable service. The figure assumes that electricity generated during the peak tariff period also has higher carbon content, but if that is not the case, then the strategies can be deployed during times when grid electricity carbon content is higher.

The last three strategies involve onsite DERs. The Energy/Carbon Impact column assumes the building has both onsite solar and some kind of storage device, a battery or an EV with a V2G-capable EVSE. The Discharge Scheduling for Peak Tariff/Carbon Period row illustrates how reductions in peak period energy and indirect carbon emissions are accomplished by charging the storage device during the off-peak period from the solar array and discharging it during the peak period, reducing the building's net load on the grid. In the Discharge Scheduling to Reduce Monthly Demand Charge row, the building manager sets a monthly demand to limit the demand charge in the BEMS, calculated based on the capacity of the solar and storage DERs and the typical load profile of the building. When the building load is away from that limit, the solar array charges the storage, but as the load approaches the limit, the solar-generated electricity is switched to servicing the building load, and the storage begins to discharge to match the load, in order to keep the load under the preset monthly demand. Additional strategies utilizing storage that also allow the building to act as a grid resource are discussed in the next section.

1.8 GEBs as a Grid Resource

The infamous "duck curve", shown in Figure 1.6, is a textbook example of what happens to a grid designed for a dispatchable fossil-fueled generation when utility scale, variable renewable generation is introduced to decarbonize without simultaneously deploying technology to compensate for renewables' non-dispatchability. The figure shows graphs of the lowest net load day each spring March to May from 2013 to 2023 in gigawatts on the CAISO transmission network [23]. The midday net load decreases every year, with a steep drop in net load during the morning when the sun moves higher and a steep ramp in net load when the sun sets in the evening. By 2023, the net load is slightly below zero between 12 PM and 2 PM, indicating that utility scale solar needs to be curtailed.

The name comes from the shape, reminiscent of a duck. This shape is due to a large increase in solar generation during the morning, but no load to utilize it in the afternoon, since temperatures in California's Central Valley are not high

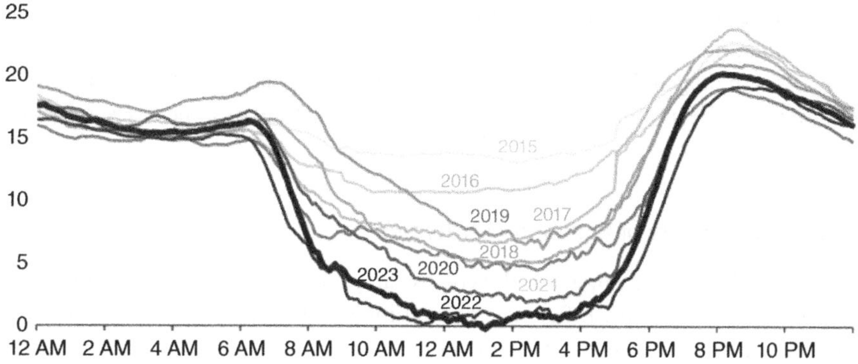

Figure 1.6 The Duck Curve. *Source:* EIA [23]/U.S. Energy Information Administration/ Public domain.

enough in the spring for air conditioners to come on. Without measures to mitigate the variability, the impact on the wholesale market is that real-time wholesale prices of electricity occasionally go negative in the spring during the afternoon, and CAISO must request that solar generation be curtailed. Conversely, during the evening, real-time wholesale prices increase substantially as people come home from work, the load ramps up, and solar generation goes offline as the sun sets.

GEBs offer a decentralized approach to mitigating the variable nature of renewables. Onsite batteries and EVSE with attached EVs can ramp up their load in the morning and afternoon, drawing renewably generated power with low carbon content from the grid. Then in the late afternoon and evening, the batteries and the EVSE (if they are V2G capable) can discharge to the grid to help service the load in the duck's neck and head. This strategy differs from that discussed in Section 1.6.2, in that no onsite solar panels are needed to utilize it. If the building has solar panels, the solar panels can charge the battery and EVs during the morning and afternoon, and the building loads can be serviced from the grid. Then in

More on Virtual Power Plants

This short video from the Energy Research Institute at the Nanyang Technological University (NTU) in Singapore provides a more detailed introduction to virtual power plants and their benefits to grid customers, renewable energy asset owners, the utility, and other stakeholders in the energy ecosystem while reducing carbon emissions by reducing fossil fuel-based generation.

the evening, the BEMS can switch the building loads to draw from the battery and EVs, mitigating the impact of the building on the grid net load curve. If a collection of GEBs is managed this way through a virtual power plant aggregator, substantial resources can be brought to bear toward grid stabilization, in some cases, upwards of 550 MW [24].

1.8.1 Grid Impact of Flexibility Strategies

In Figure 1.7 [25], the impact of the GEB strategies on the grid load curve is illustrated. The first strategy, Efficiency, doesn't involve any active control; it involves passive measures taken to reduce overall energy use, such as reinsulating the building walls with high-efficiency insulation, replacing the windows with high-performance windows, and replacing HVAC appliances with high-performance equipment. For heating applications, high-performance HVAC equipment achieves a coefficient of performance (COP, ratio of heat energy output in Kwh to electrical energy input in Kwh) of 3 or higher, while for cooling applications, the equipment achieves an energy efficiency ratio (EER, ratio of cooling energy output in BTUs to electrical energy input in Kwh) of 10 or higher [26]. The overall impact of efficiency measures is to reduce the building load on the grid by reducing the energy needed to maintain building services within the occupants' satisfaction range, since less energy is lost to the environment and less electricity is needed to deliver satisfactory service.

The rest of the strategies in Figure 1.7 involve active control of GEB systems using the control strategies discussed in Section 1.6. The Reduce Load strategy reduces overall grid load by utilizing the ability of appliances that allow modulated power draw, such as smart lighting systems, to reduce energy use when the grid load curve is peaking. The Schedule Load strategy uses the ability of appliances that support Deferral, such as smart dishwashers, and Schedulability, such as smart hot water heaters, to move their energy use away from peak periods to periods when the overall grid load is less. Note that these strategies assume that the service delivered by the appliance in question is maintained within the occupants' range of satisfactory performance.

The Ancillary Services and Generate strategies specifically address grid stability issues. Ancillary services are services other than energy delivery that are technically required to ensure grid performance and stability. Examples are maintaining the frequency of the AC voltage and current at 60 Hz in the United States and 50 Hz in Europe and elsewhere, or maintaining the power factor [27] close to 1. These measures are necessary to ensure the grid is functioning optimally, as deviations from the frequency for which appliances are designed can damage motors and deviation of the power factor below 1 can cause the grid to operate less efficiently, requiring more energy to service loads than if the power factor was

Control Strategy	Load Impact	Example	Benefit
Efficiency	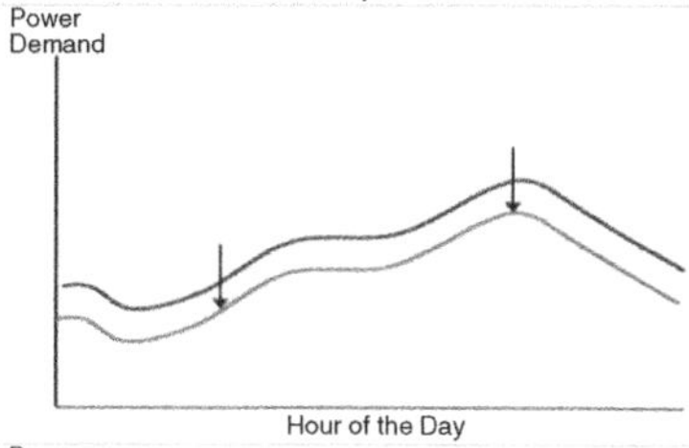	Building walls are tightly insulated and windows are high efficiency. HVAC system is high efficiency (COP of 3 or higher or an EER of 10 or higher).	Reduction in electricity cost and for HVAC, reduction in carbon emissions from gas "peaker" plants during times of grid stress. If the building heating uses fossil gas directly, direct reduction carbon emissions.
Reduce Load	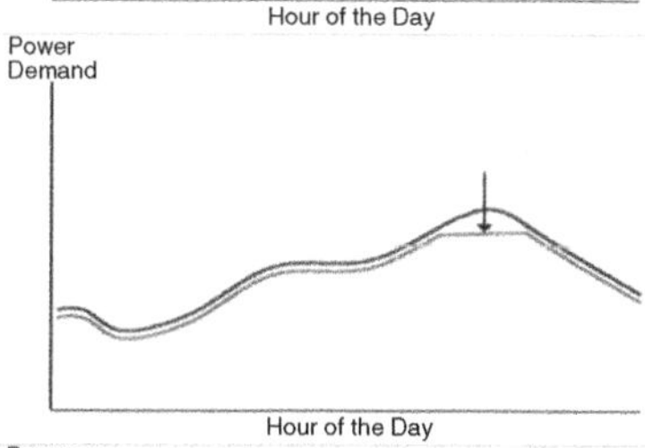	Building lights dimmed in response to a demand reduction signal from DSO while maintaining illumination within the occupants' preferred service level.	Reduction in generation and transmission equipment investment for the DSO, reduction in carbon emissions from gas "peaker" plants during times of grid stress.
Schedule Load	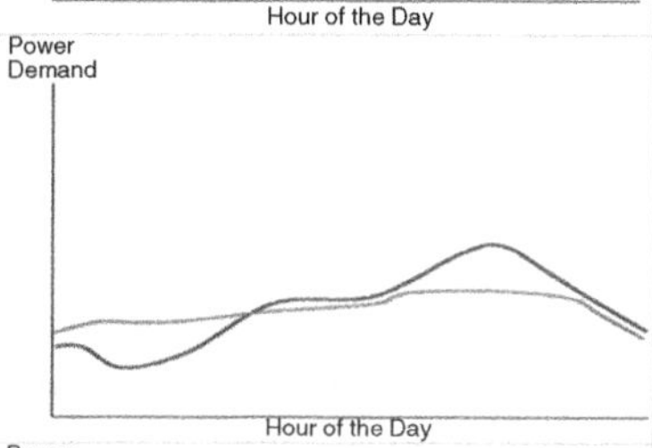	Water heater preheats water during off-peak periods in response to grid signals.	Reduction in energy costs and carbon emissions due to shifting energy consumption to a cheaper time period where carbon content of electricity is lower. Avoid need to curtail renewable generation.
Ancillary Services	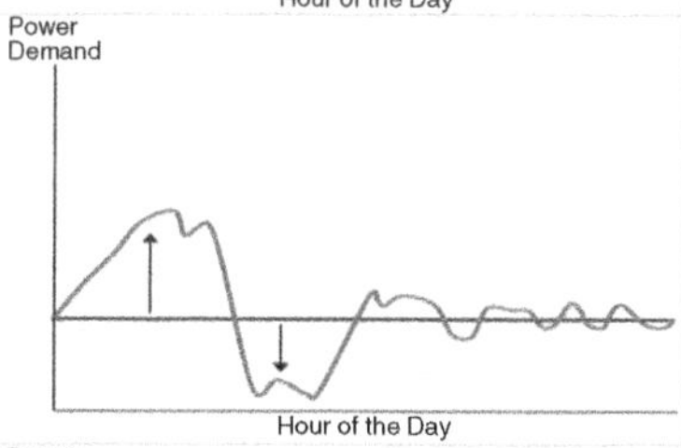	Batteries and V2G EV chargers modulate real and reactive power draw to control grid power factor or frequency.	Reduction in ancillary costs, improvement of distribution network efficiency through better power factor, improvement in integration of variable renewable generation resources.
Generate	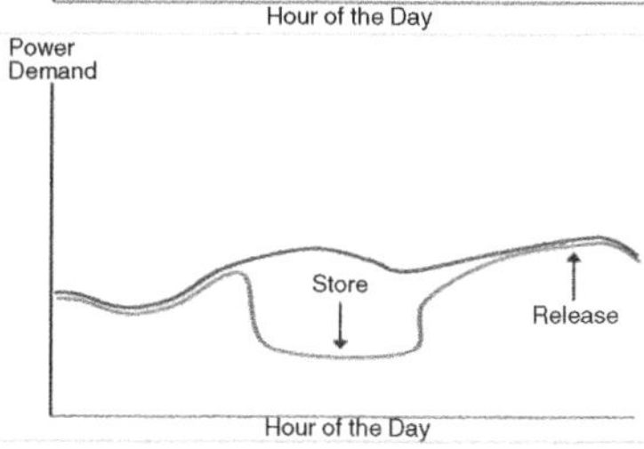	Building installed solar generated electricity stored during afternoon hours and releasd to grid or consumend during evening.	Reduction in need for grid infrastructure to handle variable generation renewable resources, improved building carbon emissions from consumption during peak period, reduced transmission and distribution losses through local consumption.

Figure 1.7 GEB Management Strategies and Their Impact on the Grid Load Curve. *Source:* Adapted from Satchwell et al. [25].

optimal. The Generate strategy involves shifting battery and EV loads to utilize more renewable energy during the day to charge and releasing it to the grid in the late afternoon when the net load begins to ramp up, or using building-mounted solar with EVs and batteries in a similar fashion, as discussed above. Building occupants are unlikely to experience any service degradation due to these strategies, but they need to be implemented in a way that either makes their net cost impact neutral or results in savings on building's energy and demand charge bills.

1.9 Building Energy Efficiency Standards

Starting with the global energy crisis of 1974–1980, national and regional governments in developed countries initiated programs to define energy efficiency standards for buildings. These standards and rating systems were repurposed in the 2010s as the scope of the climate crisis became clear and the need for reducing carbon-based energy use became urgent. The approach differed depending on the country. In Europe and Japan, energy efficiency standards for buildings were developed on a national level. In the United States, regulation of the built environment was a long-standing function of state and local governments, so energy efficiency standards for the building envelope and property were left up to them with the federal government providing guidance and model building codes for emulation. Industry groups and nonprofits were instrumental in developing rating systems for energy-efficient buildings as well. Appliance energy efficiency standards, on the other hand, were judged to be a federal responsibility, and a set of voluntary efficiency standards (Energy Star) for appliances was developed at the federal level. In some cases, the standards include additional sustainability goals such as the use of recycled materials in constructing the building and water use efficiency for landscaping. As examples of building rating systems and appliance standards, the sections below describe the nonprofit US Green Building Council's (USGBC's) LEED rating system [28], which has been widely adopted in the United States, Canada, and elsewhere, and the US Energy Star [29] energy efficiency standard for appliances, with specific focus on the building decarbonization provisions of both standards.

1.9.1 Leadership in Energy and Environmental Design

LEED development started in 1993 led by the Natural Resources Defense Council (NRDC), an environmental organization in the United States and supported by USGBC. LEED expanded between 1994 and 2015 from a single standard covering new construction to a whole series of interrelated standards covering the design and construction of buildings, maintenance, remodeling, and operation, and additionally, standards for landscaping, neighborhoods, and community involvement in the planning process [30]. LEED certifications are primarily oriented toward the commercial building sector.

The overall goals of LEED are to foster a built environment with the following characteristics [28]:

- Reduces buildings' contributions to climate change;
- Provides more healthy spaces for people to work, live, and play in;
- Helps to protect water resources;
- Enhances biodiversity and ecosystems;
- Promotes sustainable "cradle-to-cradle" cycles for materials;
- Enhances the quality of life for people who live in the community.

USGBC provides LEED certification based on the number of LEED points the building project earns. There are four levels of certification depending on the number of points: Certified for 40–49 points, Silver for 50–59 points, Gold for 60–79 points, and Platinum for more than 80 points. To earn points, a building must meet certain criteria in different categories.

With respect to building decarbonization, LEED v4.1 [31], the most recent release of LEED, was developed with the majority of credits related to reducing embodied carbon in the building during construction and operational carbon emissions [32]. Points in the Energy and Atmosphere category are awarded for building operations that increase building energy efficiency by reducing overall energy use, and that reduce carbon emissions through using energy from low-carbon energy sources. Potent non-carbon greenhouse gases used in refrigerants are also targeted by LEED v4.1. Renewable energy selection using various types of contracts such as corporate power purchase agreements is awarded points in the Energy and Atmosphere category, with priority going to contracts that encourage more utility development of renewable power rather than utilizing existing utility renewables.

Finally, the LEED v4.1 Building Design and Construction category awards points in the Optimize Energy Performance criterion, which has two metrics: energy cost and GHG reduction. The rating system doesn't specify particular technologies, but rather points are awarded for a percent reduction below the American Society of Heating, Refrigeration, and Air-conditioning Engineers (ASHRAE) 2016 standard. For example, 1 point is awarded for a 15% reduction in lighting power consumption. Up to 1 point is awarded if 75% of the rated power of appliances comes from appliances with an Energy Star rating.

1.9.2 Energy Star

The Energy Star program was started in 1992 by the United States federal Environmental Protection Agency (EPA) and is jointly run by the EPA and DOE. The EPA estimates that over the lifetime of the program, Energy Star has saved five trillion kilowatt-hours of electricity and reduced greenhouse gas emissions by four billion metric tons as well as helping United States consumers and businesses avoid more than $450 billion in energy costs [33]. Parts of the Energy Star program have also been adopted by Taiwan, Switzerland, and Canada, and in the past, the EPA has had agreements regarding Energy Star with other countries and the EU, that have since expired [34]. In addition to appliance efficiency standards, Energy Star also issues performance ratings for buildings, including residential buildings.

The Energy Star program has 75 different certified product categories, including for home appliances such as HVAC systems, office equipment such as computers and photocopiers, consumer electronics such as TVs, and residential building products such as high-efficiency windows. The procedure for setting the specifications differs depending on the product category and is determined by the EPA. Product certification must be done by an outside EPA-recognized laboratory and certified by an EPA-recognized certification agency, which are subject to oversight by an accreditation body, like ASHRE. A certain percent of the products are retested annually to be sure that they continue to meet their original specifications. Once a product has earned an Energy Star certification, it is allowed to display a distinctive Energy Star label on the packaging. The Energy Star building performance standards have also been adopted as mandatory reporting requirements into the building codes of some US cities, such as San Francisco, Seattle, and New York [34].

1.10 Summary

World energy use urgently needs to be decarbonized by 2050 in order to mitigate the causes of climate change. Decarbonizing the built environment is an essential part of the transition to a decarbonized energy system, since building operations are the fourth largest source of direct carbon emissions in developed countries. They are also responsible for indirect emissions from their electricity consumption since building operations consume around 70% of the generated electricity. While reducing the embedded carbon generated by constructing new buildings, including reducing the embedded carbon from materials and processes used to construct them, is important, the vast majority of buildings in place today will still be standing in 2050. Therefore, building decarbonization is primarily a retrofit problem, involving upgrading existing buildings with new technology to improve their energy efficiency and reduce their direct and indirect carbon emissions.

GEBs is a framework developed by the US DOE to achieve building decarbonization through deploying a combination of efficiency and control technologies.

A GEB is a building that has been optimized for energy efficiency and in which IoT-based information and control technologies have been deployed to reduce energy use and carbon emissions. Passive energy efficiency upgrades, such as installing improved insulation and high-performance windows, are applicable both to buildings with direct carbon emissions from fossil fuel use and to fully electrified buildings. But IoT technologies are primarily applicable to buildings that are fully electrified. GEBs are interactive in that their control systems can receive signaling from the building's DSO to reduce energy use during times of grid stress. In addition, their control systems allow buildings to apply control strategies to appliances for load flexibility and onsite utilization of energy generated and stored by solar and storage DERs.

Appliances have three different dimensions of constraints on load flexibility: fixed vs. schedulable loads, constant vs. modulated power draw, and occupant preferences regarding the level of service. Appliances that allow schedulable loads can have their operation either deferred or rescheduled to a time that would optimize carbon emissions or cost. Deferral involves postponement of consumption to a time outside a peak tariff period or peak carbon emissions period, for example, by scheduling a dishwasher to run later in the evening after the peak period is over. Rescheduling involves utilizing the building structure or water as a thermal mass to store energy for utilization later, for example, pre-heating a hot water tank above the desired setpoint prior to the start of the peak period, then turning off the hot water heater and allowing the thermal mass to provide water at the proper temperature during the peak period. Modulating power draw involves reducing power consumption by the appliance during the peak period, for example, reducing the lighting level during the peak period for a smart lighting system. These load flexibility strategies need to be coupled with an understanding of user preferences for the service the appliance is providing, to ensure that the service stays within the satisfactory range.

Similar strategies apply to how a GEB can utilize onsite solar and storage resources. Reductions in carbon emissions and cost can be achieved by serving the building loads using energy generated by solar and stored in batteries or EVs during peak periods and allowing building loads to draw from the grid during off-peak periods. For commercial buildings under a demand charge tariff, solar and storage energy can be scheduled to serve building loads to avoid having the overall load on the grid exceed a preset monthly demand charge maximum. EVs come into play as onsite DERs when the EVSE is bidirectional, having V2G functionality, thereby allowing the EV's battery to discharge as well as charge. Without a V2G capable EVSE, EV charging can be scheduled according to the flexible load strategies.

The BEMS serves a central role in implementing GEB flexibility strategies. The BEMS collects data from various sensors and appliances throughout the building using various network protocols and combines it with data obtained from various real-time Internet feeds, then calculates an optimized energy consumption profile for a projected time period. The BEMS uses various algorithms to calculate the energy consumption profile, guided by strategies preprogrammed by the building manager or occupants. Control actions are then scheduled during the projected time period and applied to the appliances to realize the calculated energy consumption profile. BEMS can be either completely onsite or utilize a gateway to communicate with a cloud-deployed service. A distinction should be drawn between a BEMS and such systems as smart home controllers or building security systems. These systems may use the same network protocols as the BEMS to communicate with sensors and control devices, but for a different purpose. While a smart home system may have some BEMS functionality, for example, being able to control smart thermostats from a smartphone, the primary purpose of a smart home system is lifestyle enhancement. Similarly, the purpose of a building security system is to secure a building against unauthorized access.

GEBs can serve as grid resources in a variety of ways. By enabling load flexibility, GEB control systems can shift load to times when the grid is experiencing less net load, thereby reducing the need for carbon-based energy from fossil-fueled "peaker" plants or to times when there is more renewable energy on the grid, thereby reducing indirect carbon emissions. GEB control systems can utilize data from a real-time Internet feed about the carbon content of grid electricity to schedule the consumption of energy generated and stored by onsite solar and storage DERs in a way that minimizes indirect carbon emissions from grid electricity. By reducing overall energy use, GEBs can free up grid capacity for the rapidly advancing electrification of transportation, as well as for beneficial electrification of buildings which today generate direct emissions through consumption of carbon-based fossil fuels. This can reduce the investment needed for upgrading grid infrastructure, helping make the transition to a decarbonized grid more affordable. GEBs can help stabilize the grid by shaving net peak loads, by providing ancillary services, and by consuming more renewable energy during the day when it is abundant, thereby reducing the need for solar curtailment.

Energy efficiency standards coming out of the energy crisis in the 1970s have been repurposed in the 2010s to address global climate change. National governments have had different approaches to developing and deploying standards. In the United States, the LEED standard for building environmental impact was developed by the USGBC and is primarily applicable to commercial buildings. LEED v4.1, the most recent version of LEED, was revised to emphasize the reduction of carbon emissions. Energy Star standards for appliances are developed and administered by the US federal EPA and DOE. Energy Star awards recognition to products in 75 different categories and also has categories for buildings, including residential buildings. Products that are awarded an Energy Star rating are allowed to display the distinctive Energy Star label on their packaging. The Energy Star building efficiency assessment and reporting standards have been incorporated into the building codes in some US cities.

The remainder of this book explores the BEMS architecture and the communication technologies which, if broadly instituted in a built environment that features full electrification, could result in a substantial reduction in energy use and therefore substantial savings for building owners and a reduction in investment needed for grid infrastructure. Combined with the rapid decarbonization of the grid that is now underway, the result will be a decarbonized built environment, with buildings essentially piggybacking on grid decarbonization. However, as discussed above in Section 1.3, if these measures are only applied to the existing fully electrified buildings, the carbon reduction potential is minimal. It is therefore imperative that beneficial electrification goes hand in hand with the deployment of the ICT technologies described in this book in order to achieve decarbonization of the built environment by 2050.

References

1 United Nations. High-level Dialogue on Energy, New York, September 2021 [Internet]. [cited 2022 Jan 24]. Available from: https://www.un.org/en/conferences/energy2021.

2 International Energy Agency (IEA). Buildings: A Source of Enormous Untapped Efficiency Potential [Internet]. [cited 2022 Jan 24]. Available from: https://www.iea.org/topics/buildings.

3 US Energy Information Administration (EIA). How Much Energy Is Consumed in U.S. Buildings? [Internet]. [cited 2022 Jan 24]. Available from: https://www.eia.gov/tools/faqs/faq.php?id=86&t=1.

4 Satchwell A., Piette M.A., Khandekar A., et al. A National Roadmap for Grid-Interactive Efficient Buildings. Berkeley, CA: Lawrence Berkeley National Lab (LBNL); 2021. Available from: https://gebroadmap.lbl.gov.

5 Architecture2030.org. Why the Building Sector? [Internet]. [cited 2022 Jan 26]. Available from: https://architecture2030.org/why-the-building-sector.

6 US Environmental Protection Administration (EPA). Sources of Greenhouse Gas Emissions [Internet]. [cited 2022 Jan 26]. Available from: https://www.epa.gov/ghgemissions/sources-greenhouse-gas-emissions.

7 Potter B. Every Building in America – an Analysis of the US Building Stock [Internet]. [cited 2022 Jan 26]. Available from: https://constructionphysics.substack.com/p/every-building-in-america-an-analysis.

8 EIA. What Is U.S. Electricity Generation by Energy Source? [Internet]. [cited 2024 Jan 8]. Available from: https://www.eia.gov/tools/faqs/faq.php?id=427&t=3.

9 US Department of Energy (DOE). In 2023, Non-Fossil Fuel Sources Will Account for 86% of New Electric Utility Generation Capacity in the United States [Internet]. [cited 2022 Jan 31]. Available from: https://www.energy.gov/eere/vehicles/articles/fotw-1304-august-21-2023-2023-non-fossil-fuel-sources-will-account-86-new.

10 Mir-445511. Wikipedia. Levelized Cost of Energy [Internet]. [cited 2024 Jan 8]. Available from: https://commons.wikimedia.org/w/index.php?curid=137272269.

11 Environmental and Energy Study Institute (EESI). Beneficial Electrification: An Access Clean Energy Savings Program [Internet]. [cited 2022 Jan 31]. Available from: https://www.eesi.org/electrification/be.

12 NREL. Grid-Interactive Efficient Buildings Technical Report Series [Internet]. [cited 2022 Jan 27]. Available from: https://www.nrel.gov/docs/fy20osti/75478.pdf.

13 Deign J. So, What Exactly Are Virtual Power Plants? [Internet]. [cited 2022 Jan 27]. Available from: https://www.greentechmedia.com/articles/read/so-what-exactly-are-virtual-power-plants.

14 Ward L. Amory Lovins: Solving the Energy Crisis (and Bringing Wal-Mart) [Internet]. [cited 2022 Feb 1]. Available from: https://www.popularmechanics.com/science/green-tech/a2146/4224757.

15 Brant J. Grid-Interactive Efficient Buildings: Providing Energy Demand Flexibility for Utilities in the Southwest. Southwest Energy Efficiency Project. [cited 2025 Jan 8]. Available from: https://betterbuildingssolutioncenter.energy.gov/sites/default/files/attachments/grid-interactive-efficient-buildings-report.pdf

16 EIA. Residential Energy Consumption Survey (RECS) [Internet]. [cited 2022 Feb 3]. Available from: https://www.eia.gov/consumption/residential/data/2015/index.php.

17 EIA. Commercial Energy Consumption Survey (CBECS) [Internet]. [cited 2024 Jan 4]. Available from: https://www.eia.gov/consumption/commercial/data/2018.

18 FDA. Are You Storing Food Safely? [Internet]. [cited 2022 Feb 3]. Available from: https://www.fda.gov/consumers/consumer-updates/are-you-storing-food-safely.

19 Apple. Developing Apps and Accessories for the Home [Internet]. [cited 2022 Feb 5]. Available from: https://developer.apple.com/homekit.

20 Google. Nest API Reference [Internet]. [cited 2022 Feb 5]. Available from: https://developers.nest.com/reference/api-overview.

21 Li H., Wang Z., Hong T., and Piette M.A. Energy flexibility of residential buildings: a systematic review of characterization and quantification methods and applications. *Advances in Applied Energy*, vol. 3, 100054. Available from: https://www.sciencedirect.com/science/article/pii/S2666792421000469.

22 California Independent System Operator (CAISO). California ISO – Emissions, Today's Outlook [Internet]. [cited 2022 Feb 5]. Available from: https://www.caiso.com/TodaysOutlook/Pages/emissions.aspx.

23 EIA. As Solar Capacity Grows, Duck Curves Are Getting Deeper in California [Internet]. [cited 2023 Aug 27]. Available from: https://www.eia.gov/todayinenergy/detail.php?id=56880.

24 Balaraman K. Thousands of California Homes to be Linked for 550 MW Virtual Power Plant, North America's Largest [Internet]. [cited 2022 Feb 7]. Available from: https://www.utilitydive.com/news/sce-590-mw-storage-550-mw-virtual-power-plant/591786.

25 Table 2, Satchwell, et al. A National Roadmap for Grid-Interactive Efficient Buildings. U.S. Department of Energy, Building Technologies Office, 2021, p. 4. [cited 2024 Dec 4]. Available from: https://gebroadmap.lbl.gov/A National Roadmap for GEBs - Final.pdf.

26 Wikipedia. Seasonal Energy Efficiency Ratio [Internet]. [cited 2022 Feb 7]. Available from: https://en.wikipedia.org/wiki/Seasonal_energy_efficiency_ratio.

27 Wikipedia. Power Factor [Internet]. [cited 2022 Feb 8]. Available from: https://en.wikipedia.org/wiki/Power_factor.

28 US Green Building Council. LEED [Internet]. [cited 2022 Feb 14]. Available from: https://www.usgbc.org/leed.

29 Energy Star. Energy Star: The Simple Choice for Energy Efficiency [Internet]. [cited 2022 Feb 14]. Available from: https://www.energystar.gov.

30 Wikipedia. LEED [Internet]. [cited 2022 Feb 14]. Available from: https://en.wikipedia.org/wiki/LEED.

31 US Green Building Council. LEED v4.1 [Internet]. [cited 2022 Feb 15]. Available from: https://www.usgbc.org/leed/v41.

32 Knox N. The Role of LEED in Climate Change Mitigation [Internet]. [cited 2022 Feb 15]. Available from: https://www.usgbc.org/articles/role-leed-climate-change-mitigation.

33 EPA. ENERGY STAR Impacts [Internet]. [cited 2022 Feb 15]. Available from: https://www.energystar.gov/about/origins_mission/impacts.

34 Wikipedia. Energy Star [Internet]. [cited 2022 Feb 15]. Available from: https://en.wikipedia.org/wiki/Energy_Star.

2

Architecture of Building Energy Management Systems

A key component in delivering on the decarbonization promise of GEBs is the BEMS). Many commercial and industrial buildings have legacy building automation systems (BASs), which include energy management as one function among many, to measure and control the building's power consuming devices. In most legacy BASs, however, the energy management function requires considerable human intervention. In contrast, except for multistory apartment complexes in large cities, most residential buildings have no BASs, and their BEMS in most cases consists of one or several thermostats with a few setbacks, which must be manually set.

This situation is changing in the commercial/industrial sector as building managers realize the value of controlling their building's energy consumption and carbon emissions and utilities start to introduce new tariffs like time of use rates and increase existing rates like demand charges. Investing in a BEMS to control power and energy consumption makes economic sense because the investment can result in thousands of dollars in annual operating expense savings and therefore has a short payback time. In the residential sector, "smart home" platforms from large technology vendors like Samsung, Google, Apple, and Amazon often contain an energy management function which may include integration of a smart thermostat. But most "smart home" platforms do not include control of solar photovoltaic system (PV), hot water, residential battery, and EV charging, do not have integration with utility demand response, and provide limited automated support for HVAC and lighting. They mostly shift manual control from a dedicated device like a thermostat to a smart phone with some minimal, rule-based automation or provide remote control from a smart phone where none previously existed like for controlling lighting. In many cases, utilities and third-party energy

Automating Building Energy Management for Accelerated Building Decarbonization: System Architecture and the Network Layer, First Edition. James Kempf.

Published 2025 by John Wiley & Sons, Inc.

providers offer substantial rebates to commercial/industrial and/or residential customers that can help defray the costs of purchasing and installing these systems.

An understanding of the underlying ICT system architecture is an important starting point for designing a BEMS. The BEMS ICT architecture is the topic of this chapter. The approach presented in this chapter divides the discussion into a functional architecture section, which is independent of actual deployment, and a deployment architecture section, which deals with how the functional architecture is deployed. This approach neatly separates concerns between functionality, interfaces, and protocols that need standardization for interoperability and the actual deployment of devices, software stacks, and applications, and has been used successfully in other ICT application areas, such as mobile telecommunications, web services, and enterprise information systems. Section 2.1 describes a notional BEMS functional architecture, including the most important functional blocks and how they interact. In Section 2.2, three phases of the BEMS deployment architectural evolution are presented. This evolution is of course not complete, and most existing BAS/BEMS will be somewhere on this spectrum. Security is a critical component of the BEMS, and Section 2.3 describes a cybersecurity architecture for the BEMS. Section 2.4 discusses the convergence of operational technology (OT) and information technology (IT) systems, and particularly their networks. This convergence has led to the possibility of radically simplifying the BEMS network architecture. Finally, Section 2.5 summarizes the chapter.

Introduction to Cloud Computing

This video provides some basic information about cloud computing. It presents a high-level description of cloud computing components, the basic categories of components in any cloud service, and the infrastructure components available for constructing cloud services.

2.1 BEMS Functional Architecture

This section presents a functional architecture for the BEMS. No assumptions are made about the deployment architecture in the following discussion. The abstractions provided by the functional components constitute their functionality, and the interfaces provide the means by which other functional components can access that functionality. The abstractions and interfaces are described for each functional component at a high level. In addition, the security architecture is not discussed in this section but rather in Section 2.3.

Figure 2.1 illustrates the overall BEMS functional architecture. The figure assumes that the data storage, analytics, and control

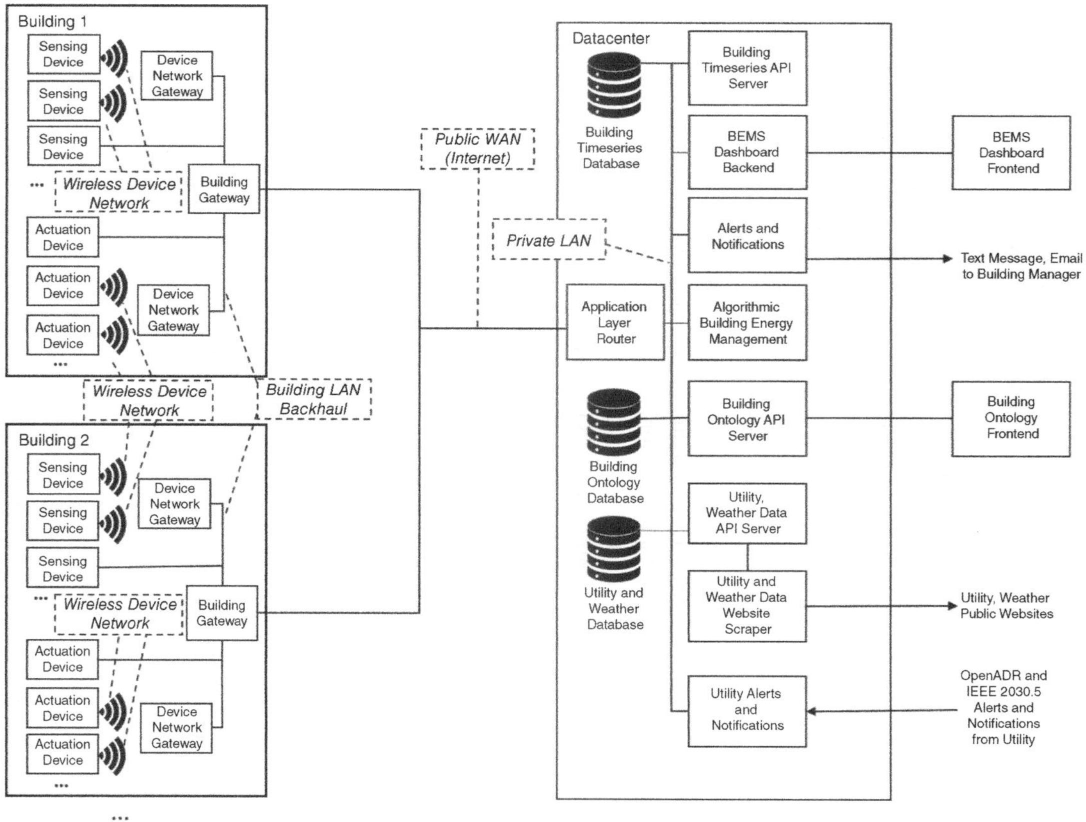

Figure 2.1 Functional Architecture for a BEMS

functional components reside in a data center since they will likely require considerable processing and storage resources, but for a single building system, they could also reside on a server in the building. Deploying these components in a data center also allows them to be shared between multiple building owners for offering BEMS as a service. The following sections discuss each of the functional components in more detail.

2.1.1 The Network

The BEMS depends on the network to communicate between the functional components of the system. Most discussions of the BAS network architecture start with the seven-layer Open Systems Interconnect (OSI) networking stack architecture [1], consisting of the physical layer, data link layer, network layer, transport layer, session layer, presentation layer, and application layer. The OSI model was developed in the early 1980s prior to the development of the Internet and was superseded by the Internet network architecture model in the 1990s when Internet networking technology was commercialized and began spreading. Today, Internet networking technology is inexpensive and ubiquitous and, through Internet of Things (IoT), can be and is used to implement BEMS networking. The Internet architecture simplified the OSI model down to four layers: the link layer, containing the OSI physical and data link layers; the network layer, containing the OSI network layer; the transport layer, containing the OSI transport, session, and presentation layers; and the application layer, containing the OSI application layer. The Internet architecture is described in Request for Comments (RFCs) 1122 [2] and 1123 [3].

The legacy standardized BAS communication protocols (Building Automation and Control Networks or BACnet, Local Operating Network or LONworks, and KNX which is an abbreviation for "Konnex," a German word that means "connection") defined full stacks from the physical layer up through the application layer because when they were developed in the 1990s, Internet hardware was expensive and viewed as primarily for IT use. Today, each of the legacy protocols features a mapping onto the Internet protocol (IP) stack, and going forward, BEMS will increasingly be deployed on IP since the hardware has become considerably cheaper and, through the use of IoT protocols, simpler to deploy for monitoring, control, sensing, and actuation use. Every building today has an IP network and a connection to a wide area network (WAN) through an Internet service provider (ISP), or can be connected through a cellular provider. Despite the fact that some older BAS networking systems do not utilize the protocols and hardware of the Internet architecture, from an architectural standpoint, their networking stacks must provide the same abstractions as the IP stack and can therefore be analyzed using the same functional architecture as the IP. Therefore, the following discussion focuses on the Internet architecture.

Figure 2.2 provides an overview of how the seven-layer OSI architecture maps onto the four-layer Internet architecture. OSI layers bundled into a single Internet layer are colored the same. The Internet stack link layer bundles the OSI data link layer and physical layer which are colored light gray; the Internet transport layer bundles the functions from the OSI transport, session, and presentation layers which are colored dark gray. The OSI network layer is colored gray and maps to the Internet network layer, and the OSI application layer is colored black and maps to the Internet application layer. Because networking connects two functional components and enables them to communicate, the abstractions and interfaces for a particular layer must reside on both sides of the communicating connection, though the two sides are really only directly connected by the physical medium.

As data moves down the stack from one layer to another, it is sectioned into fixed size chunks with a fixed size header and sometimes a footer. These chunks with attached headers and footers are called frames at the link layer, packets at the

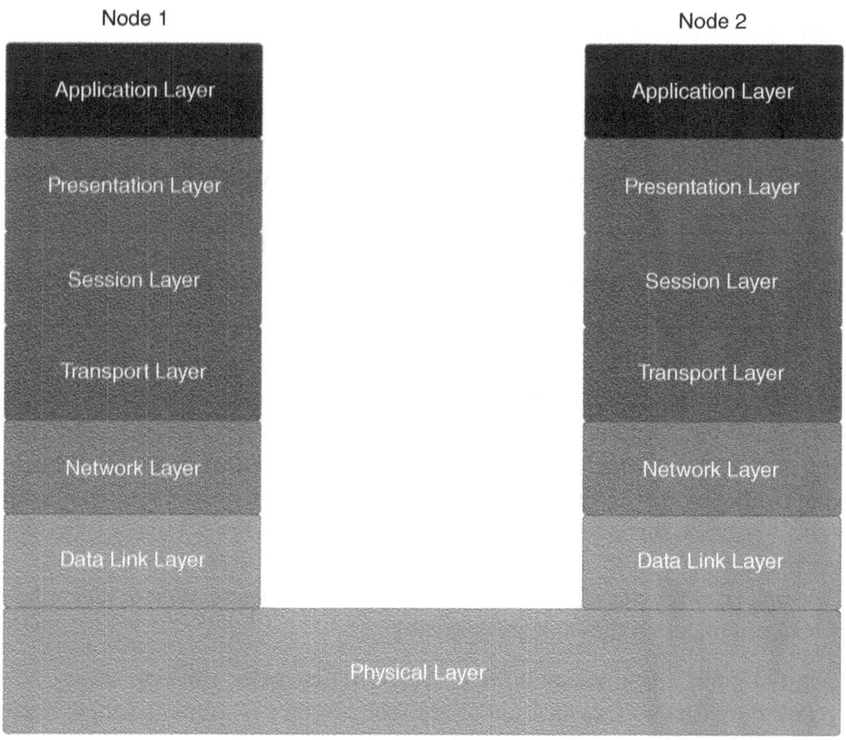

Figure 2.2 Mapping of OSI Seven-Layer Network Architecture to Four-Layer Internet Architecture

network layer, and segments by some transport layers. The header contains control parameters for the protocol implementing that layer on the sending and receiving side of the connection, for example, the addresses of the sender and recipient at the network layer. The control parameters allow the sending side to tell the receiving side what to do with the data. In the following sections, the four layers of the Internet architecture are examined from the point of view of the functions they provide to higher layers and how they utilize those functions from lower layers to provide functions to higher layers.

2.1.1.1 Link Layer

The physical and data link layers from the OSI model were condensed down into a single layer, simply called the link layer, in the Internet architecture. The physical layer is the physical communication medium used to transport data from one component in the BEMS to another. Examples of physical media are optical, electrical, and wireless (radio). The fixed-sized data chunks sent over the physical medium by the link layer are called frames.

The link layer implements a protocol that performs the following functions with respect to the physical layer:

- schedules access to the physical medium for sending and receiving chunks of binary data called frames to avoid collisions between two nodes sending data at once,
- places frames onto the physical layer, the frames consisting of a header with control information for the link layer followed by a frame data body containing the network, transport, and application layer data, and
- implements an addressing scheme allowing a node on a link to send data to another node, a subset of nodes, or all nodes on the link.

The link layer provides connectivity between network endpoints on the same physical network, for example, a local network inside a building or a WAN between the building router and the ISP's router. In many cases, the same link layer protocols are used over quite different physical layers, but the software implemented on top of the link layer mostly need not be concerned about the physical layer, as the details are abstracted away by the lower levels of the link layer protocol.

The link layer provides the following communication abstractions to higher layers between nodes on the same physical link:

- point-to-point transmission of frames between two nodes, called unicast;
- point-to-multipoint transmission of frames between a node and multiple other nodes, called multicast;
- transmission of frames between one node and all nodes on the physical link, called broadcast.

These services require the link layer to handle both transmission and reception through the physical layer upon which it is implemented.

2.1.1.2 Network Layer

The network layer provides connectivity between nodes running on different link layer networks. The networks may utilize the same or different physical media. For example, a thermostat with connectivity through the Wi-Fi protocol (IEEE 802.11 [4]) can communicate with a controller in a cloud data center running on the cloud provider's Ethernet (IEEE 802.1 [5]) network. The network layer in the Internet architecture runs the IP, either IP version 4 (IPv4 [6]) or IP version 6 (IPv6 [7]). The networking layer serves as a common point of reference for two network endpoints potentially on different link layers to address each other and communicate. In principle, two nodes anywhere in the world can communicate with each other using IP.

With respect to the data link layer, the network layer performs the following functions:

- maps individual packets onto one or several frames, depending on the packet and frame sizes;
- maps network layer addresses to data link layer addresses on the same link or, if the address is not on the same link, maps the address to the address of a router;
- to maintain communication between nodes on different physical links, distributes information about the reachability of different networks using a route distribution protocol. The route distribution protocol is handled by routers and is normally not of concern to individual nodes.

The network layer provides the following abstractions to the higher layers:

- an addressing scheme that spans across multiple physical layers and provides a node address for a communicating node, as well as addresses for multicast and broadcast;
- a communication abstraction called a packet implemented on top of the data link layer framing protocol. Data from higher layers is packaged into packets for transmission, and received data is handed off to higher layers as a stream of packets.

Because communication to and from network layer addresses can cross link layer boundaries, a router is required where one link layer network connects to another. If the two link layer networks run over different physical media, for example, optical and electrical, then the router needs physical interfaces that can handle the media on both link layers. A router may also be required if the two networks utilize the same physical medium but are for some other reason considered separate, for example, if they are administered by different organizations.

The router on the link of a sending node finds the next hop network layer address for a packet though the use of routing tables maintained using a route distribution protocol and sends the packet to the next hop address. The next hop address may be the address of the receiving node, or it may be the address of a router that can transport the packet closer to the receiving address. Routing is discussed briefly in Section 4.1.1, and reference [8] provides a well-written, easy-to-understand discussion of how Internet routing works.

2.1.1.3 Transport Layer

The next layer up in the stack is the transport layer. The Internet transport layer combines the functions of the OSI transport, session, and presentation layers and provides the following abstractions to the higher layers:

- an application endpoint, called a port, where applications on the node using an application layer protocol can send or receive data;
- a service model for communication between the sender and the receiver that implements a communication channel abstraction;
- a data container called a segment. Depending on the communication channel abstraction, the transport layer may segment larger chunks of data and insert them into the transport layer payload field together with the transport layer header for packing into network layer packets. Returned segments from the receiver are then reassembled.

The most commonly used transport protocols in the IP stack across a broad variety of applications are the User Datagram Protocol (UDP [9]) and the Transmission Control Protocol (TCP [10]). UDP simply reflects the packet-based nature of the communication channel presented by IP out to the application layer and presents a service model of an unreliable, connectionless communication channel between the sender and receiver. In contrast, TCP presents a service model of a reliable, connection-based channel. TCP is widely used for client–server applications, and other applications where streams of data larger than a single packet are exchanged over longer periods, while UDP is used for applications where most or all data can fit into one packet and communication is one-off.

2.1.1.4 Application Layer

Finally, at the top of the stack is the application layer in both the OSI and Internet models. The abstractions supported by the application layer protocol reflect the application for which the protocol is designed. An application layer protocol uses a designated port on the transport layer to receive traffic. The sending node must also supply a port on which it expects to receive return traffic, but the sending node's port need not be the same as the receiving node's port and typically is not. In total, port numbers run from 0 through 65535. According to RFC 1700 [11], ports 0 through 1023 are reserved, ports 1024 through 49151 are available for applications to register, and ports 49152 through 65535 are dynamic ports and are often used by the sender as their return port. An example of an application protocol for BEMS is the BACnet application protocol, which has registered port number 47808 (0xBAC0 in hexadecimal) [12].

Some application layer protocols provide vital system services or other system capabilities which are not specific to particular use cases. For example, the Domain Name System (DNS [13, 14]) in the Internet architecture maps human-readable names for nodes into the network layer addresses of the nodes, among other functions. DNS provides the translation of the server's name in web browser links, like www.example.com, into the IP address of the server. DNS mostly uses UDP port 53, but some functions also use TCP port 53. Other examples of general-purpose application protocols are the Hypertext Transport Protocol (HTTP [15, 16]) and Message Queuing Telemetry Transport (MQTT) [17, 18] protocol. HTTP is the fabric that ties together the World Wide Web. It is used for transporting web pages between machines connecting the user interface for services to the backend server, and for encoding web APIs allowing clients to access specific functionality and data on servers. MQTT is a protocol that supports a lightweight type of interaction called publish-subscribe (pub-sub). Pub-sub allows for the BEMS to subscribe to messages from sensors or for control points to receive commands from the BEMS. Both protocols are important in more recent vintage BEMS architectures based on IP protocols and the Internet architecture. HTTP uses TCP port 80 or, if HTTP is run over TCP with Transport Layer Security (TLS) port 443. MQTT uses TCP port 1883 or port 8883 with TLS.

2.1.1.5 Private vs. Public Networks

While not part of the network stack on the node, network coverage scope is an important part of the overall network architecture and how it is used in BEMS. There are two basic network scopes, private and public. Private or local networks cover a building, multiple buildings on a site, or an enterprise, customer account, or project on a cloud provider. If these networks are confined to a single building, they are called local area networks (LANs). Nodes on these networks have IP addresses assigned from a private IP address space which means they are only directly addressable from nodes within the same private network unless some form of network address translation (NAT) is used. NAT translates an address in the private network address space to a shared address in the public network address space. Some large private networks can span multiple physical sites across countries and continents. Public networks or WANs, often called the public Internet, connect LANs together and nodes on public networks have IP addresses assigned from the public IP address space. Public networks are run by ISPs or mobile wireless operators.

Routers on the border between private and public networks will drop any packets with a receiver address in the private address space coming from the public Internet and will translate any sender address on outgoing packets to a shared public address using some form of NAT, allowing nodes within the private address space to access nodes in other private address spaces and on the public Internet. IP addressing is different in public and private networks and is discussed in more detail in Chapter 4.

2.1.2 Sensing and Actuating Devices

Within the building, sensing and actuating devices are responsible for providing points of measurement and control for the system. These devices are associated with electrical and mechanical equipment in the building, for example, thermostat monitoring and controlling the HVAC system or automated switching for lighting. The functions these devices provide are the following:

- For sensing devices, measuring environmental parameters such as temperature, CO_2 concentration, light, occupant presence, etc., according to the type, and transmitting the data over the network using an appropriate network and application protocol to the building timeseries API server component for insertion into the building timeseries database;
- for actuating devices, presenting to the algorithmic building energy management component a control protocol or interface that the algorithmic building energy management component can use to control the equipment.

If the sensing and actuating devices utilize a legacy BAS network protocol [19, 20], the communication between the control and management components and the devices may require protocol translation through the building gateway or a protocol translation gateway. However, the sensing and actuating devices may utilize a message bus protocol such as MQTT and IP or may utilize Wi-Fi and IP to communicate directly with the control and management components in the data center.

2.1.3 Device Network and Device Network Gateway

As shown in Figure 2.1, the device network connects sensing and actuating devices up to the building LAN backhaul through a device network gateway. In some cases, devices may be directly connected into the building LAN, for example, Wi-Fi thermostats. While Figure 2.1 shows the device network and device network gateway as wireless, legacy devices may be connected to a serial bus that requires a gateway to translate the legacy building automation protocol packets into packets on the building LAN. However, many sensors and actuators may have connectivity through a wireless mesh or point-to-point wireless personal area network (WPAN) protocol like 802.15.4 [21]. These devices require a wireless device network gateway to translate the link layer from the mesh protocol to one of the building LAN link layer protocols and to route packets from the wireless device network to the building LAN and building gateway. The device network gateway has the following functions:

- manages the device network for sensing and actuating devices that communicate through a device protocol not supported by the building LAN;
- forwards packets between devices on the device network and the building LAN on which the building gateway sits.

Not shown in the figure, a wireless wide area protocol such as a cellular IoT protocol may be used to connect directly between the device and either a manufacturer website or the cloud BEMS network in some cases through a wireless ISP's network. For such devices, no wireless gateway is required since the gateway function is part of the wireless service provider's network.

2.1.4 Building Gateway

The building gateway straddles the border of the private site local network and the public Internet and performs the following functions:

- acts as a router and translator between building automation networks using legacy BAS network and data link protocols and the building's private IP network, translating to and from the legacy network protocol format to the legacy protocol's IP profile;
- provides security functions such as firewalls and distributed denial of service (DDoS) attack prevention to and from the Internet. Cybersecurity architecture is discussed in Section 2.3;
- provides routing to and from the device components in the building, including message bus routing if a message bus protocol is used to connect the building device components to the control and management components.

The device network gateway and building gateway functional components may be combined into a single device; for example, a router that connects into the building ISP's public network may also support an interface card that acts as a gateway for 802.15.4 device traffic; or they may be deployed as multiple physical devices.

2.1.5 Application Layer Router

The application layer router component sits in the data center and acts as an application layer (Layer 4) router for traffic coming into and leaving the control and management components in the data center, and between components in the data center. The functions it provides are the following:

- accepts building measurement traffic from sensing devices in the buildings under management arriving from the Internet via the data center network;
- routes the traffic to the building timeseries data API server for insertion into the building timeseries database;
- accepts building actuator device control traffic from the BEMS dashboard API server and algorithmic building energy management components and routes it to the appropriate building;
- routes the application layer traffic between control and management components in the data center.

An example of a routing function between control and management components is a query from the BEMS dashboard API server for a particular set of data points in a building temperature timeseries. The application layer router may use a message bus protocol for routing. Utilization of an intermediary for routing in this fashion ensures that if any components are upgraded or replaced, the routing information just needs to be changed in the application layer router, not in all the components.

2.1.6 Databases

The BEMS requires three databases: the building timeseries database, the building ontology database, and the utility and weather database. Data from the databases is not accessed directly by clients, but rather though an API server component that offers an HTTP API.

2.1.6.1 Building Timeseries Database

The building timeseries database stores timeseries data on the building. Timeseries data consisting of measurements made by sensors in the building and control actions initiated by the building manager through the BEMS dashboard or by the algorithmic building energy management component together with the date and time at which the data was collected. In addition, strategies for energy management within the building generated by the algorithmic building energy management component are recorded when they are changed. Historical timeseries data measured prior to the commissioning of the BEMS may be included in the database if available since such data is often useful for training machine learning models. This component also stores any alerts or notifications generated by the alerts and notifications component. The building timeseries database does not, however, store timeseries data involving the weather or carbon intensity of grid-supplied electricity since this data is not directly generated nor consumed by a building under management.

2.1.6.2 Building Ontology Database

Measuring and controlling energy use in buildings requires information on the architecture and physical characteristics of the building structure and on installed equipment. This information is known as the building's ontology or metadata and is stored in the building ontology database. A clearly defined system architecture and radically simplify how the measurement and control data on the building's energy consumption and carbon emissions is stored and accessed.

2.1.6.3 Utility and Weather Database

Finally, the utility and weather database stores all information pertinent to energy control but not strictly being generated or consumed by the building itself. Examples of stored data include the following:

- utility tariffs for controlled buildings,
- weather timeseries data including forecasts,
- timeseries of carbon emission intensity for grid-generated electricity, and
- utility alerts and notifications coming from the utility alerts and notifications component.

The data may come from other sources than the utility or weather; for example, timeseries of carbon emissions intensity may be available from the wholesale market operator rather than directly from the utility.

2.1.7 Algorithmic Building Energy Management Supervisory Control

The algorithmic building energy management component is responsible for exercising supervisory control functions over the buildings' actuators to accomplish goals set by the building manager. The goal may be to minimize building carbon emissions from grid-supplied electricity, reduce monthly demand charges below a particular amount, etc. The algorithmic building energy management component provides the following functions:

- accepts commands from the building energy manager dashboard API server for setting goals of the BEMS with respect to particular buildings;
- issues requests for data to the building timeseries database via the building timeseries database API server, the building ontology database via the building ontology API server, and the utility and weather database via the utility and weather data API server to make predictions and calculate optimal actions to carry out the goals set by the building energy manager;
- writes control actions into the building timeseries database when they are exercised via the building timeseries database API server;

- writes strategies for executing control actions into the building timeseries database via the building timeseries database API server;
- issues control actions to the actuator devices in the building via the application layer router.

2.1.8 Utility and Weather Data Scraper

The utility and weather data scraper serves as an automated front end to the utility and weather API server component. Its functions include the following:

- utilizes public APIs from utility, weather, and other relevant websites to periodically obtain data on topics such as utility tariffs, weather forecasts and historical weather data, carbon emissions intensity of grid-generated electricity at particular times, etc., as needed to keep the BEMS up to date;
- stores data obtained from these sources through the utility and weather data API server;
- offers a configuration API to the BEMS dashboard backend allowing BEMS administrative users to configure which websites should be queried for what data, how often the query should be repeated, and the buildings to which the data applies.

2.1.9 Utility Alerts and Notifications

The utility alerts and notifications component allows the BEMS to be informed by the utility of upcoming grid events, like peak times signaled by the wholesale market operator or utility when demand reduction requests are in force. Its functions include the following:

- receives messages from the utility using the OpenADR protocol [22] or IEEE 2030.5 [23] about grid events for buildings under BEMS control and returns information about the ability of buildings to respond to the events;
- writes information about such events into the utility and weather database through the utility and weather API server;
- offers a configuration API to the BEMS dashboard backend allowing BEMS administrative users and building managers to configure which events buildings should respond to and how they should respond.

2.1.10 API Servers

API servers handle the API for a particular database. An API server can serve together with a frontend to implement a particular service, for example, the building ontology service. Standalone, an API server provides an abstraction layer by which other components can access data from a database without having to utilize the database query language directly. API servers are implemented using HTTP APIs based on standards such as OpenAPI [24] and GraphQL [25]. By using a standardized API rather than directly querying the database, differences in query languages are hidden, allowing different types of databases to be utilized depending on what makes sense for the application, without proliferating code for database query languages. The result makes the BEMS more modular and allows easier modification if a particular database needs to be replaced. In addition, the data in the BEMS can be utilized more easily for other building management use cases, such as asset management and building access control. Note that the term "server" does not imply any deployment or implementation technology. The BEMS requires three API servers: the building timeseries API server, the building ontology API server, and the utility and weather data API server.

2.1.10.1 Building Timeseries API Server

The building timeseries API server supports an API for the building timeseries database. The component provides the following functions:

- accepts building timeseries measurement data routed from the building by the application layer gateway and stores the measurements in the building timeseries database;
- accepts control actions from the algorithmic building energy management component and stores the control action records in the building timeseries database;
- accepts alerts and notifications issued for a building and records them in the building timeseries database;
- accepts building energy management strategy changes from the algorithmic building energy management component and stores the new strategies in the building timeseries database;
- provides control, measurement, strategy, and alert and notification timeseries data from the building timeseries database for specific buildings to the algorithmic building energy management component and the BEMS dashboard API server.

2.1.10.2 Building Ontology API Server

The building ontology API server supports an API for the building ontology database. The component provides the following functions:

- accepts modifications to the building ontology through the building ontology frontend component and stores them in the building ontology database;
- provides data on the building ontology to the algorithmic building energy management, BEMS dashboard API server, and building ontology frontend components.

2.1.10.3 Utility and Weather Data API Server

The utility and weather data API server supports an API for entering and retrieving data from the utility and weather database. It supports the following functions:

- accepts utility and weather data from the utility and weather data scraper and stores it into the utility and weather database;
- accepts utility alerts and notifications from the utility alerts and notifications component and stores it into the utility and weather database;
- provides data from the utility including on alerts and notifications and on weather including forecasts to the algorithmic building energy management component and the building energy manager dashboard backend.

2.1.11 BEMS Dashboard Backend

The BEMS dashboard backend provides the backend for the BEMS dashboard frontend. The two components together implement a service whereby building managers and administrative users of the BEMS can view building performance data and configure various components. This component supports the following functions:

- provides an API that combines data from the building timeseries API server and the building ontology API server to allow the frontend to display the performance of the BEMS. This may include calculations using the returned data;
- provides an API that allows users to modify algorithmic building energy management calculated strategies based on occupant preferences, etc.;
- provides a push style API that pushes alerts and notifications from the alerts and notifications component to the BEMS dashboard frontend in real time;
- utilizes the configuration APIs provided by the utility and weather data scraper and the utility alerts and notifications components to provide an API for the BEMS dashboard frontend allowing building managers and BEMS administrative users to configure how these components function for particular buildings.

2.1.12 Frontends

Frontends are components that provide a graphical user interface (GUI) for a web application that runs on a mobile phone or in a browser, allowing both building managers and BEMS administrators to access data on and configure BEMS components. Permissions allowed to different users like building manager vs. the BEMS administrator are determined by their credentials and roles. Their credentials are handled using standard web identity and access management (IAM) systems.

2.1.12.1 BEMS Dashboard Frontend

The BEMS dashboard frontend provides access to all BEMS data and functions involving the building and its connection to its environment through the utility and weather. The functions provided by this component include the following:

- provides displays allowing building managers and BEMS administrative users to view the performance of the BEMS, building data, energy management strategies and control actions, and both historical and real-time alerts and notifications;
- provides controls allowing building energy managers to change energy management strategies based on occupant preferences and covey these changes to the algorithmic building energy management component via the API provided by that component;
- provides displays allowing building managers and BEMS administrative users to view utility and weather-related data and events through an API provided by the utility and weather data API server;
- provides controls allowing building energy managers and BEMS administrative users to configure how buildings respond to alerts from the utility through a configuration API provided by the utility alerts and notifications component;
- provides controls allowing building energy managers and BEMS administrative users to configure where alerts and notifications generated by the alerts and notifications component are routed to, using a configuration API provided by the alerts and notifications component;
- provides controls allowing BEMS administrators to modify what external websites are queried, which data should be fetched, and what buildings the data applies to by the utility and weather data scraper using a configuration API provided by the utility and weather data scraper;
- provides controls allowing BEMS administrative users to add or remove utility notification services for buildings through a configuration API provided by the utility alerts and notification component.

2.1.12.2 Building Ontology Frontend

The building ontology frontend provides a way for BEMS administrators to onboard new buildings and modify the metadata for existing buildings if, for example, a building is remodeled and energy efficiency improvements are incorporated. Together with the building ontology API server and the building ontology database, the building ontology frontend implements a building ontology service, which incorporates all information about the architecture and equipment of the building, essentially everything related to the building as a long-lived asset, including HVAC equipment, PV systems, etc. No information specifically related to the building energy management is covered by this service, though some of the information may be of use to the BEMS. As such, the building ontology service may be provided by another system or the BEMS building ontology service may be utilized by other applications implementing other use cases. The functions provided by the building ontology frontend include the following:

- provides graphical displays allowing building managers and others to view building metadata information including architecture, equipment, wall composition, window location, etc. This information is retrieved from the building ontology database via the building ontology API server;
- provides controls allowing metadata from new buildings to be onboarded and the metadata of old buildings to be modified based on events in the building's history. These controls are implemented through the building ontology API server.

2.1.13 Alerts and Notifications

The final functional component is the alerts and notifications component. The alerts and notifications from this component are generated by the building itself, not by the utility or any outside entity. An example is a notification that a particular air-handling unit is offline. This component has the following functions:

- takes alerts and notifications routed from the building gateway by the application layer router and sends them out via text message and email to subscribed users, and via push to the BEMS dashboard frontend in real time;
- writes the alerts and notifications to the building timeseries database via the building timeseries API server;
- provides a configuration API whereby BEMS administrative users and building managers can configure where text messages and emails should be sent if an alert or notification is received.

Practical Considerations on the Horizontal Architecture

James Dice, founder of Nexus Labs, has been at the forefront of the movement toward the horizontal architecture for BASs in smart buildings. The code above links to a podcast in which James and Rosy Khalife discuss practical considerations around the horizontal architecture, such as the benefits for building owners, occupants, and other stakeholders and difficulties in implementing it.

2.2 BAS Generations

BASs have been installed in many existing commercial buildings since the 1980s, and commercial BEMSs evolved out of and in many cases are still offered as a function of the BAS. Except for large apartment buildings, most residential buildings have no integrated BAS/BEMS deployed since "smart home" technology is newer and has been difficult to install and maintain. At a high level, in the commercial/industrial sector, BASs have evolved through the following three generations of architecture patterns [26]:

- Vertically integrated systems or point solutions in vendor and equipment specific silos with connectivity and software for specific classes of devices provided by the vendor with little interoperability between different vendor silos. Some protocols were developed that provide common communication between devices and services, but there is no single protocol standard such as the web protocol HTTP by which devices and services can communicate and most deployed systems feature a high number of vendor-specific extensions [27].
- Integration overlays in which different third-party vendors offer separate software platforms as a service that integrate different vendor silos up to and including the application layer. Typically, these platforms require time consuming, bespoke development of the underlying infrastructure on the part of the vendors for each building to accommodate the legacy building equipment and software [28].
- Horizontal integration of the infrastructure across different stack layers based on common standards that allow building managers to choose applications that best meet their needs from a service marketplace like the mobile application "app stores" [29].

Integration overlays and especially horizontal integration have been supported by the convergence of IT and OT systems discussed in Section 2.4.

While the vertical silos pattern is well established and has been the default since large commercial buildings began being outfitted with BASs in the late twentieth century, deployment of the other two patterns – integration overlays and horizontal integration – is currently ongoing in parallel. The integration overlay pattern works for existing buildings even with vertically integrated silos since it does not require replacement of existing devices or, in most cases, networks. However, integration overlays have their own architectural problems which are addressed by horizontal integration. For example, it is not possible for applications addressing other use cases such as building asset management systems to access data within the vendor's system. With a horizontal integration architecture, all interfaces and components within the management system are standardized. But horizontal integration requires agreement on standards between device vendors, BAS software vendors, data aggregators, and others. Such agreement is often difficult to obtain, requiring many years of work, and as a result, the horizontal integration pattern is still a work in progress. The horizontal integration pattern holds the most promise for establishing an ecosystem of third-party BEMS applications like the app stores for mobile smart phones.

In the residential sector, vendor and equipment specific silos also exist, but smart home systems that feature integration overlays emerged early in the development of IoT devices for the home, driven by large cloud companies (Amazon, Google, and Apple) with IoT platform services. Third-party device vendors often offer the ability to integrate with multiple smart home integration overlay platforms. This neatly separates concerns between device vendors, who specialize in hardware innovation, and platform vendors, whose expertise is in complex software platform integration. In most cases, the overlay platform vendors selected existing IP standards for their solutions on a mix and match basis, but there was little or no thought given to interoperability between the different platforms despite their use of Internet standards. However, horizontal standards are beginning to emerge in the smart home space, driven by the need for security and a simple process around device installation and configuration.

In the following sections, we review each of these generations of architectural evolution.

2.2.1 Vendor and Equipment Specific Silos

Figure 2.3 illustrates the vertically integrated BEMS architecture. Each set of vendor equipment has a separate set of proprietary communication protocols, databases, services, and application software. In some cases, a specific vendor may offer BEMS services that encompass multiple types of devices, but there is little interoperability between different vendors' stacks. If the vendor has opted to use a standard building automation protocol such as BACnet/ASHRAE SSPC 135 [20], Lonbus/ISO/IEC 14908 [30], KNX/ISO/IEC 14543 [31], or Modbus [32], often proprietary extensions exclude interoperability with other vendors' services. These standardized building automation protocols are rarely used in residential buildings since the equipment to implement networks built on them is typically

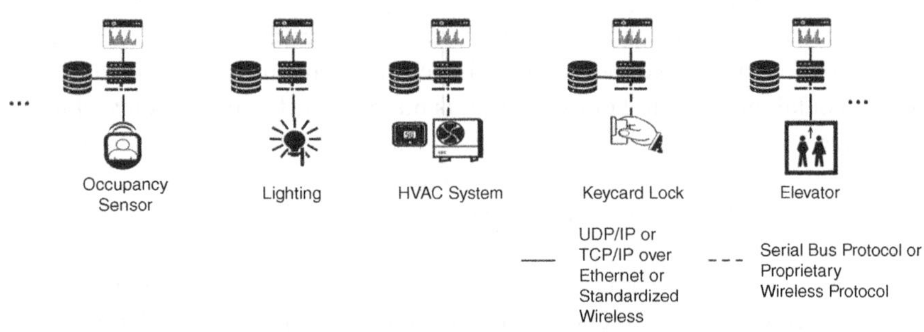

Figure 2.3 Vertically Integrated BEMS Architecture Organized Around Equipment Types/Vendors

more expensive. In addition, since the standardized building automation protocols were developed prior to the widespread use of Ethernet and IP in buildings, they specify a complete stack down to the physical layer in addition to an Ethernet/IP profile which was added later. The requirement for specialized network hardware increases the complexity and expense of using a standardized building automation protocol.

The lack of integration between different device control systems means that data from one device silo cannot be used in making control decisions for another. For example, data from occupant sensors and keycard locks cannot be used to determine when people typically enter or leave the building and what rooms are occupied to control the HVAC system. In fact, there is rarely one control application where building operations can be optimized for energy efficiency and carbon emission reduction, and progress on these goals can be tracked. The lack of a controller means that no single point exists where automated tradeoffs about what devices to throttle in response to a utility grid alert or high-tariff period can be made. In this situation, demand management – if it is done at all – must be purely manual with the building manager or home owner responding to text messages from the utility to throttle their energy use the next day. In addition, for buildings where some of the silos utilize legacy building automation protocol standards, if different vendors have chosen protocol specific stacks down to the physical layer, the building technicians may need to deal with multiple legacy networking protocols, including specialized hardware for routing and interconnection with the building's Ethernet network.

Silos are a particular problem because commercial and residential buildings are long-lived assets that are often remodeled for new tenants or owners, but BEMSs are rarely touched during a remodel. A new tenant in a commercial building may want a different type of keycard system to address their particular access concerns, but they are typically not interested in changing the HVAC system controls. They just want the HVAC system to work and maintain occupant comfort without running up a large utility bill. Similarly, a new homeowner is not going to change the HVAC controls unless the existing HVAC system malfunctions and needs to be replaced. As a result, a large collection of buildings exists with legacy energy management systems that complicate the engineering problem of implementing an integrated BEMS.

2.2.2 Integration Overlays

Figure 2.4 illustrates a cloud-hosted deployment of the integration overlay architecture [28]. The former device silos are not broken down but simply integrated through a building device network gateway. Devices communicating with legacy building automation protocols are integrated through the gateway or protocol

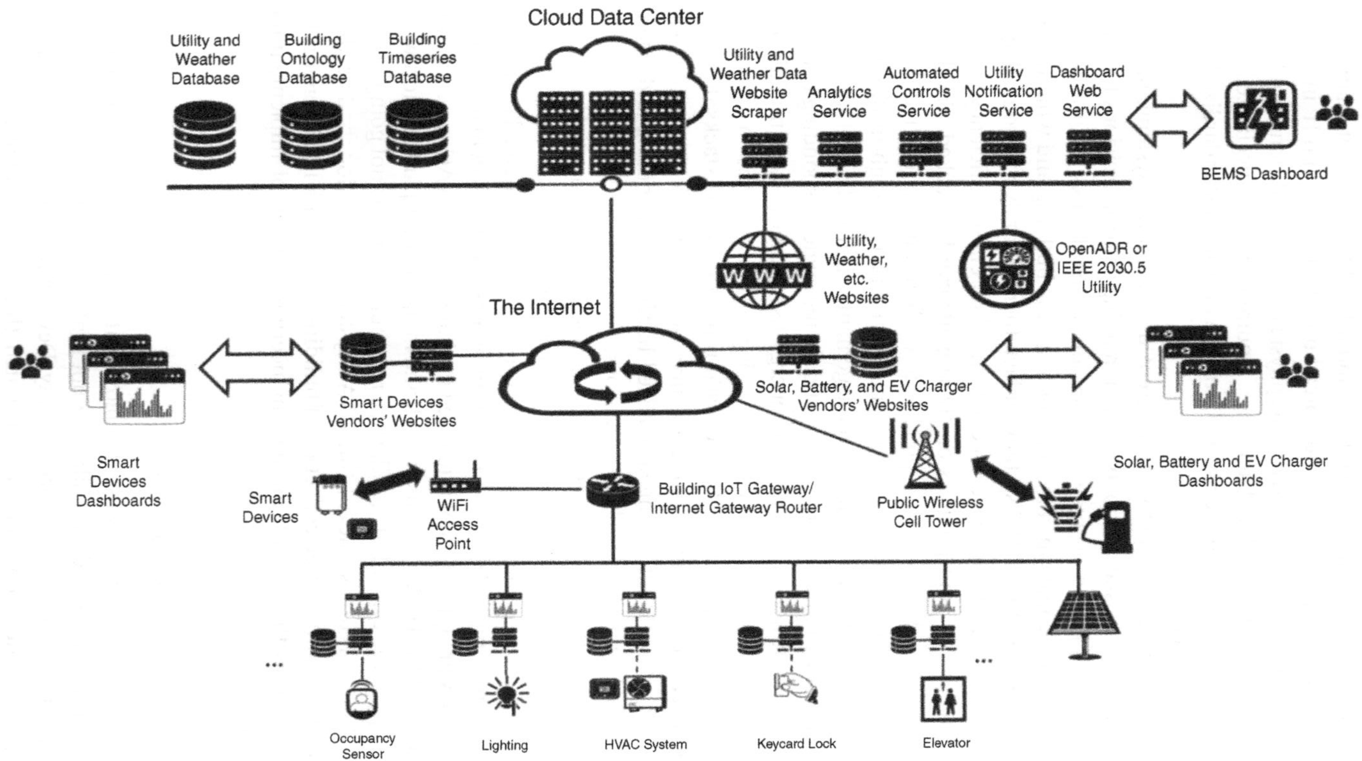

Figure 2.4 Integration Overlay BEMS Architecture with Cloud Hosted Integration Service

converters that convert the legacy protocols to run over IP. Network connectivity within the building is provided by Ethernet, a standardized wireless protocol, or a legacy proprietary wired or wireless protocol. Wide area Internet connectivity is provided through the building's Internet gateway router to the building ISP or through a public access cellular connection with a cellular provider that offers an IoT data plan. Data is sent over the Internet to a cloud service where the data is stored in a timeseries database. A building ontology database contains records about the building architecture, location of sensor and actuator points, etc. An external website scraper gathers data on the weather, utility tariffs, carbon intensity of grid electricity, etc., for use by the automated control service, and is stored in a third database.

An analytic service runs analytics on the data, and an automated control service may use rule-based systems, a model constructed with machine learning combined with occupant preferences, or other algorithmic techniques to exercise automated supervisory control. A web service hosts the BEMS dashboard, so building managers can monitor the performance and change occupant preferences if needed through a web page or smart phone app. The utility notification service accepts the notification of grid events like upcoming peak periods through the standardized utility integration protocols (OpenADR and IEEE 2030.5). A service designed in accordance with the integration overlay pattern can alternatively be deployed locally on a dedicated server in the building itself, or in a corporate data center.

New device silos have been introduced, however. "Smart" devices like Wi-Fi thermostats and lighting controls connect through a local Wi-Fi access point over the Internet to the device vendors' websites directly rather than using one of the legacy building automation protocol standards, completely bypassing the BEMS. Building-mounted solar inverters, batteries, and EV chargers likewise only communicate with their vendors' websites. However, many vendors for smart devices and building-mounted DERS also support a HTTP REST API [33] which is available to external developers, so the BEMS integration overlay automated supervisory control service can pull data and exercise control through API calls on the device vendors' websites rather than controlling devices through the device vendors' dashboards or mobile apps.

While integration overlays are a considerable improvement over vendor silos, they still have few problems. The building owner is dependent on the BEMS integration overlay vendor for introducing new applications and features and continued support. In some cases, a new BEMS app may become available for one BEMS integration overlay vendor's system but may not be available for others until much later. And if the BEMS vendor goes bankrupt or otherwise stops supporting customers, the building owners are left without energy management for their buildings [34]. For devices that utilize one of the legacy building automation protocols like BACnet, the BEMS integration overlays can provide a good solution, but the amount of work needed to integrate the legacy protocols into one coherent system is considerable, and often custom work is required for every new building.

Another problem involves non-BEMS services that need access to some of the underlying data that the BEMS is managing or data from other BAS services that the BEMS itself could use. Asset management services, for example, could use information from the building ontology database since information on devices and architecture will be the same for both. Having this information duplicated in two separate databases requires updating in two different places if something changes, a recipe for accumulating errors. If the building owner needs to install multiple integration overlays that address different service use cases, each overlay vendor needs to duplicate the integration at the infrastructure level. In addition, installation and configuration of multiple integration overlays may run into network conflicts and can run up the cost.

2.2.3 Horizontal Integration

Figure 2.5 illustrates the third generation in BEMS evolution, the horizontal integration architecture [26, 35]. The horizontal integration architecture evolved out of a realization that the value of BEMS systems to building owners lies in the applications. Devices, connectivity, and data are necessary as infrastructure to support the services, but the services are what building owners are willing to pay for. Having multiple different types of infrastructure layers for different BAS and other services is not cost effective in the end. The infrastructure layer serves the same function as the operating system on a smart phone: the basis on which services can be developed and deployed. An agreed-upon standard set of protocols, APIs, and data schemas below the service layer can lead to more rapid and cost-effective innovation at the service and application layer. The horizontally integrated infrastructure layer is sometimes called a "building operating system" [35].

The horizontally integrated architecture consists of the following layers integrated with standardized APIs and protocols:

- the Devices layer at the very bottom with the sensors and controls at the building site;
- the Network layer including the building site network, the Internet, and the cloud network. Standard protocols are used at all layers of the stack, except by legacy devices in the building;
- the Data layer based on standardized data schemas that are used for the timeseries database, the building ontology database, and any data input from external websites such as utility tariffs, carbon intensity of grid electricity, weather information, etc.;

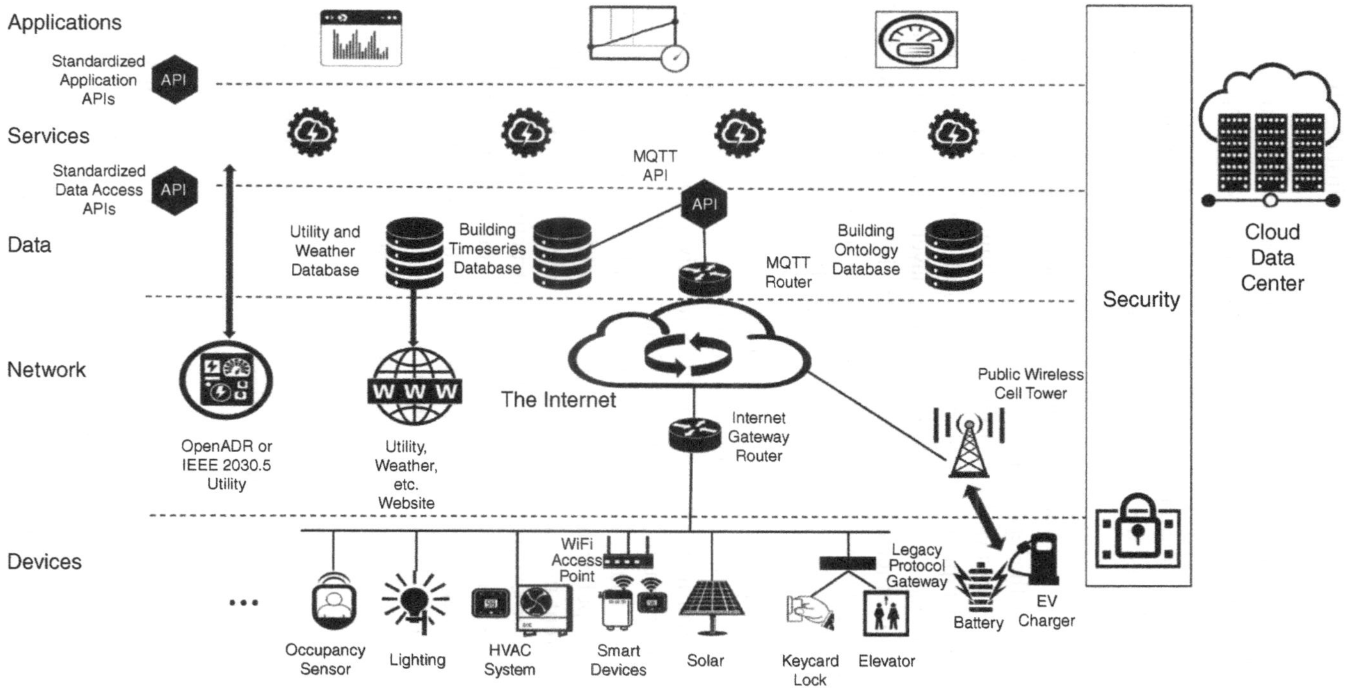

Figure 2.5 Horizontal Integration Architecture or "Building Operating System"

- the Services layer consisting of various BEMS services installed by the building manager or by the BEMS service provider for the building;
- the Applications layer where mobile or web applications make use of services to provide some energy management functions and allow building managers and others to control the services through their GUIs;
- Security which is not confined to a single layer but, as discussed in Section 2.3, cuts across all layers, from the devices to the network protocols to the applications and services.

Since buildings are long-lived assets, support for older, already installed devices that speak the legacy building automation protocols is provided through legacy protocol gateways. In addition to legacy building automation protocols, the gateway must support legacy physical and link layer networking protocols if the devices also use legacy protocols at lower layers rather than standardized protocols such as Ethernet (IEEE 802.1) and Wi-Fi (IEEE 802.11). Newer devices and the legacy gateways all have a private IP address on the building's local network, or a public IP address on the building's ISP network or a public wireless carrier's network.

Within the building, a single, standardized physical and link layer protocol is unlikely to emerge because different deployment scenarios require different types of low-level network connectivity. Similarly, connection to the Internet may take place through the building's Ethernet and Wi-Fi network to a wired Internet gateway router connecting with the building's wired ISP, or it may take place through a public access cellular provider with an IoT plan. These options all serve different use cases. For example, if the building's wired Internet service is spotty or expensive, the BEMS infrastructure provider may instead choose to use a reliable and affordable cellular connection. At the application layer, newer devices and the legacy gateway all speak the MQTT protocol directly with the MQTT gateway router in the cloud. MQTT acts as the "narrow waist" [36] through which devices integrate with the Data and Service layers in the cloud.

Standardized data schemas for storing building ontology, timeseries data, and information scraped from external websites are accessed through standardized data access APIs. Since one of the primary goals of a standardized schema is to address a broad variety of building information needs and not just BEMS or BAS, a standardized building ontology schema seems the most difficult of the three data types on which to obtain agreement. The building description in the ontology should incorporate information on the initial design ("as designed"), the actual construction ("as built"), and any remodeling modifications made to the building, as well as information on the various devices, sensors, and controls installed when the building was initially built and subsequently. Currently, information from these areas is not well integrated and often might not be available. For example, a complete set of "as built" and "as designed" information may be available from construction plans and other documents, but information on subsequent remodeling work may be lacking or in a separate set of documents on a hard drive somewhere.

Standardized APIs between the Service layer and the Data layer provide a uniform and simplified way for services to access the Data layer databases. Similarly, standardized APIs between the Service layer and Application layer allow applications to compose different services in innovative ways. REST APIs built upon an HTTP standard such as OpenAPI or GraphQL are candidates for specifying the HTTP APIs. An MQTT API based on a standardized sensor and control data schema like the Universal Device Management Interface (UDMI) [37] provides services and the timeseries database access to and from the on-site devices.

2.3 Cybersecurity Architecture

Security is an important aspect of any information system architecture, and the BEMS architecture is no exception. Information system security, in the context of OT systems like the BEMS, is sometimes called "cybersecurity" to distinguish it from physical security; for example, ensuring that physical access to critical assets is only available to authorized personal. Information system architecture diagrams often show security as a column on the right or left side, spanning top to bottom of the diagram. This is because security touches the entire stack, from the applications and services themselves through data stored by the applications, to the application layer protocols communicating between different distributed application components, down to the link layer protocols responsible for basic communication across the network. Naturally, security applies to the physical layer as well, but physical layer assets are covered by physical security measures and not cybersecurity.

2.3.1 Threat Modeling

Threat modeling is an important first step in determining what components are needed in a cybersecurity architecture. Threat modeling starts by assessing the BEMS and deciding how cybercriminals could break in and what they would target. Following an explicit methodology is important when performing threat modeling. Several different methodologies have been developed, each with specific strong points and weak points. More details on threat modeling methodologies can be found in Shostack [38]. Threat modeling should not just involve thinking about the problem. The team that will actually implement and deploy the system needs to be engaged, and the results need to be documented as part of the system architecture documentation.

Threats are sometimes confused with risks, but the two are somewhat different though related. A threat is a promise of future harm, while a risk is a probability of a negative consequence given a particular action taken to mitigate or not mitigate a threat. Risks allow managers to make tradeoffs between different mitigation strategies. A particular mitigation strategy may be excellent at reducing risks but, compared to another, may require too much time to implement and deploy or be too costly in terms of resources to run. On the other hand, the alternative may leave residual risks that are unacceptable. Knowing the risks, management can make the tradeoff and give engineering guidance as to which strategy to implement.

IoT Threat Modeling Example

This short video discusses step by step threat modeling for IoT systems with an example of a parking garage and presents an example of specific methodologies used for threat modeling. It primarily uses the Microsoft method but also briefly discusses another method called the Octave-Allegro method which was developed at Carnegie-Mellon University. Both methods offer a collection of tools for helping conduct threat modeling.

2.3.2 Trust Boundaries, Threat Boundaries, and Attack Surfaces

When threat modeling is complete, a picture of the trust boundaries for the BEMS should be clear. A trust boundary is an interface where two entities with potentially different interests interact. The entities can be human users or services. An example of the latter is the boundary between the building timeseries database server and the building timeseries database, while an example of the former is the BEMS dashboard and employees of different tenant companies that are responsible for managing their space in a multibuilding BEMS service. Whenever there is a trust boundary, entities on both sides of the boundary must prove their identity and prove their authorization to access resources using an IAM system. The threat boundary or attack surface is the overall collection of interfaces where an attacker can compromise a system. A trust boundary is often an attack surface since an unprotected trust boundary offers an attack vector to compromise the BEMS.

2.3.3 The Zero-Trust Architectural Principle

Zero trust is a cybersecurity architectural principle that eliminates any inherent or implied trust often associated with traditional security models [39]. The main principle behind the zero-trust principle is "never trust, always verify." In other words, the architecture is built on the assumption that threats can come from both outside and inside the network, and therefore, no user, device, application, or service should be trusted by default. The zero-trust model emerged as a response to the limitations of traditional perimeter-based security approaches that have become clearer with the increase in cloud-based service deployment, remote work, and the increase in sophisticated cybersecurity threats. The perimeter approach essentially sets the trust boundary and attack surface at the connection between the organization's network and a public network, or at the connection between a departmental network and the larger organization's network. If an attacker breaches that perimeter, the attacker can move laterally within the network looking for target services to compromise. The zero-trust architecture establishes the trust boundary at a much finer grained level, often at the level of an application or service, or an individual server or virtual machine.

Zero-trust architecture starts with an IAM system. The IAM performs authentication and authorization and applies not just to users but also to applications and services that perform some kind of action autonomously. The goal is to ensure that any entity claiming to be a particular identity and requesting access to resources is, in fact, the identity they claim to be and is authorized to access the resources. A related component is the principle of least privilege. The principle of least privilege means that entities only have the minimum level of access authorization necessary to perform their tasks. This helps limit the potential attack surface and prevents unauthorized access to sensitive data and systems. An additional component is encryption. Data is encrypted both at rest and in transit to ensure confidentiality and protect against unauthorized access. This is particularly important for sensitive information such as personal

More on the Zero-Trust Architecture

This short video from Microsoft provides a broad-based introduction to the motivation for the zero-trust architecture. The three basic principles are as follows: explicit verification, least privileged access, and assume a breach. The latter principle implies monitoring network traffic between services and well-defined operational procedures when (not if) an attack is identified so that the right people can swing into action to isolate the damage.

information of users and operational data such as access data for BEMS control and measurement devices like electric meters; but in a zero-trust architecture, encryption should be applied to all data.

From a network management standpoint, microsegmentation ensures that attackers who do gain access to the network cannot move around and cause more damage. Microsegmentation divides networks up into smaller segments with security controls in place between the segments and can be applied to both network traffic and applications. Finally, zero trust relies on real-time monitoring and analytics to detect and respond to threats. By collecting data on user behavior, network traffic, and device activity, security teams can quickly differentiate between normal user behavior and anomalies and take appropriate action.

2.3.4 Security Services

The network security architecture is based on a collection of security services. Security services include the following:

- Confidentiality protection – ensuring that the transmitted information remains private and is only accessible to the intended recipients of the communication,
- Integrity protection – verifying that the data exchanged between two parties has not been altered or tampered with during transmission,
- Data origin authentication – verifying that the data was indeed sent by the claimed sender by verifying the identity of the originator of the data,
- Non-repudiation – ensuring that both parties cannot deny their involvement in the communication or transaction,
- Availability – protecting against and recovering from volumetric traffic attacks and other interruptions so that the service remains available when users need it,
- Forward secrecy – ensuring that even if a private key is compromised, previous communication sessions remain secure,
- Identity authentication and access management – requiring the communicating parties to identify themselves in a provable manner and limiting their access to resources that they are authorized to access,
- Anonymity and privacy – protecting the identity of the communicating parties and preventing third parties from intercepting or tracking their communication. Anonymity and privacy are achieved by using virtual private network (VPN) technology, which is discussed in Chapter 5.

Security services are implemented in network protocols by combining the cryptographic operations and security technologies discussed in the following sections. Secure protocols are the topic of Chapter 5 and provide the foundation upon which secure BEMS services are implemented, starting with the link layer, and moving up through the network, transport, and application layers.

2.3.5 Cryptographic Operations

There are a few basic cryptographic operations that act as building blocks for implementing security services. Cryptographic operations are procedures that have some security intent. These operations depend on specific security technologies which are typically the implementation of mathematical algorithms with provable security properties. There are three basic types of cryptographic operations: authentication, encryption, and key exchange. Note that descriptions of authentication and encryption are typically worded as if an exchange is taking place between the sender and the receiver, but that need not be the case. The sender and receiver could be the same entity, with the sender storing the content, for example, a file, and the receiver being the same entity retrieving it at a later time. The cryptographic operations ensure that the file was not altered during the time it was in storage, and that no unauthorized entity gained access to it.

2.3.5.1 Authentication

Various authentication cryptographic technologies are used to implement integrity protection, non-repudiation, identity authentication, and access control. For authenticating data in messages and documents, authentication allows the receiver of a document or message to verify that the content was not modified in transit. The sender uses a cryptographic algorithm to calculate a verifier on the content and sends the verifier along with the content. The receiver recalculates the verifier and compares it to the verifier received from the sender. If the two match, then the content was not modified. This type of authentication is sometimes called data origin authentication. Depending on the algorithm, security services provided can include integrity protection or both integrity protection and non-repudiation. For identity protection and access management, authentication verifies an entity's (person's or service's) identity by verifying a security credential like a password and the receiver performs an authentication procedure on the credential to verify the identity. The two types are related in that the same cryptographic operations can be used for both, though authentication of an identity typically involves a more capable IAM system that can, in addition, establish the permissions allowed to the entity requesting access.

2.3.5.2 Encryption

Confidentiality protection is implemented with encryption. Encryption converts content to be communicated between the sender and receiver from plain text into cyphertext using cryptographic material called a key. Cyphertext is a stream of bytes that is unintelligible to eavesdroppers or other unauthorized parties. The sender uses a key plus the plain text input to be encrypted together with an encryption algorithm to convert the plain text to cyphertext. On the receiver side, the receiver uses a matched key to achieve the reverse transformation, making the content legible again.

2.3.5.3 Cryptosystem Types

Authentication and encryption are often covered by cryptographic algorithms that share a mathematical basis, called cryptosystems. There are two basic types of cryptosystems in common use:

- Symmetric (shared) key cryptosystems – both both parties to the transaction share the same key and use the same algorithm for the cryptographic operation.
- Asymmetric (public) key cryptosystems – Two parties to the transaction generate two keys per party: one that is publicly disclosed to the opposite side and another that is kept secret. Depending on the cryptographic operation and algorithm, one or the other key may be used by the sender or receiver.

Symmetric or shared key cryptosystems ensure confidentiality and secure communication by converting plaintext into ciphertext using a shared secret key. The recipient possesses the same key and can decrypt the ciphertext back into its original plaintext form. The primary strength of symmetric key cryptosystems lies in their simplicity and speed, which makes them suitable for encrypting large volumes of data. However, key distribution and management is more complex with symmetric key cryptosystems since securely exchanging and storing the shared secret key is crucial to maintaining the integrity of the encrypted data. The key must be kept confidential during the exchange and not disclosed to other parties. The Advanced Encryption Standard (AES), the winner of a competition sponsored by the National Institute of Standard and Technology (NIST) in 2001 [40], is used worldwide in all Internet standards and elsewhere for symmetric key encryption. AES can be deployed with key sizes of 128, 192, and 256 bits, but 256-bit keys are recommended for the strongest security.

Asymmetric or public key cryptosystems are based on "hard" (difficult to solve) problems in number theory and employ a pair of distinct but mathematically related keys, with one key kept private and one key publicly disclosed. The private key is known only to its owner, while the public key can be freely shared with others. Asymmetric key cryptosystems not only ensure confidentiality but also enable authentication, integrity protection, and non-repudiation using digital signatures. Though asymmetric key cryptosystems are computationally more intensive than symmetric key cryptosystems, they solve the key distribution problem because the public key can be sent by the receiver to the sender without any concern for whether the exchange is overheard. One commonly used asymmetric key system is the Rivest–Shamir–Adleman (RSA) algorithm, which is based on the difficulty of finding prime factors for large integers [41]. Another is elliptic curve cryptosystems (ECCs), which is based on the algebraic structure of elliptic curves (polynomials cubic in x and quadratic in y) [42, 43]. RSA key sizes tend to be very large, 2048 bits is the minimum recommended since computers have become much more capable since RSA was originally published in 1978 and able to factor large integers more quickly. The equivalent level of security for ECCs only requires a 256-bit key.

2.3.5.4 Key Generation and Exchange

Generating a key for use in authentication or encryption requires use of the appropriate cryptographic algorithm with selected parameters and an input of a randomly generated number. The length of the key is important for the security of the algorithm and should not be below the minimum length recommended for the algorithm. As a rule of thumb, the longer the key, the stronger the security. Some cryptosystems have known weak keys due to their particular mathematical properties. These keys should be avoided. AES has no known weak keys, as long as the random number generator used to derive the key generates numbers with sufficient entropy. If the parameters used to generate an RSA key pair are not sufficiently robust or the random number generator does not have sufficient entropy, weak keys can result [44]. The ECC algorithms in common use do not have weak keys as long as a sufficiently long key is used (e.g. 256 bits), but there are elliptic curve algorithms that are known to have weaknesses [45]. Many protocol standards specify the parameters to be used in key generation to ensure that weak parameters are not used, making the random number entropy the only possible source of difficulty. Forward secrecy can be achieved by generating unique session keys for each new session and limiting the session length by periodically regenerating the keys. This limits the damage if an attacker manages to compromise one session key.

With asymmetric key cryptosystems, exchanging a key between two parties that want to communicate confidentially is straightforward. The parties just exchange their public keys. Note that this does not allow either party to establish an authenticated identity for the other, but it does allow them to exchange encrypted messages and to calculate data origin authentication verifiers on the messages, called digital signatures, ensuring that the message was not altered in transit. However, with private key cryptosystems, some cryptographic operation is needed to transfer the shared key between the two parties.

One common technique is to encrypt the private key with the receiver's public key and send it, allowing the recipient to decrypt with their private key. Once the shared key is established, the two sides can use the much faster symmetric key encryption. The alternative is writing the key down on a piece of paper and hand delivering it to the other party. Another technique is the Diffie–Hellman key exchange algorithm. Diffie–Hellman key exchange is based on the same number theory approach as asymmetric cryptosystems but a different hard problem, called the Diffie–Hellman problem. Diffie–Hellman key exchange is widely used in IP standards to bootstrap a shared private key from a public key pair [46].

2.3.5.5 Security Associations

To implement security services for two communicating parties, the two sides must share agreement on the cryptographic algorithms to be used and a collection of parameters. The parameters must either be provisioned in a way that ensures any communication of confidential material is not overheard, or the confidential material must be kept private on the node where it is generated. A security association (SA) is a set of cryptographic parameters and keys for a collection of cryptographic algorithms that are established between two network entities to implement the security services. The SA defines the algorithms used for encryption and authentication, the keys to be used by the algorithms, and other essential security information. SAs are required for link, network, and transport layer protocol security, and a separate SA is typically required for communication with different endpoints. Therefore, each SA must be uniquely identified. The establishment, maintenance, and deletion of SAs are crucial for ensuring the correct implementation of security services.

2.3.6 Security Technologies

A collection of technologies has been developed to help in the design and implementation of secure system architectures. These technologies are based on algorithms derived from mathematical theorems in number theory and other areas, and the algorithms typically make provable statements about the security of exchanges between participants that use them. The technologies function as a kind of toolkit which, when properly applied, can substantially enhance the cybersecurity of a system. This section provides an overview of those technologies, and details on the actual algorithms as well as open-source implementations can be found in the references.

2.3.6.1 Hash Functions

A hash function is a one-way function that takes a digital document or message and creates a short digital fingerprint, called a digest, which allows a receiver to verify that the contents have not been modified. The digest is attached to the content before sending or storage. An important property of hash functions is that it is impossible to recreate the original text from the digest. This property makes hash functions a good tool to use in verifying passwords. The hash of the password is stored in a database rather than the password itself, ensuring that the password is not compromised if an attacker somehow gets access to the database. Comparing the hash of a password taken during a login attempt with the hash in the database verifies the identity of the entity attempting to login. NIST has developed recommendations about secure hash functions [47, 48]. SHA-256 and SHA-3 are recommended, and these functions are widely used in IP standards. SHA-1 is also used in some earlier protocol security standards. Specialized functions are used for hashing passwords before they are stored in an IAM system database. There are many hashing algorithms designed for keeping passwords confidential, but Argon2 [49] was the winner of the Password Hashing Competition which ran from 2013 to 2015 and has many operating system and language bindings.

2.3.6.2 Digital Signatures

Digital signatures are a security technology used to ensure the authenticity, integrity, and non-repudiation of digital documents, messages, and transactions. To create a digital signature, the sender computes a digest of the message using an appropriate hash function, for example, SHA-256, and then utilizes a public key cryptosystem like RSA or the elliptic curve digital signature algorithm (ECDSA) to sign the digest with their private key, creating a unique signature that is tied to their public key. The recipient can then calculate the digest and verify the signature using the sender's public key, confirming the integrity of the data and that the data did originate from sender. Since the private key is only known to the sender, digital signatures prevent tampering, forgery, or denial of involvement in the communication.

2.3.6.3 Hashed Message Authentication Codes

A shared key system such as AES can be used to calculate a message authenticator, but the authenticator does not have the non-repudiation property of digital signatures. A common type of algorithm used in the Internet to establish data origin authentication which does include the non-repudiation property is a hashed message authentication code (HMAC). HMAC [50] is a widely used keyed cryptographic hash requiring both parties to share a key. HMAC algorithms are paired with a hashing algorithm, like HMAC-SHA1, HMAC-SHA256, or HMAC-SHA3. HMAC is immune to length extension attacks if the hashing algorithm is. A length extension attack occurs when the hacker takes the original message and the HMAC authenticator, extends the message, and calculates a new HMAC authenticator without knowing the key. SHA1 and SHA256 are vulnerable to a message extension attack, but the newer SHA3 algorithm is not [48].

2.3.6.4 Open-Source Implementations

Open-source implementations of many cryptographic algorithms are available through the OpenSSL project [51]. A wide variety of algorithms are implemented in useful tools and libraries that can be incorporated directly into the application code. In addition, an open-source bcrypt implementation is available from SourceForge [52]. As with any software, however, implementations of cryptographic algorithms are subject to bugs, though typically the code is scrutinized much more closely than other code before release. If open-source implementations of cryptographic algorithms are used in the implementation of a BEMS, the network management team must stay on top of notices for upgrades, and promptly apply upgrades when they are released, to ensure that the BEMS remains protected against attacks.

2.3.7 Security Reference Architecture

A security reference architecture serves as a comprehensive blueprint for designing, implementing, and managing the BEMS cybersecurity infrastructure. It provides a structured approach to address cybersecurity by outlining best practices, guidelines, and principles for building a resilient and secure BEMS. A security reference architecture additionally facilitates communication and collaboration among stakeholders involved in commissioning and operating the BEMS by presenting a common language and framework for discussing security-related matters. A well-defined and well-documented security reference architecture helps organizations to align their security strategies with business objectives and stay ahead of emerging risks.

2.3.7.1 Network Protocol and Data Security

The foundations of the security reference architecture rest on secure communications and data storage. Security technologies for communications are incorporated into the BEMS in the network stack. Standardized protocols particularly at the link, network, and transport layer include specifications for security measures based on the cryptographic operations and the security technologies discussed in Sections 2.3.5 and 2.3.6. For zero-trust architectures, these protocols should be utilized wherever network communication between two entities occurs, without exception. They should be used when a human user wants access to the BEMS through a dashboard, when two services within the BEMS communicate with each other, and when an internal BEMS service communicates with an external site, or vice versa. Chapter 5 discusses protocol security in more detail.

Security for data requires ensuring that all data stored in databases is encrypted if possible. Sensitive data like protected personal information such as user account information should always be encrypted in any case. Other sensitive data may involve specific settings for HVAC machinery or electrical meters that would pose a health and safety risk if compromised. Passwords should never be stored in plain text, but should always be hashed.

2.3.7.2 Firewalls, Security Groups and Access Control Lists

The next level up in a security reference architecture involves controlling communications between different entities to ensure that only entities with authenticated identities are allowed access within the scope of their authorized permissions. An important component for controlling access is the firewall. A firewall is a network security component that monitors and controls incoming and outgoing network traffic based on predetermined security rules. It helps control access so that unauthorized parties are excluded. Firewalls are pervasive in modern cloud service deployments and can be implemented as hardware, software, or a combination of both. They function by examining incoming and outgoing data packets and determining whether to allow or block them based on a set of predefined rules, effectively filtering, and managing network traffic. The rules typically include fields in the protocol headers at different layers of the stack, like the IP address at the network layer and the port number at the transport layer, that are allowed or denied.

In modern cloud deployments, every virtual or physical machine instance has an associated firewall called a security group. A security group controls inbound and outbound network traffic for one or more virtual instances. Security groups act as a set of rules that determine which network traffic is allowed to flow in and out of the associated instances. These rules are defined based on fields in the protocol headers, such as protocol numbers, ports, and source or destination IP addresses. When network traffic tries to reach an instance, the security group rules are evaluated, and only the traffic that meets the specified criteria is allowed through. Security groups are typically stateful, meaning that if a request is allowed to flow from a source to a destination, the return traffic is automatically permitted, even if there is no explicit rule for it. They also include default rules; for example, to deny all inbound traffic by default while allowing all outbound traffic. The rules are evaluated on a case-by-case basis and modified where appropriate. Security groups help to isolate and protect instances from unauthorized access and attacks. When configuring security groups, it is essential to follow the principle of least privilege and only allow the necessary traffic to minimize the attack surface.

A related functional component is a firewall associated with networks, called an access control list (ACL). An ACL is a virtual firewall that operates at the IP subnet level within a cloud environment, typically implemented in subnet routers. It filters network traffic entering or exiting a subnet based on a set of rules similar to those defining a security group. ACLs are applied to control traffic to and from entire subnetworks. However, ACLs are not stateful. Inbound and outbound traffic is controlled by a separate set of rules and the rules must be defined for both. ACLs and security groups also have different ways of evaluating rules. Security groups typically evaluate rules based on the most permissive set when multiple security groups are associated with a single instance or resource. In contrast, ACLs evaluate rules in a specific order (usually determined by rule numbers), and the first rule that matches the traffic is applied.

2.3.7.3 Identity and Access Management

IAM systems provide a foundation for managing and controlling access to an organization's resources, applications, and data. By ensuring only authenticated entities have access to just the resources they need to accomplish their job, IAM systems help maintain the confidentiality, integrity, and availability of information, while also mitigating the risk of unauthorized access, data breaches, and insider threats. IAM systems apply not just to human users but also to services and applications accessing other services and applications in an autonomous fashion. Such accounts associated with software and system entities are called service accounts.

IAM systems cover authentication, authorization, access control, and lifecycle management of the accounts associated with entities needing access to BEMS data and services. Authentication verifies the identity of entities by validating their credentials. The nature of the credentials depends on the system component being accessed. For human users, most IAM systems require a username and password and issue an authentication token after the user has successfully navigated the authentication process. The token is then used for authenticating with various BEMS services. For services, a public/private key pair tied to the service with a public key certificate may be required, though some IAM systems use tokens for service-to-service IAM. Authentication tokens and public key certificates are discussed in Chapter 5. Authorization, on the other hand, involves defining and enforcing access policies that determine the level of access an authenticated identity has to specific resources or applications. Account lifecycle management encompasses the entire process of creating, managing, and removing accounts and access rights within the IAM system.

2.3.7.4 Access Control

In a zero-trust architecture, services need to provide a verifiable identity to access resources on another service, and their access rights need to be established through an IAM systems. Most cloud services use one of the two access control approaches: role-based access control (RBAC) or policy-based access control.

RBAC establishes a collection of permissions for accessing resources with particular roles and assigns roles to authenticated identities. The roles were originally associated with job functions in an organization when the authenticated identities only included human users, but as RBAC uses expanded to encompass authorization of services, roles now include functions that are accomplished by services autonomously, such as writing sensor data into the building timeseries database through its API server. Permissions are assigned to roles based on the principle of least privilege, and an authenticated identity may have more than one role assigned to it if it requires access to multiple sets of resources. Permissions should be as granular as possible to ensure that only sufficient privilege is granted for the entity to accomplish its tasks.

Policy-based access control allows a more fine grained determination of what services and users can access specific resources by establishing policies that govern which authenticated identities can access resources and under what conditions. The architecture consists of three functional components: policies, which encompass the rules and conditions for access; a policy decision point (PDP), which evaluates access requests against these policies; and a policy enforcement point (PEP), which enforces the decision of the PDP by either granting or denying access based on the policy evaluation. The PEP is typically located at or near the service's connection with other services, so it can intercept incoming traffic, while the PDP is part of the IAM service where identities and policies are recorded. Most policy-based access control systems also include a policy definition language for expressing resource access permissions.

2.3.8 Commissioning Security

Installing and configuring a device so that it is connected and functioning in the building network, called commissioning, require that the device and network undergo a procedure to establish trust. For IT systems, the installation is typically done by system administrators running software tools through a command line interface or shell to bring a laptop or switch into the network. Unlike IT systems, IoT devices in OT systems like a BEMS are typically not installed by people with access to a laptop running a shell in a window, but rather technicians with a practical knowledge of how to install and configure the hardware, having access to a cell phone. For "smart home" devices, the installer is typically the end user and they may have little or no idea about how to manage IT systems much less commission them securely into their home network. The commissioning of IoT devices introduces an important point where a security architecture is necessary to safeguard data and operational integrity. Cryptographic protocols, secure key management systems, and identity verification mechanisms allow a device to establish trust with the network and thereby become an operational member of the in-building device network.

The security services involved in commissioning security are the same as for security of the operating network, but are often deployed in different ways. IoT devices need to be pre-provisioned with certificates to act as a trust anchor, allowing the device to identify itself to the network. In a zero-trust architecture, the network, which may be represented by a commissioning service, also needs to identify itself to the device. Some commissioning protocols go further and allow verification of the supply chain by which the device was purchased. Since devices require commissioning before any network, transport, or application level code other than the commissioning code is run, commissioning protocols and methods are often defined at the link layer. An important consideration with respect to certificate management is that the same certificate not be used for commissioning as for identifying the device-to-end nodes during the execution of network or TLS protocols. This ensures that the operational certificate can be renewed periodically, something that is typically not possible with the commissioning certificate. In addition, secure commissioning protocols need to be defined in a way that does not require running software from the command line as most IT security configuration procedures do since installers may only have a cell phone handy during the commissioning process.

2.4 The Convergence of OT and IT Systems

Traditionally, IT and OT systems have been kept separate with separate networks operating in silos and dedicated workstations for supervisory control of OT devices. IT systems have been primarily concerned with data processing, communication, and the overall management of enterprise networks. OT systems have been focused on the direct control and monitoring of physical devices and processes within buildings, such as HVAC, lighting, and other energy-related systems. The historical separation of IT and OT networks was partly due to differing technical requirements, priorities for reliability and security, and the specialized nature of OT systems. However, with the maturation of the IP stack and the advent of "smart" IoT devices, the barriers to integrating these networks have diminished. The falling cost of IT equipment and the widespread deployment of IT networks in buildings of all types make deploying an OT network on top of IT infrastructure a much cheaper option than deploying a bespoke OT network. The integration can achieve enhanced efficiency, centralized management and can unlock advanced analytics and control strategies. Yet, it also requires careful consideration of the unique characteristics and requirements of OT use cases.

For example, commissioning plays a crucial role in OT systems, differing significantly from the process of integrating new devices or systems within IT networks. Section 2.3.8 discusses commissioning security, but there are other aspects to commissioning that are important. Commissioning of OT systems like a BEMS involves rigorous testing and validation of control systems and their components to ensure they operate correctly, meet the required performance standards, and can effectively manage the building's energy systems. This process is far more complex than simply bringing a new laptop or network switch online in an

IT environment and is similar to the kind of rigorous testing done during the development of a software product. For OT systems, commissioning includes calibrating sensors, validating control logic, and ensuring that automated systems can reliably interact with physical processes. This meticulous approach ensures that energy management systems can efficiently regulate building operations and reduce carbon emissions.

Leveraging IT capabilities in OT environments can lead to more sophisticated control strategies, improved energy efficiency, and better overall management of building systems through data analytics and integrated operational views, in addition to lower cost due to the commodified nature of most IT equipment. However, this integration requires addressing the unique security, reliability, and commissioning requirements of OT systems. As buildings become more connected and reliant on advanced technologies for energy management, understanding the distinct requirements OT systems impose on the underlying IT systems used for their implementation becomes increasingly important. This understanding is essential for designing, implementing, and maintaining systems that not only enhance building operational efficiency but also contribute to the goal of eliminating building carbon emissions.

2.5 Summary

The functional architecture presented in Section 2.1 provides the basis for agreement on the components in the BEMS, what functionality these components must provide (the role they play in the overall system), and the interfaces between the components. Agreement on the functional architecture is the foundation on which standards for the protocols at the interfaces are based. In turn, standardized protocols ensure interoperability between devices and applications from different vendors. From a deployment perspective, the horizontal integration deployment architecture presented in Section 2.2.3 removes most of the problems with the vertically integrated and overlay deployment architectures, discussed in Sections 2.2.1 and 2.2.2 respectively, by allowing the building owner to purchase devices and applications developed by different vendors that interoperate. In particular, data collected through one set of devices and applications can be utilized by a variety of other applications. In addition, having a well-defined, well-documented, and robust security architecture is essential for ensuring the cybersecurity of the BEMS. Section 2.3 discusses some architectural concepts for developing a strong cybersecurity architecture. Finally, the convergence of OT and IT systems, especially for networking, has been driven by the falling cost of IT equipment and the widespread deployment of IT networks in buildings of all types.This, in turn, drives deployment of horizontal architecture systems.

Section 2.4 briefly discusses the IT/OT convergence, but this topic is a theme that will return periodically throughout the book.

Achieving agreement on standards between different stakeholders in the BEMS business ecosystem remains an outstanding challenge however. The horizontal integration architecture is very much a work in progress, and many of the necessary standards have yet to be firmly established. In some cases, there are multiple technical approaches for a technology component at a particular layer, or legacy protocols that must be accommodated. Additional design work may be needed on the hardware side to accommodate legacy BAS networking protocols. In coming chapters, the focus turns to the details of networking technologies at all layers of the stack, and how they can be utilized in a BEMS. Technologies that are candidates for including in a horizontal integration architecture are explored, in addition to the legacy technologies, such as the specialized standard building automation protocols, that will continue to play a role in existing buildings' BEMS.

References

1 Wikipedia. OSI Model [Internet]. [cited 2022 Aug 7]. Available from: https://en.wikipedia.org/wiki/OSI_model.

2 Internet Engineering Task Force (IETF). Requirements for Internet Hosts – Communication Layers, RFC 1122 [Internet]. [cited 2023 Jan 13]. Available from: https://datatracker.ietf.org/doc/html/rfc1122.

3 IETF. Requirements for Internet Hosts – Application and Support, RFC 1123 [Internet]. [cited 2023 Jan 13]. Available from: https://datatracker.ietf.org/doc/html/rfc1123.

4 Wikipedia. IEEE 802.11 [Internet]. [cited 2022 Jun 16]. Available from: https://en.wikipedia.org/wiki/IEEE_802.11.

5 Wikipedia. IEEE 802.1 [Internet]. [cited 2022 Jun 16]. Available from: https://en.wikipedia.org/wiki/IEEE_802.1.

6 Postel J. Internet Protocol, RFC 791 [Internet]. [cited 2022 Feb 13]. Available from: https://www.rfc-editor.org/rfc/rfc791.

7 IETF. Internet Protocol, Version 6 (IPv6) Specification, RFC 8200. July 2017. Available from: https://www.rfc-editor.org/rfc/rfc8200.txt.

8 Schuler R. How Does the Internet Work? [Internet]. [cited 2022 Aug 8]. Available from: https://web.stanford.edu/class/msande91si/www-spr04/readings/week1/InternetWhitepaper.htm.

9 Wikipedia. User Datagram Protocol [Internet]. [cited 2022 Aug 8]. Available from: https://en.wikipedia.org/wiki/User_Datagram_Protocol.

10 Wikipedia. Transmission Control Protocol [Internet]. [cited 2022 Aug 8]. Available from: https://en.wikipedia.org/wiki/Transmission_Control_Protocol.

11 IETF. Assigned Numbers, RFC 1700. October 1994. Available from: https://www.ietf.org/rfc/rfc1700.txt.

12 TCP-UDP-Ports.com. TCP UDP Ports: Port 47808 [Internet]. [cited 2022 Aug 8]. Available from: http://tcp-udp-ports.com/port-47808.htm.

13 IETF. Domain Names: Concepts and Facilities, RFC 1034. November 1987. Available from: https://www.ietf.org/rfc/rfc1034.txt.

14 IETF. Domain Names: Implementation and Specification, RFC 1035. November 1987. Available from: https://www.rfc-editor.org/rfc/rfc1035.

15 IETF. Hypertext Transfer Protocol (HTTP/1.1): Message Syntax and Routing, RFC 7230. June 2014. Available from: https://www.rfc-editor.org/rfc/rfc7230.

16 IETF. Hypertext Transfer Protocol (HTTP/1.1): Semantics and Content, RFC 7231. June 2014. Available from: https://www.rfc-editor.org/rfc/rfc7231.

17 Organization for the Advancement of Structured Information Standards (OASIS). MQTT Specification [Internet]. [cited 2022 Nov 10]. Available from: https://mqtt.org/mqtt-specification.

18 International Organization for Standardization/International Electrotechnical Commission (ISO/IEC). ISO/IEC 20922:2016 Information Technology — Message Queuing Telemetry Transport (MQTT) v3.1.1 [Internet]. [cited 2022 June 21]. Available from: https://www.iso.org/standard/69466.html.

19 Neptronic.com. BACnetModbus [Internet]. [cited 2022 Oct 8]. Available from: https://www.neptronic.com/controls/PDF/EVC/BACnetModbus/BACnet%20MSTP%20Overview%20Manual-160405.pdf.

20 ASHRE. BACnet: Official Website of ASHRAE SSPC 135 [Internet]. [cited 2022 May 25]. Available from: http://www.bacnet.org.

21 Institute of Electrical and Electronics Engineers (IEEE). IEEE Standard for Low-Rate Wireless Networks. In: IEEE Std 802.15.4-2020 (Revision of IEEE Std 802.15.4-2015). pp. 1–800, 2020 Jul 23. doi: https://doi.org/10.1109/IEEESTD.2020.9144691.

22 OpenADR Foundation. OpenADR Specifications [Internet]. [cited 2022 May 24]. Available from: https://www.openadr.org/specification.

23 IEEE. IEEE Standard for Smart Energy Profile Application Protocol [Internet]. [cited 2022 May 24]. Available from: https://standards.ieee.org/ieee/2030.5/5897.

24 OpenAPI Initiative. OpenAPI Specification v3.1.0 [Internet]. [cited 2022 Jun 20]. Available from: https://spec.openapis.org/oas/v3.1.0.

25 GraphQL Foundation. GraphQL specifications [Internet]. [cited 2022 Jun 21]. Available from: https://spec.graphql.org.

26 Dice J. Horizontal Architecture 101: The Building Blocks of a Smart Building. [cited 2024 Jan 8]. Available from: https://www.nexuslabs.online/whitepapers/nexus-lore-the-smart-building-blocks (Accessed 2024 Jan 08).

27 Dice J. (Smart) building blocks Part 1: The Silos, Nexus Labs [Internet]. [cited 2022 Jun 9]. Available from: https://www.nexuslabs.online/smart-building-blocks-the-silos-nexus-newsletter-125.

28 Dice J. Part 2: The Overlays, Nexus Labs [Internet]. [cited 2022 Jun 10]. Available from: https://www.nexuslabs.online/part-2-the-overlays.

29 Dice J. Part 3: The Horizontal Architecture. [Internet]. [cited 2022 Jun 14]. Available from: https://www.nexuslabs.online/part-3-the-horizontal-architecture.

30 ISO/IEC. ISO/IEC 14908-1:2012: Information Technology – Control Network Protocol – Part 1: Protocol Stack [Internet]. [cited 2022 Jun 10]. Available from: https://www.iso.org/standard/60203.html.

31 ISO/IEC. ISO/IEC 14543-4-1:2008: Application Layer for Network Enhanced Control Devices of HES Class 1 [Internet]. [cited 2022 Jun 10]. Available from: https://www.en-standard.eu/iso-iec-14543-4-1-information-technology-home-electronic-system-hes-architecture-part-4-1-communication-layers-application-layer-for-network-enhanced-control-devices-of-hes-class-1.

32 Modbus Organization. The Modbus Organization [Internet]. [cited 2022 Jun 10]. Available from: https://modbus.org.

33 Wikipedia. Representational State Transfer [Internet]. [cited 2022 Jun 10]. Available from: https://en.wikipedia.org/wiki/Representational_state_transfer.

34 Amando R. Shameful: Insteon Looks Dead – Just Like Its Users' Smart Homes [Internet]. [cited 2022 Jun 14]. Available from: https://arstechnica.com/gadgets/2022/04/shameful-insteon-looks-dead-just-like-its-users-smart-homes.

35 Lam S. How Building Digitalization Is Changing The Master Systems Integrators Focus Area [Internet]. [cited 2022 Jun 15]. Available from: http://automatedbuildings.com/news/may21/articles/sabine/210418015701sabine.html.

36 Katz R., Brewer E., and McCanne S. The Evolution of Internet Services [Internet]. [cited 2022 Jun 20]. Available from: https://www.researchgate.net/publication/267972931_The_Evolution_of_Internet_Services.

37 Universal Device Management Interface (UDMI). UDMI Overview [Internet]. [cited 2022 Jun 21]. Available from: https://github.com/faucetsdn/udmi.

38 Shostack A. Threat Modelling: Designing for Security. Wiley, New York; 2014.

39 Wikipedia. Zero Trust Security Model [Internet]. [cited 2022 Jun 23]. Available from: https://en.wikipedia.org/wiki/Zero_trust_security_model.

40 National Institute of Standards and Technology. Specification for the Advanced Encryption Standard (AES). FIPS PUB 197. Information Technology Laboratory, Gaithersburg, MD; 2001.

41 Rivest R.L., Shamir A., and Adleman L. A method for obtaining digital signatures and public-key cryptosystems. *Communications of the ACM*, vol. 21, no. 2, pp. 120–126, 1978.

42 Koblitz N. Elliptic curve cryptosystems. *Mathematics of Computation*, vol. 48, no. 177, pp. 203–209, 1987.

43 Miller V.S. Use of elliptic curves in cryptography. In: Advances in Cryptology—CRYPTO'85 Proceedings. Lecture Notes in Computer Science, vol. 218, pp. 417–426; 1985.

44 Langkemper S. Attacking RSA Keys [Internet]. [cited 2023 Apr 2]. Available from: https://www.sjoerdlangkemper.nl/2019/06/19/attacking-rsa.

45 Wikipedia. Elliptic-Curve Cryptography [Internet]. [cited 2023 Apr 2]. Available from: https://en.wikipedia.org/wiki/Elliptic-curve_cryptography.

46 Rescorla E. Diffie–Hellman Key Agreement Method. RFC 2631 [Internet]. [cited 2023 Apr 2]. Available from: https://www.rfc-editor.org/rfc/rfc2631.html.

47 National Institute of Standards and Technology. Secure Hash Standards (SHS). FIPS PUB 180-4. Information Technology Laboratory, Gaithersburg, MD; 2015.

48 National Institute of Standards and Technology. SHA-3 Standard: Permutation-Based Hash and Extendable-Output Functions. FIPS PUB 202. Information Technology Laboratory, Gaithersburg, MD; 2015.

49 Dinu D., Khovratovich D., Aumasson J.P., and Neves S. Argon2 [Internet]. [cited 2023 May 31]. Available from: https://github.com/p-h-c/phc-winner-argon2.

50 Krawczyk H., Bellare M., and Canetti R. HMAC: Keyed-Hashing for Message Authentication, RFC 2104 [Internet]. [cited 2023 May 23]. Available from: https://www.ietf.org/rfc/rfc2104.html.

51 OpenSSL. Welcome to OpenSSL! [Internet]. [cited 2023 Mar 31]. Available from: https://www.openssl.org.

52 Shelley J. and Stolarczky P. bcrypt [Internet]. [cited 2023 Mar 31]. Available from: https://bcrypt.sourceforge.net.

3

Link–Layer Protocols

Communication between BEMS and devices performing sensing and control for energy-consuming devices on site in the building requires a communication network. As discussed in Chapter 2, the architectural component of the BEMS embodying network communication is founded on a physical layer. The physical layer provides the primary physical medium upon which the software stack for communication between sensors, control points, computation nodes, gateway nodes, and the cloud is built. In the Internet architecture, the physical and data link layers are condensed into the link layer. The link layer connects the software stack of different nodes on the same physical link and provides services coordinating multiple access to the physical layer from variou snodes. Different data link layers can be used to connect device nodes and the building gateway and between the building gateway and the BEMS in a data center.

Within the building, local network data link layers connect nodes to the building gateway. Data from sensors is aggregated onto the building's LAN before being sent to the building gateway or local BEMS server and the BEMS databases, and control actions from the BEMS are sent over the LAN to actuators. This chapter discusses physical and link–layer technologies for the on-site building network between devices and the building gateway. Section 3.1 introduces the two data link layers for LANs that are the most widely deployed, specifically Ethernet for wired LANs, discussed in Section 3.2, and Wi-Fi for wireless LANs (WLANs), discussed in Section 3.3. In addition to LANs, device nodes are often connected through mesh wireless networks that cover relatively small areas. These wireless local device networks are sometimes called WPANs. In Section 3.4, link layers for WPANs are introduced. The 802.15.4 WPAN protocol is presented in Section 3.5, while Bluetooth Low Energy (BLE) is presented in Section 3.6. Unlike Wi-Fi, most devices that use the 802.15.4 protocol do not implement the Internet stack directly

Automating Building Energy Management for Accelerated Building Decarbonization: System Architecture and the Network Layer, First Edition. James Kempf.

Published 2025 by John Wiley & Sons, Inc.

on top of the link layer. If they implement a full Internet stack, they implement an adaptation layer between the 802.15.4 data link layer and the Internet layer. Some 802.15.4 devices implement alternative stack layers up to the application layer. The IoT stacks built on top of 802.15.4 are examined in Chapter 7.

Connecting devices within the building to services outside the building requires a WAN. Wired WAN link-layer protocols are not discussed here because the BEMS engineer usually has little choice in a wired link-layer protocol. If the building has access to a wired Internet service, the building manager or owner typically has already chosen an ISP that provides 50 Mbps to 1 Gbps Internet service IT services through an onsite router. The building network never sees the link-layer protocol since the interface is entirely through the IP network layer. The data rate and latency of the service are usually perfectly adequate for BEMS, but congestion, resiliency, and reliability may, in some cases, be of concern, and cost will naturally vary.

Often a more economical or reliable option is to use a wide-area wireless protocol for the device network or aggregated BEMS traffic. Section 3.7 discusses the Long RAnge (LoRA) wireless wide area sensor network protocol, an option widely implemented in inexpensive sensors that are easy to deploy. Section 3.8 discusses cellular protocols, specifically fourth-generation long-term evolution (4G/LTE) CAT M1 and Narrowband-Internet of Things (NB-IoT) . These protocols are relatively recent additions to the options for WAN connectivity and can be used to bypass a wired ISP if the building's wired Internet connection is unreliable or frequently congested. In Section 3.9, the topic of how to select link-layer technologies for a BEMS is discussed. Finally, Section 3.10 summarizes the chapter.

3.1 LAN and WLAN Link-Layer Protocols

Commercial and residential buildings today typically have a private LAN installed that provides communication on the site with WAN access to the public Internet through an Internet router provided by an ISP. In most cases, the Internet connection is to a wired ISP of some type, but a more recent trend is to use a high data rate 4G or 5G cellular wireless router for the link to the Internet. As discussed in Chapter 2, devices involved in measuring and controlling energy use are connected to the building backhaul network, perhaps through a gateway from a legacy serial bus or an intermediate WPAN. A building gateway aggregates traffic and sends it through the public Internet to the BEMS in the cloud or an on-premises data center. Wired and wireless LAN networks are the middle piece between the devices or device networks and the Internet. In this section, the dominant wired and wireless link-layer protocols for commercial and residential BEM backhaul networks are discussed.

3.2 IEEE 802.3 (Ethernet)

The link–layer protocol for wired private LANs, both commercial and residential, is overwhelmingly IEEE 802.3 or Ethernet [1]. The Ethernet protocol was initially developed in the early 1970s at the Xerox Palo Alto Research Labs [2] and required bulky coaxial cables for physical connectivity. Since then, the protocol has been adapted to multiple different types of physical media suited to a wide variety of physical implementations and network deployment types. The most common physical connectivity deployed in commercial and residential buildings today consists of the ubiquitous RJ 45 cable connector fastened to a twisted pair or multi-conductor flat cable, although deployment through an optical cable is also available in some cases. The Ethernet physical layer runs at data rates from 1 Mbps through 400 Gbps, although the most common data rates in buildings are 10 Mbps (10BASE-T), 100 Mbps (100BASE-TX), and 1 Gbps (1000BASE-T). The maximum cable length for these data rates is 100 m.

3.2.1 Ethernet Medium Access Control

Prior to the development of switched Ethernet in the 1990s, Ethernet used an algorithm called Carrier Sense–Multiple Access with Collision Detection (CSMA/CD). CSMA/CD is specified in the 802.3 Ethernet Standard. CSMA/CD requires every node on the same link to receive every frame, even though the node's address may not match the destination address. The following is an outline of the CSMA algorithm:

- When the node wants to transmit a frame, it monitors the channel for a specified period to see if the channel is busy.
- If the channel is busy, the node waits until the channel is free.
- If the channel is free, the node transmits the frame.
- While transmitting, the node monitors the channel to detect a collision.
- If a collision is detected, the node begins the collision resolution algorithm.
- When the node successfully completes frame transmission, it resets the retransmission counter.

If a collision is detected, the node runs the following steps to resolve the collision:

- The node continues the transmission of the frame for a specified period and includes a jam signal to ensure the collision is detected by other nodes.
- The node increments the retransmission counter.
- If the maximum number of retransmissions is reached, the node terminates the retransmission.

- If not, the node waits for a backoff interval that is a function of the number of retransmissions and retries the transmission.

CSMA/CD was replaced by switched Ethernet in the late 1990s and early 2000s. The forwarding algorithm for switched Ethernet was initially specified in IEEE 802.1D, but the newest version of the standard is IEEE 802.1Q [3].[1] With switched Ethernet, each node is connected into the LAN through a separate switch port with its own forwarding queue, and forwarded frames are switched by hardware. The switch learns the link–layer address of a node by noting the link–layer source address of the first frame transmitted by the node after it is connected and associates the node's link–layer address with the port from which the packet arrived in a forwarding table. Incoming frames for the node are then switched to that port for delivery if the frame destination address matches the address associated with the port in the forwarding table. If the switch does not have an association between a port and a destination address for an arriving frame, the switch floods the frame out of all ports except the one over which the frame arrived. When the destination node replies, the switch records the port to address mapping. Since each node has a separate connection to the switch, nodes only receive frames with their destination address. Connections in switched Ethernet are also full duplex, meaning that both sides can transmit and receive at the same time, and since there are only two nodes on the link (the node and the switch) no frame collisions can occur, allowing much higher throughput.

Ethernet switches provide network coverage over a broader area, but forwarding loops must be avoided; otherwise, the bandwidth could be consumed by infinitely looping packets. Switched Ethernet eliminates forwarding loops by running the spanning tree protocol. The spanning tree protocol takes an arbitrary graph of interconnected switches, which may contain loops, and converts it into a tree with a single root and one path through the tree from every switch to the root, eliminating redundant forwarding paths that might cause loops. The most recent version of the spanning tree protocol is called rapid spanning tree. It was standardized in 2004 by 802.1D and later merged into 802.1Q.

To calculate the spanning tree, the switches use the following algorithm [4]:

- Periodically (every two seconds in the rapid spanning tree protocol), the switches exchange dedicated frames called bridge protocol data units (BPDUs) with other switches. These frames are sent to a dedicated multicast link layer address to which all switches are listening, called the system medium access control (MAC) address.

1 The 802.1Q standard describes the protocols and algorithms for an Ethernet *bridge*, which provides a subset of the features provided by a switch. Although the two are different types of nodes, most building LANs today use Ethernet switches and so this term is used throughout the following discussion.

- The switches automatically elect a root switch by using a switch priority value recorded in the BPDU and the system MAC. If the switch priority value (default 32,768) is the same for all switches, then the switch with the lowest system MAC is elected the root. Otherwise, the switch with the lowest switch priority value wins. The switch priority provides a mechanism to select the highest data rate switch as the root regardless of its system MAC by configuring the priority value to something less than 32,768 on the highest data rate switch.
- Each nonroot switch calculates the cost for each of its ports that have a direct or indirect connection to the root switch using the BPDU messages. The root switch sets its cost to 0 in the BPDU messages it sends out. A switch calculates the path cost for a port by taking the accumulated cost from the arriving BPDU frame and adding a port cost for each of its ports. The switch then records the ports and cost for the ports in a new BPDU frame and transmits it. The cost for a port is based on the port data rate, with slower ports having higher costs. The total cost is the sum of the costs of all the ports on the path to the root.
- The switch assigns the port with the least cost path back to the root as the root port and blocks other ports back to the root to avoid loops.
- The switch calculates a designated port for each connection to other switches based again on the lowest cost. Other ports that could reach the destination switch are marked nondesignated and are blocked.
- The switch listens for 15 seconds to determine what nodes are accessible through the root and designated ports, recording the nodes' source addresses and the incoming port in the forwarding table as described above but does not forward the frames.
- Finally, the switch ports enter the forwarding state, where they forward frames based on the destination address.

3.2.2 Ethernet Frame Format

Figure 3.1 illustrates the Ethernet frame format [5]. The fields are:

- Preamble – consists of 7 bytes of alternating 1's and 0's that allow nodes to synchronize their receiver clocks at the bit level.
- Start Frame Delimiter (SFD) – terminates the Preamble and starts the actual frame contents. It consists of the 1-byte bit pattern 10101011 or 0xd5 in hexadecimal.

Preamble (56 bits)	SFD (8 bits)	Destination Address (48 bits)	Source Address (48 bits)
VLAN Tag (optional) (32 bits)	Length or Frame Type (16 bits)	Payload (variable)	CRC (32 bits)

Figure 3.1 Ethernet Frame Format

- Destination and Source Address – the Ethernet address of the destination node and source node.
- Virtual Local Area Network (VLAN) Tag – an optional field defined in the 802.1Q standard that can be used to designate virtual LANs (VLANs); see the following for further discussion.
- Length or Frame Type – contains either the size of the Payload field in bytes if the size is less than 1,500 or a specific payload type if the value is greater than 1,536. If the field is used as a Frame Type field, the length is determined by the number of bytes between the initial interframe gap and the cyclic redundancy check (CRC). The Frame Type is sometimes called the EtherType.
- Payload – contains the frame contents.
- CRC – a 4-byte CRC. The CRC is used at the receiving Ethernet interface to detect data corruption within the frame.

After the frame is complete, there is an interframe gap, a minimum of 12 bytes of idle time, before transmitting another frame.

The minimum size for the entire frame is 64 bytes, with 42 bytes being the minimum payload size if the VLAN tag is present and 46 bytes if it is absent. If the actual payload is less than the minimum, the remainder is filled out with padding. The maximum payload size is 1,500 bytes and all IP packets are sent with this length, so the Length or Frame Type field in IP networks functions strictly as a frame-type designator. For example, the IPv4 frame type code is 0x800. Other IP networking suite protocols, such as the Address Resolution Protocol (ARP), discussed in Chapter 4, and IPv6, have their own type codes. If more than 1,500 bytes must be sent, the content must be segmented into multiple frames unless the Ethernet switches in the network support nonstandard jumbo frames with larger sizes.

A VLAN is a virtual slice of the local area network that runs over the same physical links as other VLANs but whose traffic is isolated from other traffic at the data link–layer level. Since frames in a VLAN behave as if they were traversing a completely separate physical infrastructure from frames in other VLANs, deploying separate physical networks for individual use cases is unnecessary. VLANs are commonly used in enterprise networks for isolating different IP subnets from each other. VLANs can be used together with IP subnets in building networks for isolating OT traffic for the BEMS from IT traffic, and could also be used to restrict access to only authorized users and devices with the 802.1X port-based access control standard [6]. Chapter 4 discusses IP subnets, and Section 5.4 discusses port-based access control in more detail.

3.2.3 Ethernet Addressing

Ethernet uses a flat 6-byte (48-bit) address space, for a total of roughly 2×10^{14} addresses, with the Ethernet MAC address assigned to the network interface. The addresses are formally called Extended Unique Identifiers-48 (EUI-48) addresses [8].

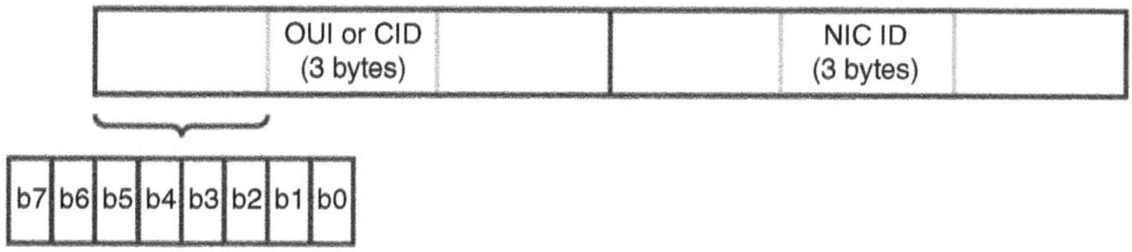

Figure 3.2 Ethernet Address Structure. *Source:* Adapted from Wikipedia [7].

The scheme is designed to ensure that addresses are globally unique so there should be no collision between frames issued from two different Ethernet interfaces. EUI-48 addresses are also used for wireless LAN and Bluetooth links. As shown in Figure 3.2, the 6 byte address is divided into two fields of 3 bytes each. The leftmost 3 bytes constitute the organizationally unique identifier (OUI) or company identifier (CID), designating the administrative authority for the address, while the rightmost 3 bytes designate the network interface controller (NIC) identifier.

The two least significant bits (bit 0 and bit 1) in the first byte transmitted in the OUI indicate the type of address. Bit 0 is the Individual/Group (I/G) bit. If the I/G bit is set to 1, then the address is a multicast address, an address on which multiple interfaces can listen and receive the same frame. If the I/G bit is set to 0, then the address is a unicast address on which a single interface receives a frame. Bit 1 is the unique/local (U/L) bit. If the U/L bit is set to 1, then the address is locally generated and administered, possibly with a CID in the first 3 bytes, minus B0 and B1 of the first byte, indicating the administration authority. If the U/L is set to 0, then the rest of the first 3 bytes transmitted contain an OUI which has been assigned by the Institute of Electrical and Electronics Engineers (IEEE) Registration Authority to the organization that manufactured the hardware or produced the software (for virtual machines [VMs] and other software with virtual Ethernet interface controllers). The last 3 bytes transmitted are assigned to the NIC and are allocated by the organization whose OUI or CID appears in the first 3 bytes. It is up to that organization to ensure that there is no conflict between different addresses.

EUI-48 addresses are written in hexadecimal form, with the hexadecimal digits grouped into bytes and the bytes separated by a character, typically the colon character (:) or a dash character (–) or sometimes a space, to make the addresses easier to read. For example:

- ac-de-48-23-45-67
- ac:de:48:23:45:67
- ac de 48 23 45 67

All OUIs assigned by IEEE Registration Authority have their U/L bit set to 0 and their I/G bit set to 0; hence, EUI-48 addresses formed by concatenating an OUI with the 3-byte suffix are only unicast addresses and are guaranteed to be globally unique if the organization ensures the 3-byte suffix is unique. All CIDs assigned by the IEEE Registration Authority have their U/L bit set to 1 and their I/G bit set to 0; hence, EUI-48 addresses formed by concatenating a CID with a 3-byte suffix are only unicast addresses and are local addresses not guaranteed to be globally unique. Local addresses typically do not need to be unique on any network further than the IP router between one IP subnet and another. The EUI-48 standard does not call out any specific multicast addresses, with the exception of the all-stations Ethernet link-layer broadcast address ff:ff:ff:ff:ff:ff. The broadcast address is not, strictly speaking, an EUI-48 address because it does not follow the EUI-48 address format standard, but is still used for broadcast in Ethernet networks.

Power over Ethernet

Power over Ethernet (PoE) is a collection of IEEE standards that allow injecting power over a four wire Ethernet cable. This avoids having to run a separate power line to a sensing or control device, and is especially handy in new buildings. The above QR code links to a video with an introduction, and more information can be found at the following link: http://tinyurl.com/y22jcmvs.

3.2.4 Ethernet over Power Line

An alternative to twisted pair or optical cable as the physical layer for Ethernet is Ethernet over power line. Ethernet over power line uses a building's existing electrical wiring for physical connectivity. Converters with standard RJ 45 Ethernet jacks plug into electrical outlets, and twisted pair cables connect to devices with Ethernet interfaces. The converters use high-frequency signaling to transmit data over the power line, so the 60 or 50 Hz power signal does not interfere with data transfer. Ethernet over power line is useful for deploying devices with Ethernet interfaces in parts of a building where running new cable is difficult or impossible or for connecting Wi-Fi access points or switches to a building gateway. The twin of Ethernet over power line is power over Ethernet, and the video link box to the right provides more information on that.

Ethernet over power line equipment is available compatible with the following standards:

- HomePlug – HomePlug is a set of standards developed by the HomePlug Powerline Alliance, an industry consortium dedicated to promoting power line communication. HomePlug is primarily designed for home and small business deployments and includes several different specifications for specific use cases with varying rates of data transmission.

- G.hn – G.hn is a standard developed by the ITU for high-speed power line communication. G.hn is designed to work over a wide range of frequencies and supports data transfer rates of up to 1 Gbps [9].
- IEEE 1901 – IEEE 1901 is a standard developed by the IEEE for medium-speed power line communication. IEEE 1901 is designed for use in both residential and commercial deployments and supports data transfer rates of up to 500 Mbps [10].
- Powerline Intelligent Metering Evolution (PRIME) – PRIME is a standard developed by the PRIME Alliance for power line communication in smart grid deployments. PRIME supports data transfer rates of up to 600 Mbps and is designed to work in residential and commercial deployments [11].

An Ethernet over power line communication link can run up to a maximum of 1,500 ft or up to the nearest transformer if the transformer is closer than 1,500 ft.

HomePlug is the most widely adopted standard and is compatible with IEEE 1901 . One of the main features of HomePlug is that it uses a frequency band between 2 and 86 MHz, allowing data to be transmitted at high speeds while avoiding interference with other networking devices that may be using the same electrical wiring. The first specification for HomePlug 1.0 was introduced in June 2001 and featured a data transfer rate of 14 Mbps. HomePlug 1.0 was superseded by HomePlug AV, which was introduced in August 2005. The HomePlug AV specification provides high-speed data transfer rates of up to 200 Mbps and is designed for high-definition video and other bandwidth-intensive applications. HomePlug AV2 improved on HomePlug AV with data rates up to 2 Gbps and was introduced in January 2012. Finally, the HomePlug Green PHY specification is targeted specifically at smart energy applications and provides data transfer rates of up to 10 Mbps [12]. It is designed for inclusion into smart meters and devices such as HVAC thermostats, home appliances, and plug-in electric vehicles so that data can be shared over a home network and with the power utility. GreenPHY uses up to 75% less energy than AV. The international electric vehicle charging standard Combined Charging System (CCS) [13] specifies the use of HomePlug GreenPHY.

Ethernet over power line has some disadvantages that can limit its effectiveness and usability:

- Interference from other devices using the same wiring can reduce the quality of the signal and the data rate.
- Circuit breakers can cause disruptions in power line communication signals when they are tripped. Additionally, circuit breakers can filter out high-frequency signals used by power line communication converters, reducing the data rate.
- Isolation transformers are designed to isolate different parts of a power system and can block power line communication signals from passing through,

limiting the range. Transformers can also cause a voltage drop which can affect the signal quality.
- Power line communication converters can be relatively expensive compared to other options, such as Wi-Fi.
- Power line communication is dependent on the power quality. Factors such as voltage sag, voltage swell, harmonic distortion, and the presence of high-frequency noise can affect the performance of power line communication devices.

3.3 IEEE 802.11 (Wi-Fi)

A popular option for private LANs in residential and commercial buildings is the WLAN protocol, IEEE 802.11. The IEEE 802.11 standards define both the physical layer, that is, the frequency band, modulation scheme, and data rates, as well as the link layer for WLAN. Most of the IEEE 802.11 standards are defined to operate in the internationally designated industrial, scientific, and medical (ISM) spectrum. These frequency bands were initially allocated by the International Telecommunication Union (ITU) to ISM use cases, such as microwave ovens and microwave diathermy [15], but have been increasingly used in standards for license-free network communication protocols such as 802.11 and Bluetooth. They are not licensable to telecommunications providers, so any network communication standards developed in these bands must be prepared to deal with interference from other use cases.

The first release of the 802.11 protocol was in 1997 and featured 1–2 Mbps throughput in the 2.4 GHz ISM band. In 1999, 802.11b increased the maximum throughput to 11 Mbps in the 2.4 GHz band and added 802.11a in the 5 GHz ISM band with 6–54 Mbps throughput. The latest version of the base standard 802.11-2020 [14] updates the legacy 2.4 GHz and 5 GHz ISM band protocols and folds in security enhancements. An extension for high performance, mesh WLAN, 802.11.ax [16], was released in 2021 and features 574 Mbps to 9.608 Gbps in the 2.4, 5, and 6 GHz bands, along with other new features.

In 1999, the Wi-Fi Alliance was founded as a nonprofit industry trade organization to promote the 802.11 standard under "Wi-Fi," a brand name with which consumers can more easily connect. Since 2019, the Wi-Fi Alliance has designated the different 802.11 standards with easier-to-remember names. The 802.11ax standard was designated "Wi-Fi 6" and "Wi-Fi 6E" for the 6 GHz standard, while previous releases of the standard were retroactively named back to 802.11n, released in 2008, which was designated "Wi-Fi 4." The Wi-Fi Alliance runs a certification program through which manufacturers can test their nodes and receive a certification sticker [17].

While the 2.4, 5, and 6 GHz bands cover the bulk of the deployed 802.11 nodes, there are few standards in other bands. For example, the 802.11y amendment was passed in 2008 and runs in the licensed 3.7 GHz spectrum. There are also 802.11 standards for Wi-Fi over infrared and standards existing and in development for Wi-Fi over light and millimeter wave frequencies. More relevant for BEMS systems, the 802.11ah amendment, also known as Wi-Fi HaLow, was passed in 2017 and runs in the 700, 800, and 900 MHz bands, depending on the prevailing regulatory regime. Wi-Fi HaLow is designed for low-power, long-range use cases such as smart grid, industrial automation, and smart cities. Wi-Fi HaLow also has the ability to operate in environments with high levels of interference. However, it has a lower data rate than other Wi-Fi standards, such as Wi-Fi 6, and is not intended to be used for high-data-rate applications. Wi-Fi HaLow is not nearly as widely deployed as Wi-Fi 4, 5, and 6 since there are other wireless protocols, notably 802.15.4, that operate in the same frequency bands and are designed for the same use cases but have been available much longer.

A significant attraction of Wi-Fi is that it removes the need to run cables in existing buildings that do not have preinstalled cabling, which can vastly simplify the deployment of a backhaul network for the BEMS. With the new Wi-Fi 6 and 6E standards, the transmission speed available rivals that of low-end wired Ethernet. The coverage area of an 802.11 access point runs from about 150 ft in the 2.4 GHz band to 50 ft for the 5 and 6 GHz bands. Unlike wired Ethernet, Wi-Fi signals propagate through free space, and any obstacles will cause the signal to degrade and attenuate. A general rule of thumb is that 2.4 GHz Wi-Fi can penetrate two walls while 5 and 6 GHz can penetrate only one. The wall material also influences the propagation characteristics. Walls made of wood and drywall, which are commonly used in single-family residences in the United States, typically allow much better propagation than walls in commercial buildings, which may have steel or concrete incorporated into them, or stone and masonry walls such as are more common in European residences and large commercial buildings.

3.3.1 Wi-Fi Network Deployment Architectures

Wi-Fi 6 features three different network deployment architectures, depending on the deployment use case:

- An ad-hoc network architecture where nodes communicate in a peer-to-peer fashion, with no access point required unless they wish to communicate out to the Internet;
- An infrastructure network architecture where nodes communicate with a wired access point that connects back to the Internet through wired Ethernet;
- With Wi-Fi 6E, a mesh network architecture where nodes communicate through access points that are meshed through a wireless backhaul, with at least one access point connecting to the wired Ethernet for Internet access.

3.3.1.1 Ad-hoc Architecture

Figure 3.3 illustrates the ad-hoc network architecture. In an ad-hoc Wi-Fi network, nodes communicate directly with each other without the use of an access point unless they need to communicate with the nodes on a wired network or with the Internet. Frames destined for other nodes in the ad-hoc network are routed in a peer-to-peer fashion directly between the nodes. If the destination node is on the wired LAN or the Internet, the traffic is routed through an access point. Ad-hoc nodes can use the same frequency to communicate with the access point as with other nodes in the network, but the access point needs to be configured to accept ad-hoc connections. The advantage of an ad-hoc network is that no wired connections are needed except for one access point, which can be placed conveniently near the Internet router. Ad-hoc networks are particularly applicable for use cases where the nodes are highly mobile within a small area.

Note that in the figure, no sensors or actuators are shown. Wi-Fi ad-hoc networks are almost never used for device networks. Most devices that run ad-hoc networks tend to use WPANs, such as BLE or one of the 802.15.4 stacks, rather than ad-hoc Wi-Fi. These link–layer protocols use the same hop-by-hop routing protocol as ad-hoc Wi-Fi, but their power consumption is considerably lower, increasing

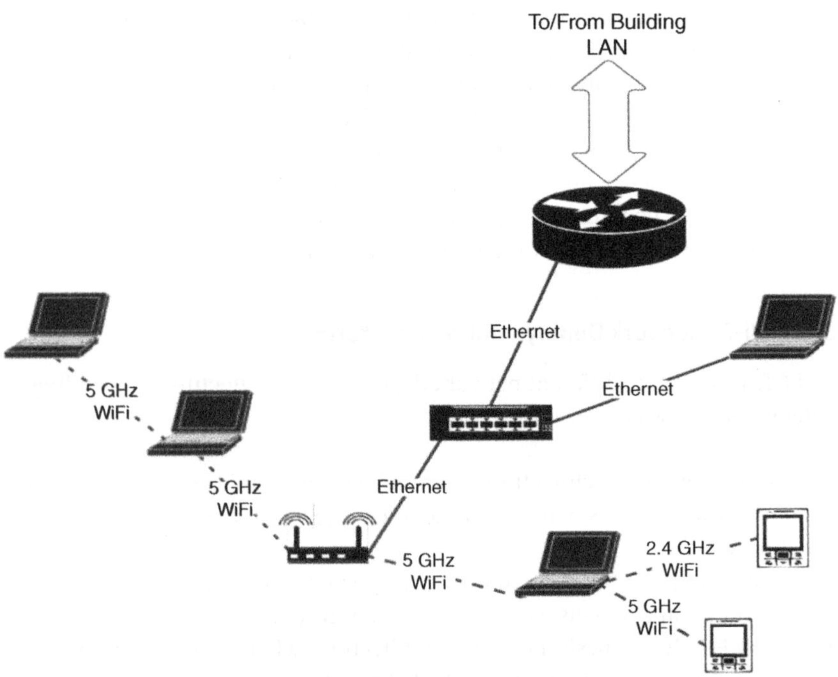

Figure 3.3 Wi-Fi Ad-hoc Network Architecture

the battery life or even allowing the nodes to be powered by environmental energy harvesting. Although Wi-Fi ad-hoc networking was used in the past for commissioning devices to connect them to an infrastructure network, the series of steps needed to set up a Wi-Fi ad-hoc network is much more complicated than connecting a Wi-Fi device to an infrastructure or mesh network. In addition, security is much more difficult to configure in a Wi-Fi ad-hoc network. The introduction of the mesh network architecture in Wi-Fi 6E has largely eliminated the need for ad-hoc architecture since security on the mesh backbone can be configured as part of the network infrastructure, yet the access points can be positioned to achieve optimal wireless reception for devices.

3.3.1.2 Infrastructure Architecture

Figure 3.4 illustrates the infrastructure network architecture. In the infrastructure architecture, a Wi-Fi access point with a wired Ethernet LAN port connects to Ethernet switches and through the switch to the Internet router. In this architecture, the access point serves as the hub of the Wi-Fi network. The access point periodically transmits a beacon management frame so unassociated nodes can find it, and unassociated nodes can additionally send out a probe request

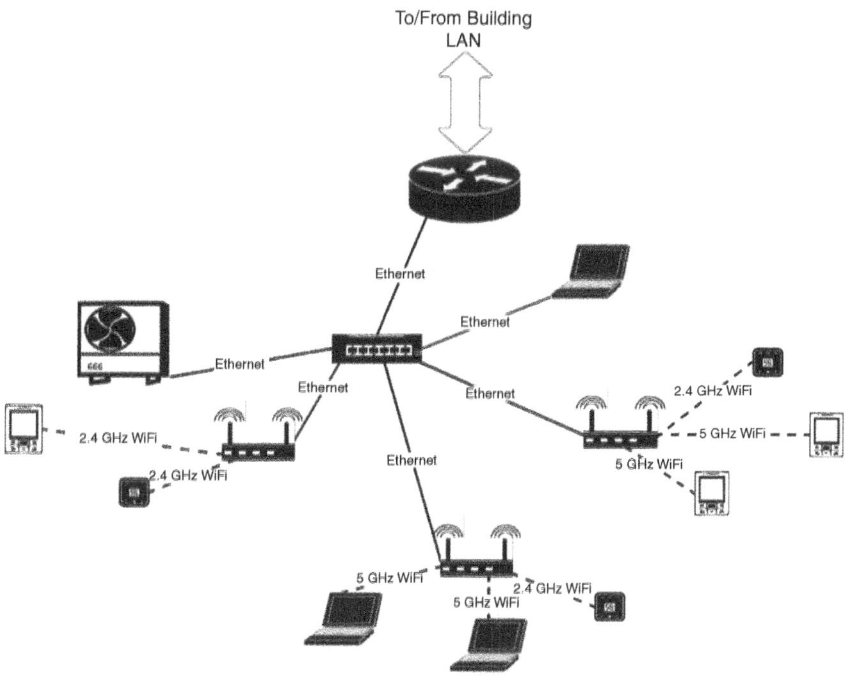

Figure 3.4 Wi-Fi Infrastructure Network Architecture

management frame to which the access point replies with a probe reply management frame containing network information. Nodes associate with the access point and communicate exclusively through the access point rather than directly with each other, as in the ad-hoc architecture. The access point is connected to the building's LAN through an Ethernet switch and connects the wireless nodes with wired nodes on the LAN and, through the building's Internet router, to the public internet. In larger buildings, multiple access points may provide connectivity to different areas of the building to ensure good wireless reception.

The infrastructure network architecture is widely used in Wi-Fi deployments, and prior to the introduction of the mesh architecture in Wi-Fi 6E, it was the only alternative to the ad-hoc network architecture. As shown in the figure, devices using Wi-Fi are very compatible with infrastructure mode networks. The figure shows smart Wi-Fi thermostats connected to the access point on the 2.4 GHz band, a common deployment pattern. One drawback of the infrastructure architecture is the need for wired Ethernet connections wherever an access point is deployed. The access point must immediately connect to the wired LAN for backhaul. This requirement can make deployments difficult in older buildings without cabling for a wired LAN built into the walls.

3.3.1.3 Mesh Architecture

Another way to deploy Wi-Fi 6E is to use a mesh architecture, in which multiple access points create a wireless mesh network for backhaul rather than connecting directly to the wired LAN. Figure 3.5 shows an example of the mesh architecture. Nodes on the Wi-Fi network are associated with access points as in the infrastructure architecture, but the access points route the frames through a wireless backhaul network between themselves to an access point that either has the destination node associated if the destination node is on the Wi-Fi network or has a wired connection into the building's LAN if the destination node not on the Wi-Fi network. The wireless backhaul is run over the 6 GHz band because it has higher capacity and is less subject to interference than the 2.4 and 5 GHz bands, allowing higher transmission speeds and longer range for the aggregated traffic on the backhaul.

As shown in the figure, the mesh architecture is also very compatible with device deployment use cases. An attraction of the mesh architecture is that a wired LAN connection is not necessary for each access point, so it is a good choice for older buildings without wired Ethernet cabling installed and where installing cable would be expensive. Unlike the ad-hoc architecture, the wireless mesh backhaul can be configured with security as part of the building's networking infrastructure. One downside of the mesh architecture is that since the wireless backhaul cannot use the 2.4 GHz band, the access points must be carefully placed so that they have sufficient wireless reception with at least one other mesh access

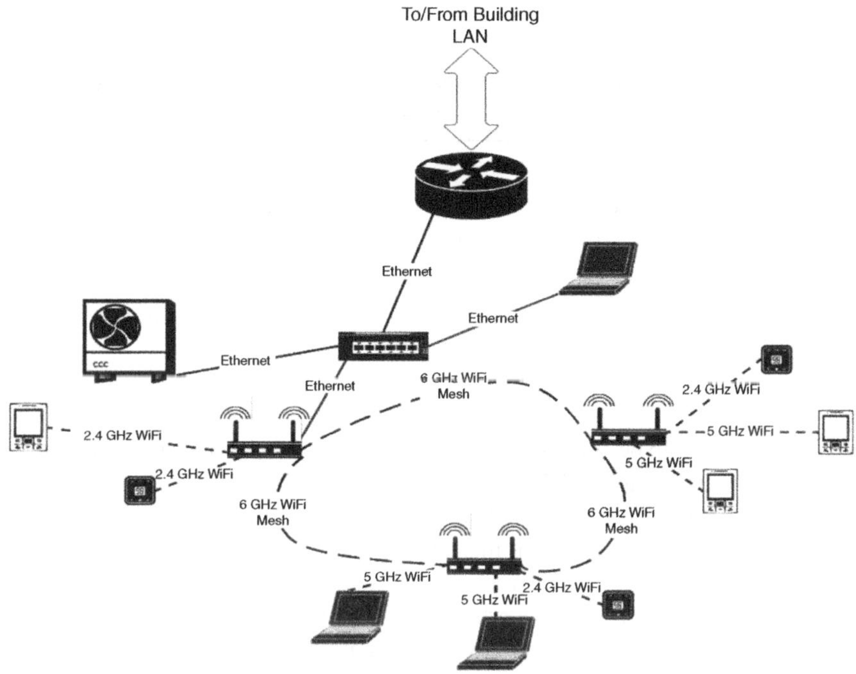

Figure 3.5 Wi-Fi 6E Mesh Network Architecture

point. If the building's geometry is such that walls occlude the wireless reception, additional access points may need to be installed to improve reception.

3.3.2 Wi-Fi Physical Layer

Wi-Fi channels confine traffic within the span of one access point or one network to specific, narrowly defined frequency bands within the broader frequency bands assigned to Wi-Fi in the regulatory jurisdiction. The channels mostly do not conflict with Wi-Fi traffic on other channels. Nodes associated with the access point share the channel and must utilize a MAC algorithm to handle channel contention. Although Wi-Fi channels are not considered formally part of the MAC mechanism in Wi-Fi, they do enable the isolation of traffic to certain frequencies, thereby reducing interference and channel contention [14]. Table 3.1 below contains some channel allocations from two regulatory jurisdictions representing the largest markets for Wi-Fi equipment. In some jurisdictions, certain channels may be inaccessible due to carve-outs for previous users of the spectrum. The exact channel structure in jurisdictions other than the United States and Europe can be different [18]. As shown in the table, while the base channel size is 20 MHz,

Table 3.1 Wi-Fi 6 and 6E Channel Structure

Band	Regulatory Jurisdiction	Band Endpoints (GHz)	Number of 20 MHz Channels	Primary Channels	Channel Bandwidth (MHz)
2.4 GHz	United States	2.400–2.473	11	1, 6, and 11	20 or 40
	Europe	2.400–2.4835	13		
5 GHz	United States	5.150–5.895	45	36, 40, 44, and 48	20, 40, 80 or 160
	Europe	5.150–5.875	23		
6 GHz	United States	5.945–7.125	59	Every fourth channel	20, 40, 80, 160
	Europe	5.945–6.425	24		

several channels can be bonded, essentially doubling the channel size and thereby increasing the data rate available.

Access points must be configured to operate on specific channels to avoid interference with other Wi-Fi access points and networks or with other protocols, such as Zigbee and Bluetooth, that overlap on certain frequency bands. In addition, some adjacent Wi-Fi channels overlap in the 2.4 GHz band so care must be exercised in allocating channels if many overlapping Wi-Fi access points and networks are in the vicinity. In the 5 and 6 GHz bands, channels do not overlap, but some channels are close together; for example, channels 36, 40, and 44 in the 5 GHz band. Allocating channels to ensure wide separation can result in less probability of interference. While there is no guarantee that traffic confined to a particular channel will not experience interference since the ISM bands Wi-Fi runs in are unlicensed, careful attention to channel allocation can reduce the probability of interference occurring from other Wi-Fi traffic.

3.3.2.1 OFDMA Algorithm

Wi-Fi 6 introduced the orthogonal frequency division multiple access (OFDMA) algorithm for allocating frequency and transmitting and receiving data over the wireless medium. The OFDMA algorithm is also used in recent cellular standards such as the 4G and 5G. OFDMA provides close to the maximum theoretical efficiency for communicating data over a wireless medium. Earlier versions of Wi-Fi used different modulation and access control algorithms, but they did not allow multiple users on the channel without the support of a MAC protocol. In OFDMA, multiple nodes transmit and receive data concurrently on the same channel by dividing the channel into smaller subchannels called subcarriers.

Wi-Fi 6 divides the 20, 40, 80, and 160 MHz channels into 256, 512, 1,024, and 2,048 subcarriers, respectively. Each subcarrier can be assigned to a different node by the access point, or a node can have multiple subcarriers assigned to it depending on its data communication needs. This allows nodes to communicate in parallel with the access point rather than sequentially, as is the case when nodes share a single channel, increasing the overall network throughput. Channel contention is prevented by carefully assigning subcarriers to nodes and scheduling transmissions in advance. OFDMA can be used in both the uplink (from node to access point) and downlink (from access point to node) directions. OFDMA is an alternative to the basic MAC mechanisms discussed in the next section.

3.3.3 Wi-Fi Medium Access Control

Within a channel, Wi-Fi nodes use two basic MAC mechanisms:

- The distributed coordination function (DCF) requires nodes to coordinate to enable fair, contention-free communication without requiring a central coordinator. The DCF is the primary MAC algorithm in all 802.11X amendments of Wi-Fi and applies to all network architectures.
- The point coordination function (PCF) in which an access point acts as a central coordinator to schedule communication between the access point and associated nodes. The PCF applies to the infrastructure and mesh network architectures.

In addition, the DCF has an extension called the enhanced distributed coordination function (EDCF) that allows certain nodes and types of traffic to be prioritized for channel access.

3.3.3.1 Distributed Coordination Function

The DCF utilizes two different MAC algorithms:

- The Carrier Sense Multiple Access with Collision Avoidance (CSMA/CA, which is similar to the CSMA/CD algorithm used in Ethernet, except the sending node must avoid collisions. CSMA/CA is the basic Wi-Fi MAC algorithm and has been part of all Wi-Fi standards since the original 802.11 standard in 1997.
- Exchange of Request-to-Send/Clear-to-Send (RTS/CTS) messages between the sending node and receiving node giving the sending node permission to send. RTS/CTS was introduced in 802.11b in 1999 [19].

The CSMA/CA algorithm allows nodes to transmit data onto the network only when the wireless medium is idle. The CSMA/CA algorithm is illustrated in Figure 3.6.Unlike on a wired network, a node cannot monitor the medium while it is transmitting because its signal overwhelms the signals from other nodes, so it cannot perform collision detection. In CSMA/CA, the sending node monitors the channel until the medium becomes idle. The time interval that the node monitors is called the distributed interframe space (DIFS) interval. In the basic CSMA/CA

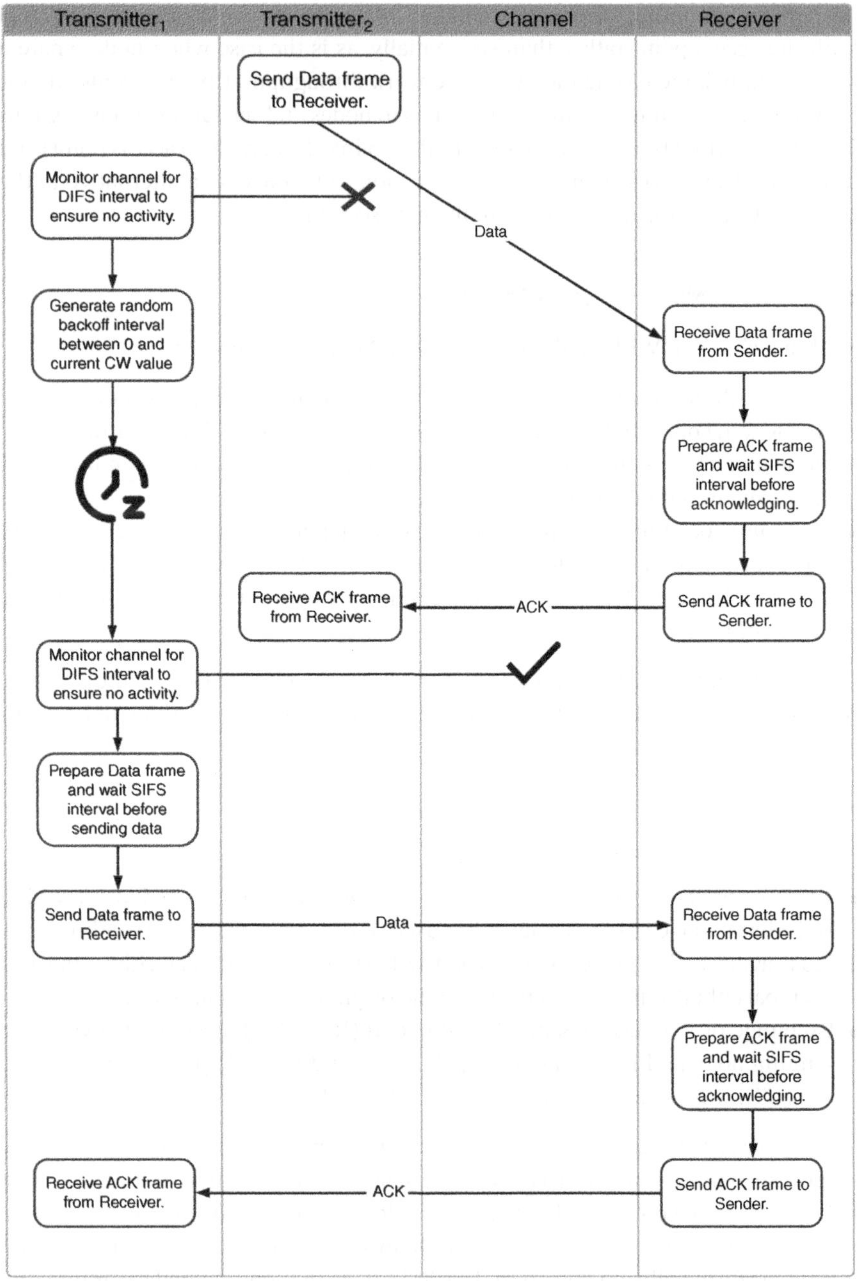

Figure 3.6 Wi-Fi CSMA/CA Protocol with Channel Contention

algorithm, the DIFS is the same for every frame and every node, ensuring fairness. If the channel is not busy, the node can transmit the data immediately.

If the channel is busy, the node randomly generates a backoff time value between zero and a contention window (CW) upper bound before attempting to transmit again. When the backoff time expires, the node checks the channel to see if it is still busy. If the channel is still busy, the node exponentially increases the CW bound and generates a new random backoff time, then checks again when that backoff time expires, waiting for the DIFS interval before deciding whether to transmit or not. The CW value is increased at every retransmission attempt up to a maximum value. Once the channel becomes available, the node resets the CW upper bound to its minimum value, transmits the data, and waits for an acknowledgment (ACK) from the intended recipient. If the node reaches the maximum CW upper bound without being able to transmit, it can drop the frame.

If the recipient receives the data successfully, it sends an ACK frame back to the sender. The receiver waits for a time interval called the Short Inter-Frame Space (SIFS) interval before sending the ACK. If the sender does not receive an ACK within the SIFS interval, the sender assumes that the frame was not received successfully and retries the transmission. The duration of the SIFS is equal to the sum of the physical layer delay and processing delay in the link–layer required to prepare a reply. The SIFS functions as an inter-frame spacing period during which the channel is idle and typically lasts around 10 μsec.

In the RTS/CTS algorithm, the sending node explicitly signals to the receiving node that it wants to send data. Figure 3.7 illustrates the algorithm. A node wanting to transmit data first waits for the DIFS interval to ensure the channel is clear, then sends an RTS message to the intended recipient, including the size of the data frame that it wants to send, as well as the duration of the transmission. The receiving node responds with a CTS message to indicate that it is ready to receive the frame. When the sending node receives the CTS, it proceeds to send the frame. Both sides wait for the SIFS interval before replying. If the receiving node does not respond with an ACK within the SIFS interval after the frame is sent, the sending node waits for a random amount of time and then resends the frame.

The RTS/CTS protocol is used to prevent collisions caused by hidden nodes. Hidden nodes are nodes that are within range of the receiver but not the sender. Because the hidden node can only hear the receiver, it cannot detect when the channel is clear and may try to transmit at the same time as the sender if the sender simply sends data according to the CSMA/CA algorithm, causing a collision. By using RTS/CTS, the hidden node is made aware of the communication since it will receive the CTS from the receiver and can defer its transmission until the sender has finished transmitting. RTS/CTS is especially useful in ad-hoc networks since the movement of nodes can cause hidden nodes to occur more frequently. However, it imposes a significant overhead on transmission since the RTS/CTS exchange is required prior to sending, unlike the CSMA/CA protocol.

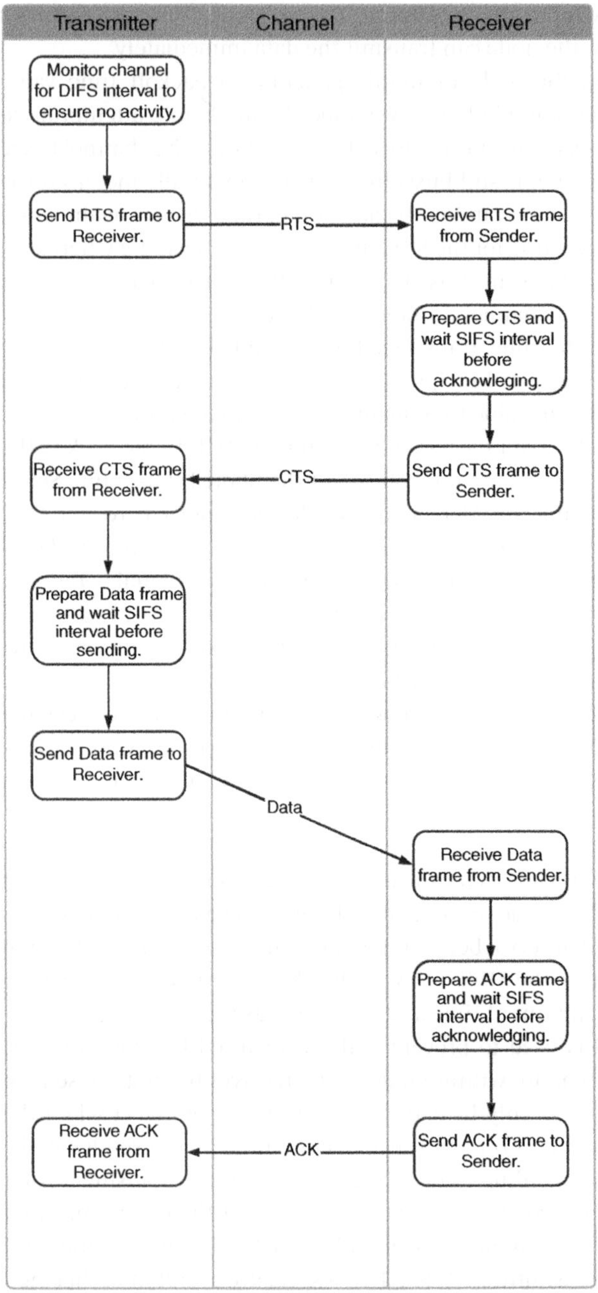

Figure 3.7 RTS/CTS Protocol

3.3.3.2 Enhanced Distributed Coordination Function

The DCF is designed to allow nodes to share the wireless channel efficiently and fairly without requiring a central coordinator. A modified version of the DCF protocol is called the EDCF and is a more efficient MAC protocol when network traffic and nodes have different latency bounds and data rate requirements because it allows access points to prioritize certain types of traffic and certain nodes over others, at the expense of fairness [20].For example, real time audio and video traffic, which must be delivered within a maximum latency bound, can be prioritized higher than web browsing. In a BEMS, EDCF may be useful for high-priority traffic, such as alerts, or low-priority traffic, such as bulk-data-like logs. Traffic prioritization is often called quality of service (QoS) or differentiated services in standards because it takes the latency and data rate requirements of the traffic into account when making channel contention resolution and forwarding decisions.

EDCF predicates the DIFS value and the CW bound values on the traffic priority classification of the node and frame. Instead of assigning all nodes and traffic the same DIFS value and a single minimum and maximum CW bound, EDCF defines four different access priorities, each with its own defer time, called the arbitrary distributed interframe space (AIFS) interval, and its own minimum and maximum CW bounds. Within a node, frames with the same access priority share the same queue. This sets up two levels of traffic contention: between frames of different priority within the node and between nodes of different priority externally. Collisions of channel access requests can happen on both levels and are resolved such that higher priority traffic has a higher probability of obtaining the channel before lower priority traffic.

3.3.3.3 Point Coordination Function

PCF divides medium access into a contention-based period during which the DCF is used and a contention-free period during which the PCF is used. The access point acts as a centralized coordinator for the medium during the PCF, and it gains access to the medium during the contention-based period in the usual way, using DCF; once it has access to the channel, it transmits a beacon frame starting the PCF period. The beacon frame contains information about the maximum expected length of the contention-free period and must additionally be transmitted periodically, even if a PCF period is not scheduled. Nodes note the expected length of the contention-free period from the beacon frame in their MAC layer and will defer restarting the DCF until after the contention-free period has expired.

During the contention-free period, the access point monitors the channel for the priority interframe space (PIFS) interval to ensure the channel is free. Because the PIFS is shorter than the DIFS, the PCF suppresses the start of the DCF until the contention-free period is completed, in case any nodes miss the beacon frame. Once the PCF interval starts, the access point polls associated nodes whether they have data for the access point or informs nodes that it has data for them. The data is then exchanged, with the receiving side replying with an ACK when the transmission is complete. The access point announces the end of the contention-free period by transmitting a contention-free end frame. Nodes then reset their MAC to restart the DCF. Because the PCF was not included in the Wi-Fi Alliance interoperability standards, it has been implemented by very few access point products [17].

3.3.4 Wi-Fi Frame Format

Figure 3.8 illustrates the common frame format for all 802.11X amendments and revisions of Wi-Fi. Several fields are optional fields, and whether they are included or not depends on the frame control field contents. The Sequence Control through the Payload fields are also absent from control frames since control frames have no payload. In addition, many amendments, such as 802.11ah and the more recent amendments, such as 802.ax, feature alternate header and frame formats optimized for specialized use cases, such as very high-performance media streaming over OFDMA or aggregation of multiple wired Ethernet frame payloads into one wireless frame payload.

As shown in Figure 3.8, the frame includes the following fields:

- Frame Control – information about the frame, such as the type of frame (e.g., data, control, management), the duration of the frame, and the protection mechanism (e.g., encryption) used. The Frame Control field is discussed in more detail in Section 3.3.4.1.
- Duration/ID – serves two purposes:
  - For data and management frames, it contains the duration of the transmission in microseconds.
  - For control frames which are all the same length, it contains a unique identifier for the frame.
- Address 1, Address 2, Address 3, and Address 4 – addresses for the receiver, transmitter, source and destination for the frame. The receiver and transmitter addresses are needed because an access point may serve as an intermediary

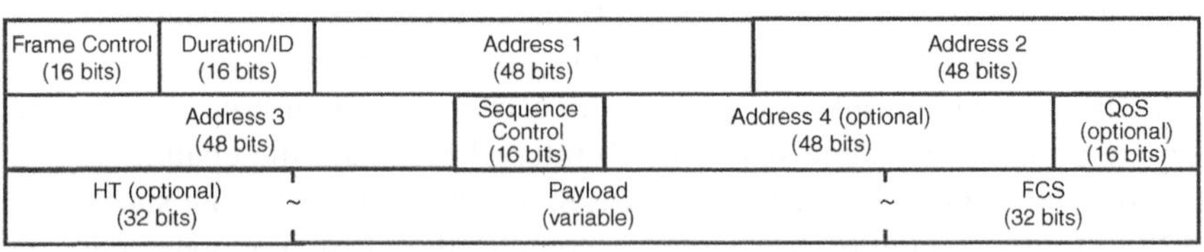

Figure 3.8 Common Wi-Fi Frame Format

between the ultimate source and destination nodes. Only the Address 1, Address 2, and Address 3 fields are required in all frames. The exact number and content of the address fields are determined by the Control Header field setting and are discussed in Section 3.3.5.

- Sequence Control – the sequence number of the frame and the fragment number if the frame is being transmitted in fragments. The Sequence Control field is described in more detail in a following subsection.
- QoS – the priority of the frame and its service class (e.g., real-time, best effort, etc.). The QoS field is optional and only present if the frame is being transmitted with equipment utilizing an 802.11e-compatible protocol [21].
- HT field – conveys information about the high throughput capabilities of the transmitting and receiving nodes, as well as indicating the presence of any high throughput information in the frame. The HT field is optional and only present for protocols that feature high throughput, such as 802.11ac (Wi-Fi 4) [21].
- Payload – the actual data being transmitted. The size of the Payload field can vary from 0 to a maximum value. In the original 802.11 standard, the maximum size was 2,312 bytes but the maximum payload size has increased over the years in later amendments with the addition of frame aggregation.
- Frame Check Sequence (FCS) – a cyclic redundancy checksum value that is calculated on the contents of the frame by the sender. The receiver calculates the cyclic redundancy value and compares the calculated value to the FCS field in the frame to determine if any data corruption occurred in transit.

The fields from the Frame Control field through the Sequence Control field (or, optionally, the HT field) constitute the Wi-Fi frame header.

3.3.4.1 Frame Control Field

The Frame Control field, illustrated in Figure 3.9, is a bit field that provides information on the rest of the frame, and in particular, the rest of the header fields. The receiving node uses this field to direct processing through the link layer. The Frame Control field is divided into the following subfields:

- Version Number – indicates the protocol version number. Despite the many amendments and changes over the years, the 802.11 version number remains at 0.

<table>
<tr><td>Version Number
(2 bits)</td><td colspan="2">Frame Type
(2 bits)</td><td colspan="4">Frame Subtype
(4 bits)</td></tr>
<tr><td>Transfer Direction
(2 bits)</td><td>More Fragments
(1 bit)</td><td>Retry
(1 bit)</td><td>Power Mgmt.
(1 bit)</td><td>More Data
(1 bit)</td><td>Protected Frame
(1 bit)</td><td>Order
(1 bit)</td></tr>
</table>

Figure 3.9 Wi-Fi Frame Control Field

- Frame Type – indicates the frame type. The following are the possible values of the Frame-Type field:
  - 00 – Management
  - 01 – Control
  - 10 – Data
  - 11 – Reserved
- Frame Subtype – indicates the frame subtype together with the Frame Type field. For example, if the Frame Type field is 10 and the Frame Subtype field is 0000, the frame is the Data frame subtype (10000). Some other examples are a Beacon frame (001000), which is a Management frame subtype, and an RTS frame (011011) which is a Control subtype frame. Altogether, there are 13 Control frame message subtypes, 16 Data frame message subtypes, and 12 Management frame message subtypes in Wi-Fi 6.
- Transfer Direction – indicates the transfer direction of the frame used only in data frames. In all other frames, the 2 bits are set to 0. The first bit indicates if the frame is moving toward the wired LAN or other backhaul, called the distribution system in the standard. The second bit indicates if the frame is coming from the distribution system. These bits are used in combination to indicate the number of address fields and the mapping from the address fields to their logical function and are discussed in more detail in Section 3.3.5.
- More Fragments – set to 1 if more data frame fragments are coming; otherwise, set to 0.
- Retry – set to 1 if the frame is the retry of a previous transmission; otherwise, set to 0.
- Power Management – set to 1 if the node is transitioning into power saving mode; otherwise, set to 0. Only set on frames sent from a node to its associated access point. This allows the access point to buffer frames that arrive while the node is asleep and send them again when it wakes up.
- More Data – set to 1 if there is more buffered data for a node that has just exited power saving mode. This ensures the node stays awake until all its buffered data is transmitted. In the last buffered frame sent, this bit is set to 0. Only set on frames from the access point to an associated node.
- Protected Frame – set to 1 if the Payload field in the frame is encrypted; otherwise, set to 0. Wi-Fi security is discussed in Section 5.5.
- Order – set to 1 if the frames must be processed in order; otherwise, set to 0.

A complete list of Frame-Subtype field values can be found in reference [22].

3.3.4.2 Sequence Control Field

The Sequence Control field is a 2-byte field divided into two subfields: a 4-bit Fragment Number subfield followed by a 12-bit Sequence Number subfield. The Sequence Control field is intended to allow a node to eliminate duplicate received frames and to track content from higher layers in the stack that must be fragmented into multiple link-layer frame payloads because it is too large to fit into a single frame. Only data and management type frames have the Sequence Control field in the header, control frames omit it because they have no payload. The Sequence Control field is divided into two subfields: the Fragment Number subfield and the Sequence Number subfield.

The Fragment Number subfield is a 4-bit number that is set in a sequence of frames containing content fragments when content is too large to be transmitted in a single frame. Oversized content is divided into smaller fragments, each of which is given a unique fragment number, starting with 0 for the first fragment sent, and transmitted in separate Wi-Fi frames. The receiver uses the fragment number to reassemble the fragments in the correct order. The fragment number is not changed if the frame must be retransmitted.

The Sequence Number subfield contains a 12-bit sequence number assigned by the sending node to each piece of oversized content that is larger than can be transmitted in a single Wi-Fi frame. The sequence number is incremented after all the frames containing the oversized content fragments are transmitted and wraps back to 0 at 4095. The sequence number for a set of frames containing the oversized content fragments remains the same across different frames and retransmissions of individual frames containing parts of the content.

3.3.5 Wi-Fi Addressing

Because wireless networks can be deployed in such a way that multiple link layer networks overlap in the same physical area, some identifier is needed to differentiate between networks operated by different organizations or that have been deployed for different purposes. For example, in a building housing multiple companies, each company may have its own individual Wi-Fi network and the building owner may additionally have a Wi-Fi network for managing the building infrastructure. A 30-byte identifier called the service set ID (SSID) is used in such cases to differentiate between networks and is typically a name in a natural language indicating to whom the network belongs or for what the network should be used. Prior to 2012, no fixed encoding was specified for the SSID, but the 2012 update of the 802.11 specification defined a flag in the beacon and probe reply management frames indicating whether the SSID was in Unicode [23]. The SSID is broadcast in clear text in the beacon advertisements and probe replies. A single access point may advertise SSIDs for multiple networks as long as the access point is configured to send and receive traffic for the separate networks on different channels or frequency bands. The SSIDs then appear in a list of networks available to an unassociated node, for example, when a user browses through to the Wi-Fi association menu on their mobile phone or laptop to connect with a new Wi-Fi network.

Wi-Fi uses the same EUI-48 addresses discussed in Section 3.2.3, that wired Ethernet uses. The access point provides a set of wireless LAN services to associated nodes called a basic service set (BSS) in the standards, and the access point's EUI-48 link–layer address is therefore called the basic service set ID (BSSID). In an ad-hoc network, since there are no access points providing the BSS, the BSSID is a random string of numbers generated by the node that initiated the ad-hoc network. The collection of all Wi-Fi access points offering service on the same logical network that all advertise the identical SSID is sometimes called the extended service set (ESS).

As mentioned in Section 3.3.4, Wi-Fi frames have four address fields, not two as with Ethernet frames. The four address fields play different roles depending on the direction the frames are being transmitted. Two roles vary depending on the direction of the frame transmission (i.e. to the distribution system or from the distribution system). They are:

- Receiving Address – the address of the receiver to which the frame is sent over the wireless link.
- Transmitting Address – the address of the sender which transmitted the frame onto the wireless link.

Three roles are fixed and assigned to specific link–layer addresses of nodes:

- Destination Address – the node address of the frame's end destination.
- Source Address – the node address of the frame's original source.
- BSSID – the address of an access point.

The Address 1 field always contains the receiving address, and the Address 2 field always contains the transmitting address, but the specific link–layer addresses in the Address 1 and 2 fields can be any of the three fixed roles. The roles of the other two address fields and whether the fourth field is present or not depend on the contents of the Transfer Direction subfield. Table 3.2 shows a map of the Transfer Direction subfield contents to the contents of the address fields.

The two Transfer Direction fields have been labeled "to distribution" and "from distribution" to indicate the direction of the frame transfer to or from the distribution network. The distribution network could be either a wired LAN or a wireless mesh backhaul between access points. An empty cell in the Address 4 column means the field is omitted from the frame header.

The first row indicates that the frame is not destined for any distribution network and covers some special circumstances. In ad-hoc networks, for example, there is no distribution network and so both Transfer Direction bits are set to 0. In addition, if both the source node and destination node are associated with the same access point, or the node or access point is sending control or management frames, then no distribution network is involved, and the bits are set to 0.

Table 3.2 Address Fields Contents Depending on Transfer Direction Subfield

Transfer Direction					
To Distribution	**From Distribution**	**Address 1 (Receiving Address)**	**Address 2 (Transmitting Address)**	**Address 3**	**Address 4**
0	0	Destination address	Source address	BSSID	
1	0	Destination address	BSSID	Source address	
0	1	BSSID	Source address	Destination address	
1	1	Receiving access point BSSID	Sending access point BSSID	Destination address	Source address

The second and third rows cover the cases in infrastructure and mesh networks where the access point is acting as an intermediary between the node and the distribution network, either wired or wireless. The bits are set to indicate the direction of the transfer. In the third case, the frame is being sent from one access point to another over a wireless mesh backhaul distribution network. In this case, both bits are set to 1, and the receiving and transmitting addresses are the link-layer addresses of the receiving and transmitting access points. Address 3 and 4 are both required in this case for the destination and source address.

3.3.6 Wi-Fi Association Protocol

Unlike Ethernet, Wi-Fi nodes must undergo an exchange of messages between the node and the access point prior to establishing the link-layer connection, in which the two sides exchange management frames to learn about the capabilities of each other. Over the years as more features have been added to Wi-Fi and the protocol has expanded to new frequency bands and to more use cases, more information has been added to the initial association protocol. Security parameters for setting up a robust security network are included in this exchange of capabilities and are discussed in more detail in Section 5.5. The capability parameters are specified in elements defined as key-value pairs, called Information Elements in the standard, and are included in the management frame Payload field.

Figure 3.10 illustrates a generic management frame and information element. The Element ID field identifies what parameters the information in the Information field contains, while the Length field provides the length in bytes, excluding the Element ID and Length fields. The Information field contains the

Element ID (8 bits)	Length (8 bits)	Element ID Extension (optional, 8 bits)	Information (variable)

Figure 3.10 Generic Wi-Fi Management Frame Information Element

actual parameters being specified. When the element id field is set to 255, the Element ID Extension field identifies the parameters, since so many parameters were included as the standard expanded that space in the 8-bit field ran out. The Element ID and Element ID Extension codes are listed in Table 9-92 of the 802.11-2020 Standard [24].

3.3.6.1 Association and Disassociation Management Message Types

The management message types involved in the association protocol are listed in Table 3.3 along with their type codes and a brief description of their functions. The management frame type codes are included in the Frame Subtype field of the Wi-Fi header Frame Control field described in Section 3.3.4.1 and the Frame Type

Table 3.3 Association and Disassociation Management Messages

Message	Code	Function
Beacon	8	Periodically broadcast by an access point to announce its presence and relay information and capabilities about the network.
Probe Request	4	Sent by an end node to discover an access point. Either broadcast or unicast to a specific access point if the access point SSID is known.
Probe Response	5	Unicast from the access point to the end node with similar information as in the Beacon message.
Association Request	0	Unicast from the end node to the access point with information about the client's capabilities.
Association Response	1	Unicast from the access point back to the end node in response to the Association Request indicating whether the association was successful and providing additional information needed for communication.
Authentication Request/ Authentication Response	11	Unicast from the end node to the access point and from the access point to the end node. Includes a code for the authentication algorithm and, depending on the algorithm, may also perform mutual authentication. See below.
Disassociation	10	Unicast by either the client or the access point when they want to end the association.
Deauthentication	14	Unicast by either the client or the access point to end the security association.

field is set to 00, indicating a management frame. The table also lists two messages involved in breaking an association (Disassociation and Deauthentication). Note that the Authentication Request/Authentication Response management messages use the same Frame-Subtype code.

The end node sends an Authentication Request with the Status field set to 0, while the access point replies with the Status field set to 0 if the request was processed successfully, or an error code if the message failed. Depending on the authentication algorithm, the authentication message exchange may simply covey a code for the authentication algorithm to be used, or it may perform a bilateral authentication of the end node and access point. If bilateral authentication is not performed, an extension of 802.1X, an identity authentication and access control protocol for Ethernet and Wi-Fi, is used. 802.1X is discussed in Section 5.4.

3.3.6.2 Example Wi-Fi Association Exchange

Figure 3.11 shows an example of the association protocol between the end node and the access point [25]. Note that this is an example of a successful association, details differ depending on the outcome of the association and the authentication algorithm in use.

The access points may optionally periodically broadcast a Beaconmessage, as shown in the figure, which the end node receives. Based on the SSID and the connectivity parameters in the Beacon message, the end node selects one access point and sends a Probe Request message to the access point's BSSID, which is its link-layer address, including a few connectivity Information Elements. The access point responds with a Probe Response repeating the SSID and including Information Elements for connectivity and security.

The end node acknowledges the Probe Response with an ACK message, which is not entirely shown in the figure. The ACK message is a control message with Frame Type field code 01 for control and Frame Subtype field code 13. The ACK message is very simple and consists of a 2-byte Frame Control field, a 2-byte Duration field which is 0 for the ACK messages sent during the association protocol, the 6-byte Return Address field, which is the link-layer address of either the access point or the end node, depending on whose message is being acknowledged, and the FCS field.

After the probe exchange, the end node needs to signal what algorithm it wants to use for identity authentication, access control, and setting up the security association for encryption and data origin authentication. The end node sends an Authentication Request message with a code for the desired algorithm, and the access point sends an Authentication Reply with a status code indicating if the requested algorithm is acceptable. In this example protocol exchange, the Authentication Request/Authentication Reply exchange does not actually perform bilateral authentication, it simply conveys which algorithm the two sides will use for identity authentication and access control.

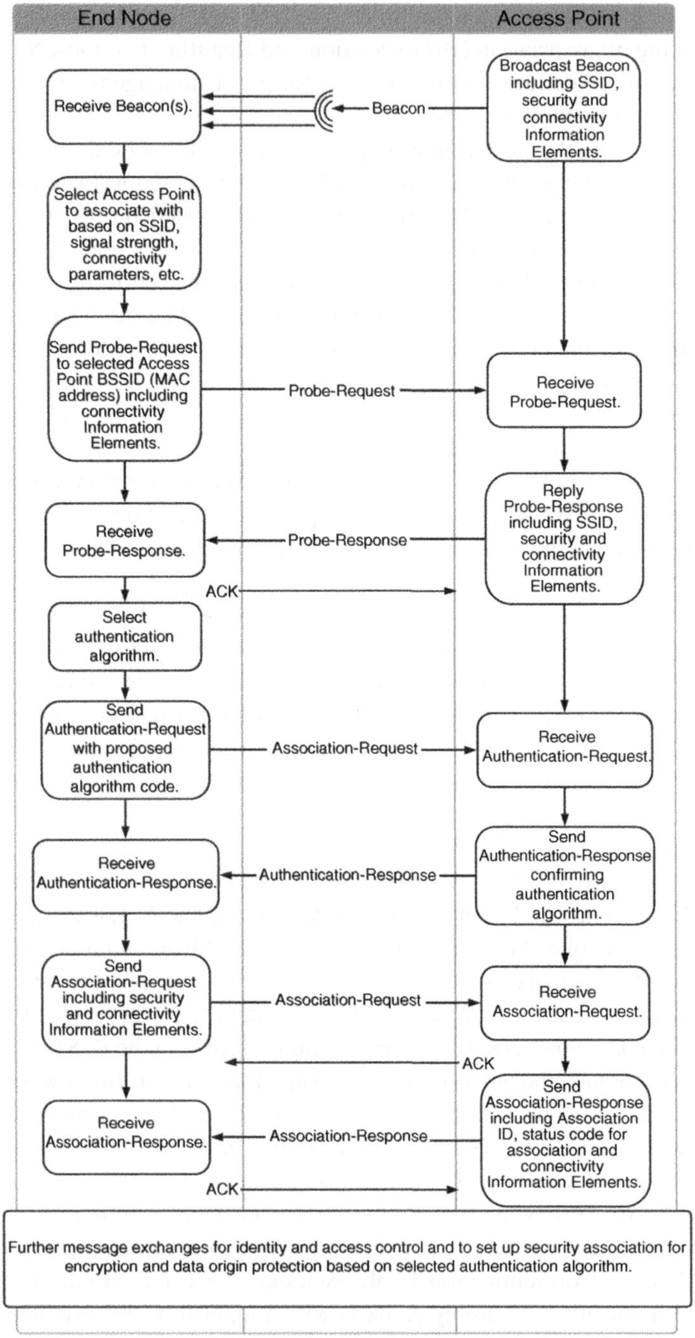

Figure 3.11 Wi-Fi Access Point Association Protocol

Once the authentication exchange is complete, the end node sends an Association Request message with a complete set of security and connectivity Information Elements. The access point responds with an ACK message and an Association Response message with the Association ID assigned to the end node and a status code indicating the association was successful. The end node responds with an ACK message, and the association is complete. However, the end node cannot send any network layer traffic over the link until completing identity authentication and access control and setting up the security association for encryption and data origin authentication.

3.4 WPAN Link-Layer Protocols

WPANs are short-range wireless networks that connect devices over a small area, such as within a single room or building. WPANs are typically used for low-data rate, low-power applications, such as smart energy, home automation, and other residential and office applications. WPANs connect nodes over a range of up to 32 ft, and are not intended to provide the same level of coverage or performance as a WLAN or a cellular network. The advantage of WPAN over WLAN for BEMS use cases is that the nodes do not require much, if any, additional infrastructure to deploy, consume ultralow power, and are inexpensive to develop and purchase. Because power consumption is low, nodes can be powered with batteries that in some cases last for 10 years or even do not ever need charging if the batteries recharge through ambient energy harvesting from photocells or power from other wireless signals. Since, in many cases, the devices will be installed in locations that are difficult to access, low power consumption can radically reduce the amount of maintenance by increasing the time between battery changes or even eliminating it if ambient energy harvesting is used. In addition, installing dense device networks covering large areas, such as multiple floors in an office building, is less expensive because the nodes are cheaper. The tradeoff is that the wireless data rate is relatively low but in most cases, if a high data rate is needed for aggregating sensor traffic toward the building network and Internet, a Wi-Fi or Ethernet gateway can be installed. WPANs operate in the unlicensed ISM frequency bands. The two most common WPAN networks for the BEMS use case are IEEE 802.15.4 [26] and BLE [27].

3.5 IEEE 802.15.4

The IEEE 802.15.4 standard defines the characteristics of a wireless link–layer protocol for devices to communicate over a WPAN. Unlike Wi-Fi, many different mutually incompatible implementations of the network and transport layers have been built on top of the base 802.15.4 link-layer protocol. Two relevant for building energy management applications are Zigbee [28] and Thread [29]. Gateway nodes are necessary and are available that run separate stacks, allowing nodes running incompatible network layer protocols to interoperate, and to interoperate with the building LAN running Wi-Fi and Ethernet. Zigbee and Thread are covered in Chapter 7, while this section covers the basic 802.15.4 WPAN standard.

3.5.1 802.15.4 Network Architecture

The 802.15.4 standard defines two node types: a full-function device and a reduced-function device, and two types of network architectures: the star architecture and the peer-to-peer architecture. Node types and the two architectures are described in the following subsections.

3.5.1.1 Node Types

Full-function devices can perform the following functions:

- Coordinate the creation of a network and invite other nodes to join.
- Send periodic beacon messages to announce the presence of the network and allow nodes to associate with the network.
- Forward data between two nodes that are not in direct communication range with each other using hop-by-hop routing.
- Set up encryption and authentication for the network to protect the network from unauthorized access.
- Control the power usage of the network to optimize battery life for low-power nodes.

Full-function devices can act as network coordinators for nodes within their local area. One coordinator in the network is designated as the personal area network (PAN) coordinator and is responsible for managing the entire network. Other coordinators act as routers connecting reduced-function devices and other coordinators that are not within range of each other. Full-function devices can additionally perform the same functions sensing and actuation functions as reduced-function devices, in addition to originating and receiving transmissions.

Reduced-function devices can perform the following functions:

- Associate with the WPAN through a local coordinator or the PAN coordinator.
- Send and receive data to or from a coordinator for further routing by the coordinator, if necessary.
- Receive commands from a coordinator and execute them.

Some reduced-function devices may be transmit-only or receive-only to reduce node complexity and power consumption. Reduced-function devices can also perform sensing and actuation.

3.5.1.2 Star Architecture

Figure 3.12 illustrates the star architecture. In the star architecture, all nodes establish communication with a single PAN coordinator. The PAN coordinator acts as an access point in 802.11 networks and is responsible for establishing the network. As shown in the figure, the PAN coordinator often features a gateway back into the building's LAN, so data from the sensor nodes can reach the BEMS and commands from the BEMS can reach the actuator nodes. Because the PAN coordinator needs to perform lots of communication, it usually will be powered by a grid power connection and not by a battery. Star networks operate independently from each other and connect only through a gateway in the PAN coordinator. The first full-function device to start sets itself up as the PAN coordinator. It then chooses a unique PAN ID, like the SSID in Wi-Fi, that differs from that of other PAN networks within the reception vicinity. The PAN ID is included in the beacon. Higher layers use a set of procedures defined in the standard [30] to set up the star network and invite other nodes to join the network.

3.5.1.3 Mesh Architecture

Figure 3.13 illustrates the peer-to-peer or mesh architecture. The mesh architecture is similar to the ad-hoc architecture in Wi-Fi. One node can optionally be nominated as the PAN coordinator for the network because it was the first node to start or because it has access to a grid power connection. As in the star network, if the network has a PAN coordinator, it chooses a PAN ID and broadcasts the ID in

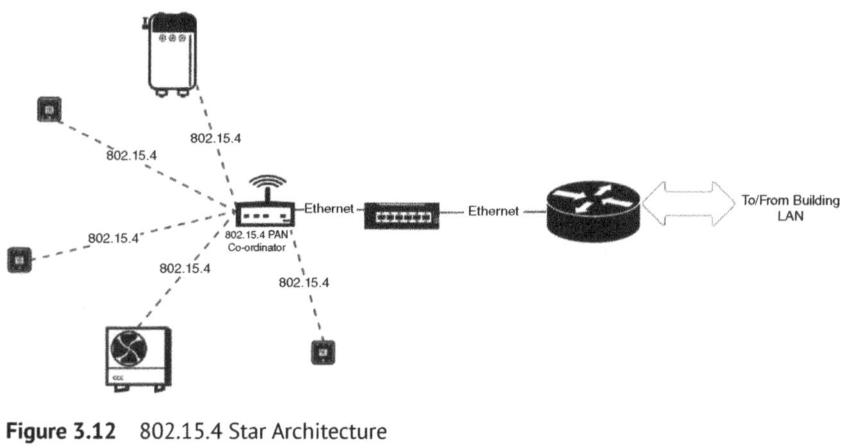

Figure 3.12 802.15.4 Star Architecture

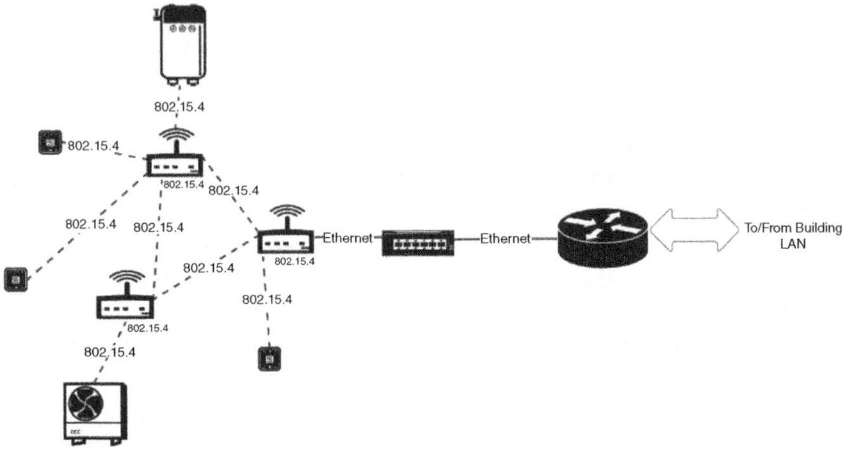

Figure 3.13 802.15.4 Mesh Architecture

its beacon. However, nodes are not constrained to communicate strictly with the PAN coordinator. If two or more nodes attempt to establish themselves as PAN coordinator, some contention resolution mechanism is required, but the standard does not specify any algorithm.

Alternatively, the network can operate without a PAN coordinator, as shown in the figure. Each node can communicate with any other node within the wireless range. If the network has no PAN coordinator, then a node wanting to join a network must probe for a beacon frame. A node receiving a beacon frame from a coordinator can request to join the network. If the coordinator allows the node to join, it adds the new node as a child in its neighbor list, and the newly joined node adds the coordinator as its parent in its neighbor list. If the new node is a full-function device, other candidate nodes can then join the network through the new coordinator. On the other hand, if the node is a reduced-function device, the node can then just send and receive data through the coordinator. If the node is not able to join the network at the coordinator for some reason, it can search for another parent node, which could be a coordinator in the same PAN or another PAN.

3.5.2 802.15.4 Physical Layer

Table 3.4 shows the basic 802.15.4 frequency bands and channels for Europe, the United States, and many other regulatory jurisdictions [26]. The 2.4 GHz band is the primary band utilized by 802.15.4 internationally. As mentioned in Section 3.2.2, the 802.15.4 2.4 GHz band partially overlaps with the Wi-Fi 2.4 GHz band, so even though both protocols have mechanisms built in to operate in the

Table 3.4 802.15.4 Channel Structure

Band	Regulatory Jurisdiction	Band Endpoints	Number Channels	Channel Bandwidth (MHz)	Data Rate (kbps)
2.4 GHz	International	2.400–2.835 GHz	16	5	250
868 MHz	Europe, other regulatory jurisdictions	868.0–868.6 MHz	1	2	20, 40, and 100
915 MHz	United States, other regulatory jurisdictions	902–928 MHz	10	2	20, 40, and 100

presence of interference, care should be exercised when deploying overlapping 802.15.4 and 802.11 networks as excessive interference can degrade network throughput. This is especially a concern for 802.15.4 because the basic throughput is low. 802.14.5 uses either frequency hopping or direct sequence spread spectrum to modulate the wireless medium. These modulation mechanisms were also used in older Wi-Fi versions prior to 802.11a. More recent amendments of the 802.14.5 standard have incorporated additional frequency bands and, like the 802.11 standard, OFDM, but these new features are primarily designed for low-power, long-range outdoor applications such as smart energy utility networks and rail communications and control.

3.5.3 802.15.4 Medium Access Control

The 802.15.4 standard specifies two MAC mechanisms for coordinating access to the wireless channel:

- Slotted mode in which a PAN coordinator periodically sends out beacons announcing the start of a communication slot, called a superframe. Slotted mode is used in the star network architecture.
- Unslotted mode in which there is no PAN coordinator and no periodic beacons. Nodes need to send out probes to discover full-function devices that can act as coordinators. Unslotted mode is used in the mesh network architecture.

Within these two mechanisms, the 802.15.4 standard specifies the same basic MAC algorithms as Wi-Fi, namely, CSMA/CA.for contention-based access and a reserved access time, called a guaranteed time slot, for contention-free access, but with differences in the details. Slotted mode utilizes both algorithms, while unslotted mode uses only CSMA/CA.

3.5.3.1 Medium Access Control Parameters

The 802.14.5 CSMA/CA algorithm has a collection of parameters. that govern the backoff time, CW size, number of backoffs, etc. These are:

- Number of Backoff Attempts – set to 0 at the beginning of a transmission attempt, incremented by 1 at each backoff until a configurable maximum between 0 and 5 with default 4.
- Backoff Period Unit – minimum time granularity for backoff. The default is 320 µsec in the 2.4 GHz band. All time periods are denominated in units of this basic dimension.
- Backoff Exponent – Exponent used in calculating the number of backoff periods for a particular transmission retry. The backoff exponent is set to the minimum backoff exponent for the first backoff period incremented by 1 before each calculation of the backoff period for a new attempt to transmit until it reaches the maximum backoff exponent.
- Minimum Backoff Exponent – exponent used to calculate the first backoff period length. Configurable within a range is 0 to the maximum backoff exponent with default 3.
- Maximum Backoff Exponent – maximum value of the backoff exponent. Configurable within a range is 3–8 with default 5.
- Number of Backoff Periods and Hence the Length of Time before a Retransmission attempt – this is calculated as $2^{(\text{backoff exponent})} - 1$.
- CW length – the number of backoff periods that need to be clear of channel activity before the transmission can start. Set to two before each transmission attempt and reset to two every time channel monitoring determines the channel is busy. This parameter is only used in the slotted MAC method.

These parameters are used during CSMA/CA to control the backoff time and number of retransmission attempts.

3.5.3.2 Unslotted Medium Access Control

Figure 3.14 illustrates the unslotted CSMA/CA algorithm.When a node wants to transmit data, it first randomly generates a backoff period and waits until the period expires unless the backoff exponent is 0, in which case collision is possible. When the backoff period expires, the node performs a clear channel assessment to check if the channel is clear. If the channel is clear, the node begins transmitting its data. If the channel is busy, the node increments the backoff exponent and generates a new backoff period, then sleeps until the backoff timer expires. When the backoff period has expired, the node rechecks the channel and transmits if it is free.

The node continues to perform backoffs until the maximum number of backoff attempts has been reached, in which case it discards the frame. The receiver sends an ACK after waiting for a SIFS interval, which for the 2.4 GHz band is

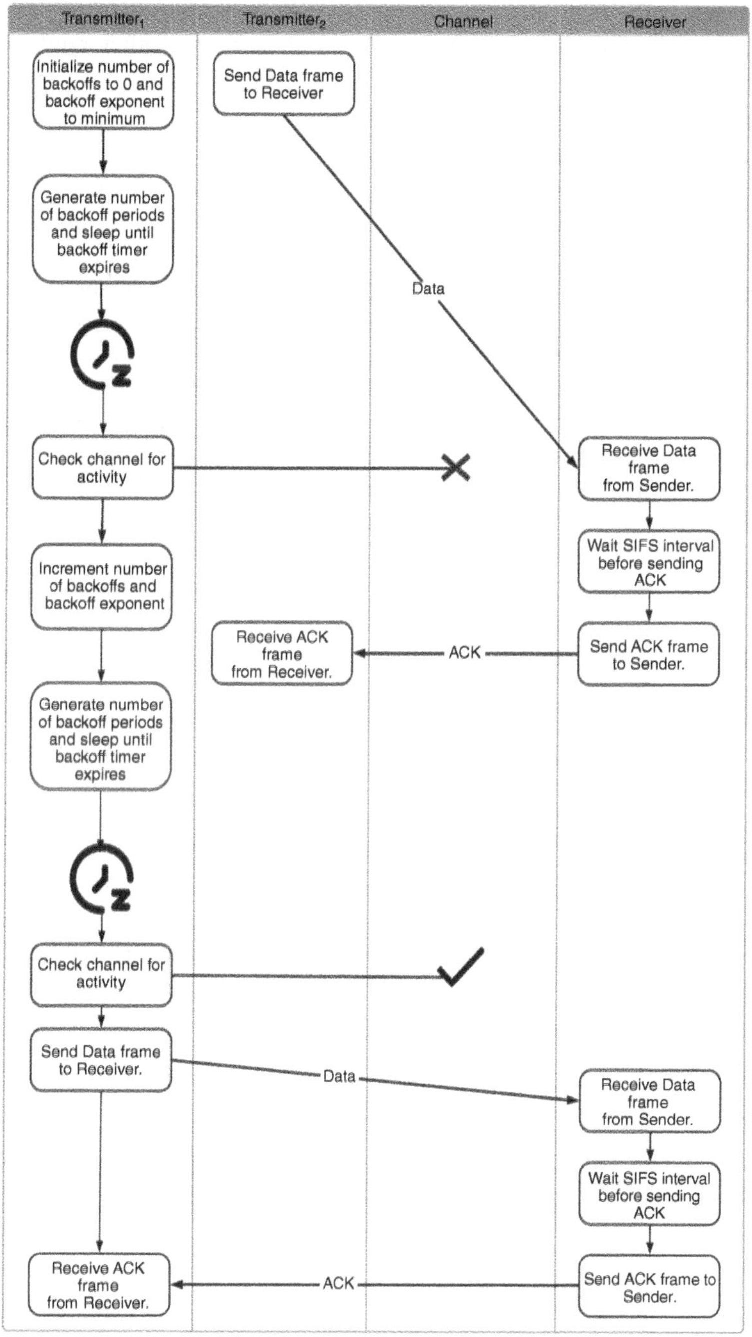

Figure 3.14 802.15.4 CSMA/CA Algorithm with Channel Contention

around 192 μsec. If the node does not receive an ACK, it assumes that the data was not successfully received and retransmits the data after another random backoff period.

3.5.3.3 Slotted Medium Access Control

Figure 3.15 illustrates the superframe structure for the slotted MAC [26]. The superframe is divided into two parts: an active interval and an inactive interval. During the inactive interval, the PAN coordinator and nodes can sleep, reducing power consumption.During the active interval, the PAN coordinator sends the beacon, after which nodes can exchange traffic with the PAN coordinator. The active interval is divided into up to 16 equally sized basic time slots of size 960 μsec, during which nodes can communicate data with the PAN coordinator. The time slots are shared between the contention-active period and the contention-free period. The PAN coordinator sends the beacon during slot 0. The first part of the active interval is the contention-active period, during which nodes use the CSMA/CA algorithm to contend for the channel. The second part is the contention-free period during which nodes have reserved time slots when they are guaranteed channel access. The contention-free period is optional; if no nodes have reserved any time slots, then it is dropped. As shown in the figure, guaranteed time slot reservations can include several basic time slots if the data transmission requires one.

The PAN coordinator signals the start of a superframe by sending out a beacon. One of the fields included in the beacon frame is the superframe specification. The superframe specification contains information about the superframe structure, including the length of the active and inactive intervals, the timing of the beacon period, and the length of the contention access period. The beacon also contains a Guaranteed Time Slot Information field, with a list of addresses for nodes that have requested a guaranteed time slot, if any, along with the starting slot, length of the slot, and an indicator of whether the direction is transmitting or

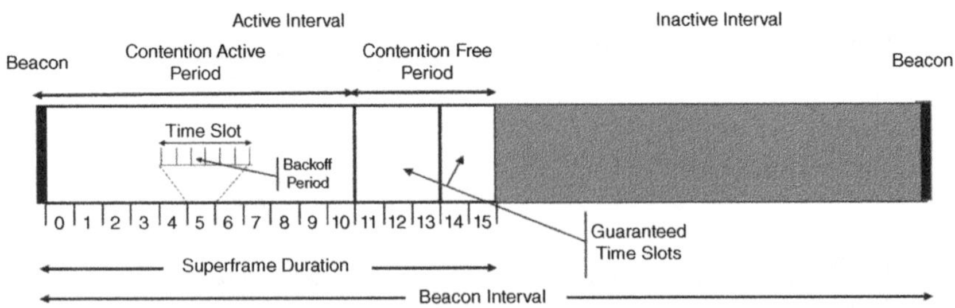

Figure 3.15 Example Superframe Structure

receiving, from the node's standpoint. A superframe can support up to seven guaranteed time slots per superframe.

The CSMA/CA algorithm for the slotted MAC is similar to the unslotted mechanism, except that a few additional steps have been added to synchronize the utilization of the superframe between the PAN coordinator and other network nodes. In Figure 3.15, an exploded view of a contention-active time slot is shown above slot 5, with the backoff period designated. In slotted mode, before the initial backoff period, the nodes must align with the backoff boundary. This ensures that all nodes are contenting for the channel on the same schedule. Since the unslotted mode is strictly peer-to-peer, the backoff periods are uncorrelated, and this step is unnecessary. The other difference is that after the backoff timer expires, instead of performing a clear channel assessment once before transmitting, it performs the assessment twice, using a CW to count the number of clear channel assessments. If, on the second clear channel assessment, the channel is occupied, the node resets the CW to 2 and returns to the backoff procedure, incrementing the number of backoffs again.

To gain access to a guaranteed time slot, nodes send a guaranteed time slot request command to the PAN coordinator. The command contains the desired length of the guaranteed time slot, the direction of communication (transmit or receive), and the node's address. The PAN coordinator then assigns a guaranteed time slot to the node if one is available and communicates the slot assignment to the node and other nodes in the network through the beacon signaling the start of the next superframe. The node can then send or receive data during its guaranteed time slot.

3.5.4 802.15.4 Frame Format

The 802.15.4 frame format has considerably more variability than the Wi-Fi or Ethernet frame formats, as shown in Figure 3.16. The 802.15.4 standard has undergone three major version changes since it was first issued in 2003, during which many new features were added. As new features were added, the frame format has been optimized to remove fields that were unnecessary for particular types of frames. For example, the acknowledgment frame, sent by a

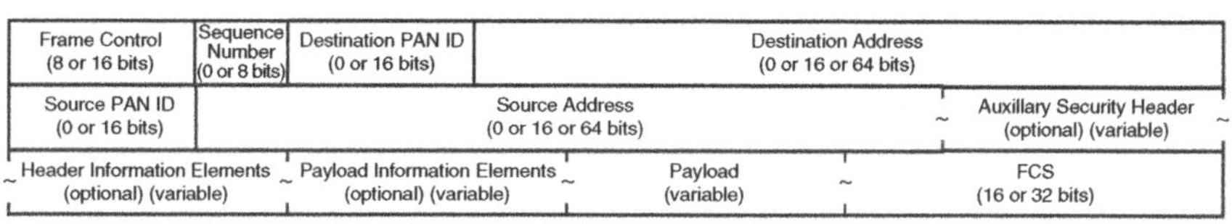

Figure 3.16 Common 802.15.4 Frame Format

receiver to acknowledge receipt of data, can be as short as 5 bytes with no payload for the Immediate acknowledgment type or as long as 21 bytes for the header plus additional data in the information element and payload fields for the enhanced acknowledgment. Because the base standard contains so much flexibility, the wireless network specifications that build on top of 802.15.4, such as Zigbee and Thread, can tailor their frame formats for their particular network architectures and use cases. Consequently, the discussion below only summarizes the frame fields at a high level. More details can be found in the latest 802.15.4 specification [26] or one of the device networking wireless specifications that build on 802.15.4.

The frame format includes the following fields:

- Frame Control – information about the frame, such as the type of frame, security on the frame, the addressing mode (e.g. 16-bit or 64-bit addressing), and other details about the frame. The Frame Control field is discussed in more detail later in this section.
- Sequence Number – a unique identifier for the packet to allow the receiver to detect duplicate packets.
- Source and Destination PAN ID – the PAN ID for the source and destination networks. This field is discussed in more detail in Section 3.5.5 since it is part of the frame addressing.
- Source and Destination Address – depending on the addressing mode used in the packet, the header may include these addressing fields to provide the source and destination address of the packet. This field is also discussed in more detail in Section 3.5.5.
- Auxiliary Security Header – information about the security applied to the packet. This field is only present if the Security Enabled bit is set in the Frame Control field.
- Header and Payload Information Element Fields – a list of key-value pairs of Information Elements that can modify the processing of the frame. The payload Information Element field is part of the payload and is encrypted if the frame is encrypted. More information on these can be found in [31].
- Payload – the packet payload. The maximum Payload field size depends on the physical layer.
- FCS – a 2- or 4-byte cyclic redundancy checksum value that is calculated on the contents of the frame by the sender. The receiver calculates the cyclic redundancy value and compares the calculated value to the FCS field in the frame to determine if any data corruption occurred in transit.

The fields from the Frame Control field through the Header Information Element field (if present) constitute the frame header.

3.5.4.1 802.15.4 Frame Control Field

As in Wi-Fi, the Frame Control field in 802.15.4 contains information about the format of the frame. The Frame Control field is shown in Figure 3.17. The Frame Control field is divided into the following subfields:

- Frame Type:
  - 000 – Beacon
  - 001 – Data
  - 010 – Acknowledgment
  - 011 – MAC command
  - 100 – Reserved
  - 101 – Multipurpose
  - 110 – Fragment
  - 111 – Extended
- Security Enabled – set to 1 if the Auxiliary Security Header is present; otherwise, set to 0.
- Frame Pending – set to 1 if the sender has more data for the receiver; otherwise, set to 0.
- ACK Required – set to 1 if the receiver must send an acknowledgment frame to the sender; otherwise, set to 0.
- PAN ID Compression – version-specific indication about whether one or both of the Destination PAN ID or Source PAN ID fields are omitted. See [32] for details;
- Reserved – Set to 0.
- Sequence Number Suppressed – set to 1 if the Sequence Number field is not present in the header; otherwise, set to 0. If the Frame Version field is 00 or 01, always set to 0.
- Information Elements Present – set to 1 if either the Header Information Elements or Payload Information Elements fields are present; otherwise, set to 0. If the Frame Version field is 00 or 01, always set to 0.
- Destination Addressing Mode, Source Addressing Mode:
  - 00 – PAN ID and address fields are not present
  - 01 – Reserved

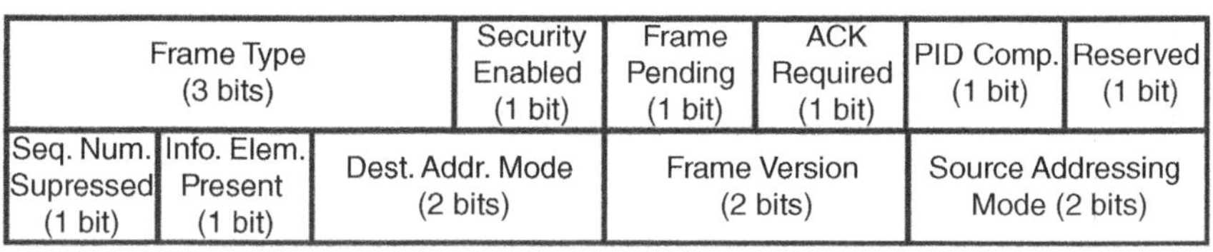

Figure 3.17 802.15.4 Frame Control Field

  - 10 – Address field contains a 16 bit short address
  - 11 – Address field contains 64 bit extended address.
- Frame Version – indicates the version of the 802.15.4 specification the frame format follows. For the Beacon, Data, Acknowledgment, and MAC Command type frames, 00 indicates the 2003 version, 01 indicates the 2006 version, 10 indicates the 2020 version, which is currently active, and 11 is reserved. For the Multipurpose type frame, 00 indicates the 2020 version; the other values are reserved. In fragment and extended frames, this field is not present.

3.5.5 802.15.4 Addressing

The extended address type for 802.15.4 nodes is an EUI-64 address type, an extension of the EUI-48 address type from 6 to 8 bytes [8]. The EUI-64 address is assigned to the hardware interface of the device. The main difference is the EUI-64 addresses have two additional bytes for the node-specific address, which could be a benefit for sensor networks in which there are considerably more devices than in a network used for IT purposes only. However, because 802.15.4 networks have a relatively low data rate, the reduction of unnecessary header fields can help to increase network throughput overall, so the Frame Control header has indications if the Destination or Source PAN ID and Destination or Source Address fields can be dropped. In addition, the star architecture network allows reduced-function devices that only transmit and do not receive to optionally omit their address from the frame header.

Another way 802.15.4 networks reduce header size is by using a short address. The short address is a 16- bit (2-byte) address that is only valid within the local PAN. If the destination or source is outside the PAN, the extended address must be used. In star networks, the PAN coordinator assigns the short address when a node joins the network. In peer-to-peer networks where there is no PAN coordinator, short addresses can still be used, but the 802.15.4 standard does not provide any guidance about how to generate them. The specifications that are built on top of 802.15.4 provide methods for how to generate short addresses. For example, a node can self-generate addresses and use address conflict resolution to determine if another node has claimed the address, or it can obtain an address from the coordinator with which it connects to join the network.

No particular structure is defined for the PAN ID other than that it is a 2-byte number, typically represented in hexadecimal. The PAN ID is not expected to be utilized in user interface tools for humans, such as the SSID in Wi-Fi. The 802.15.4 standard also reserves some addresses for broadcast. The broadcast PAN ID is defined to be 0xff:ff. The broadcast short address is defined to be 0xff:ff, while the broadcast extended address is 0xff:ff:ff:ff:ff:ff:ff:ff [33].

3.6 Bluetooth Low Energy

The first Bluetooth specification was released in 1999 by the Bluetooth Special Interest Group (SIG). The specification was created by Ericsson, Intel, IBM, Nokia, and Toshiba and was intended to provide a wireless alternative to RS-232 serial cables. Version 1 of the specification provided a way for devices to communicate with each other at a distance of up to 33 ft. The first Bluetooth-enabled products, such as mobile phones and headsets, became available on the market in 2000. Since then, the specification has been updated multiple times, with the latest version being Bluetooth 5.4, which was released in January 2023 [27].

The name "Bluetooth" comes from the 10th-century Danish king Harald "Bluetooth" Gormsson. At an initial meeting in 1996 of the team working on developing the technology, Jim Kardach from Intel suggested Bluetooth as a temporary code name because the technology aimed to unify communication between various devices, similar to how Harald Bluetooth united various tribes to form a kingdom in Denmark. The team intended to replace the name with something else, but a full trademark search could not be completed before the launch so the name stuck. The Bluetooth logo is also a combination of the two letters "H" and "B" rendered as Danish runes from King Harald Bluetooth's initials [35].

The first BLE specification, also known as Bluetooth Smart, was released in 2010 with Bluetooth 4.0. The latest version of the specification is included as part of the Bluetooth core specification. BLE is intended to provide low-power, low-cost communication for short-range devices, similar to 802.15.4, with a range of around 330 ft. Designed as an extension of the classic Bluetooth technology, BLE provides similar functionality while consuming considerably less power. It also operates in the 2.4 GHz ISM band. In 2019, the Bluetooth Special Interest Group released a new specification for a mesh architecture called Bluetooth Mesh [34] in a separate specification as an alternative to the original BLE star architecture. Bluetooth Mesh consists of an entirely different stack that sits directly on top of the Bluetooth link and physical layers.

BLE devices can communicate with other BLE devices or with devices that support classic Bluetooth, but they cannot communicate directly with 802.15.4, Wi-Fi, or Ethernet devices, except through an interoperation gateway having a BLE stack and a stack from one of the other link–layer protocols with which BLE is connecting. One advantage of using BLE over the 802.15.4-based protocols for device communication is that all smartphones have Bluetooth built in, so smartphone apps can provide an easy-to-use interface for commissioning and monitoring. A smartphone can also provide convenient uploading of sensor data to the Internet through the cellular wireless data connection without the need for a wired Internet gateway. Note that the BLE protocol is very complex, with many optimizations, some of which are at the physical layer and some of which have only been recently introduced. This book only covers a BLE at a fairly high level; for more details, please see the latest Bluetooth core specification [27].

3.6.1 BLE Stack Architecture

As discussed in Chapter 2, since BLE was designed for WPAN and device networks, the stack architecture does not follow the standard Internet stack architecture. WPAN networks typically do not need a networking layer because they only span one link layer hop, so transport layer functions can sit directly on top of the link layer. Figure 3.18 illustrates the BLE stack architecture.

The stack has three layers:

- The Application layer containing code for the particular application written by the developer. This code interfaces to functions in the Hhost layer through a software development kit (SDK) provided by the device manufacturer. The code at this layer is not, strictly speaking, part of the BLE stack but rather an independent application using BLE for communication.
- The Host layer contains a collection of functions for managing the connection and structuring data transfer between BLE nodes. The functions are the Generic Access Profile (GAP), Generic Attribute Profile (GATT), Attribute Profile (ATT), Security Manager (SM), and the Logical Link Control and Adaptation Protocol (L2CAP). These functions cover the functionality provided by the transport

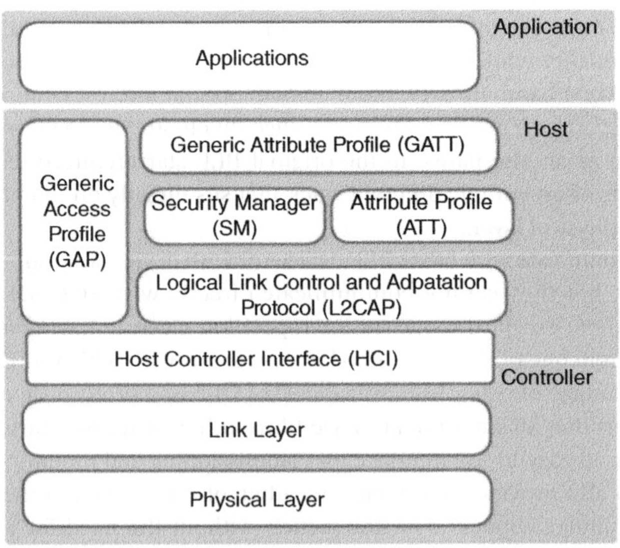

Figure 3.18 BLE Stack Architecture

layer and certain parts of the application layer in a typical Internet stack application, The SM is discussed in Section 5.7, and the GAP and GATT are discussed in Section 7.1.

- The Controller layer contains the link layer and physical layer. These functions are covered in this section.

The Controller layer is structured so that it can be implemented either on the same device as the Host Layer or on a separate hardware device, such as a networking card or a plug-in Universal Serial Bus (USB) device, with the Host layer hardware communicating over a wired connection. The interface between the Controller layer and the Host layer is the Host Controller Interface (HCI).

3.6.2 BLE Network Architecture

BLE supports two different network architectures: a star architecture and a mesh architecture. The star architecture is similar to the 802.15.4 star architecture shown in Figure 3.12 or the Wi-Fi infrastructure architecture, while the mesh architecture is similar to the 802.15.4 mesh architecture and the Wi-Fi ad-hoc architecture. Unlike the Wi-Fi mesh architecture, however, the BLE mesh architecture does not use the mesh for a wireless backhaul, but rather to build a basic peer-to-peer network. The relationship between the BLE mesh architecture and the basic BLE link–layer protocol is similar to the relationship between the 802.15.4 link–layer protocol and Zigbee. BLE mesh is an application built with the functions in the Host Layer, but it adds additional functionality, just as Zigbee does for 802.15.4. BLE mesh does not add any additional functionality to the link layer but only changes how it is used. In this section, only the basic star architecture and link–layer functions are discussed.

One important difference between BLE, 802.15.4, and Wi-Fi is the way nodes join the network. In 802.15.4 and Wi-Fi, controllers or access points advertise, and devices wanting to join the network respond to those advertisements to join the network. In some cases, devices can solicit advertisements, but the responsibility for advertising rests solely with the controller or access point. The BLE architecture is exactly the opposite. It is the responsibility of the devices wanting to join the network to advertise their presence, and the controller nodes respond with an offer to connect. As such, advertising plays a more fundamental role in the BLE protocol than simply making known an access point for WPAN, LAN, or Internet access. More details on advertising in BLE are discussed in Section 3.6.4.

3.6.2.1 Star Architecture and Node Types

In the star architecture, a central device organizes the connectivity and other devices advertise to connect with the central device and communicate with it in a one-to-one fashion. BLE calls the central device responding to advertisements the

Central. A Central is a BLE node that listens and discovers BLE devices that are advertising. It is also capable of establishing a connection with other devices and transferring data to and from them, and it can normally establish connections with multiple devices at the same time. A subtype of a Central is the Observer. An Observer does not establish any connections with devices but just monitors advertisements. The Peripheral is a device that sends out advertisements and accepts a connection from a Central for bidirectional data transfer. A subtype of the Peripheral is the Broadcaster. A Broadcaster is a Peripheral that only advertises. It does not form connections to transfer data. Because devices of the Observer or Broadcaster types have reduced functionality, they also have reduced power requirements and only need to implement a subset of the BLE hardware and software stack, allowing them to be much simpler to implement and cheaper.

Typically, the Central is in a mobile phone, laptop, or gateway device and has some source of grid power or a large rechargeable battery, so it can always be listening. Peripherals only need to advertise at periodic intervals, and the period is configurable so that they can sleep between advertisements if they are not connected and transfer data to a Central. The Central performs the more power-consumptive functions of connection management, time management, and processing, helping to reduce power consumption and processing power requirements on the Peripheral. This allows the Peripherals' batteries to have a longer lifetime or they can use energy harvesting for power. In some cases, a Central may be paired with a Peripheral back-to-back in the same device. For example, a mobile phone may play a Central role in home energy monitoring devices such as thermostats but play a Peripheral role in a laptop or home computer. This allows the mobile phone to connect with the devices and upload their data either to the home computer or to the Internet through the mobile phone's cellular data connection.

3.6.3 BLE Physical Layer

BLE operates only in the 2.4 GHz ISM band, from 2.4 to 2.4835 GHz. The frequency band is divided into 40 channels, 2 MHz wide from 2.402 to 2.480 GHz. BLE channels are twice as wide as classic Bluetooth channels. BLE uses frequency hopping modulation to avoid interference from other devices that may be operating on the same frequency. Frequency hopping involves rapidly switching between different frequency channels in a predefined pattern, making it difficult for an interfering signal to cause problems. The latest version of BLE has three physical layers which support three different throughput rates. One of the physical layers supports a mandatory 1 Mbps rate, which was part of the older BLE specifications, while the second supports an optional 2 Mbps. Neither of these uses error-correcting codes.

The third physical layer utilizes error-correcting codes. Error-correcting codes add extra bits to the transmitted data so errors can be detected and corrected, and are especially useful in environments where there is a lot of wireless noise or for longer range communication to counteract signal attenuation. There are two types of coding used in BLE: 1/3 coding and 1/9 coding, and the number of extra bits in both is always less than if the transmission was duplicated entirely:

- 1/3 coding adds two coding bits for every three data bits, resulting in a data rate of 1/3 or approximately 33% of the original rate, reducing the throughput to 500 kbps.
- 1/9 coding adds eight coding bits for every nine data bits, resulting in a data rate of 1/9 or approximately 11% of the original rate, reducing the throughput to 125 kbps.

The coding physical layer can obtain a range of up to 2,600 ft but at significantly reduced throughput. Note that the Bluetooth 1.0 mesh network specification does not support the coded physical layer.

3.6.4 BLE Medium Access Control

BLE uses channels and different modulation techniques to coordinate medium access. Table 3.5 summarizes the channel structure of BLE BLE dedicates 3 of the 40 channels to primary advertising (37, 38, and 39), with the other 37 channels for secondary advertising and general-purpose data transmission. The advertising channels fall between or outside the main frequencies used for 802.11 channels

Table 3.5 BLE Channel Structure

Channel Type	Number of Channels	Medium Access Control (MAC)/Modulation Type	Channel Bandwidth (MHz)	Data Rate
Primary Advertising Channel	3	Adaptive frequency hopping spread spectrum	2	125 kbps
Secondary Advertising and Data Channel	37	Adaptive frequency hopping spread spectrum	2	125 kbps, 500 kbps, 1 Mbps, 2 Mbps
Periodic Advertising Channel	1	Time-division multiplexing	83.5	1 Mbps
Isochronous Physical Channel	1	Continuous wave	83.5	2 Mbps

allowing for better coexistence with Wi-Fi. The advertising channels are used for device discovery and connection establishment, while the data channels are used for secondary advertisement and data transfer after a connection has been established. The data rate on the primary advertising channels is restricted to 125 kbps, while the rate on the secondary advertising and general-purpose data transmission channels is 1 or 2 Mpbs if the uncoded physical layers are used.

Devices use adaptive frequency hopping to limit channel contention. With adaptive frequency hopping, the sender and receiver rapidly switch the channel on which the devices are communicating, using a pseudorandom sequence agreed upon during connection establishment to determine to which channel to hop next. This helps to avoid interference from other devices that may be using the same channel and provides a measure of security by making it difficult for an attacker to intercept the communication. In addition, the Central determines the channels included in the hopping sequence by assessing which channels are most heavily used by other devices and avoids them in the proposed hopping sequence. For advertisement, the Peripheral similarly uses a hopping pattern across the three primary advertisement channels. Adaptive frequency hopping controls medium access by avoiding overloading congested channels, improving communication reliability.

In addition to the three primary advertisement channels, BLE supports secondary advertisements on the 37 data channels. The advertisement frames sent on the primary channels contain information to discover advertisements sent on the secondary channels. Secondary advertisements allow a Peripheral to send more data than primary advertisements since secondary advertisement frames may contain up to 255 bytes as opposed to 31 bytes in primary advertisements. Note that the 2 Mbps physical layer may only be used on the 37 data channels, so a Peripheral must use either 1 Mbps or the coded physical layer for primary channel advertisements. In addition, Bluetooth Mesh 1.0 does not support secondary advertisements.

BLE 5.1 introduced the Periodic Advertising Channel (PACH) as an additional channel for sending periodic advertising packets. The PACH is a one-way advertising channel that operates on the same 2.4 GHz band as the primary advertising channels. The PACH uses time-division multiplexing for controlling medium access rather than adaptive frequency hopping. The PACH channel is shared between multiple devices and is divided into time slots; each time slot is 6.25-msec long, and each device is allocated a specific time slot based on its advertising set identifier (SID) value. The advertising SID is a unique number assigned to each device that is used to identify the device. The PACH can support a maximum data rate of 1 Mbps. The PACH also supports a higher transmit power than the standard BLE advertising channels, which means that the PACH can support longer range advertising. The PACH is intended to be used in conjunction with the standard BLE advertising channels for use cases such as location-based services and proximity services that require a higher data rate and longer range advertising.

BLE 5.2 added the Isochronous Physical Channel to support higher data rate, low-latency, and time-sensitive data transfer, such as might be required for control systems, while still maintaining low power consumption. Instead of frequency hopping, the Isochronous Physical Channel uses continuous wave modulation. The standard does not specify the channel width, but it is expected to be very narrow. The Isochronous Physical Channel is expected to provide 2 Mbps transfer.

3.6.5 BLE Frame Format

Figure 3.19 illustrates the BLE frame format. This frame format is used on the uncoded physical layers. The fields are as listed in the following:

- Preamble – contains a sequence of bits used for synchronization. For the 1 Mbps physical layer, the preamble contains 1 byte, while in the 2-Mbps physical layer frames, the preamble contains 2 bytes. The bit values alternate between 0 and 1 and start with the same value as in the first bit in the access address field.
- Access Address – contains an access address for the connection or advertisement, see Section 3.6.6 for more on BLE addressing.
- Payload – contains the frame content. The format differs depending on whether the frame is a Data frame or an Advertising frame; see the following subsection.
- CRC – contains a CRC, calculated over the frame contents by the sender and checked by the receiver to detect corruption.

The coding frame format is different and is described in Volume 6, Part B, Section 2.2 in the Bluetooth core specification [36].

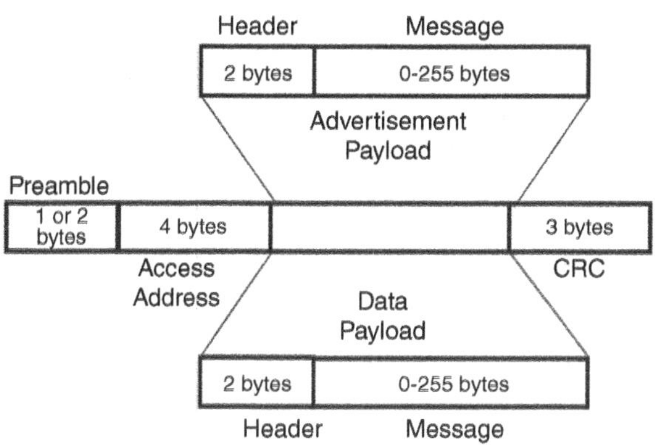

Figure 3.19 BLE Frame Format

3.6.5.1 Advertising Payload Format

The Advertising Payload contains data for a Peripheral's advertisements as they appear on the primary and secondary advertising channels. The header consists of a 16-bit (2-byte) field with the format shown in Figure 3.20 and a variable Message field, as shown in Figure 3.19. The header bitfields are listed in the following:

- Frame Type – the type of advertising message.
- RFU, ChSel, TxAdd, RxAdd – these are flags specific to the different types of advertising messages. More information on them can be found in the Bluetooth Core specification, Volume 6, Part B, Section 2.3.1 [37].
- Length – the length of the Advertising Payload field, including the header. If the advertising frame is sent on one of the three advertising channels, the Message field has a length of between 0 and 37 bytes. If the frame is sent on a secondary advertising channel, the length is 0–255 bytes.

As mentioned above, BLE uses advertisements extensively for establishing presence and connecting as well as for transmitting periodic data updates, partially as a means of reducing power consumption since a node can simply issue data in advertisements periodically instead of maintaining a connection. The BLE specification lists 15 separate advertising types that can appear on primary advertising, secondary advertising, and periodic channels. More information on advertisement types and message bodies can be found in the Bluetooth Core specification [37].

3.6.5.2 Data Payload Format

The Data Payload contains the data a Peripheral is exchanging with its Central. The payload consists of a 16-bit (2-byte) header with an optional 8-bit (1-byte) extension shown in Figure 3.21 with a variable-sized Message field shown in Figure 3.19. The header bitfields are:

- Frame Type – there are two data frame types:
  - 01 or 10 – the Message field contains application data from the L2CAP layer.
  - 11 – The Message field contains data to control the link–layer connection.
- NESN, SN – used in link–layer acknowledgment and flow control.
- More Data – set to 1 to indicate that the device has more data to send so the connection should be kept active, otherwise 0.
- CTE Field Present – set to 1 to indicate that the CTE Extension field is present, otherwise 0.

Frame Type (4 bits)	RFU (2 bits)	CHSel (1 bit)	TxAdd (1 bit)	RxAdd (1 bit)	Length (6 bits)

Figure 3.20 BLE Advertising Payload Header Format

<table>
<tr><td>Frame Type
(2 bits)</td><td>NESN
(1 bits)</td><td>SN
(1 bit)</td><td>More
Data
(1 bit)</td><td>CTE
Present
(1 bit)</td><td>Reserved
(2 bits)</td></tr>
<tr><td colspan="6">Length
(8 bits)</td></tr>
<tr><td colspan="6">Constant Tone Extension (CTE, optional)
(8 bits)</td></tr>
</table>

Figure 3.21 BLE Data Payload Header

- Length – length of the Data Payload field, including header.
- Constant Tone Extension (CTE) – an optional field containing parameters for the CTE. The CTE is a physical layer optimization that allows the receiver to locate the direction of the sender more accurately.

3.6.6 BLE Addressing

BLE uses three types of addresses in the Access Address field. The address on all broadcast frames is fixed and set to 0x8e89bed6. For data frames, the node has an EUI-48 public address assigned by the manufacturers like Wi-Fi and 802.15.4. A node can also generate a random address. Random addresses come in two types, static addresses and private addresses. Static addresses are generated at boot time and cannot change until the node power cycles. They can also remain the same for the life of the node. Static addresses can be used in place of public addresses.

Private addresses are split into two subtypes: nonresolvable private addresses and resolvable private addresses. Nonresolvable private addresses are temporary but are not commonly used in applications. Resolvable private addresses are used for privacy and are cryptographically generated using a random number and a key. Other BLE nodes can resolve these addresses if the two nodes exchange the key during connection initiation. Resolvable private addresses can change even during the lifetime of a connection and are primarily used to avoid tracking by unknown scanners.

3.6.7 Establishing a Connection in BLE

In Figure 3.22, the basic connection process between a Peripheral and a Central is illustrated. The Peripheral sends an advertising message out to one or several of the primary advertising channels. Not shown in the figure is an optional scan request that the Central can send to the Peripheral, to which the Peripheral replies with a scan response. The scan message exchange allows the Peripheral and Central to exchange additional information before initiating a connection.

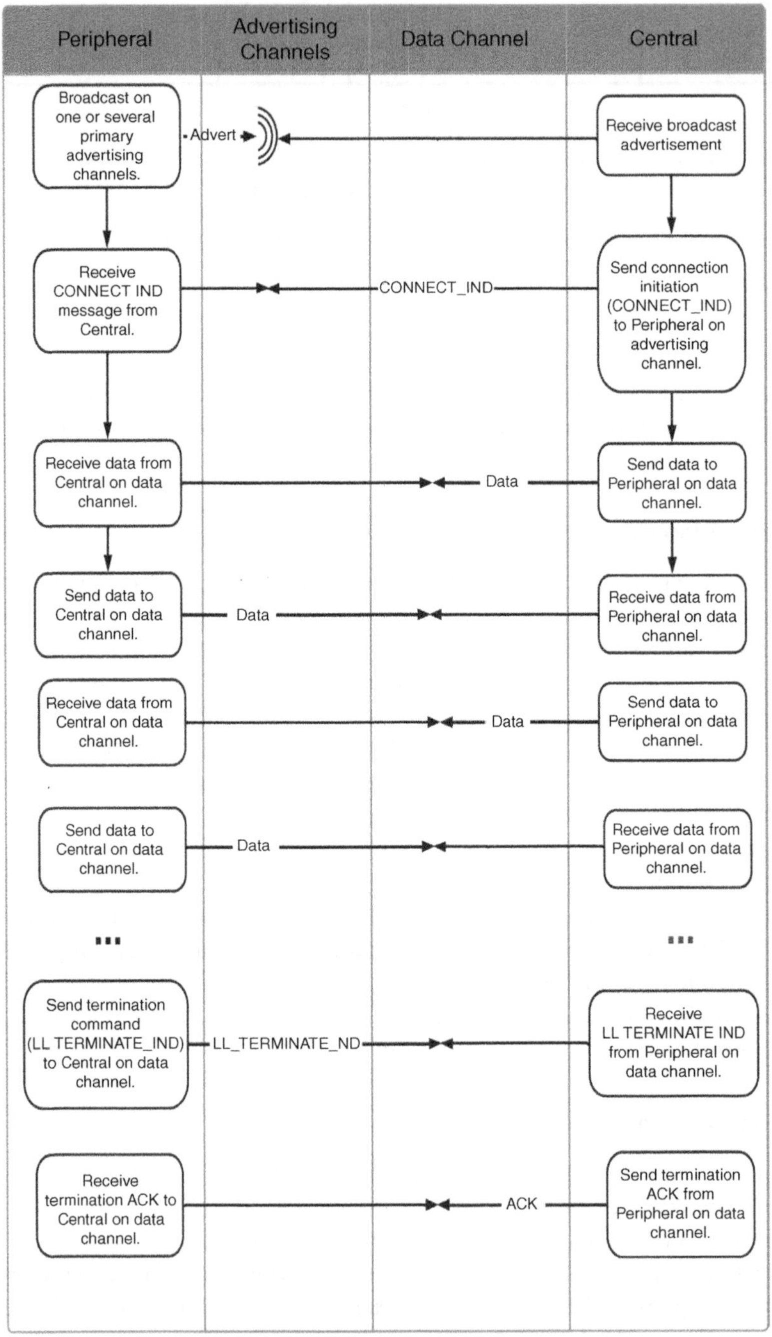

Figure 3.22 BLE Connection Protocol

Once the Central has enough information to initiate the connection, it sends a CONNECT_IND message to the Peripheral. Both sides then set up a state to maintain the connection. At this point, the connection is complete. The two sides exchange data frames until one or the other is finished or needs to disconnect for some reason. The terminator then sends the LL_TERMINATE_IND to the other party to request that the connection be dismantled. The receiver of the LL_TERMINATE_IND message sends an ACK message to the sender, and the connection state is discarded by both sides, dismantling the connection.

3.7 LoRa and LoRaWAN

LoRa is a wireless communication protocol that uses the ISM frequency bands for extremely low data rate, long-range communications between wireless devices and a communication server. The Long Range Wide Area Network (LoRaWAN is a communication protocol built on top of LoRa that enables the creation of low-power, wide-area wireless networks for IoT applications. LoRa provides communication up to 3 mi. in urban areas where devices may be located deep within a building and up to 10 mi. or more in rural areas when line-of-sight communications with a gateway is possible. LoRa devices have substantially reduced maintenance requirements since they can be powered by a battery for up to 10 years or use energy harvesting for power and can have their firmware updated over the air, both for the LoRaWAN stack itself and any applications running on devices.

In most countries, public LoRaWAN networks are run by service providers that act as ISPs for devices. Public access LoRaWAN provides an alternative to running a device network onsite composed of Wi-Fi, 802.15.4, or BLE devices and device network gateways. Rather than hiring IT staff to run the device network, a BEMS service provider or a building owner can simply subscribe to a LoRaWAN service and let the LoRaWAN service provider handle the network. If your BEMS service is deployed in a public cloud, most large public cloud providers (Amazon, Microsoft, Google, IBM, etc.) have LoRaWAN servers that can be deployed into VMs or provide a LoRaWAN service as part of their service offerings. The public cloud server takes traffic from the gateways in the WAN and routes it to their server and then to the BEMS virtual private cloud network. LoRaWAN service from public cloud providers is available at low cost. Private LoRaWAN gateways are also available, but many LoRaWAN device network deployments utilize public access service providers due to ease of use and convenience.

The LoRa protocol was developed by a French company named Cycleo, which later became part of Semtech, a US-based semiconductor company. The technology was introduced in the early 2010s, and the first public demonstrations were conducted in 2012. LoRa by itself is not very useful since it just defines the physical layer, so several groups in the early 2010s began working on a link–layer protocol for LoRa. The LoRaWAN protocol was developed by the LoRa Alliance, a nonprofit industry association that was formed in 2015. The LoRa Alliance was established to provide a link–layer standard for wide-area wireless device networks based on LoRaWAN and to promote LoRa and LoRaWAN among chip vendors, device manufacturers, and other players in the sensing and actuating device industry ecosystem. The LoRaWAN protocol was standardized in 2015 and quickly gained traction among developers and service providers [39]. In December 2018, LoRaWAN became the official ITU standard Y.4480, including specifications for the physical layer and link–layer protocols [40]. Today, LoRaWAN networks are deployed in over 100 countries and are used by thousands of companies and organizations worldwide. LoRaWAN is particularly attractive for smart city applications such as street and traffic lighting control due to its long range, but a wide range of sensor and actuator products are also available for building energy management and control applications as well.

This section provides a high-level overview of LoRa and LoRaWAN. The discussion is not as detailed as the previous sections on Wi-Fi, 802.15.4, and BLE since the consumer of a LoRa service has no need to know about the details of the link–layer protocol, even if running a private LoRaWAN network. The LoRaWAN network acts as a black box for communicating data and commands between devices in the building and the BEMS in the cloud. For more detailed information on the LoRa physical layer and the LoRaWAN link layer, see the LoRa Alliance website for the latest specification (version 1.1) [38].

3.7.1 LoRaWAN Stack Architecture

Figure 3.23 illustrates the LoRaWAN device node stack. The layers are as follows:

- The Application layer is not really part of the LoRaWAN stack, as with BLE. Unlike BLE, however, the bottom layers of the LoRaWAN stack do not provide any additional application support. Applications manage connections and send and receive data through an SDK in a programming language that provides programmatic access to the MAC layer. At least 64 open-source LoRaWAN application SDKs are available on Github [41], in addition to proprietary application stacks. The bottom layers are similar to an IP stack, except the SDK code may not reside in the operating system kernel but rather may be a program library. In addition, the "application" could actually be the network and transport layers of the IP stack rather than an end-user application if the end-user application resides in the cloud and is designed to use the IP stack.
- The MAC layer, or data link layer, contains the code for the MAC algorithm, framing, and addressing. The MAC layer supports three different device node classes: Class A, Class B, and Class C. These are discussed in more detail in Section 3.7.4.

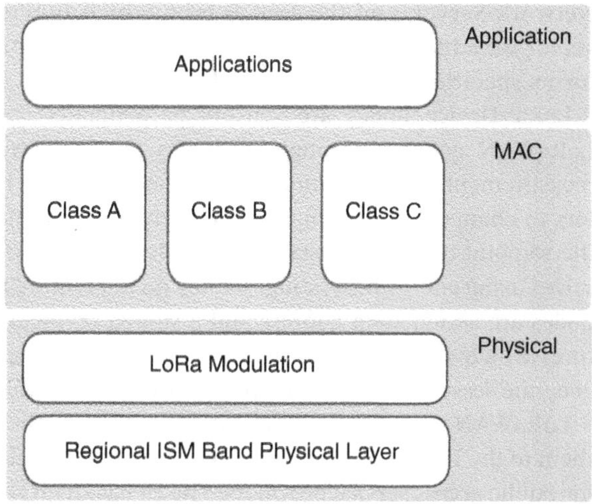

Figure 3.23 LoRaWAN Stack Architecture

- The Physical layer contains the LoRa modulation protocol and access to the regional ISM band physical medium and is discussed in more detail in Section 3.7.3.

3.7.2 LoRaWAN Network Architecture and Network Node Types

LoRaWAN has a modified star architecture, in the sense that nodes do not communicate peer-to-peer but rather to fixed gateways, as shown in Figure 3.24. In addition to Device nodes, the LoRaWAN network architecture contains three other node types: the LoRaWAN Gateway, the LoRaWAN

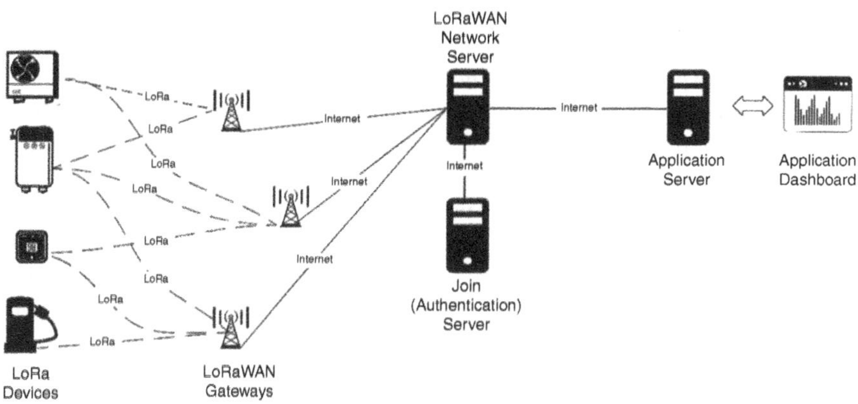

Figure 3.24 LoRaWAN Network Architecture

Network Server, and the Join Server, which functions as an authentication server for admitting nodes to the LoRaWAN network [42]. Each of these performs specific functions.

LoRa Device nodes are sensors or actuators connected wirelessly to the LoRaWAN network through the LoRa physical layer protocol. They perform measurements on environmental parameters such as temperature or act as actuators to change some settings on an energy-consuming device, such as changing the setpoint on a thermostat. Most devices can run on batteries or power themselves using energy harvesting, so they do not require much maintenance. Device nodes are assigned an address and a shared cryptographic key by the manufacturer which are used to authenticate when the device joins the network and to generate session keys for encrypting application traffic.

LoRaWAN Gateway nodes receive frames from the Device nodes and forward them to the LoRaWAN Network Server over an IP network, typically the Internet for public access service providers. The IP backhaul can run over any appropriate link layer. The contents of the LoRa frames from the Device nodes are copied directly from the wireless physical layer into IP packets which are then sent over the Internet connection. Unlike Wi-Fi, Device nodes do not associate with a single LoRaWAN Gateway. Instead, any LoRaWAN gateway node that can hear a device node's transmissions sends packets to the LoRaWAN Network Server with the contents of the same frame.

The LoRaWAN Network Server performs forwarding of join requests to the Join Server, frame authentication, frame deduplication, routing of uplink application packets to the application, and forwarding of downlink frames to an appropriate Gateway for the Device node. The Network Server performs deduplication by checking the quality of the frames from a Device sent up from the Gateway nodes. It judges the quality using a received signal strength indicator sent by the Gateway in the IP packet along with the frame. Frames from the Gateway with the best-received signal strength indicator are then selected for routing to the Application Server. The Gateway node with the best-received signal strength indicator is also selected as the transmitting gateway for downlink packets sent by the Application Server to the device since it has the best reception for the Device node.

The Join Server acts as the authentication agent for the LoRaWAN network. It receives join request frames forwarded from the Network Server and sends back join replies, which the Network Server then forwards to the Device node. It also connects the Application Server to the Device node via the Network Server, performs the cryptographic session key derivations for the communication link between the Application Server and the Device node, and communicates these keys to the appropriate parties.

Finally, the Application Server hosts the Application Dashboard, whereby users can view sensor data and change actuator settings. The Application Server is also responsible for managing sensor data coming from the Device nodes and generating application packets for changing actuator settings.

3.7.3 LoRaWAN Physical Layer

Table 3.6 contains the LoRaWAN frequency bands and channel structure in Europe and the United States. Other jurisdictions have different regulations about what bands are used; see the LoRa Alliance regional parameters specification for more information [43]. Worldwide, LoRa uses exclusively subgigahertz spectrum, ensuring that signal propagation will not degrade as quickly due to obstacles such as walls and distance, making LoRaWAN well suited for BEMS device network deployments deep inside buildings.

The LoRa physical layer uses a modification of the chirp spread spectrum algorithm for modulation. LoRa uses six different spreading factors, numbered 7 through 12, which help to reduce the impact of interference; although, because the spreading factors are not orthogonal, frame collisions are still possible depending on the signal strength of the sender [44]. LoRaWAN also uses an adaptive data rate scheme to reduce power consumption. The Device nodes that are closest to the Gateway transmit at the lowest spreading factor, thereby minimizing their transmission time and prolonging battery life. Devices that are further away from the Gateway use a higher spreading factor. LoRa data rates vary between 980 bps and 21.9 kbps depending on the spreading factor and frame payload size [42].

Table 3.6 LoRaWAN Channel Structure

Band (MHz)	Regulatory Jurisdiction	Band Endpoints (MHz)	Number Channels	Channel Bandwidth (kHz)	Direction
433	Europe, other regulatory jurisdictions	433.175–434.665	3	125, 250	Uplink and Downlink
868	Europe, other regulatory jurisdictions	863.0–870	3	125	Uplink and Downlink
915	United States, other regulatory jurisdictions	902–928	64	125	Uplink
			8	500	Uplink
			8	500	Downlink

3.7.4 LoRaWAN MAC Layer and Device Node Types

The LoRaWAN MAC layer supports three different Device node types. The three types have different MAC layers and are designed to support different classes of applications depending on their power and signaling requirements. All devices must support Class A operation, but the other two classes are optional. Class B devices must support Class A operation, and Class C devices must support Class A and Class B operation [42].

The Class A device has the lowest power consumption requirement. The Class A device MAC layer is designed for devices that require infrequent communication, such as environmental sensors. The LoRa network interface remains in sleep mode most of the time while the device is performing environmental sensing. When the monitored environmental parameter changes, it activates the LoRa network interface and transmits a report to the uplink, then sleeps for a default period of one second. When it wakes up, it listens for a response during a receive window. If there is no response, the device sleeps again and then wakes up again for the second receive window. If again there is no response, the device deactivates the LoRa network interface until it has something to report again. The length of sleep periods is configurable, and the delay between the first and second receive windows is configured based on the delay from the endpoint of the uplink transmission. Since there is no way for a Class A device to receive a downlink transmission while in sleep mode, Class A devices cannot be used for actuators.

The Class B MAC algorithm uses time-division multiplexing to control medium access. A Class B device is given a regularly scheduled, fixed time slot when the device can receive downlink traffic from the network. The Class B device MAC is designed for applications that require more frequent communication than Class A devices while still maintaining low power consumption and is suited for both sensor and actuator applications. In order to maintain time synchronization, Class B devices must align their internal clocks with the Gateway nodes. The Gateway nodes periodically send out beacons containing time information which the Class B devices receive and use to perform this alignment. A Class B device can send an uplink packet to the network at any time, but if there is no data to send, it can deactivate the LoRaWAN network interface until the next receive window opens, thereby reducing power consumption.

Class C devices are basically always listening for downlink traffic, and therefore the Class C MAC is primarily suited for devices that have access to grid power. A Class C device has the lowest latency from the Dpplication Server to the Device node. The Class C MAC has the same two receive windows as a Class A device, but it does not close the second window as the Class A device does. It also continues the second window for a fixed interval after transmitting an uplink packet.

After the second receive interval is closed, the first receive interval opens for a fixed period, but the second receive window is opened immediately after the first window closes.

3.8 Cellular IoT Link-Layer Protocols

Another WAN alternative for a building's wireless device network is to utilize one of the IoT protocols offered by licensed cellular providers. Everyone is familiar with cellular service from using a mobile phone, but in addition to offering the full Internet and voice service powering smartphones, wireless service providers also offer a variety of services targeted at device networks. Like LoRaWAN, a device network that communicates using one of the cellular IoT protocols frees the BEMS service provider or building manager from having to build, configure, and manage their own private wireless device network based on the WLAN and WPAN protocols. However, the monthly LTE subscription cost is more than the monthly cost of LoRaWAN service since cellular IoT protocols run in licensed spectrum, and the service provider has to pay for the spectrum license. A subscription to a cellular IoT service typically comes with a limit on the number of devices and amount of data that can be transmitted per month.

IoT Protocols in 5G

5G is the latest cellular standard developed by 3GPP, the international standards body for cellular. In many countries, 5G is still in the process of rollout, but in others, deployment is well underway. The video link above provides an introduction to 5G IoT and describes how NB-IoT evolves in 5G.

One lower cost option is to deploy one of the WLAN or WPAN protocols for the network inside the building but deploy an Internet gateway with a cellular IoT protocol for the Internet service, rather than using the building's wired ISP. As long as the expected data rate and expected traffic volume do not exceed the limits set by the service provider's subscription plans, a cellular Internet connection can often perform better and be more reliable than a wired Internet connection, especially in buildings with multiple tenants or where the wired Internet service is haphazardly maintained. With a wired connection, the device network must share the link with other traffic and could experience congestion if the data volume of other traffic increases. With a dedicated device network, the link is exclusively available for IoT traffic.

For 4G LTE, the two most widely deployed and utilized services are Cat M [45] and NB-IoT [46], although other protocols are available. In this section, a high-level overview of Cat M and NB-IoT in LTE is presented, comparing the two protocols. In the 5G standards development, machine-to-machine communication was one of the top use cases considered, and 5G standards have four different protocols designed for device network use, including NB-IoT. The QR code in the video link box introduces IoT in 5G.

3.8.1 LTE Cat M

3GPP, the worldwide standardization organization for cellular protocols, developed a protocol based on the standard LTE link layer and a physical layer called Category M or LTE-M for "machine-to-machine" communication. The standard was included in Release 13, which was approved in June 2016. In Release 14, which was approved in mid-2017, 3GPP included Cat M2, an enhanced version of Cat M1. Many international wireless service providers have deployed Cat M service. The main benefit to wireless network providers of Cat M is that Cat M devices utilize smaller-sized frequency bands compared to standard LTE cell phones and modems. This allows carriers to pack more devices into the cell site spectrum. For example, a typical LTE cell site can support 10x the number of Cat M devices as regular LTE devices.

Cat M1 is a variation of the standard LTE physical and link layers and uses a subset of LTE features to reduce power consumption and cost while still providing wide area coverage and high reliability. Cat M1 utilizes a bandwidth of 1.08 MHz and includes the following features:

- Lower power consumption compared to the LTE protocol utilized in regular LTE devices.
- Wide coverage area, up to 3 mi in urban areas and 22 mi in rural areas.
- Peak data rate of 800 kbps downlink and 1 Mbps uplink, average data rates of 300 kbps downlink and 375 kbps uplink.
- 5–10x less latency than regular LTE devices. Cat M1 has 10–15 msec vs. 50–100 msec for regular LTE devices.
- Full or half duplex communication.
- Lower device complexity and cost than is typically the case for standard LTE devices.

Because Cat M1 devices are designed to be low-power and energy-efficient, they have much longer battery life and reduce the need for battery replacement or recharging. In a regular LTE device, the device goes into idle mode which can reduce power consumption somewhat. Cat M1 devices can negotiate an extended sleep period, from hours to days, when they are not listening for pages from the network. If any traffic arrives for the device, it is partially buffered in the network until the device wakes up. Cat M1 devices can also request a longer paging cycle, a 10.24-second paging cycle, as opposed to the 1.28-second paging cycle, used by regular LTE devices. These optimizations lead to a battery life of up to 10 years for Cat M1 devices.

Cat M1 devices are not designed for mobility, nor do they support fallback to the older 2G and 3G networks when LTE coverage is lacking. Cat M2 is similar to Cat M1 in terms of data rate and power consumption but additionally supports mobility and fallback to 2G/3G networks if an LTE network is not available. Average data rates are the same for Cat M2 as for Cat M1, but peak data rates are higher: 4-Mbps downlink and 7-Mbps uplink.

Worldwide, many wireless service providers have deployed Cat M1 service. In the United States, Cat M1 service is available for prices that start at around $1.00 for 500 kB of data per device per month and run to $6.00 for 10 Mb of data per device per month, plus an initial one-time connection charge. Very few carriers have deployed Cat M2.

3.8.2 NB-IoT

In 2016, 3GPP approved Cat NB1, or NB-IoT, as part of Release 13 of LTE. NB-IoT is an alternative to Cat M for low-power, low-cost IoT devices. In 2018, 3GPP release an updated version of NB-IoT in Release 14, called Cat NB2 that added mobility and fallback to 2G/3G networks.

For individual devices, NB-IoT offers a slightly different feature set from Cat M1. NB-IoT is a wide-area cellular communication protocol designed for indoor coverage, low cost, and low power. It uses a subset of the LTE standard that limits the bandwidth to 180 kHz and can use the standard LTE guard bands in addition to the main band. NB-IoT includes the following features:

- Lower power consumption compared to the LTE protocol utilized in regular LTE devices.
- Wide coverage area, up to 0.5 mi in urban areas and 6 mi in rural areas.
- Peak data rate of 170 kbps downlink and 250 kbps uplink, average data rates of 26 bbps downlink and 62 kbps uplink.
- Latency of 1.6–10 second.
- Half duplex communication.
- Lower device complexity and cost than is typically the case for standard LTE or Cat M1 devices.

In addition, NB-IoT supports the same paging power optimizations as Cat M1. As with Cat M1, the NB1 protocol does not support mobility or failover to 2G/3G if LTE deployment is lacking, while NB2 does support these features.

As with Cat M1, many wireless service providers worldwide have deployed NB-IoT, primarily NB1. In the United States, prices for NB-IoT service start at $2.50 per month for unlimited data and unlimited devices, with a $32 initial payment for an account and a $0.70 per device charge for a SIM card. NB-IoT is thus substantially less expensive than Cat M1, partially because the carrier can utilize the guard bands and therefore does not have to accommodate IoT devices at the expense of regular LTE devices.

3.9 Selecting Link-Layer Technologies for a BEMS Network

Planning the design and implementation of the communication network for a BEMS deployment requires considering a number of factors. The communication network for a BEMS breaks down into the following parts:

- The device network for sensor and actuator devices.
- The backhaul network for transporting data from the devices to the BEMS analytics and control software, typically hosted in a cloud or an on-premises data center.
- The WAN network over which data from the building is sent to the BEMS through the Internet.

Single Pair Ethernet

An alternative to wireless technologies for new buildings and remodels is single pair Ethernet, a new IEEE standard for running Ethernet over a single pair of wires rather that two pairs as in the standard twisted pair cable. In addition to data, power can be transmitted over the wires but this is not the same as the PoE standard discussed in Section 3.2.4. The above QR code links to a video with more information. Single pair Ethernet has been explicitly designed for industrial IoT applications, but it also includes building use cases.

3.9.1 Selecting a Device Network

First and foremost, a decision about which device network technology to install depends to a large part on whether devices are available that support the device network technology and which can measure the important parameters and control the equipment in the buildings. If, for example, the only devices available run Wi-Fi, then a Wi-Fi network must be installed. Usually, however, different options at varying price points are available, and a choice must be made.

The advantage of WLAN and WPAN devices such as Wi-Fi, the upper layer 802.15.4 networking technologies (Zigbee and Thread), and BLE is that once the device is purchased and installed, there is no ongoing subscription charge for data transfer. The coverage area of these networking technologies is roughly comparable. With respect to data rate, Wi-Fi has the highest (574 Mbps–9.608 Gbps), BLE the next highest (125 kbps–2 Mbps), and 802.15.4 the lowest (20–250 kbps). The amount of data in a measurement or command may, therefore, influence the choice of network technology. Since power utilization for Wi-Fi is higher than for 802.15.4 and BLE, Wi-Fi sensors, such as smart thermostats, mostly need access to grid power. If grid power is not available, an 802.15.4 or BLE device is usually a better option, other considerations being equal. Wi-Fi devices connected to grid power need no physical maintenance, but 802.15.4 and BLE devices that are battery powered typically require changing the battery periodically, usually no more than once every 10 years. If the device uses energy harvesting, even changing the battery is not required. The disadvantage of using these protocols is the costs for ongoing network administration and maintenance, in particular security software upgrades, which requires hiring qualified IT staff. To the extent that such staff is needed for the building LAN anyway, an additional expense may not be necessary.

The advantage of LoRaWAN, NB-IoT, and Cat M1 is that the service provider handles the network administration, including network security upgrades, although the device owner is sometimes responsible for device firmware upgrades depending on the conditions of the service plan. LoRaWAN, NB-IoT and Cat M1 have different data rates and coverage areas, with LoRaWAN having 980 bps–21.9 kbps data rate and 3–10 mi coverage area, NB-IoT having 26 kbps downlink/62 kbps uplink data rate and 0.5–6 mi coverage area, and Cat M1 having around 300 kbps data rate and 3–22 mi coverage area, although their power utilization requirements are quite similar. Again, the maximum amount of data in a measurement or command might suggest one technology over the others, given the equal availability of devices in an affordable price class. Different service providers structure their service offerings differently and in some cases, even cover device replacement if a device malfunctions or is somehow damaged.

3.9.2 Selecting a Building Backhaul LAN Technology

In many cases, the building backhaul technology selection decision has already been determined because the building is already wired for Ethernet. This will typically be the case for most large commercial and government buildings, and also some small commercial buildings, multifamily dwellings, and residential buildings. VLANs can be used to isolate BEMS traffic from other traffic. By setting up a separate VLAN for the BEMS traffic with access controls on what devices and people can access it, OT and IT traffic in the building can be kept separate. Devices can communicate through the device network and device network gateway, if necessary, to the building gateway through the BEMS VLAN. If an Ethernet drop is missing in certain rooms due to a lack of twisted pair Ethernet cable, Wi-Fi is a low-cost and convenient option to fill in the gaps, especially with the introduction of mesh Wi-Fi in Wi-Fi 6E. A mesh Wi-Fi network can cover a larger area and provide excellent throughput over the wireless backhaul until an Ethernet drop or the building gateway is reached.

If the building does not have wired Ethernet, then the options for backhaul technology depend primarily on how well the building's walls allow Wi-Fi wireless signal propagation. If the building's walls allow reasonable wireless propagation, then a Wi-Fi network with mesh backhaul is probably the best option. The Wi-Fi mesh backhaul can be terminated at an access point directly connected to the building gateway or Internet router. If the wireless reception is poor, then Ethernet over power line is an option to fill in the gaps where the Wi-Fi mesh propagation is inadequate, although it may be more expensive than Wi-Fi. In either case, if a WPAN mesh is used for the device network, the device network gateway for the WPAN mesh network needs to be deployed as well. Combining the device network gateway with the building network gateway may be possible for smaller buildings with good wireless reception.

An alternative if the building does not have wired Ethernet is to use LoRaWAN or one of the cellular technologies for the device network instead, removing the need for having a building backhaul that uses the building LAN. In that case, no backhaul network is necessary, as the BEMS can access the devices in the device network directly through their service provider-supplied IP address or another identifier rather than through a building gateway. For devices such as smart thermostats, where the thermostat's Wi-Fi talks with a manufacturer's website rather than directly to the BEMS, however, a Wi-Fi access point and backhaul network to the building's Internet router are still required.

3.9.3 Selecting a WAN Technology

In general, the networking technology used for the WAN between the building and the cloud or on-premises data center will be out of the BEMS engineer's control. Usually, a building has access to a few ISPs through a wide area wired, fiber optic, or coax (TV cable) wide area connection, and the decision about which to use may already have been taken by the building's existing Internet users. If the building is part of an enterprise campus, the WAN connectivity may be provided by the enterprise IT department.

If the choice of ISPs is open, the service provider with the best reliability and lowest cost service plan for the data rate expected from the BEMs and associated devices should be selected. An alternative is to use a wireless ISP. If a wireless ISP is used, the building gateway needs to be outfitted with an appropriate wireless modem. Wireless Internet connectivity is particularly attractive if the wired connection is unreliable due to maintenance issues or heavily loaded by high volume traffic such as video and voice by other users in the building, which would interfere with the device network traffic.

If the data rate is expected to be less than 300 kbps, then a Cat M1 modem with a monthly data plan matching the expected amount of data transferred per month may be a very cost-competitive option since Cat M1 plans tend to be much cheaper than wired Internet and regular LTE data plans. Data should be buffered at the building gateway if the data rate exceeds 300 kbps, but real-time response should not be expected in any case. If the data rate is considerably higher, more data need to be transferred per month, and real-time response is needed, such as in a larger building or a campus complex, a regular LTE modem with a higher data rate plan can be used.

3.10 Summary

A wide variety of link–layer technologies are available for the device network, the building backhaul LAN, and the WAN. Most of the link technologies are quite mature. Ethernet and Wi-Fi are the primary LAN technologies. For the building backhaul network, Ethernet is the dominant link–layer technology in wired networks offers versatility in physical media and data rates, and provides a wide range of deployment options. Wi-Fi offers high-speed networking appropriate for situations where wired Ethernet is not available or would be challenging to deploy.

For the device network, IEEE 802.15.4 and BLE are both options with a wide variety of available devices. Documentation for sensor and actuator devices typically would not advertise that they support IEEE 802.15.4 directly but rather will mention that they support one of the adaptation protocols, such as Zigbee or Thread, which are built on top of the 802.15.4 physical and link layers. BLE provides excellent coverage for short-range communication and is well-suited for devices with limited energy resources, providing efficient, low-power connectivity within a limited radius. LoRa, LTE Cat M, and NB-IoT are wide-area wireless communication protocols where connectivity is typically supplied by a wireless ISP, freeing the building manager from having to manage the device network. LoRa is suitable for sensor networks that require limited bandwidth. LTE Cat M and NB-IoT are cellular-based technologies that run over the same wireless networks as smartphones but require more limited bandwidth. These protocols are ideal for applications needing reliability and moderate data rates over larger distances. In some cases, LoRa, LTE Cat M, and NB-IoT may be appropriate choices for connecting the building gateway to the BEMS in a public cloud or private data center rather than using a wired ISP.

Selection criteria for the building backhaul LAN, device network, and WAN depend on factors such as range, data rate, power consumption, and deployment context. Ethernet and Wi-Fi are preferred for high-data-rate, infrastructure-centric environments because they are fairly ubiquitous in commercial and government buildings. For energy-efficient, low-data-rate applications, WPAN technologies such as IEEE 802.15.4 and BLE are suitable. LoRa and cellular technologies cater to scenarios requiring wide-area coverage and robust connectivity. The choice of technology for these three networks should align with the cost and performance goals of the building management.

References

1 IEEE. IEEE Standard for Ethernet. In: IEEE Std 802.3-2022 (Revision of IEEE Std 802.3-2018). pp. 1–7025, 2022 Jul 29.

2 Wikipedia. Ethernet [Internet]. [cited 2022 Dec 15]. Available from: https://en.wikipedia.org/wiki/Ethernet.

3 IEEE. IEEE Standard for Local and Metropolitan Area Network--Bridges and Bridged Networks. In: IEEE Std 802.1Q-2018 (Revision of IEEE Std 802.1Q-2014). pp. 1–1993, 2018 Jul 6.

4 ComputerNetworkingNotes. STP – Spanning Tree Protocol Explained With Examples [Internet]. [cited 2022 Dec 19]. Available from: https://www.computernetworkingnotes.com/ccna-study-guide/stp-spanning-tree-protocol-explained-with-examples.html.

5 Wikipedia. Ethernet Frame [Internet]. [cited 2022 Dec 19]. Available from: https://en.wikipedia.org/wiki/Main_Page.

6 IEEE. IEEE Standard for Local and Metropolitan Area Networks – Port-Based Network Access Control. In: IEEE Std 802.1X-2020 (Revision of IEEE Std 802.1X-2010 Incorporating IEEE Std 802.1Xbx-2014 and IEEE Std 802.1Xck-2018). pp. 1–289, 2020 Feb 28.

7 Wikipedia. MAC Addresses [Internet]. [cited 2022 Dec 21]. Available from: https://en.wikipedia.org/wiki/MAC_address.

8 IEEE. Guidelines for Use of EUI, OUI, and CID [Internet]. [cited 2023 Jan 13]. Available from: https://standards.ieee.org/wp-content/uploads/import/documents/tutorial/eui.pdf.

9 ITU-T. Applications of ITU-T G.9960, ITU-T G.9961 Transceivers for Smart Grid applications: Advanced Metering Infrastructure, Energy Management in the Home and electric vehicles [Internet]. [cited 2023 Jan 28]. Available from: https://www.itu.int/dms_pub/itu-t/opb/tut/T-TUT-HOME-2010-PDF-E.pdf.

10 IEEE. IEEE Standard for Broadband over Power Line Networks: Medium Access Control and Physical Layer Specifications. In: IEEE Std 1901-2020 (Revision of IEEE Std 1901-2010). pp. 1–1622, 2021 Jan 19.

11 PRIME Alliance. Specification [Internet]. [cited 2023 Jan 28]. Available from: https://www.prime-alliance.org/alliance/specification.

12 HomePlug Alliance. HomePlug Green PHY Version 1.1.1 Specification [Internet]. [cited 2023 Jan 28]. Available from: https://web.archive.org/web/20170920023510/http://www.homeplug.org/tech-resources/spec-acknowledgement/?spec=gp.

13 Wikipedia. Combined Charging System [Internet]. [cited 2022 Jan 28]. Available from: https://en.wikipedia.org/wiki/Combined_Charging_System.

14 IEEE. IEEE Standard for Information Technology – Telecommunications and Information Exchange Between Systems – Local and Metropolitan Area Networks – Specific Requirements – Part 11: Wireless LAN Medium Access Control (MAC) and Physical Layer (PHY) Specifications. In: IEEE Std 802.11-2020 (Revision of IEEE Std 802.11-2016). pp. 1–4379, 2021 Feb 26.

15 Wikipedia. ISM Radio Band [Internet]. [cited 2022 Dec 21]. Available from: https://en.wikipedia.org/wiki/ISM_radio_band.

16 IEEE. IEEE Standard for Information Technology – Telecommunications and Information Exchange Between Systems Local and Metropolitan Area Networks – Specific Requirements Part 11: Wireless LAN Medium Access Control (MAC) and Physical Layer (PHY) Specifications Amendment 1: Enhancements for High-Efficiency WLAN. In: IEEE Std 802.11ax-2021 (Amendment to IEEE Std 802.11-2020). pp. 1–767, 2021 May 19.

17 Wi-Fi Alliance. Certification Programs [Internet]. [cited 2023 Jan 1]. Available from: https://www.wi-fi.org/certification/programs.

18 Wikipedia. List of WLAN channels [Internet]. [cited 2022 Dec 31]. Available from: https://en.wikipedia.org/wiki/List_of_WLAN_channels#2.4_GHz_(802.11b/g/n/ax).

19 IEEE. IEEE Standard for Information Technology – Telecommunications and Information Exchange Between Systems – Local and Metropolitan networks – Specific Requirements – Part 11: Wireless LAN Medium Access Control (MAC) and Physical Layer (PHY) Specifications: Higher Speed Physical Layer (PHY) Extension in the 2.4 GHz Band. In: IEEE Std 802.11b-1999. pp. 1–96, 2000 Jan 20.

20 IEEE. IEEE Standard for Information Technology – Local and Metropolitan Area Networks – Specific Requirements – Part 11: Wireless LAN Medium Access Control (MAC) and Physical Layer (PHY) Specifications – Amendment 8: Medium Access Control (MAC) Quality of Service Enhancements. In: IEEE Std 802.11e-2005 (Amendment to IEEE Std 802.11, 1999 Edition Reaff 2003). pp. 1–212, 2005 Nov 11.

21 IEEE. IEEE Standard for Information Technology – Telecommunications and Information Exchange Between Systems – Local and Metropolitan Area Networks – Specific Requirements – Part 11: Wireless LAN Medium Access Control (MAC) and Physical Layer (PHY) Specifications – Amendment 4: Enhancements for Very High Throughput for Operation in Bands Below 6 GHz.

In: IEEE Std 802.11ac(TM)-2013 (Amendment to IEEE Std 802.11-2012, as Amended by IEEE Std 802.11ae-2012, IEEE Std 802.11aa-2012, and IEEE Std 802.11ad-2012). pp. 1–425, 2013 Dec 18.

22 Wikipedia. 802.11 Frame Types [Internet]. [cited 2023 May 20]. Available from: https://en.wikipedia.org/wiki/802.11_Frame_Types.

23 IEEE. IEEE Standard for Information Technology – Telecommunications and Information Exchange Between Systems Local and Metropolitan Area Networks – Specific Requirements Part 11: Wireless LAN Medium Access Control (MAC) and Physical Layer (PHY) Specifications. In: IEEE Std 802.11-2012 (Revision of IEEE Std 802.11-2007). pp. 1–2793, 2012 Mar 29.

24 Table 9-92, IEEE 802.11-2020. pp. 939–948.

25 Dionicio, R. Station Authentication and Association [Internet]. [cited 2023 May 20]. Available from: https://netbeez.net/blog/station-authentication-association.

26 IEEE. IEEE Standard for Low-Rate Wireless Networks. In: IEEE Std 802.15.4-2020 (Revision of IEEE Std 802.15.4-2015). pp. 1–800, 2020 Jul 23.

27 Bluetooth-SIG. Bluetooth Core Specification, Version 5.4 January 2023.

28 Connectivity Standards Alliance. Zigbee Specification [Internet]. [cited 2023 Jan 5]. Available from: https://zigbeealliance.org/wp-content/uploads/2019/11/docs-05-3474-21-0csg-zigbee-specification.pdf.

29 Thread Specification, v1.1.1, Thread Group. 348 pp., February, 2013.

30 Clauses 6.3 and 6.4, IEEE 802.15.4–2020. pp. 75–93.

31 Clause 7.4, IEEE 802.15.4 Standard 2020. pp. 179–237.

32 Clause 7.2.2.6, IEEE 802.15.4 Standard 2020. pp. 163–164.

33 IEEE. IEEE Standard for Local and Metropolitan Area Networks: Overview and Architecture. In: IEEE Std 802-2014 (Revision to IEEE Std 802-2001). pp. 1–74, 2014 Jun 30.

34 Bluetooth-SIG. Mesh Profile. Bluetooth Special Interest Group, Mesh Working Group, Version 1.0.1, January, 2019.

35 Bluetooth-SIG. Origin of the Bluetooth Name [Internet]. [cited 2023 Jan 14]. Available from: https://www.bluetooth.com/about-us/bluetooth-origin.

36 Volume 6, Part B, Section 2.2, Bluetooth Core Specification, Version 5.4. pp. 2687–2688.

37 Volume 6, Part B, Section 2.3.1, Bluetooth Core Specification, Version 5.4. pp. 2691–2698.

38 LoRa Alliance. LoRaWAN® Specification v1.1 [Internet]. [cited 2023 Jan 23]. Available from: https://lora-alliance.org/resource_hub/lorawan-specification-v1-1.

39 Wikipedia. LoRa [Internet]. [cited 2023 Jan 23]. Available from: https://en.wikipedia.org/wiki/LoRa.

40 ITU-T. Recommendation Y.4480 (11/21) [Internet]. [cited 2023 Jan 23]. Available from: https://www.itu.int/rec/dologin_pub.asp?lang=e&id=T-REC-Y.4480-202111-I!!PDF-E&type=items.

41 Github. [Internet]. [cited 2023 Jan 24]. Available from: https://github.com/topics/lorawan-application.

42 Semtech.com. LoRa Developer Portal [Internet]. [cited 2023 Jan 24]. Available from: https://lora-developers.semtech.com/documentation/tech-papers-and-guides/lora-and-lorawan.

43 LoRa Alliance. LoRaWAN 1.1 Regional Parameters [Internet]. [cited 2023 Jan 25]. Available from: https://lora-alliance.org/resource_hub/lorawan-regional-parameters-v1-1ra.

44 Benkhelifa F., Bouazizi Y., and McCann J.A. How orthogonal is LoRa modulation? *IEEE Internet of Things Journal*, vol. 9, no. 20, pp. 19928–19944, 2022 Oct 15.

45 Wikipedia. LTE-M [Internet]. [cited 2023 Dec 22]. Available from: https://en.wikipedia.org/wiki/LTE-M.

46 Wikipedia. Narrowband-IoT [Internet]. [cited 2023 Dec 22]. Available from: https://en.wikipedia.org/wiki/Narrowband_IoT.

4

The IP Stack

As discussed in Chapter 3, the communication system for a BEMS consists of multiple links potentially running different physical and link-layer protocols. Communication must span from the devices, through the device network gateway and the building backhaul network, to the building gateway and router and over the WAN to the cloud or on-prem data center. If the devices in the device network use one or more of the wireless WAN device network protocols (LoRaWAN, LTE/Cat M1, or LTE/NB-IoT), communication must span from the sensors and actuators through the wireless service provider's network to the BEMS in the cloud. All these disparate link layers are unified under a single network layer running the IP that routes packets between the sensors and actuators in the device network, the building gateway (if one is used) in the building LAN, and the BEMS in the cloud. Since the early 2000s, the IP stack has been the go-to platform for the Internet of Things (IoT), a term that has been popularized to describe integrated, unified communications between distributed devices and applications. The IP stack and the IP networking protocol are primarily responsible for making the Internet the ubiquitous communication platform of the twenty-first century.

The IP stack originated in the late 1970s and early 1980s as the system software for packet-switching hardware. Prior to the development of packet switching, data (primarily voice communication) was sent as analog signals through telephone lines. Packet switching, in contrast, encodes data in binary form as 8-bit bytes and bundles the bytes into small packets, which are then sent to the destination. The transmitted data is not restricted to voice or any other specific analog format input. At the destination, the data is reassembled from the individual packets. Packet switching is much more efficient and resilient than switching analog signals and is also more resilient to corruption in transit. In the 1970s, the U.S. Department of Defense began a research project called the Advanced

Automating Building Energy Management for Accelerated Building Decarbonization: System Architecture and the Network Layer, First Edition. James Kempf.

Published 2025 by John Wiley & Sons, Inc.

Research Projects Agency Network (ARPANET) to investigate how packet switching could be applied. The U.S. National Science Foundation (NSF) joined as a sponsor in the 1980s. During the 1970s and 1980s, the ARPANET was only available to researchers at academic institutions and research labs, but in the early 1990s, the U.S. government spun it out to private industry, and it became available for commercial users, renamed the Internet [1].

With the growth of the Internet in the late 1980s and early 1990s, a standards organization was needed to guide architecture and protocol development. The Internet Engineering Task Force (IETF) was founded in 1986 to manage development and standardization of the Internet. The IETF is a nonprofit standardization group concerned with the evolution of the Internet architecture and the engineering work for Internet protocols and operational standards [2]. All IETF standards are available on-line as Request for Comments (RFCs) without cost.[1] The management of the two principal name spaces on the Internet defined by IETF standards, the IP address space and the Domain Name Service (DNS), is the responsibility of another nonprofit, the Internet Corporation for Assigned Names and Numbers (ICANN), which delegates responsibility to five regional Internet registries (RIRs) operating in different international regions [3]. The Internet Assigned Numbers Authority (IANA) is a department within ICANN responsible for the global coordination of various Internet identifiers, including IP addresses and protocol parameters. The IETF and ICANN operate under the guidance of the Internet Society, a nonprofit organization that provides leadership in Internet-related standards, education, and policy.

In this chapter, we discuss the IP stack: network, transport, and system-level application layer protocols like DNS available for use in BEMS communication systems. Section 4.1 is an introduction to the IP network layer. The IP networking layer has two versions: the universally deployed IP version 4 (IPv4) and the more recent IP version 6 (IPv6), which is just beginning to see wider deployment, especially for smart home and home energy management use cases. Section 4.2 discusses IPv4, Section 4.3 discusses IPv6, and Section 4.4 discusses IPv6 for Low-Power Wireless Personal Area Networks (6LoWPAN), an IPv6 adaptation layer for lower-power mesh networks like 802.15.4. Section 4.5 introduces the IP transport layer. In Sections 4.6 and 4.7, User Datagram Protocol (UDP) and Transmission Control Protocol(TCP), the predominant transport layers for applications written on the IP stack, are discussed. In fact, TCP is so commonly used that many descriptions of applications implemented on the IP stack say they run on the TCP/IP stack.

System-level application protocols are introduced in Section 4.8. Section 4.9 describes the DNS protocol, a system-level application protocol that maps

1 RFCs are available at the following link: https://www.ietf.org/standards/rfcs.

human-readable service and node names into IP addresses, in addition to providing other system-level information. Section 4.10 describes the Dynamic Host Configuration Protocol (DHCP), which is used widely in IPv4 networks and in some IPv6 networks to configure a node with an IP address and other network information when it enters the network. In Section 4.11, an open-source tool, Wireshark, for tracing packets is described. Finally, Section 4.12 briefly summarizes the chapter.

The importance of the IP stack in making cloud and on-prem data center-based BEMS possible cannot be overemphasized. If the IP stack were not available or if the networking layer were split into multiple noninteroperable protocols, some of which were proprietary – which was the situation prior to the early 1990s – device networks and unified communication to data centers would not exist. Implementing a BEMS to say nothing of offering it as a service would be exponentially more difficult. The protocols of the IP stack are the ground on which the communication for the BEMS use case is built, so a thorough understanding of these protocols is required of BEMS engineers.

4.1 The IP Network Layer

The IP network layer is primarily responsible for routing packets between networks running different physical and link layer technologies and ensuring that packets reach their destination. IP provides a connectionless, best effort packet forwarding communication service upon which various higher-level communication models can be implemented at the transport layer. The basic traffic model is unicast: one node sending traffic to another. IP also supports broadcast and multicast models. In the broadcast model, a node sends traffic to all nodes on the local link. A more restricted form of broadcast is offered by the multicast traffic model. In the multicast model, nodes subscribe to a multicast group address, and any node within the group can send traffic to all the other nodes in the group by using the multicast group address. Multicast reduces the amount of unwanted traffic a node needs to process compared to broadcast by allowing the node to only receive traffic from the multicast groups it joins. The network layer also handles mapping of packets into and out of link-layer frames, and for IPv4 and 6LoWPAN, fragmentation and reassembly of packets that are too large to fit into a single frame.

4.1.1 IP Routing Overview

The network layer implements an address space for end nodes, in which each end node is assigned an address. The addresses are organized into subnets, grouping end nodes that are close together from a routing standpoint. The subnet prefix is encoded in the leftmost bits of the address, while the rightmost bits contain the identifier of the node. A router is a node with network interfaces on two or more links which may run different physical and link layer technologies or may run the same technologies on the two interfaces. In the latter case, the subnets are primarily to improve routing efficiency. Each interface is assigned to a different subnet, and the router is responsible for forwarding packets from one interface/subnet to another. On each interface, the router maintains one or more forwarding queues. If packets come into an interface faster than they can be forwarded, they are queued on one of the queues associated with the outbound interface. If the queues fill up and packets arrive faster than the router can forward them, the network is experiencing congestion, and newly arrived packets must be dropped. Maintaining multiple queues allows routers to provide different levels of service, for example, prioritizing video and voice traffic above web browsing traffic by using different queues for different traffic.

To determine the outbound interface, the router reads the destination address from the IP header, looks up the next hop in the routing table or routing information base (RIB). The next hop is the address of a node that would move the packet closer to its destination. Each subnet prefix (or the full IP address if the destination is directly connected) in the RIB has associated with it the IP address and link-layer address of the next hop. This allows a router intermediate on the path to construct a frame with the next hop link-layer address as the frame destination address and its own link-layer address as the frame source address.

To forward the packet, the router constructs a link-layer frame with the next hop link-layer address in the header and the packet as the payload, and writes the frame out on the outgoing interface or into the outbound interface queue. The router primarily uses the subnet prefix of the destination address to determine the next hop. If however the subnet prefix exactly matches a subnet assigned to one of the router interfaces, then the destination is directly connected to the link served by that interface, and the packet can be forwarded directly to its destination. The packet is copied into the frame payload field without modification. One route in the RIB is typically designated as the default route, which is applied if no other route matches. The default route usually points to an Internet gateway router or a router with more information about destination reachability than the local router holds.

To configure their RIB, routers require information about destination reachability. This is obtained by exchanging information with other routers using a routing protocol. There are two general classes of routing protocols:

- Interior gateway protocols, which only exchange information between routers within a routing domain under a unified administration,
- An exterior gateway protocol, which exchanges information between Internet gateway routers on the border between routing domains.

There are several interior gateway protocols but there is only one exterior gateway protocol, the Border Gateway Protocol (BGP) [4]. BGP distributes routing information between the border routers that route packets over the public Internet.

Routing protocols are based on different algorithms. For example, BGP is based on the distance vector algorithm [5], as are several interior gateway protocols used in various applications. The distance vector algorithm is relatively simple and routes packets based on the metric of the number of hops to the destination. Some distance vector protocols may use other information such as traffic latency. Each router periodically sends its RIB to other routers on the subnets configured on its interfaces to ensure all routers have up-to-date reachability information. The heart of distance vector routing is the Bellman–Ford algorithm [6], for computing the path with the least number of hops to a particular destination. While there are other types of routing algorithms used in interior gateway protocols, for configuring and managing a BEMS, distance vector protocols are the most important.

4.1.2 Internet Network Architecture

Figure 4.1 contains a high-level overview of the Internet architecture. A routing domain under a single administrative entity like an ISP or company is called an autonomous system. Each autonomous system has a number assigned by IANA. In the figure, Autonomous System A's equipment is labeled. A border router sits on the edge of the routing administrative domain and connects to other border routers.

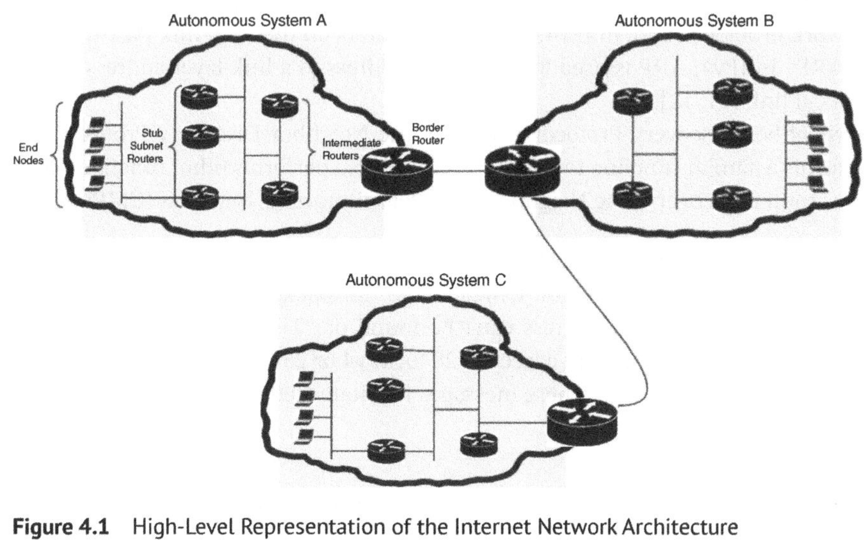

Figure 4.1 High-Level Representation of the Internet Network Architecture

BGP routes traffic between autonomous systems, while an interior gateway protocol routes within the autonomous system.[2] Stub subnet routers are connected through intermediate routers to the border router, and end nodes are connected to the stub subnet routers. Stub subnets are subnets from which packets originate on end nodes and to which packets are delivered, but they do not serve as transit networks to further subnets. The acyclic graph of routers and nodes within an autonomous system constitutes a routing topology. The public Internet consists of a collection of ISPs woven together through the BGP protocol, with private routing domains connecting into the public Internet through ISPs. These ISPs are known as the Internet backbone.

Subnets are used within larger networks to organize routing in a hierarchical fashion, with one or a small number of routers connected to the public Internet, and intermediate nodes provide routing to stub subnet routers where end nodes are connected through the stub subnets. Within larger networks, the interior gateway protocol Open Shortest Path First (OSPF) [7] is used to distribute routes. OSPF is based on a different mechanism than BGP, and uses the link state routing algorithm [8]. The link state routing protocols are more complex than distance vector protocols. End nodes typically use a routing protocol based on a distance vector algorithm.

4.1.3 ICMP, ARP, and Neighbor Discovery

Three other protocols provide a management support to the IP network layer:

- Internet Control Message Protocol (ICMP) – used to diagnose and resolve network problems, and, in IPv4, to discover routers on the local link [9, 10],
- ARP– in IPv4, ARP is used to map an IP address to a link-layer address on the local link [11, 12],
- Neighbor Discovery Protocol – in IPv6, the Neighbor Discovery Protocol performs a similar function to ARP in IPv4 and also performs other functions such as router discovery. The Neighbor Discovery Protocol is based on ICMPv6 [13].

The primary function of ICMP is to inform nodes of error conditions. It returns messages, such as "Destination Unreachable" for packets that can't be delivered because the destination address can't be found or "Time Exceeded" if the hop count during packet delivery exceeds 128 for IPv4 or 255 for IPv6, to the sender if an error condition occurs. These messages facilitate troubleshooting connectivity

2 BGP can be used within autonomous systems too, especially for large routing domains spanning multiple countries and continents.

problems or diagnosing network congestion. In addition, ICMP is used to discover routers on the local link that connect to other subnets. Both the `ping` command, which provides information on whether a node is alive and how long the latency is between the source and destination nodes, and the `traceroute` command, which generates a list of routers required to reach a destination node, also use ICMP.

Since the IP protocol does not handle link-layer addresses, a node needs some means to map the IP address of a destination node on the local link to the link-layer address, so its link layer can set the destination link-layer address correctly on frames. This function is performed in IPv4 by the ARP protocol. Strictly speaking, ARP is not an IPv4 layer protocol, it operates at the link layer. ARP packets are sent directly in link-layer (nominally Ethernet) frames without an IP protocol header, and they have a separate frame type code (0x806) from IPv4. ARP Request messages are broadcast to the local link all-nodes broadcast address (0×ffffffff on Ethernet). The broadcasting node includes the IP address for which a link-layer address is requested in the message. The node with the matching IP address sends an ARP Reply, allowing the sending node to determine the link-layer address from the link-layer address in the frame. If no node on the link has a matching IP address, the subnet router replies with its link-layer address so packets can be routed off the link. Typically, the sending node caches the IPv4 address to link-layer address mapping for a period of time, to avoid burdening the local link with unnecessary broadcast traffic. For link layers that are not Ethernet or Wi-Fi, other solutions may be required.

For IPv6, the Neighbor Discovery protocol performs a similar function to ARP, except it utilizes ICMPv6. This unifies the base protocol used for IP address to link-layer address resolution with router discovery and allows the Neighbor Discovery protocol to perform some additional functions. It also makes address resolution more easily portable to link layers that are not Ethernet and allows the use of IP authentication and security mechanisms in securing the packets. In addition, Neighbor Discovery uses multicast instead of broadcast. Section 4.3.2.1 describes how an IPv6 node configures its address and router using Neighbor Discovery.

4.2 IPv4

IPv4 was the first version of the IP protocol. The original design for IPv4 dates from 1981 [14]. While the underlying features such as the packet header and address format have remained largely the same, many changes have been made to accommodate the exponential growth in the number of nodes connecting to the Internet.

4.2.1 IPv4 Packet Header

Figure 4.2 illustrates the basic IPv4 packet header [15], totaling 20 bytes. The fields are:

- Version – the IP version number. In IPv4 packets, this is 4,
- IP Header Length – specifies the number of 32-bit words in the header, including header options,
- Diffserv Code Point – may contain a differentiated services code value for a different type of forwarding service, for example applications, that need low latency like streaming. This field allows routers to identify packets that need to use different queues with different routing characteristics, like video and audio traffic which needs low latency forwarding vs. web browsing traffic. RFC 2474 [16] describes differentiated services,
- Explicit Congestion Notification (ECN) – an optional feature that allows routers to inform end nodes and other routers about congestion without actually dropping packets, if both the network and end nodes support it. This field is defined in RFC 3168 [17],
- Packet Length – length of the entire packet, including the payload in bytes. The minimum size is 20 bytes and includes the header without data, while the maximum size is 65,535 bytes. The maximum link layer size on the end-to-end path is called the maximum transmission unit (MTU). Packets are fragmented if they cannot fit through the link layer having the smallest MTU on the end-to-end path. IPv4 nodes are required to reassemble datagrams of size up to 576 bytes, but most nodes handle larger packets. On Ethernet links, all frames are 1,500 bytes unless the link layer handles jumbo frames. Fragmentation in IPv4 is performed in either the sending node or in routers. Reassembly is performed at the receiving node. See [15] for more details on how IPv4 performs fragmentation and reassembly,
- Fragment Identification – identifies a group of fragments that need to be reassembled into a single source packet. All fragments of the same original packet have the same identification number. See RFC 6864 for details [18],
- Flags – controls fragmentation. The bits in order from most to least significant are:
  - bit 0 – reserved, set to zero,
  - bit 1 – if set, instructs routers to not fragment the packet,
  - bit 2 – more fragments are coming.

  If bit 1 is set and a router must fragment the packet to route it, the packet is dropped. If a packet is unfragmented, bit 2 is cleared. If a packet is fragmented, all fragments except the last have bit 2 set, but the last fragment has a non-zero Fragment Offset field, differentiating it from an unfragmented packet for which this bit is also set to zero,

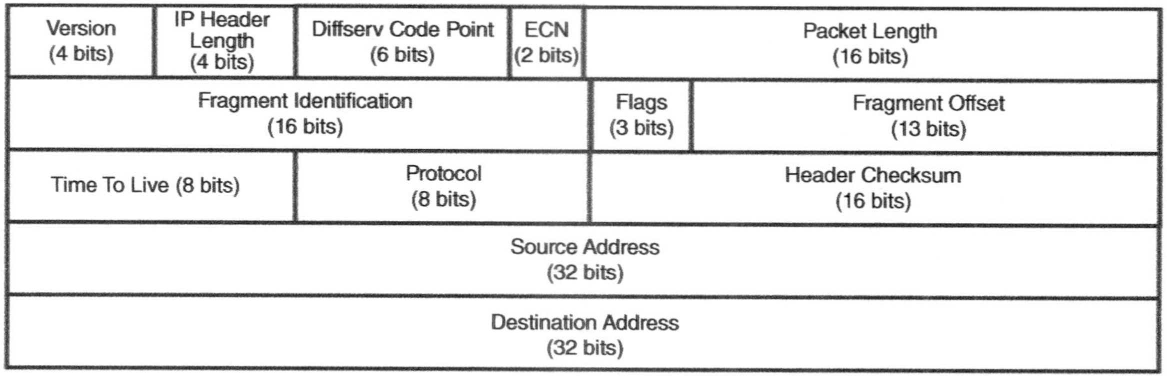

Figure 4.2 IPv4 Packet Header

- Fragment Offset – the offset of the current fragment from the start of the packet in units of 8 bytes (64 bits). The Fragment Offset value for the first fragment is 0. Subsequent fragment offsets are calculated by adding together the number of bytes previously transmitted to the previous offset and then dividing by 8. The Fragment Offset allows the receiving node to position the fragments in the proper order to rebuild the packet [19],
- Time to Live – the maximum number of hops the packet can be forwarded. On each hop, this field is decremented by the router. If it reaches zero, the packet is discarded. As mentioned in Section 4.1.3, the maximum value of this field is 128 for IPv4. The Time to Live field helps prevent routing loops,
- Protocol – the protocol number of the payload field contents. Protocol numbers are maintained by IANA [20],
- Header Checksum – a checksum value used to detect header corruption in transmission. A router calculates the checksum of the header and compares it with this field. If the two do not match, the packet is discarded. This field only covers the header, protocols further up the stack are responsible for ensuring their own integrity. The router needs to recalculate the checksum at each hop because the Time to Live field changes. See [9] for a description of how the checksum is calculated,
- Source and Destination Addresses – these fields contain the 4-byte, 32-bit IP address of the source node and destination node. Note that these addresses may be changed in transit by network address translation (NAT), see Section 4.2.2.4.

The basic header can be followed by up to 40 bytes of additional extension headers. Extension headers are almost never used in IPv4 because most routers refuse to forward packets with them.

4.2.2 IPv4 Addressing

The original Internet architecture called for transparent, end-to-end addressing with no intermediaries except for routers; however, that architecture has been modified over the years. The IPv4 address space consists of addresses having 32 bits (4 bytes) so the maximum number of addresses is 2^{32} or 4,294,967,296. IPv4 addresses are most often represented using the quad-dot notation, for example, 192.72.42.8. Each number represents a different byte in the address.

As described in Section 4.1.1, the Internet address space is divided into subnets to enable more efficient routing. A subnet is a group of nodes having the same address prefix. A node address consists of a subnet prefix and the node identifier suffix. If the node identifier is zero, the address designates the entire subnet, while a node identifier of 255 designates a broadcast to all nodes on the subnet. For example, 192.72.42.0 designates the entire subnet including the node 192.72.42.8 which has a node identifier of 8. The maximum number of nodes on the 192.72.42.0 subnet is 254, since the node identifiers 0 and 255 are reserved and there are a total of 256 (2^8) addresses on the subnet. The subnet portion of the address is often specified by a subnet mask when configuring routers and other network intermediaries. The subnet mask is a bit mask represented like a quad-dot address, but with all the bits in the mask corresponding to bits in the subnet prefix set to 1. Bits in the node identifier are set to 0. For example, the subnet mask 255.255.255.0 corresponds to a subnet prefix with 24 bits and a node identifier suffix of 8 bits.

Originally, allocation of and routing between IPv4 subnets were based on the concept of IP address classes. Three IP classes of IP addresses were defined: Class A, Class B, and Class C. Each class was defined by a range of IP subnet prefixes, with prefixes defined on byte boundaries in the address. Class A addresses had a subnet prefix consisting of the leftmost byte of the address, or a subnet mask of 255.0.0.0, Class B addresses had the two leftmost bytes, or a subnet mask of 255.255.0.0, etc. The rest of the address was left for the node identifier. This meant that Class A networks had room for 16,777,214 nodes, while Class C networks had room for only 254 nodes. ISPs were assigned an address class depending on the size of their network. However, with the explosion in nodes connecting to the Internet after the commercialization of the Internet in the 1990s and the looming exhaustion of the IPv4 address space, the IETF realized that a more efficient addressing architecture was needed.

Classless Inter-Domain Routing (CIDR) [21] was developed to provide more flexibility in subnet prefix assignment. CIDR did away with the old class structure and, in its place, assigned subnet address blocks on single-bit boundaries instead of on byte boundaries. CIDR designates the subnet prefix size by a slash followed by the number of bits in the subnet prefix. For example, the subnet prefix 192.72.42.0 is a /24 prefix, but so are others, for example, 102.168.50.0. An example of a CIDR prefix that is not on a byte boundary is a /20, for example, 192.72.15.0. The /24 has 255 node identifiers in the subnet, while the /20 has 4,095 due to having 4 fewer bits in the subnet prefix and 4 more in the node identifier. The slash notation alone is usually used to designate a group of subnet prefixes in a common routing domain, and is typically part of the routing architecture specification for the domain; for example, a /24 domain has nodes with 8-bit node identifiers and 24-bit subnet identifiers. The slash notation can also be appended to the IPv4 node address itself, to indicate the size of the subnet prefix; for example, 192.72.42.8/24 indicates that the first 24 bits of the address constitute the subnet prefix and the last 8 designate the node identifier. If the node identifier is 0, the CIDR notation designates the subnet prefix itself, for example, 192.72.42.0/24. A CIDR subnet prefix written in this form designates an address block, which is a group of addresses that are all routable through the prefix.

Occasionally, a CIDR address may be written as a /32, that is, the entire address appears to be a subnet prefix. A /32 address indicates that the subnet has only a single node on it, and all traffic will be forwarded directly between the node and the router. This can be used to prevent the node from communicating directly with other nodes on the subnet. While the Internet originally had a more fixed addressing architecture, the addressing architecture of the Internet today is more contextual. The addressing architecture within a routing domain needs to be carefully designed to ensure that the address space is efficiently used. The flexibility of the CIDR addressing architecture provides the space for efficient routing architecture design.

4.2.2.1 Broadcast and Multicast Addresses

IPv4 has reserved addresses for broadcast and multicast. The address 255.255.255.255/32 is reserved for broadcast to all nodes on the local network. The scope of the local network varies and can include multiple subnets, but the traffic having this destination address is dropped at the border router to the public Internet and may be dropped on other subnets of the local routing domain as well. Addresses in which all the node address bits are set to 1 correspond to the local subnet broadcast. For example, 192.72.42.255/24 is the subnet broadcast address for the 192.72.42.0/24 subnet. In addition, any incoming traffic to the local subnet broadcast address is dropped at the local subnet router.

The CIDR address block 224.0.0.0/4 (addresses 224.0.0.0–239.255.255.255) is reserved for IPv4 multicast on the Internet. Addresses in this range are assigned by IANA [22]. Multicast is not used as heavily in IPv4 as in IPv6 because of the restricted address space and therefore the need for IANA assignment.

4.2.2.2 Unspecified, Local Host and Local Subnet Addresses

The address 0.0.0.0 is the unspecified address in IPv4. However, it can be used by a node in certain circumstances. A node with multiple network interfaces can use 0.0.0.0 to indicate that it should listen for incoming traffic on all network interfaces, regardless of the specific IP address of each interface. If a node boots and has not yet been assigned an IPv4 address, the unspecified address can be used by the node until an IPv4 address is assigned. The unspecified address is not routable and is only used as a placeholder indicating that the node has not yet received a valid IPv4 address.

Addresses in the address block 127.0.0.0/8 are confined to the node on which they originate. If packets appear on any external interface with addresses in this block, they are dropped. The address 127.0.0.1 is assigned to the node's interface and is called the localhost, or loopback, address. Addresses in the address block 169.254.0.0/16 are confined to the local subnet on which they originate and are dropped at the subnet border router. Local subnet addresses can only be received by nodes that are directly connected to the subnet. This address block is primarily used for IPv4 address autoconfiguration [23], a technique for a node to generate an address if the standard address configuration mechanisms – DHCP and static configuration – are not available.

4.2.2.3 Private Address Space

When IPv4 was originally designed in the 1980s, 2^{32} addresses seemed sufficient for all the nodes that would ever be connected to the Internet. However, after the Internet was commercialized in the early 1990s, concern developed at the IETF that the address space would ultimately run out. Whereas in the 1980s, the only end nodes connected to the Internet were large time-sharing machines at universities and government research labs, in the 1990s more and more workstations and PCs at companies were connecting. In the 2000s, mobile phones and devices have all started connecting, resulting in accelerated address assignments. By the 2011, IANA had allocated all the available top-level address blocks to the RIRs [24]. On January 31, 2011, the address space was officially declared exhausted by IANA, and on February 3, 2011, the last address block was allocated. By 2019, all the RIRs had allocated all their free addresses. The only addresses available were recycled addresses from owners that had given up their allocations or unallocated addresses from ISPs with large blocks.

One response to IPv4 address space exhaustion was the development of IPv6 in the late 1990s, which has an address space consisting of addresses having 2^{128} bits. However, transitioning to IPv6 has been slow, and it has mostly served as a solution for new applications like mesh networks. Another response was the development of private address space blocks. Table 4.1 illustrates the private address space blocks in IPv4. Routing domains can reuse addresses from the private address space blocks internally without requiring coordination with IANA. Routers on the public Internet drop packets that appear from the public Internet with addresses in a private address space block, but routers within a routing domain can freely route them.

IPv4 has three private address space blocks, the largest of which can host over 16 million nodes. For example, private addresses from the 10.0.0.0/8 block are

Table 4.1 IPv4 Private Address Blocks

CIDR Address Block	Address Range	Number of Addresses
10.0.0.0/8	10.0.0.0–10.255.255.255	16,777,216
172.16.0.0/12	172.16.0.0–172.31.255.255	1,048,576
192.168.0.0/16	192.168.0.0–192.168.255.255	65,536

commonly used in cloud computing networks to provide each account with its own private network. Private address space blocks allow portions of the address space to be reused in different routing domains, thereby effectively eliminating the cap on IPv4 address space for communication within the domain. Another advantage of private address space blocks is that they obscure the internal routing topology of a routing domain, making it harder for attackers to single out important nodes for attack.

4.2.2.4 Network Address Translation

If a node has an address in a private address space, it needs some way to communicate with nodes that are in other private address spaces or on the public Internet. Since private addresses are not routable on the public Internet, a solution is needed to map the addresses in the private address space into publicly routable addresses obtained from the organization's ISP. One solution is NAT [25].

NAT requires a router or network intermediary with an interface facing the public Internet to modify the packet headers when a packet is sent to or received from a node in the private address space to or from another node that is not within the private address space. The router maps the node's private IP address and UDP or TCP source and destination ports to one of the organization's publicly routable IP addresses and potentially different TCP or UDP ports. If a node running with a private address is offering a service on the public Internet that is available through a well-known port (for example, a Web server at TCP port 80), the outgoing port number must match the well-known port number of the service. Because the small pool of public addresses must be shared among all nodes in the organization's private address space, the router needs to keep a table of the mapping, so the header translation can be reversed for traffic returned from the external node. The address (source and destination)/protocol/port (source and destination) tuple serves to uniquely identify the traffic stream and allows the mapping to proceed.

NAT functionality is widely implemented and included in any Internet capable router. A NAT is typically automatically configured by the ISP when Internet service is installed. If the private network is run by an organization, the organization's IT staff configures the NAT.

4.2.2.5 IPv4 Node Configuration

When an IPv4 node joins a subnet, it needs to configure an address, determine what subnet router to use for communicating with nodes that are not part of the subnet (its default gateway), its subnet prefix, and its default DNS server. IPv4 nodes have two options for configuring: DHCP or manual configuration. Most organizationally managed networks utilize DHCP, since it allows tight control of the subnet prefix to align with the local routing domain architecture designed by the IT staff. This control is necessary because of the relatively small

IPv4 subnet prefix space. Within a local routing domain, the IPv4 address most likely comes from one of the private address space blocks, and the IT staff can configure the DHCP servers with address blocks for different subnets from the chosen private address space block. DHCP also provides a node with the other configuration information, and is discussed in more detail in Section 4.10.

If the node is manually configured, in addition to an IPv4 address, it must be configured with the additional information necessary to connect with other nodes. Manual configuration is rarely used and only in very specific circumstances. If a node needs to have a fixed permanent IPv4 address to enable it to be easily identifiable, the address may be statically configured. Examples are printers or servers that are widely used within an organization. Network administrators may also need to assign an IPv4 address temporarily if they are troubleshooting a networking issue. Finally, for security purposes, a network administrator may want to manually assign a /32 IPv4 address to limit the communication between nodes on a subnet.

4.3 IPv6

IPv6 was developed in the late 1990s in response to the rapid acceleration in nodes connecting to the global Internet after commercialization. The IETF realized that the public IPv4 address space would shortly be fully allocated. The design feature most directly relevant toward relieving address space exhaustion was to increase the number of bits in the IP address from 32 to 128, thereby increasing the available addresses to 2^{128}, or approximately 3.4×10^{38} total addresses. Many problems that arose over the 15 years since IPv4 was originally designed were addressed with acceptable if complicated workarounds in IPv4. But the designers of IPv6 took the opportunity of designing a new network layer protocol to fix these problems in the new protocol. The original design for IPv6 was published by the IETF in 1999 with the most recent update in 2017 [26].

While the IPv6 design is in many ways simpler and more elegant than IPv4, the enormous momentum behind the World Wide Web, email, and other public services running on the IPv4 Internet has led to slow adoption of IPv6 for public, Internet-facing services. IPv4 and IPv6 are not directly interoperable, so nodes need to run two stacks in order to use IPv6. Though the major operating systems all offered dual stack functionality by the early 2010s, the introduction of private IPv4 address space blocks and the widespread deployment of NATs in routers have enabled the IPv4 Internet to continue to flourish and grow.

One of the use cases that has seen the most utilization of IPv6 is device networks. The motivation for using IPv6 rather than IPv4 is that a device network deployment in a single large building could easily result in more IP address assignments than in a medium-sized enterprise IT network. IPv6 is more suitable for transparent addressing between devices and management applications running on fixed PCs, laptops, and mobile phones. Since device networks are not part of the public Internet, they do not need direct access to public Internet services. The management applications running on PCs, laptops, and mobile phones can provide any data they need from the public Internet through gateways. Thread and Matter, two device network protocols discussed in Chapter 7, both utilize IPv6. Since device networks are the foundation on which a BEMS is built, understanding IPv6 is likely to become increasingly important to the BEMS engineer.

4.3.1 IPv6 Header

In Figure 4.3, the basic IPv6 header is shown. The total number of bytes in the header is 40 bytes, double the number of bytes in the IPv4 header. As shown in the figure, the IPv6 header is much simpler than the IPv4 header, primarily because many of the fields were moved into extension headers or their function was eliminated. Extension headers are included in the packet immediately after the basic header. Extension headers are more widely used in IPv6 than in IPv4.

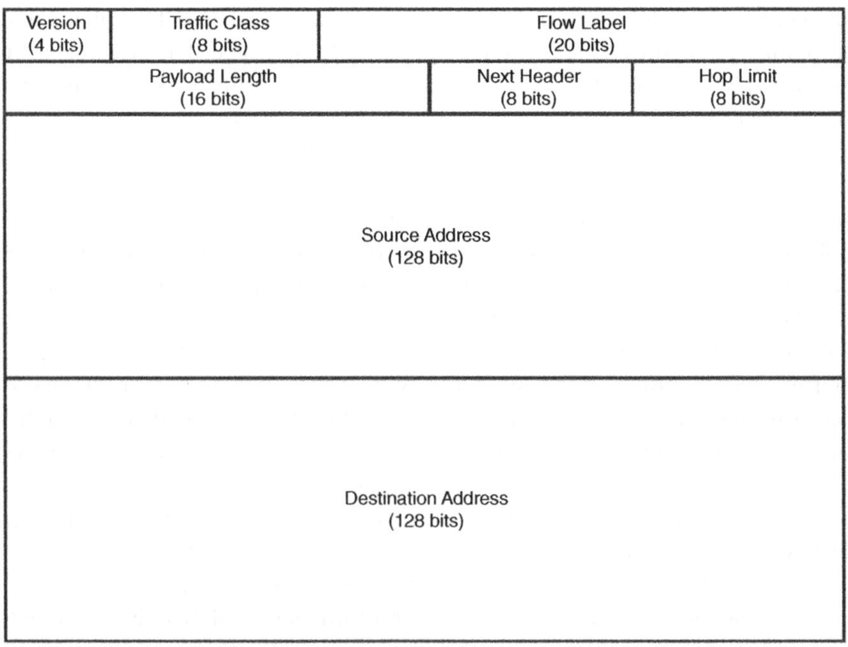

Figure 4.3 IPv6 Packet Header

The fields are as follows:

- Version – the IP version number. In IPv6 packets, this is 6.
- Traffic Class – unifies the Diffserv code point and ECN fields from the IPv4 header into one field. The most significant six bits specify the Diffserv code point for differentiated services, while the least significant two bits are used for ECN. See RFC 2474 [16] for more on differentiated services and RFC 3168 [17] for more on ECN,
- Flow Label – used by end nodes to label a sequence of packets that should be treated as a single flow. The flow label is specified in RFC 6437 [27].
- Payload Length – contains the length of the payload in bytes. The payload length includes any extension headers.
- Next Header – a code for the next header. The next header includes both the extension headers and the upper-level protocol headers, like transport. The codes for transport layer protocols are the same for both IPv4 and IPv6 [20].
- Hop Limit – equivalent to the Time to Live field in the IPv4 header. Routers decrement the field by one on every hop. If the field reaches zero or a router receives a packet with the field equal to zero, the packet is discarded As mentioned in Section 4.1.3, the maximum hop count is 255 for IPv6.
- Source and Destination Addresses – these fields contain the 16-byte, 128-bit IP addresses of the source node and destination node (see Section 4.3.2 for more on IPv6 addressing).

As with IPv4, extension headers are used for more complex options. In addition, some functions included in the IPv4 header have been delegated to other levels in the stack. There is no Header Checksum field; for example, IPv6 assumes that integrity protection for the packet is handled by the link layer or transport layer. Furthermore, IPv6 routers do not perform fragmentation. End nodes are required to perform MTU discovery and fragment packets to fit within the smallest MTU or not send a packet any longer than 1,280 bytes, which is the default MTU. RFC 8201 describes the path MTU discovery process [28].

4.3.2 IPv6 Addressing

The design of IPv6 addressing is based on a different principle than IPv4. Subnets are used in IPv4 to improve routing efficiency given the smaller IPv4 address space, so the size of the subnet prefix is variable. In IPv6, because the address space is so large, the size of the node identifier and network identifier is fixed. The network identifier is the most significant (leftmost) 64 bits, and the node identifier is the least significant (rightmost) 64 bits. The node identifier is usually derived from the 48-bit EUI-48 link-layer address on Ethernet-compatible link layers, but other node identifier schemes can be used. IPv6 addressing is described in RFC 4291 [29].

IPv6 addresses are represented as eight groups of four hexadecimal digits, each group representing 16 bits of the address, with the groups separated by colons. An example is:

2345:0425:2ca1:0567:0000:0000:5673:23b5.

In this case, 2345:0425:2ca1:0000 is the network identifier and 0000:0567:5673:23b5 is the node identifier. Because IPv6 addresses are long, some rules have been developed to shorten their representation. The first rule is that leading zeros can be dropped.

The second rule is that if a field consists of all zeros, it can be abbreviated by a single zero. For example:

2345:425:2ca1:567:0:0:5673:23b5

Finally, two groups that are all zeros can be abbreviated with a double colon, but this can only be done once within the same address:

2345:425:2ca1:567::5673:23b5

The IPv6 network identifier is also structured differently than the IPv4 subnet prefix. For unicast addresses routable on the public Internet, the network part is divided into two parts:

- Site Prefix – The leftmost 48 bits of the network identifier are the public network prefix, assigned to a site by the site's ISP. This part is used by ISPs and other organizations on the public Internet to route packets to the site border router.
- Subnet Identifier – The rightmost 16 bits of the network identifier are assigned by the local site administrator and indicate the local subnet within the site where the node is located.

In the above address, the site prefix is 2345:425:2ca1 and the subnet prefix is 0567.

CIDR notation is also used in IPv6, primarily to specify the structure of the network identifier. A site prefix with 48 bits is a /48, for example, 2345:425:2ca1::/48. Covering the entire network identifier including both the site identifier and the subnet identifier is a /64, for example, 2345:425:2ca1:567::/64. The double colons on the end of the network identifier in these two cases indicate that the rest of the address is zeros and is omitted since a network identifier is designated.

Note that if IPv6 addresses are used in URLs for Web sites, they must be enclosed in square brackets to distinguish the colon separating the port number from the address. For example:

http://[2345:425:2ca1:567::5673:23b5]:80

designates the Web page at port 80 located at the IPv6 address 2345:425:2ca1:567::5673:23b5.

4.3.2.1 Multicast and Anycast Addresses

Unlike IPv4, IPv6 does not support broadcast. IPv6 uses multicast for one-to-many addressing. Multicast addresses begin with the reserved prefix ff00::/8. The two 4-bit bitfields following the 0xff contain flags indicating the assignment status and scope of the address. The final 112 bits of the address are the multicast group identifier, which indicates the topic of the multicast group.

The flags in the first bitfield indicate the assignment status of the address. For example, the bitfield 0000 indicates that the address has been assigned by IANA and is "well-known." The bitfield 0001 indicates that the address has been dynamically assigned and is transient. The field additionally supports a few other, more specialized flags. The second bitfield indicates the scope of the address, or over what part of the network the address is valid and routable. Assigned scope values include:

- 0x1 – node local scope,
- 0x2 – link local scope,
- 0x4 – admin local scope,
- 0x5 – site local scope,
- 0x8 – organization local scope,
- 0xe – global scope.

Node local scope applies only to a single interface on the node and has the same scope as the local host unicast address. Admin local scope designates the smallest scope that must be administratively configured and cannot be automatically derived from physical connectivity or non-multicast-related configuration. Organization local scope is intended to span multiple, topologically separate sites in the same organization. See RFC 4291 [29] for more details on these two bitfields.

Anycast addresses are addresses having a unicast address format but different routing properties. They are assigned to a node's network interface in addition to the node's unicast address. A unicast address assigned to more than one interface automatically becomes an anycast address, but the node must be configured to treat the address as an anycast address. Packets sent to an anycast address are sent to the nearest node with the address configured, where the distance is measured by the routing topology. RFC 4291 defines a single anycast address, the subnet router anycast address. The subnet router anycast address is formed from the network identifier for the subnet, followed by a node identifier of all zeros. The subnet router anycast address is assigned to the interfaces on all the routers for the local subnet. It is used by nodes on the subnet to perform router discovery when they join the subnet.

4.3.2.2 Unspecified, Local Host, Link Local, and Private Addressing

The unspecified address in IPv6 consists of 128 bits set to zero, or :: in the IPv6 notation. It is used similarly to how the unspecified address is used in IPv4, namely to allow a node to send traffic until a fixed IPv6 address is allocated, and it is also not routable. The local host address in IPv6 is ::1. As with IPv4, packets with this address are not routed outside of the node, but IPv6 does not designate a fixed address block for the local host, unlike IPv4. There is a single loopback address in IPv6. Link local addresses are specified in IPv6 using a reserved link local 10-bit network prefix and a 64-bit node identifier. The link local network prefix is fe80::/10. An example of a link local address is fe80::23a1:b152, the right-most 64 bits being the node identifier. As in IPv4, link local addresses are not routed off the local subnet.

IPv6 originally did not have any scheme for private addressing because private addresses were added to IPv4 primarily to handle address space exhaustion. The amount of address space available in IPv6 is so much larger than IPv4 that address space exhaustion is unlikely to ever be a problem. However, after private address space had been deployed in IPv4, some side benefits were discovered. For example, a high-capacity border router can deal with harmful traffic from the public Internet more easily than requiring each node to handle the traffic itself if the traffic must first traverse a NAT on the router, so private address space was added to IPv6 in 2005 [30]. IPv6 addresses having the format fc00::/7 are available for private address space assignment.

4.3.2.3 IPv6 Node Configuration

As briefly mentioned in Section 4.1.3, in IPv6, the most common mechanism used for node configuration is the Neighbor Discovery protocol [13], though static configuration and DHCP are also available in IPv6 as well. Neighbor Discovery uses specific ICMPv6 messages along with multicast to replace ARP and DHCP and allows an IPv6 node to discover their subnet router, subnet prefix, and DNS server, and autoconfigure their address.

An IPv6 node joining the network with the Neighbor Discovery protocol must subscribe to the following local multicast addresses to discover the link-layer address of other nodes on the link, configure its address, and discover its subnet router:

- ff02::1 – the all-nodes multicast address, used to reach all nodes on the link,
- ff02::2 – the all-routers multicast address, used to discover routers on the link,
- ff02::1:ffxx:xxx – the solicited node multicast address, used to discover the link-layer address of an IPv6 node with xx:xxx as the rightmost 24 bits of the node identifier.

An IPv6 node configuring its addresses with address autoconfiguration must first determine a unique node identifier for its network interface, construct an address, then run duplicate address detection to see if any other node on the link has claimed the address. The node identifier must be statistically likely to be unique to avoid conflicting with other nodes on the link. Since the EUI-48 link-layer address typically has this property for Ethernet interfaces, the most common way to form a node identifier is to simply use the EUI-48 link-layer address as the rightmost bits in the node identifier and set the remaining 16 bits to zero. Another option is to use a universally unique identifier (UUID) [31]. UUIDs are 128-bit labels that are statistically unique for all practical purposes if generated using standard methods. The rightmost 64 bits of the UUID can be used for the node identifier.

Figure 4.4 contains a high-level illustration of the duplicate address detection protocol. A new node first performs duplicate address configuration on its link local address. Here is a short description of the steps in the figure illustrating how duplicate address detection uses Neighbor Discovery [32]:

- The entering node multicasts a Neighbor Solicitation ICMPv6 message with the destination address set to the solicited node multicast address.The rightmost 24 bits of the solicited node multicast address are set equal to the rightmost 24 bits of the tentative node identifier the node has selected. The source address is the unassigned address. The entering node includes the address being checked as the Target Address in the Neighbor Solicitation message payload.
- If an established node on the link has already claimed the proposed address and configured a network interface with it, the established node will receive the multicast message. It replies by multicasting a Neighbor Advertisement ICMPv6 message, with the destination address set to the all-nodes multicast address and the source address set to the established node's link local address. The replying node includes a Target Link-Layer Address option in the reply, with the link-layer address set to the address on the interface configured with the conflicting address's node identifier.
- If an entering node receives a multicast reply to its Neighbor Solicitation, it drops the tentative node identifier, generates a new one, and performs duplicate address detection on the new address. When duplicate address detection fails to detect a conflicting address, the link local address is formed from the link local prefix (fe80::/10) together with the validated node identifier.

Once the link local address is constructed, the entering node must execute router discovery and configure a routable unicast address. Figure 4.5 illustrates the router discovery protocol. The steps in Figure 4.5 use the ICMPv6 Neighbor Discovery protocol for router discovery in the following way [13]:

- The entering node multicasts a Router Solicitation ICMPv6 message with destination address set to the all-routers multicast address and the source address set to either the unassigned address (::) or the link local address of the entering node if the link local address is already configured. The entering node includes its link-layer address in a Source Link-Layer Address option with the Router Solicitation.

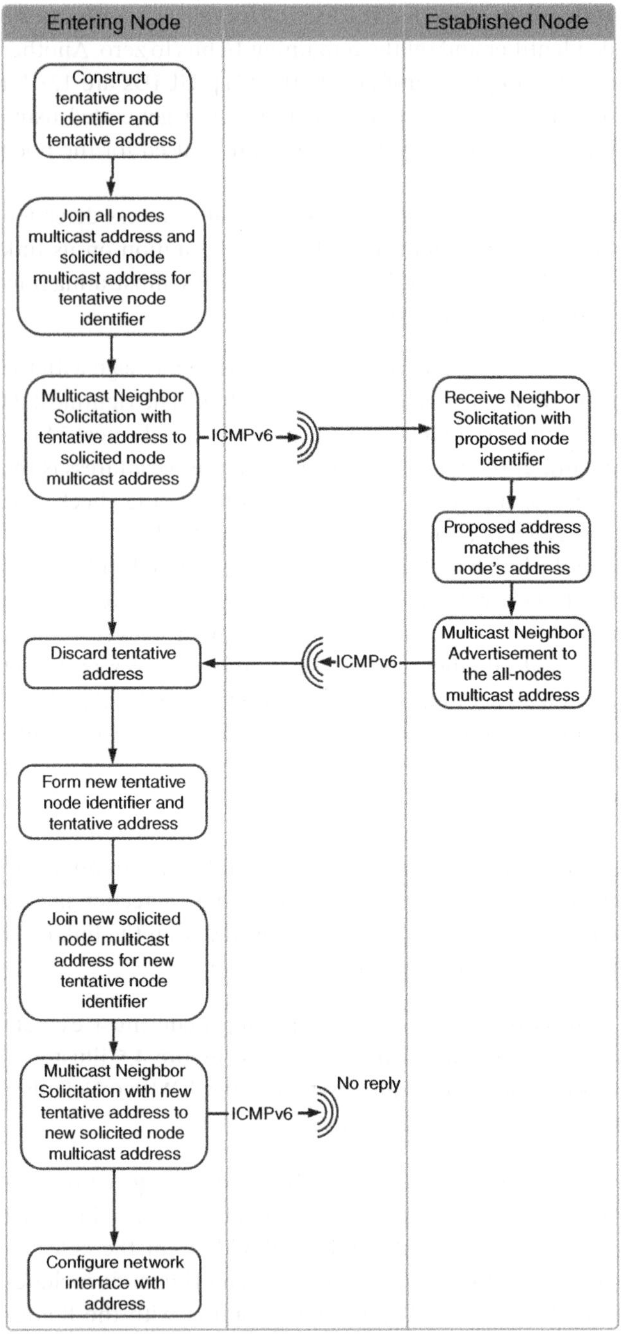

Figure 4.4 IPv6 Duplicate Address Detection Protocol

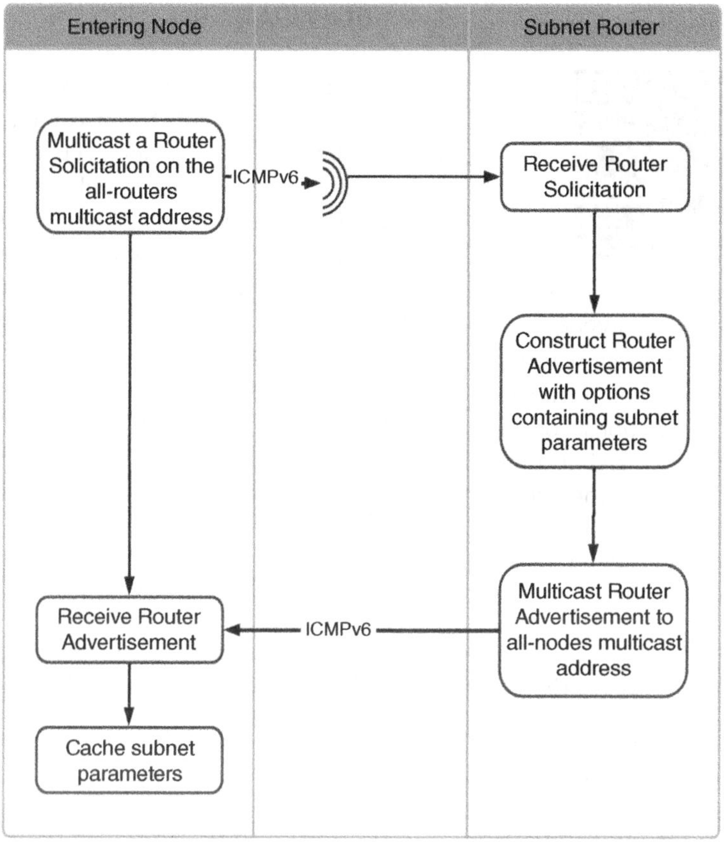

Figure 4.5 IPv6 Router Discovery Protocol

- A subnet router replies by multicasting a Router Advertisement ICMPv6 message with destination address set to the all-nodes multicast address and the source address set to the router's link local address. The router includes the following options: the Source Link-Layer Address option with the router's interface link-layer address, a Prefix Information option with network identifiers for the subnet, and a DNS Server option with DNS servers (see RFC 8106 for more information on the latter [33]).

Based on the network identifier returned in the Prefix Information option and the constructed node identifier, the node constructs a tentative IPv6 routable unicast address and performs duplicate address detection on it, as shown in Figure 4.4.

Introduction to 6LoWPAN

This video presents a short introduction to 6LoWPAN and how it is used in wireless mesh networks. The video includes graphical illustrations of a 6LoWPAN network architecture and how it relates to the building LAN and compares the network stacks of a Wi-Fi node with a 6LoWPAN node.

4.4 6LoWPAN

IPv6 is used in several device networking protocols running on WPANs for communicating between sensors and actuators and their management applications through gateways running on LANs. Matter and Thread use IPv6 running on the 802.15.4 link layer, and a standard IPv6 profile exists for BLE. The IETF developed a standard adaptation layer called 6LoWPAN allowing device networks running over WPANs to utilize IPv6. The protocol design was based on the 802.15.4 link layer and is documented in a number of RFCs [34–38]. 6LowPAN over BLE is specified in RFC 7668 [39]. The rest of this section describes 6LoWPAN for 802.15.4.

To support the smaller packet sizes, limited bandwidth, and reduced power of WPAN devices, 6LoWPAN rearchitected IPv6 by adding the following functionality:

- Modifications in the address format to reduce the size of addresses,
- Modification of multicast addressing since 802.15.4 only supports broadcast natively and not multicast,
- Header compression and reorganization of the header design to reduce the header size and remove redundant or unneeded fields,
- Adds a fragmentation and reassembly algorithm to map IPv6 packets with a minimum size of 1,280 bytes to the 802.15.4 link layer having a maximum frame size of 81–102 bytes,
- Redefines how node configuration is implemented for new nodes entering the link,
- Adds a mesh routing protocol since the mesh network architecture, or peer-to-peer architecture as 802.15.4 calls it, is an important deployment architecture for WPANs.

These additions allow IPv6 packets to be routed over WPANs but require that the packets be rebuilt into standard IPv6 packets to be routed over an IPv6 backhaul, or transferred to IPv4 if the backhaul is running IPv4.

4.4.1 6LoWPAN Network Architecture

Figure 4.6 illustrates two WPAN deployment architectural patterns that can serve as the basis for a 6LoWPAN network [40]. One pattern is the simple LoWPAN shown in the upper right corner. In this pattern, a collection of mesh WPAN routers and WPAN devices are connected to the backhaul through a single edge router. In the extended LoWPAN, two or more routers connected through the same backhaul subnet are responsible for maintaining connectivity from devices and mesh routers into the backhaul and between devices and mesh routers that have no direct forwarding path due to wireless propagation conditions. A third pattern, not shown in the figure, is based on the 802.15.4 star architecture discussed in Section 3.5.1.2. Every device has connectivity directly to the edge router, and forwarding to the edge router happens at the link layer rather than through multihop mesh routing, eliminating the need for mesh routers within the WPAN. This section focuses on the simple LoWPAN architecture since it seems the most relevant to BEMS deployments.

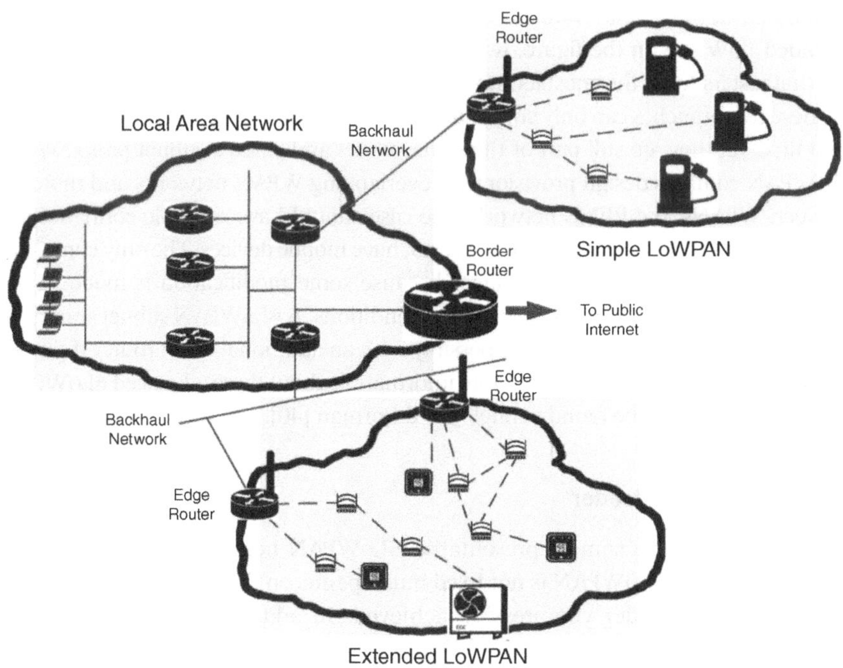

Figure 4.6 Network Architecture for 6LoWPAN Energy Management Devices

The WPAN edge router is responsible for organizing the WPAN subnet and performs the following additional functions besides simple packet forwarding between the WPAN subnet and the backhaul:

- Distributing the subnet prefix for the 6LoWPAN subnet to entering nodes and mesh routers,
- Header compression and decompression for packets entering and exiting the WPAN, and distributing compression context for stateful compression of IPv6 addresses to nodes within the WPAN subnet,
- Serving as a central point for IPv6 Neighbor Discovery and duplicate address detection by maintaining a table of IPv6 addresses for nodes in the WPAN subnet so individual WPAN nodes do not have to respond,
- Serving as the root node in the WPAN mesh routing topology.

The 6LoWPAN network architecture is considerably different from the typical IP network architecture discussed in Section 4.1.2. A WPAN network is a stub subnet that consists of an embedded island of wireless devices. In the typical IP subnet, devices in the subnet have full connectivity at the link layer without having to go through a router. In WPAN networks, some devices on the subnet may not be directly reachable from others at the link layer due to wireless propagation conditions. As shown in the extended LoWPAN in the figure, two paths of embedded wireless forwarding exist, one that just includes thermostats and one that also includes a heat pump. The devices on these two branches can only communicate through their edge routers via the backhaul link, yet they are still part of the same subnet and share a subnet prefix. While 6LoWPAN contains design provisions for overlapping WPAN networks and mobility between WPANs, the BEMS networks use case should have a simple configuration without the need for overlapping and will not have mobile devices. The only complication is if wireless connectivity changes because some modification is made in the building that changes wireless propagation conditions. A 6LoWPAN subnet should be able to recover from these changes, possibly with an additional mesh router if part of the subnet is cut off from the rest. More information about these advanced 6LoWPAN network features can be found in Shelby and Borman [40].

4.4.2 6LoWPAN Header

Figure 4.7 illustrates some representative 6LoWPAN headers [40]. Unlike IPv6, the header size in 6LoWPAN is not fixed but depends on the content of the header and amount of header compression achieved. In addition, some options that would be carried in header options following the basic header in IPv6 are instead placed in headers before the basic header in 6LoWPAN. This repositioning allows the functions in the software stack to be implemented in a way that reduces unnecessary processing on low-power, memory-constrained 6LoWPAN nodes.

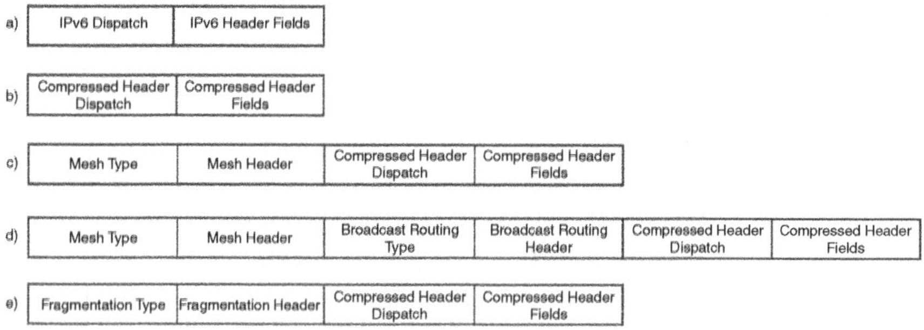

Figure 4.7 Representative 6LoWPAN Headers: (a) IPv6 encapsulated packet; (b) 6LoWPAN encapsulated packet; (c) 6LoWPAN encapsulated packet with mesh addressing; (d) 6LoWPAN encapsulated packet with mesh addressing and broadcast routing; (e) 6LoWPAN encapsulated packet with fragmentation

Basic 6LoWPAN headers begin with a dispatch field that indicates what type of compression is used in the packet. Figure 4.7a shows a 6LoWPAN header for an uncompressed IPv6 packet, while in Figure 4.7b, the header for a basic 6LoWPAN packet is shown. Figures 4.7c and d show the 6LoWPAN headers with additional preheaders added in cases where mesh routing is used. Finally, Figure 4.7e shows a 6LoWPAN header from a packet fragment. All combinations of preheaders are possible, but the preheaders must be ordered according to the following:

- Mesh addressing header,
- Broadcast Routing header,
- Fragmentation header.

This order ensures the packets can be fast and efficiently routed over the mesh network without requiring devices to process the entire packet.

4.4.3 6LoWPAN Addressing

802.15.4 nodes can have two types of link-layer addresses as described in Section 3.5.5:

- The network interface comes with a 64-bit EUI-64 link-layer address preinstalled in the hardware interface.
- Once a node has joined a PAN, the PAN coordinator can assign it a temporary 16-bit short address.

The EUI-64 address can be used directly as the node identifier in the rightmost 64 bits of the IPv6 address. The short address can also be used, but RFC 4994 cautions that, because short addresses are basically transient in nature and only valid as long as the PAN and node are associated, they could potentially become invalid if the association is dropped [34]. The tradeoff is that short addresses allow for better header compression and therefore better scalability. 6LoWPAN also assumes that the PAN corresponds to a single link and therefore will be mapped into a single IPv6 subnet.

While 802.15.4 supports broadcast, it does not natively support multicast, and IPv6 does not support broadcast, so 6LoWPAN defines a technique for performing multicast over broadcast. All 6LoWPAN frames that are to be sent to a multicast address are broadcast to the PAN broadcast address (0xffff). Multicast addresses are represented as 16-bit short addresses with the leftmost bit set to 1 and the next two bits set to 0, leaving the rightmost 12 bits available for the multicast group id. A unicast address has the first bit set to 0 and 15 bits for the node identifier. Sending to the broadcast address causes all nodes in the PAN to receive the frames holding the packet, but the packet will be dropped at the 6LoWPAN layer for those nodes that have not joined the multicast group. The standard also specifies an IANA registry for short addresses, with the unicast and multicast prefixes and a few other prefixes specified as reserved. The IANA registry allows well-known multicast addresses to be registered. In addition, the registry includes the broadcast address 0xffff and the unspecified address 0xfffe.

4.4.4 6LoWPAN Header Compression

802.15.4 frames have a maximum payload size of between 81 and 102 bytes depending on whether link layer security is enabled, considerably less than the minimum packet size for IPv6, which is 1,280 bytes. If full-sized IPv6 and transport headers were included, not much room would be left for application data. Since many of the fields in the IPv6 and transport headers can be inferred in a PAN, 6LoWPAN specifies a header compression algorithm to compress the IPv6 and transport layer headers. The base 6LoWPAN header compression algorithm in RFC 4944 is not recommended, since it was replaced by a newer algorithm in RFC 6282 [36]. The newer algorithm includes both stateless and stateful components.

RFC 6282 defines two compression algorithms, LOWPAN_IPHC for compressing the IPv6 header and LOWPAN_NHC for compressing arbitrary next headers including IPv6 option headers. The LOWPAN_IPHC compression algorithm reduces the size of the IPv6 header by making the following changes in the header:

- The version is assumed to be 6 so the Version field is removed.
- The Traffic Class and Flow Label are assumed to be zero so they are removed.
- The Hop Limit is set by the source to one of the following values: 1, 64, and 255, although it must be expanded out to one byte on the first hop out of the subnet because it is decremented on each hop.

- The Payload Length is inferred from the 6LoWPAN fragmentation header or the 802.15.4 frame header.
- The network identifier is either a link local identifier or one of a small number of routable prefixes assigned to the entire WPAN.
- The node identifier is either the EUI-64 address of the 802.15.4 network interface or a 16-bit address assigned by the PAN coordinator.

Unlike the original algorithm, LOWPAN_IPHC is not completely stateless, since compression of the IPv6 addresses requires the node to maintain state.

The LOWPAN_NHC algorithm can apply to arbitrary next headers in contrast with the original algorithm which only applied to UDP, TCP, and ICMPv6. The LOWPAN IPHC indicates whether the following header is encoded using LOWPAN_NHC. If so, the bits immediately following the compressed IPv6 header start the next header. Since UDP is the most commonly used transport protocols in WPANs, LOWPAN_NHC defines a compression algorithm for UDP that removes the UDP checksum if the application layer implements a message integrity check. Compression formats are additionally defined for commonly used IPv6 headers, such as IPv6 in IPv6 tunneling and the IPv6 extension headers.

In the best case, for link local communication between device nodes and other nodes such as the device gateway in the WPAN subnet, LOWPAN_IPHC can reduce the IPv6 header to two bytes. For packets that are routed over multiple 6LoWPAN hops, the header can be reduced to 7 bytes. If the packet is routed off the WPAN onto an IPv6 building backhaul network, the 802.15.4 edge router at the device gateway needs to expand the network and transport headers out to full IPv6 headers.

4.4.5 6LoWPAN Packet Fragmentation and Reassembly

RFC 8200 requires links that have an MTU size of less than 1,280 bytes must provide a packet fragmentation and reassembly service at a layer below IPv6 [26]. Since the maximum frame size for 802.15.4 is 81–102 bytes, 6LoWPAN is intended to be such an adaptation layer, and it therefore provides fragmentation and reassembly. Fragmentation is implemented through two preheaders as shown in Figure 4.7e. The fragmentation preheader contains the size of the packet before fragmentation, a tag value which is the same for each fragment that is part of the same packet, and the offset of the fragment in the original packet.

A destination node receiving packets with the fragmentation header uses the sender's 802.15.4 link-layer address or the originator's address if the mesh header is present and the sender is not the originator of the packet, or the destination's 802.15.4 link-layer address or the final receiver's address if the mesh header is present and the destination is not the final receiver, the packet size, and the tag to identify all the fragments that belong to the same original packet. When the destination node receives a fragment, it starts constructing the original packet, with the size of the reassembly buffer determined by the original packet size. The node uses the offset field to determine where within the original packet the fragment is located. Any error in fragmentation causes the receiver to flush the fragmentation buffer and restart, for example, if a fragment is received that overlaps one already in the buffer. If the node dissociates from the PAN, the receivers of fragments must discard all received fragments that have not been reassembled into packets, and the transmitters of fragments must discard all fragments of partially transmitted packets that have not yet been transmitted. Nodes also maintain a reassembly timer, limiting the amount of time they wait for fragments in an unassembled packet [34].

The original 6LoWPAN specification requires reassembly of packets at each hop. In a star network architecture, there is only one hop to the PAN coordinator which then reassembles the packet and converts it to a standard IPv6 packet for further delivery over a WLAN or LAN. In a mesh network, however, frames need to traverse multiple hops in the PAN, and since each full-function device intermediary acts as a router, reconstructing the 6LoWPAN packet would lead to increased latency or reduction in battery life for nodes providing forwarding. RFC 8930 [37] and RFC 8931 [38] specify an algorithm that allows fragments to be forwarded without reassembly if the next hop is a 6LoWPAN link.

4.4.6 6LoWPAN Node Configuration

A node entering a 6LoWPAN network needs to find a subnet router and configure an IPv6 address just as in a full IPv6 subnet. Because 802.15.4 and other WPAN networks are bandwidth-constrained and often subject to more packet loss than in Ethernet and Wi-Fi networks, the node configuration process has been redefined for 6LoWPAN nodes to reduce the amount of signaling and replace multicast with unicast where possible. 6LoWPAN nodes also require compression context for stateful address compression. The edge router in a 6LoWPAN subnet maintains the address compression context and a table of IPv6 addresses for nodes and their associated link-layer addresses so that multicast neighbor discovery is not needed. Modifications to the IPv6 Neighbor Discovery protocol for 6LoWPAN are described in RFC 6775 [35].

RFC 6775 defines the following extensions to Neighbor Discovery messages to implement the above optimizations:

- Authoritative Border Router Option – contains the edge router's IPv6 address and version information on the current subnet configuration, and is propagated along with all Router Advertisements sent between mesh router nodes and other nodes in the subnet.

- 6LoWPAN Context Option – contains context information for stateful header compression. The compression context can be for any IPv6 address and a Router Advertisement can contain up to 16 6LoWPAN Context options.
- Address Registration Option – included by nodes in unicast Neighbor Solicitation messages to register their interface identifiers with a router. The Address Registration option includes the EUI-64 link-layer address of the 6LoWPAN interface.

A high-level view of the node configuration protocol is shown in Figure 4.8 and consists of the following steps:

- The entering node constructs a link local IPv6 address using the link local address prefix (fe80::) and the EUI-64 link-layer address of its WPAN interface.
- The node then multicasts a Router Solicitation message with the destination address being the all-routers multicast address and the source address the node's link local address, and includes a Source Link-Layer Address option to enable the replying router to unicast a reply. The node must not use the unspecified IPv6 address (::) as the source address.
- Mesh routers and edge routers respond back with Router Advertisements unicast to the node's link local IPv6 address. The Router Advertisement includes a Prefix Information option with the subnet prefix of the subnet, a 6LoWPAN Context option with context information for stateful IPv6 address compression, and an Authoritative Border Router option with the IPv6 address of the edge router.
- The entering node selects a router based on the returned Router Advertisements and caches the stateful IPv6 address compression contexts, the default router's IPv6, and link-layer addresses.
- The node constructs a globally routable IPv6 address using the IPv6 subnet prefix from the selected Router Advertisement message Prefix Information option and its EUI-64 link-layer address as the node identifier.
- The node then unicasts a Neighbor Solicitation to its newly selected default router with the source address set to the node's newly constructed globally routable IPv6 address and includes a lifetime. The message contains an Address Registration Option with the node's EUI-64 link-layer address indicating to the router that the node is directly reachable, and a Source Link-Layer Address option. The node periodically renews the address registration prior to the expiration of the lifetime.
- The node's default router checks if there is any address conflict, makes an entry for the node's globally routable IPv6 address in its neighbor cache if not, and responds to the new node with a Neighbor Advertisement message. The message contains a status code indicating the status of the registration request. The status code is 0 if the registration request was successful, 1 if there is an address conflict with another request in the cache, and 2 if the router's neighbor cache is full. If the registration succeeds, the node is free to use the address but must periodically inform the router that it is still active.

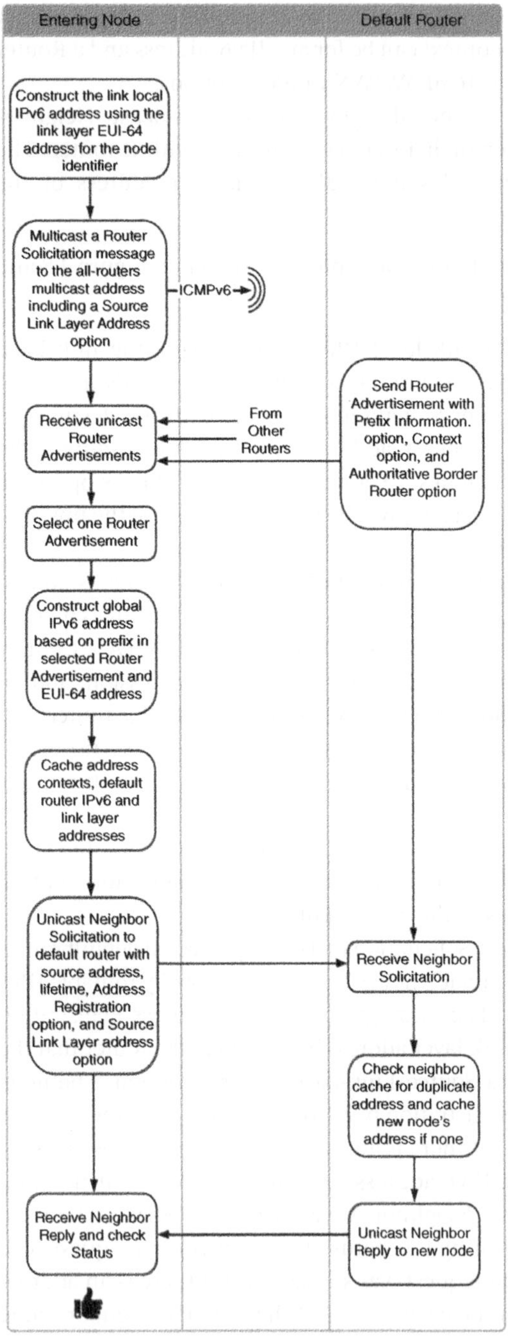

Figure 4.8 6LoWPAN Node Configuration

If the node uses its EUI-64 address to construct its globally routable address, there should be no possibility of address collision because EUI-64 link-layer addresses are uniquely assigned to 802.15.4 network interfaces. If, however, another method is used for the node identifier part of the IPv6 address, the entering node needs to construct a new IPv6 address and register it. If the Neighbor Advertisement indicates that the neighbor cache is full, the entering node needs to select another router from the returned Router Advertisements and try again.

4.4.7 6LoWPAN Mesh Routing

For 802.15.4 networks with the star architecture, devices only connect with the PAN coordinator, and it forwards packets between the WPAN subnet and the building LAN backhaul link somewhat like a Wi-Fi access point. The 802.15.4 specification defines link layer forwarding in a partial mesh network architecture in which multiple PAN coordinators associate and support a star architecture, but the specification has no provision for a full mesh architecture. Typically, an IP subnet consists of a single link with a small number of routers. Other devices on the link act as end nodes. For networks with a full mesh architecture, however, the subnet is composed of multiple links rather than a single link and multiple full-function devices act as routers. Each link also requires at least one full-function device to act as a coordinator on the link layer. The coordinator also performs peer-to-peer routing on the network layer, so there may be no PAN coordinator for the entire WPAN subnet. A packet may need to transit multiple peer-to-peer routing links between the device and a WPAN edge router that is connected to the backhaul network.

RFC 6550 [41] defines a routing protocol for mesh routing called IPv6 Routing Protocol for Low-Power and Lossy Networks (RPL). RPL is a distance vector routing protocol that uses a directed acyclic graph (DAG) to represent the network topology. In a DAG, nodes are organized into a loop-free, tree-like structure, with a single root node that initiates the formation of the DAG. The root of the global routing topology in an RPL network typically resides in the edge router responsible for routing packets between the WPAN and the backhaul link and acts as a sink for the routing topology. The edge router may not necessarily perform the PAN coordination function on the link layer; however, it may rather simply be a full-function device with two interfaces and perform IP routing on the network layer. The root maintains information about all the nodes in the WPAN and the routing topology. RPL calls the DAG a destination-oriented directed acyclic graph (DODAG). The DODAG is the topology representation used by RPL to organize forwarding nodes and provides a framework for routing packets.

The DODAG is generated through the following steps:

- The root node is selected, typically as an edge router for the global DODAG. The root node is responsible for initiating the DODAG construction and managing the topology of the network. The root node's full, 128-bit IPv6 address becomes the DODAG ID, along with an 8-bit RPL Instance ID. These IDs identify the routing topology global to the WPAN subnet.[3]
- The root node initiates the DODAG construction by multicasting out DODAG Information Object (DIO) messages to the all-RPL multicast address (ff02::1a). The DIO messages contain routing topology information about a node, such as the subnet prefix, the node's IPv6 address, its distance from the root, and the unique DODAG ID and RPL Instance ID.
- As nodes receive the DIO messages, they use the information to join the DODAG and become part of the network. They update the topology information and propagate it to their neighbors.
- The network topology may change due to changing wireless propagation conditions or if one of the nodes moves. This can disrupt routing since the DODAG will become invalid. If a link fails, the node experiencing the failure sends out a Destination Advertisement Object (DAO) message with routing information allowing other nodes to repair their routing tables. The nodes then propagate the updates.

Unlike other interior gateway protocols, WPANs have unique requirements with respect to routing metrics. As discussed in Section 4.1.1, distance vector routing protocols like BGP use the number of hops to the destination as the routing metric. Link state protocols such as OSPF use link metrics such as interface bandwidth to decide which interface to transmit a packet over. The assumption with these protocols is that the router is always available and has few constraints. In contrast, peer-to-peer routers in WPANs have more constraints than routers in WANs, LANs, and WLANs.

One major constraint is energy conservation. A peer-to-peer router has a variety of strategies to implement energy conservation. It may, for example, periodically go into sleep mode limiting its ability to receive packets, and will in any case limit its broadcast power when transmitting. These constraints are discussed in detail in RFC 6551 [42]. RFC 6551 also defines a DAG Metric Container Object to encode the constraints. The DAG Metric Container Object is included in the DIO message as an option. The constraints and metrics are used by RPL to determine the preferred path for an outgoing packet.

---

3 It is also possible to have local DODAGs that represent only part of the WPAN, see RFC 6550 [41] for details.

In addition to the constraints and metrics, RPL uses an objective function to compute an upstream path in the DODAG. An objective function specifies the objectives used in the computation of the constrained path as the rules for determining the upstream parents, how to do load balancing, etc., using the metrics and constraints included in the DAG Metric Container Object option. Two types of objective functions have been specified for RPL:

- Objective Function Zero (OF0) [43] – OF0 is the default objective function. It uses various link properties between nodes that result in a metric equivalent to hop count to form the rank.
- Minimum Rank with Hysteresis Objective Function (MRHOF) [44] – MRHOF selects routes that minimize a metric, usually the average number of transmission frames and acknowledgment frames necessary for transmission while using hysteresis to reduce churn in response to small metric changes. Without hysteresis minimization, small changes in the metric could result in large changes in the routing table. This minimizes routing table churn if the routing topology changes frequently.

4.4.8 6LoWPAN Mesh Router Configuration

After a full-function 802.15.4 node has entered an RPL mesh-routed subnet and has a globally routable IPv6 address, it may need to configure itself as a mesh router. A parent is a mesh routing node closer to the root than the entering node as determined by the objective function and the link metrics and constraints distributed with the DIO object. Parents are therefore able to progress packets toward the root. The rank is a positive integer used to measure the relative position of a node in the topology and thus its suitability to act as a forwarding node toward the root. The rank of a node is computed using the objective function for the DODAG which is determined by the DODAG ID, the RPL Instance ID, and the link metrics and constraints in the DIO messages. In general, the rank is a measure of the distance between the entering node and the root, with nodes closer to the root having a smaller rank, and the root node having rank 0.

As shown in Figure 4.9, when an entering node configures itself as a mesh router in an RPL routed subnet, it performs the following steps:

- The node first listens for topology information multicast by its neighbors in the form of DIO messages and notes the neighbors' ranks.
- The node calculates its own rank based on the DODAG information, the objective function, the ranks of its neighbors, and link metrics and constraints contained in the DIO messages.
- The node then selects the neighbor with the lowest rank as its parent.

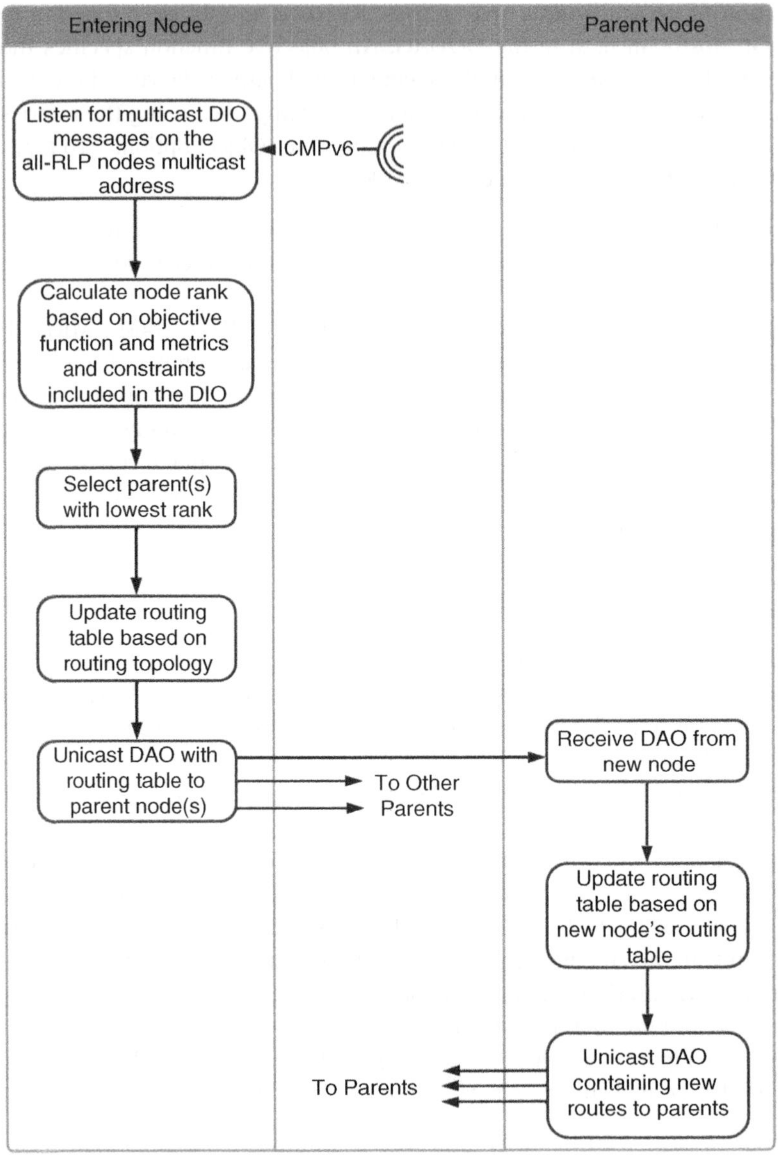

Figure 4.9 6LoWPAN Mesh Router Configuration

- The node unicasts a DAO message to the parent informing the parent that the node has chosen it as a parent. A node may have multiple parents for redundancy. The new node includes its routing table with distance vector information in the DAO message.
- The parent then updates its routing table based on the new node's information and propagates it backward toward the root through its parents.

4.5 Transport Layer

The transport layer sits immediately above the network layer in the IP stack and provides several important services for the application layer. One of these services is a collection of application endpoints, called ports. Application layer protocols can send and receive data through the ports. For sending, data from the application layer written to these ports is packaged into the transport layer payload field, a transport layer header is added, and the packet is transferred to the network layer for transmission to the receiver. For receiving, the network layer transfers received packets to a port, and the transport layer makes the data in the packets available to the respective application layer protocol listening on the port. The transport layer may optionally segment larger data chunks received from the application layer so that they fit into the transport layer payload field, and reassemble segmented chunks received from the network layer to present to the application layer. The transport layer is also responsible for presenting a communication channel abstraction toward the application layer on the ports. The communication channel abstraction specifies a collection of properties that the transport layer enforces on the communication between the sender and receiver. An example of the communication channel abstraction is a reliable data stream.

Transport layer protocols are programmatically available from a node's operating system API through the `socket()/bind()` interface. The `socket()` call opens a connection to the transport protocol type specified as a parameter in the function call and returns a socket file descriptor. The `bind()` call binds the socket to a particular port and IP address. Different programming languages implement these application programming interfaces (APIs) in different ways but the basic functionality is the same regardless of operating system and programming language. Application protocols then build on top of the `socket()/bind()` API by sending data through a socket to the receiver and listening on a socket for the reply.

While there are several transport layer protocols available for the IP stack, for all practical purposes, the only two protocols of interest for BEMS system design are UDP and TCP. UDP presents an abstraction of a connectionless, unreliable data stream that mirrors the underlying communication model of the IP layer. TCP presents an abstraction of a connection-oriented, reliable data stream with congestion control. These two protocols are discussed in Sections 4.6 and 4.7, respectively.

4.5.1 Transport Layer Ports

Ports take the form of a 16-bit, 2-byte number between 0 and 65,535. The transport layer maintains a separate port number space for each transport protocol. The port number space is divided into three ranges:

- Well-Known or System Ports – port numbers between 0 and 1,023 are assigned by IANA to protocols that are critical for system functions such as DNS for server name resolution and to routing protocols.

Table 4.2 Registered Port Numbers

Port Numbers	Transport Protocols	Application Protocol
20	UDP	File Transfer Protocol (FTP)
21	UDP	FTP control channel
22	UDP, TCP	Secure Shell (SSH)
53	UDP, TCP	Domain Name Service (DNS)
67	UDP	Dynamic Host Configuration Protocol (DHCP) for IPv4 servers
68	UDP	Dynamic Host Configuration Protocol (DHCP) for IPv4 clients
80	UDP, TCP	Hypertext Transfer Protocol (HTTP)
443	UDP, TCP	Hypertext Transfer Protocol secure (HTTPs)
546	UDP, TCP	Dynamic Host Configuration Protocol (DHCP) for IPv6 clients
547	UDP, TCP	Dynamic Host Configuration Protocol (DHCP) for IPv6 servers

- Registered Ports – port numbers between 1,024 and 49,151 have been assigned by IANA to application protocols requesting them; for example, ports 4780810 (0xbac0) to 4782310 (0xbacf) have been assigned to the BACnet protocol for building automation.
- Dynamic or Ephemeral Ports – port numbers between 49,142 and 65,535 are used by applications as needed; for example, a client sending a request to a server may use an ephemeral port as the source port and the application port on the server as the destination port.

IANA assigns port numbers for use by application protocols on nodes that need to receive communications specific to the application. IANA keeps an online registry [45] of port numbers, service names, and transport protocols for all system ports and application ports.

A few representative registered port numbers together with their respective transport and application protocols are shown in Table 4.2.

4.6 UDP

UDP provides a simple, unreliable, and low-latency mechanism for sending self-contained units of data called datagrams between nodes in a network. UDP does not establish a dedicated end-to-end connection between the sender and receiver like TCP. Instead, each UDP datagram is treated independently and is delivered to the receiver without any prior setup or negotiation. UDP does not provide any mechanisms for error correction, flow control, or congestion control, nor does it acknowledge transmissions, mirroring the IP layer. This means that UDP datagrams may be lost, duplicated, or delivered out of order depending on network conditions. Applications using UDP must be prepared to deal with these conditions if reliability is required. Because UDP does not have the overhead of establishing and maintaining a connection, it can transmit data with very low latency compared to TCP. The lightweight nature of UDP makes it a good choice for a transport protocol in a 6LoWPAN network, and many applications built on Matter and Thread, which are implemented on top of the 6LoWPAN adaptation layer, use UDP for communicating device data.

UDP datagrams have a maximum size of 65,535 bytes including the UDP header. Each datagram is independent of any others and can be routed and delivered independently. UDP does not segment datagrams, the entire datagram received from the application layer is presented to the network layer in the UDP payload field. If the total packet size including IP header, UDP header, application layer header, and application data exceeds the MTU size on the end-to-end path in IPv4, the packet will be fragmented by the IP layer. In IPv6, however, it is the responsibility of the application layer to ensure that no datagram exceeds the MTU. UDP is specified in RFC 768 [46] which is only three pages long, and, even though the protocol was originally specified over 40 years ago, the basic design remains the same.

4.6.1 UDP Header

As shown in Figure 4.10, the UDP header is just 8 bytes and only contains the source port, destination port, datagram length, and a checksum field. The small header size makes UDP an efficient protocol for transmitting small amounts of data. The Checksum field is used for error checking on the datagram and is calculated by the sender and verified by the receiver. The checksum calculation is described in RFC 1071 [47] and is optional for IPv4 but required for IPv6. If the checksum is not calculated, the field is filled with zeros. The calculation is done over the entire packet, including a partial network layer header, called a pseudo-header [48]. The pseudo-header is different for IPv4 and IPv6 and omits information that is not possible to determine at the transport layer.

Source Port (16 bits)	Destination Port (16 bits)
Datagram Length (16 bits)	Checksum (16 bits)

Figure 4.10 UDP Header

More on the QUIC Protocol

This longer video of a 50 minute talk given by Peter Doornbosch from Luminis at the DevOx conference in Belgium in 2023 describes QUIC in more detail and how it is both similar to and different from TCP. Doornbosch is the author of the Kwik open-source implementation of QUIC. The talk also discusses how HTTP/3, the latest version of the HTTP application layer protocol, is well matched for use with QUIC

4.6.2 QUIC

In response to the widespread deployment of web browsers on mobile devices, a new transport based on UDP has been developed. Quick UDP Internet Connection (QUIC) is a modern, high-performance protocol that was developed by Google to address some of the limitations of TCP and UDP. TCP provides reliable communication, but it can be slow due to its connection setup and congestion control mechanisms. UDP is faster, but it lacks reliable delivery because it does not deal with packet loss. QUIC uses UDP as its underlying transport layer protocol to provide low-latency communication but includes additional features for reliable delivery. QUIC also provides several advanced features, including stream multiplexing, encryption, and header compression to performance and security. One of the primary advantages of QUIC is its ability to establish a connection faster than TCP, as it does not require a three-way handshake. This makes QUIC well-suited for use in mobile networks and other low-latency environments. QUIC is specified in RFC 9000 [49].

4.7 Transmission Control Protocol

The TCP is one of the most widely used protocols in computer networking. It is responsible for providing a communication channel abstraction that is reliable, ordered, and error-checked. This abstraction is built on top of the IP layer, which is neither reliable, ordered, nor error-checked. TCP ensures that transmitted data is delivered without errors and in the correct order and it was originally specified in RFC 793 [50] in 1981 prior to the formation of the IETF. Over the years, as the ARPANET grew into the Internet and use of the Internet exploded with commercialization in the 1990s, the IETF made many amendments and changes to TCP to address problems that arose as use of the protocol radically scaled up. In 2022, the IETF published a comprehensive update to the TCP standard, RFC 9293 [51], that gathered all the amendments and changes into one specification.

The following are key features of TCP:

- A connection is established between two nodes before transmitting any data. It uses a three-way handshake protocol to establish a connection and a similar handshake to terminate a connection.
- Large data chunks are divided into smaller segments to transmit over the network. On the receiving end, the segments are reassembled into the original data chunk.
- Reliable data delivery is ensured by using sequence numbers and acknowledgments. These guarantee that the order in which the data segments are sent is reconstructed on the receiving side even if packets arrive out of order, and if any segment is lost, a retransmission is requested.
- A sliding window mechanism is used to implement flow control. This ensures that the receiving node can handle the same transmission bandwidth as the sending node.
- A congestion control algorithm detects and responds to network congestion, reducing pressure on routers.
- Communication between the senders and receivers is full duplex, allowing the two sides to send data at the same time.

To implement this feature set, TCP requires considerably more complex mechanisms than UDP. This section presents a short overview of some of these mechanisms. For more detailed information, see RFC 9293 [51].

4.7.1 TCP Header

Figure 4.11 illustrates the TCP header. The header fields contain the following information:

- Source Port, Destination Port – two fields containing the 16-bit source and destination port numbers on which the application protocol sends and receives data.
- Sequence Number – every byte sent over a TCP stream has a sequence number. The first byte immediately following the header is the lowest numbered, and the following bytes are numbered consecutively. The Sequence Number field holds the sequence number of the first data byte in this segment, except if the SYN flag is set. If the SYN flag is set, the sequence number is the initial sequence number and the first data byte is the initial sequence number + 1. See Section 4.7.2 for details on the connection establishment protocol.
- Acknowledgment Number – if the ACK flag is set, this field contains the value of the next sequence number the sender of this segment is expecting to receive.
- Data Offset – indicates the size of the TCP header including any options in bytes. The minimum size is 20 bytes, and the maximum size is 60 bytes.

Source Port (16 bits)			Destination Port (16 bits)
Sequence Number (32 bits)			
Acknowledgment Number (32 bits)			
Data Offset (4 bits)	Reserved (4 bits)	Flags (8 bits)	Window Size (16 bits)
Checksum (16 bits)			Urgent Pointer (16 bits)

Figure 4.11 TCP Header

- Reserved – set to zero by the sender and ignored by the receiver.
- Flags – indicates the type of message and other conditions (see Section 4.7.1.1).
- Window Size – unsigned integer indicating the maximum number of bytes that the sender is willing to accept beginning with the byte indicated in the Acknowledgment Number field.
- Checksum – a checksum value used to detect corruption in the TCP header, payload, and the parts of the IP header that are known to the transport layer.
- Urgent Pointer – if the URG flag is set, contains an offset from the sequence number that indicates the last urgent data byte. See Section 4.7.1.1 for a discussion of the URG flag.

The header field can be followed by up to 40 bytes of options: in units of 4-byte words, with the list being padded to the next four-byte boundary if the option ending the list is less than 4 bytes. The options are formatted as type-length-value (TLV) fields. The type is specified by an integer code, with a code of 0 used to terminate the list. The length of the field is determined by the Data Offset field. The following are some commonly used options [52]:

- Maximum Segment Size – during the initial connection establishment protocol, each side of the TCP connection can specify the maximum number of bytes it is willing to accept in a single segment. To avoid fragmentation in IPv4, the size should be less than the minimum path MTU size. For IPv6, path MTU discovery is required in any case.
- Window Scaling – this option allows the connection to use high bandwidth networks more efficiently by allowing both sides to set a receive window greater than 65,535 bytes during the connection establishment protocol.
- Timestamps – senders can timestamp TCP segments, and receivers can echo the most recently received timestamp.
- Selective Acknowledgments – to avoid inefficiencies when packets are lost, receivers can acknowledge correctly but discontinuously received blocks of

packets, in addition to performing the basic TCP acknowledgment of the sequence number immediately following the last sequence number of the last correctly received contiguous byte. Selective acknowledgments are only used if both sides support it and are negotiated during the connection establishment protocol.

4.7.1.1 TCP Header Flags

Figure 4.12 illustrates the TCP header flag field. Each flag is one bit. The flags indicate the following:

- CWR – if the CWR flag is set, the congestion window has been reduced by the sending node. The CWR flag is set in response when the node receives a packet with the ECN flag in the IP header set during the initial connection negotiation.
- ECE – if the ECE flag is set during the initial connection establishment protocol (SYN flag is 1), it indicates that the node can utilize the ECN protocol at the IP layer to reduce transmission in the face of congestion. If the ECE flag is set during transmission by the receiver, it indicates that the receiver has received a packet with the ECN flag set at the IP layer indicating that congestion was detected or is imminent.
- URG – set if the Urgent Pointer field is set.
- ACK – set if the Acknowledgment field is significant, and is set in all packets sent by the initiator of the connection after the initial connection establishment protocol.
- PSH – requests that buffered data be pushed to the application layer.
- RST – resets the connection.
- SYN – synchronize sequence number. This flag is set by the initiator of the connection during the initial connection establishment protocol (see Section 4.7.2).
- FIN – indicates this is the last packet sent by the sender. This flag is set by the initiator of the connection during the protocol breaking the connection (see Section 4.7.2).

The Urgent Pointer field allows the sender to signal that part of the transmitted data is urgent and requires immediate attention from the receiver. The Urgent

CWR (1 bit)	ECE (1 bit)	URG (1 bit)	ACK (1 bit)	PSH (1 bit)	RST (1 bit)	SYN (1 bit)	FIN (1 bit)

Figure 4.12 TCP Header Flags

Pointer value indicates the sequence number of the last byte of the urgent data, and the sender sets the URG flag in the header flags field to indicate that the urgent pointer is also set. This signals to the receiver that the bytes within the payload up to the urgent pointer should be processed immediately and pushed to the application. The urgent pointer is rarely used as it can lead to performance issues and security vulnerabilities and is not uniformly implemented on all operating systems [52].

4.7.2 Procedures for Establishing and Breaking a Connection

The initiator of a TCP connection is typically called the client while the party receiving the request to initiate is called the server. The server opens a socket on the application port where the client is expected to send the request and listens for the request. TCP uses a three-way SYN-ACK connection establishment protocol to establish the connection between the two nodes. In the simplest case where the client wants to initiate the connection, the protocol involves the following steps, illustrated in Figure 4.13:

- The server opens a socket on the application port and listens for requests to establish a connection.
- The client opens a socket on an ephemeral port in preparation for sending the request.
- The client sends a TCP packet with the SYN flag set to 1, the ACK flag set to 0, and the Sequence Number field set to an unsigned integer sequence number, X. The initial sequence number is generated from a 32-bit number sequence that increments roughly every 4 μs until it wraps around. This ensures that any unacknowledged packets outstanding from a previous connection that was broken will be ignored.
- The server receives the packet and responds with a packet having both the SYN and ACK flags set to 1. The Sequence Number field is set to an unsigned integer, Y, generated as for the client, and the Acknowledgment Number field is set to X + 1.
- The client receives the SYN-ACK packet and responds with the final ACK message. The Sequence Number field is set to X + 1, and the Acknowledgment Number field is set to Y + 1.

When the three-way handshake is completed, both the sender and receiver exchange data using the established connection. The server is expecting the initial data packet from the client to have sequence number X + 1 and acknowledgment number Y + 1, the same sequence and acknowledgment numbers in the final ACK. From this point on, the two sides increment their sequence numbers on each exchange.

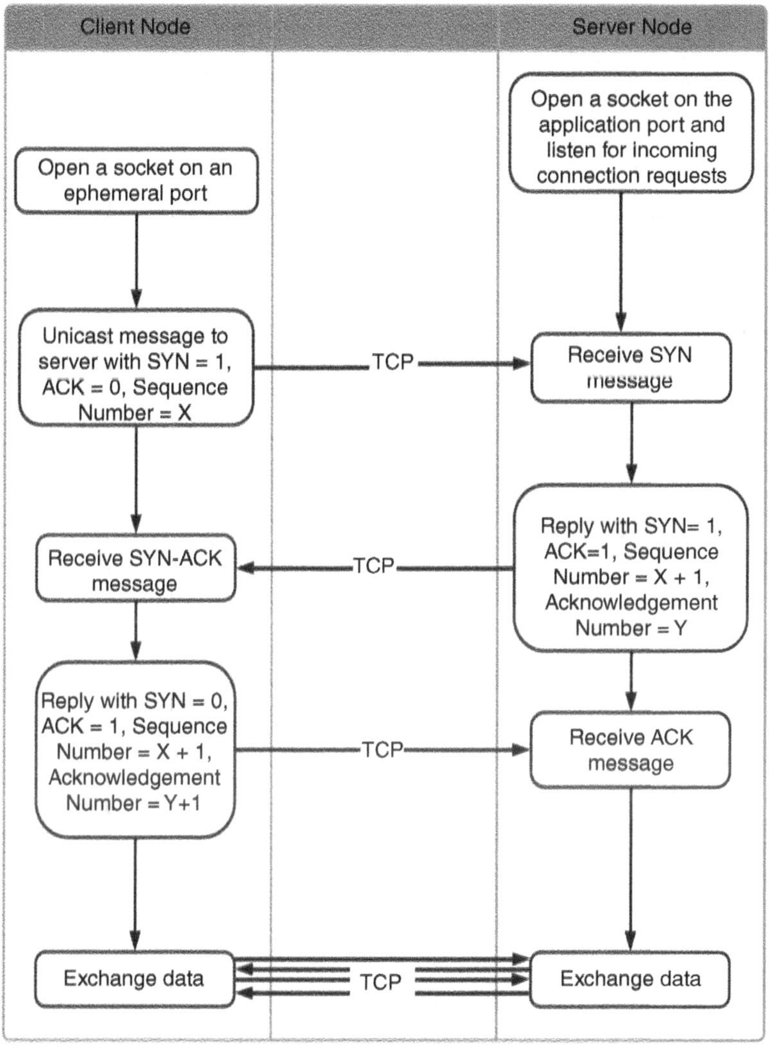

Figure 4.13 TCP SYN-ACK Protocol for Establishing a Connection

Once the two sides are finished with the communication and the client side needs to terminate the connection, a four-way FIN-ACK protocol is used. The protocol follows the steps illustrated in Figure 4.14:

- The client sends a TCP packet with the FIN bit set to 1 and the ACK bit set to 1. The Sequence Number field is set to the next number in the client sequence, say X, and the Acknowledgment Number is set to the next number in the server sequence, say Y.

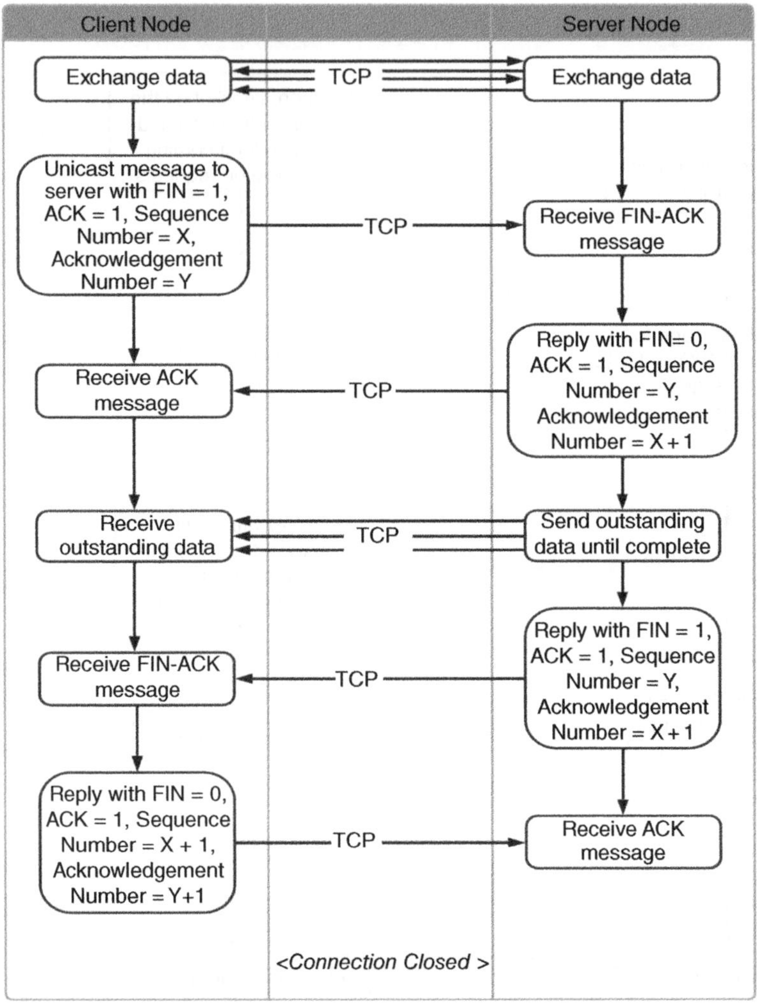

Figure 4.14 TCP FIN-ACK Protocol for Terminating a Connection

- The server sends a reply with the ACK bit set to 1, the Sequence Number field set to Y, and the Acknowledgment Number field set to X+1.
- The client continues to receive outstanding segments from the server having client sequence numbers less than X but does not initiate any further requests for data from the server.
- Once the server has completed all outstanding requests, it sends a TCP packet with the FIN bit set to 1 and the ACK bit set to 1. The Sequence Number field is set to Y and the Acknowledgment Number field is set to X+1.

- The client acknowledges the final packet by replying with a TCP packet having the ACK bit set to 1, the Sequence Number field set to X+1, and the Acknowledgment Number field set to Y+1.

When the four-way handshake is completed, the connection is terminated, and no further segments will be sent or received by either side. The TCP FIN-ACK protocol ensures that the communication is properly terminated. There are a few other cases for initiating and closing connections (see RFC 9293 [51] for details) but these are the most common.

4.7.3 Congestion Control Mechanisms

Network congestion occurs when the forwarding queues in a router fill up and the rate of packet arrival exceeds the rate at which the router can forward packets. When this happens, the router drops newly arrived packets until queue space opens. Such congestion and the resulting dropped packets can severely impact network performance, even though TCP will recover by retransmitting any packets that are lost. Consequently, TCP incorporates a collection of congestion control measures to help reduce congestion when it is detected. These mechanisms regulate the rate of data transmission and are described in RFC 5681 [53].

Each side in the TCP connection maintains a congestion window limiting the maximum number of bytes it is willing to send before receiving an acknowledgment. The sender communicates this in the TCP header Window Size field. The receiver's window is advertised in its TCP header on the reply packet. Each side maintains a sender window and a receiver window as part of the connection state. When a new TCP connection is established, the sender starts in the slow-start phase, sending at a minimal rate, and gradually increases the transmission rate until it reaches the receiver's congestion window limit. This avoids overwhelming the network and triggering congestion. Once the congestion window limit on both sides has been reached, TCP switches to the congestion avoidance phase. During the congestion avoidance phase, the sender increases its window by roughly one segment per round-trip time until congestion is detected through a timeout or lack of an acknowledgment for a segment, and then reduces its window downward. The sender recovers by entering a fast retransmit and recovery phase, where it retransmits the lost packet immediately without waiting for a timeout, and reduces its congestion window by half to avoid any further network overload.

4.8 Application Layer Protocols

Application layer protocols sit at the top of the IP stack, and in the original Internet architecture, they were thought to be where distributed applications would implement data communication protocols specific to their needs. Some application layer protocols such as the File Transfer Protocol (FTP) [54] or the Secure Shell (SSH) protocol [55] are based on the original architecture. FTP was designed to transfer files from one computer to another, while SSH was designed to allow a user on one computer to remotely log into another in a secure fashion. In the BMS application space, BACnet [56] is an application protocol designed specifically for building automation applications, and BACnet is discussed in more detail in Chapter 6. Recently, however, most distributed applications have moved to a model where their interfaces are specified as APIs using the rich text-based programming resources of the World Wide Web, such as HyperText Markup Language (HTML) [57] or specific API definition standards such as OpenAPI [58], rather than bespoke binary application protocols specifically defined for their use case. HTML is carried over the network by the HyperText Transfer Protocol (HTTP) [59] which has become the basis for connecting distributed applications built around a client–server architecture. In Chapter 8, application architectures, API standards, and protocols for distributed applications are discussed in more detail.

However, there are a few protocols that, while technically at the application level of the IP stack, are in practice fundamental to the operation of all applications built on top of the stack, and therefore are basically system protocols. One of these is DNS protocol [60]. The basic function of DNS is to resolve the name of a server to its IP address, thereby allowing developers to avoid hard coding IP addresses into their applications but it also supports other name resolution tasks. The other protocol is DHCP [61]. DHCP allows a node entering an IP network to be configured with its net mask, IP address, subnet prefix, subnet router, and DNS server, among other important properties. DHCP is the default way that node configuration is done in IPv4 networks, and is also used in some IPv6 networks, though IPv6 has a well-designed set of autoconfiguration primitives at the IP level that are used in most IPv6 networks as discussed in Section 4.3.2.3. In this section, DNS and DHCP are discussed.

4.9 DNS

The DNS is a hierarchical, decentralized system of name servers that translates human-readable domain and host names into the IP addresses of other name servers and end nodes. DNS uses a UDP-based protocol to communicate between nodes seeking to look up the IP address of another node based on the host name. It is a critical piece of Internet infrastructure, as it allows nodes to be referred to by name in applications rather than directly by an IP address. DNS also provides information on other resources associated with node and domain names such as services, mail servers, and the name servers for a domain. DNS acts as a basic naming directory for the Internet and, as such, has many components that all play a role in resolving a name to an address.

4.9.1 DNS Names

RFC 1035 specifies the syntax of DNS names [62]. A DNS name consists of a series of labels separated by periods ("."). The labels are restricted to between 1 and 63 characters and can include both upper- and lower-case letters, numbers, and a hyphen character ("-"). DNS does not, however, distinguish between upper and lower case letters in name resolution. Two labels with the same letters but with some in a different case are functionally equivalent. The overall length of a DNS name including all subcomponents and their separating periods must not exceed 253 characters. A label can be any valid combination of characters and numbers but cannot begin with a hyphen. International character sets can also be used, but are translated into ASCII characters with an `xn-` prefix indicating that the name is internationalized. An example of a DNS host name is www.example.com.

Each component of the name refers to a DNS domain, except for the leftmost component which refers to the end node itself. A domain is a scope within which names are resolved to addresses, either end node addresses or addresses of DNS servers that can further resolve a DNS name to find an end node. The rightmost label in a DNS name is called the top-level domain. So, for example, in www.example.com, the `.com` component is the top-level domain. A DNS name containing all the components in the name from the top-level domain name down to the name of the end node itself is called a fully qualified domain name (FQDN).

Top-level domains are registered by IANA. There are three different classes of top-level domains maintained by IANA:

- Country code top-level domains – two letter codes that represent a particular country, territory, or group of countries in a political association, like `.de` for Germany, `.us` for the United States, or `.eu` for the European Union. Organizations registering these names must be located in the respective region represented by the domain name,
- Generic top-level domains – the familiar `.com` and `.org` are examples of the original generic top-level domains. The `.com` domain is for use by commercial entities, while `.org` is for nonprofits, although no check is made on the non-profit status of an entity registering a `.org` domain name. Generic top-level domains that were introduced it the early 2000s include `.info`,`.biz`, `.museum`, and `.travel`. Generic top-level domains can vary in size,

- Sponsored top-level domains – to register a domain name in a sponsored top-level domain, the organization must meet certain criteria. IANA delegates the responsibility for checking the credentials of an organization registering for a sponsored top-level domain to another registrar. For example, to register a domain name in the `.edu` domain, the organization must be an accredited post-secondary educational institution in the United States. The `.edu` top-level domain is administered by Educause [63], a nonprofit organization dedicated to advancing education using information technology.

The next label to the left of the top-level domain name is the second-level domain. The second-level domain is typically the organization that registers the domain name and manages the domain. For example, in the DNS name www.example.com, `.example` is the second-level domain. Any additional labels to the left of the secondary domain name are subdomains and often represent the specific content or service, for example, in the name http://calendar.example.com, `.calendar` is a subdomain name that represents a calendar service. Finally, if a specific node is being referenced, it is the last label on the left with no period; for example, in http://my-server.example.com, `my-server` is the host being referenced.

4.9.2 DNS Architecture

DNS is based on a client–server architecture and consists of four main components:

- Recursive Resolver – a client-side service, typically a process running on the client's operating system that receives requests from applications to translate domain names into IP addresses and forwards the request to an appropriate DNS server,
- Root Server – the first step in the DNS resolution process. It is responsible for directing the recursive resolver to the appropriate top-level domain server. There are 13 root server clusters worldwide, managed by different organizations,
- Top-Level Domain Server – top-level domain servers store information about the registered top-level domains. A top-level domain server is responsible for returning information on DNS name servers for domains registered under the top-level domain,
- Authoritative Name Server – holds the actual DNS records for a specific domain. The authoritative name server is responsible for returning IP addresses and other information about names registered in the domain.

A DNS server at an upper level knows the name of the servers responsible for subdomains in zones at the level below it. An example resolution of a FQDN starting with the root server is described in Section 4.9.4.

Network administrators use a process called zone transfer to exchange information between DNS servers in the form of zone files. A DNS zone file is a text file that contains all the information about resources in a domain. Zone transfers are a way for DNS servers to share information with each other about their zones. There are two types of zone transfers: full zone transfers and incremental zone transfers. Full zone transfers are used when a new DNS server is added to the network. Incremental zone transfers are used to communicate updates to a zone file on an existing DNS server. Full zone transfers use the absolute zone transfer protocol (AXFR) [64]. AXFR replicates the entire zone file from the primary DNS server to the secondary DNS server. Incremental zone transfers use the incremental zone transfer protocol (IXFR) [65]. IXFR replicates only the changes to the zone file since the last transfer. Both AXFR and IXFR use TCP for the transport protocol.

4.9.2.1 DNS Records

DNS also supports records about other types of resources in the domain besides IP addresses. These records are known as resource records. Here is a list of common DNS resource record types:

- Address (A) – maps an FQDN to the IPv4 address of the respective node.
- Quad-A (AAAA) – maps an FQDN to the IPv6 address of the respective node.
- Canonical Name (CNAME) – contains an alias for a domain name. Canonical names are used to map a subdomain label such as www or mail to the domain hosting the subdomain's name records.
- Mail Exchange (MX) – contains the FQDN of mail servers responsible for handling email for a domain.
- Name Server (NS) – contains the FQDN for the authoritative name servers for a domain.
- Service (SRV) – contains the FQDN and port number of services such as a web service, enabling service discovery.
- Text (TXT) – can contain arbitrary text information, commonly used for spam prevention, domain ownership records, contact information for the domain, etc.

Multicast DNS Service Discovery

In this video, Stuart Cheshire, Distinguished Engineer at Apple and author of the Bonjour multicast DNS service discovery protocol, describes how service discovery is accomplished in IP networks. Bonjour is the basis of the IETF multicast service discovery protocol standard, RFC 6763 [66]. Service discovery is a key part of the BEMS since different services constituting the BEMS must find the IP address of other services they need to use without hard coding the IP address.

Resource records for other, less commonly used information are important for administering the domain and are also supported by DNS.

4.9.3 DNS Packet Format

DNS uses UDP as its primary transport protocol, although it can also use TCP when necessary for large responses or zone transfers. DNS operates on port 53 for both UDP and TCP. DNS packets are divided into five fields:

- Header – contains control information for the packet. The header fields indicate if the packet is a query or response and the number of questions in a query or answers returned in the response.
- Question – if the message is a query, contains a list of queries for the DNS server.
- Answers – if the message is a response, contains a list of resource record responses to the questions from the DNS server.
- Authority – if the message is a response, contains a list of resource records pointing to an authoritative name server.
- Additional Information – if the message is a response, contains a list of additional resource records bearing information on the query.

4.9.3.1 DNS Header

Figure 4.15 illustrates the DNS application layer protocol header format. The fields in the header are:

- Message Identifier – a unique 16-bit identifier assigned by the application that generates the query. Matching responses will have the same identifier allowing the application to match queries with responses,
- QR – a 1-bit field indicating if the message is a query (0) or response (1),
- OpCode (OC) – if the message is a query, a 4-bit field indicating the type of query, see Section 4.9.3.2 for a list of opcodes. If the message is a response, set to zero,
- Flags – a 4-bit field with flags, see Section 4.9.3.3 for a list of flags,
- Response Code (RC) – if the message is a response, a 4-bit field indicating the response results, see Section 4.9.3.2 for a list of response codes. If the message is a query, set to zero,
- Reserved – a 3-bit field set to zero and reserved for future use,
- Question Record List Length – a 16-bit field indicating the number of questions in the Questions field,
- Answer, Authority, and Additional Information Records List Length – 16-bit fields indicating the number of resource records in the respective list.

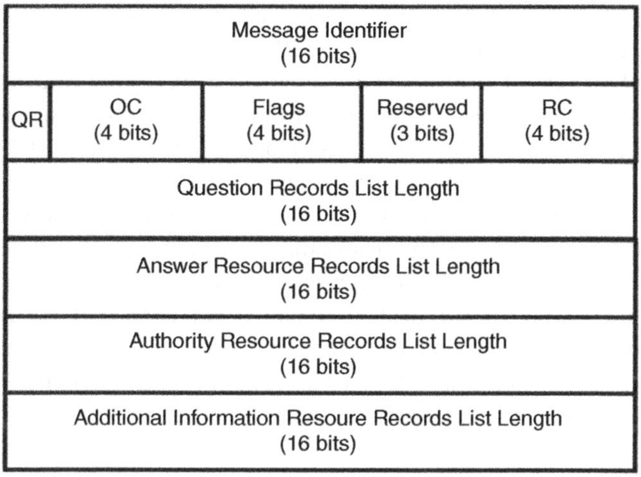

Figure 4.15 DNS Packet Header Format

4.9.3.2 DNS Opcodes and Response Codes

Table 4.3 contains a list of the DNS query opcodes set by the client to indicate the nature of the message. Most messages will have an opcode of 0, indicating a standard query. Opcode 1 is now obsolete; it was reserved for reverse queries, where the client provided an IP address and requested resolution to the FQDN but this functionality is now implemented in a different way (see RFC 2317 [67] for details). Opcode 2 is a request for the status of the server. Opcode 4 was added by RFC 1996 [68] to allow authoritative name servers to prompt secondary name servers that the records for the zone have changed and to request a zone update. Opcode

Table 4.3 DNS Query Opcodes

Code Value	Code Name	Condition Description
0	Query	Standard query.
1	Obsolete	Originally used for queries giving the IP address and requesting the FQDN in response but not used.
2	Server Status Request	Request for server status.
3	Reserved	Reserved, not used.
4	Notify	Added by RFC 1996 allowing prompt notification of zone changes.
5	Update	Added by RFC 2136 allowing dynamic update of DNS.

5 was added by RFC 2136 [69], called dynamic DNS, to allow resource records to be incrementally added, deleted, or updated. See the RFC for details.

Table 4.4 contains a list of the DNS return codes set by the server to indicate the results of processing the query. Return codes 0–5 are from the original DNS specification in RFC 1035 [62], while codes 6–10 were added by the dynamic DNS specification in RFC 2136 [69]. Code 0 indicates a successful resolution. Codes 1–4 indicate various error conditions, such as a formatting error or if the server encountered an unspecified error. Code 5 indicates that the query was refused for policy reasons; for example, if the client requests a zone transfer but is not permitted to make such a request. Codes 6–8 indicate problems with records and names within the server. Code 9 indicates that the server receiving a request is not authoritative for the zone specified, while code 10 indicates that the name in the query is not in the zone specified in the query.

4.9.3.3 DNS Header Flags

Figure 4.16 shows the DNS header flags. With the exception of the RD flag, flags are set in responses only. The flags are the following:

- Authoritative Answer (AA) – set to 1 if the server responding is the authoritative name server for the domain name specified in the Questions field; otherwise, set to 0.

Table 4.4 DNS Return Codes

Code Value	Code Name	Condition Description
0	No Error	No Error.
1	Format Error	The server could not respond due to a formatting error in the query.
2	Server Error	The server encountered an error.
3	Name Error	The name specified in the query does not exist in the zone.
4	Not Implemented	The requested type of query is not implemented by this server.
5	Refused	The server refused to process the query for policy reasons.
6	YX Domain	A name exists when it should not.
7	YX RR Set	A resource record exists when it should not.
8	NX RR Set	A resource record that should exist is missing.
9	Not Auth	The server receiving the query is not authoritative for the zone specified.
10	Not Zone	A name in the query is not within the zone specified in the query.

Figure 4.16 DNS Header Flags

AA	TC	RD	RA
(1 bit)	(1 bit)	(1 bit)	(1 bit)

- Truncation (TC) – set to 1 if the response was truncated due to UDP's packet length limit. The client can retry with TCP if the results do not include what the client was looking for.
- Recursion Desired (RD) – set to 1 by the client to indicate that it wants the server to answer the query recursively if possible. Unchanged by the server in the response.
- Recursion Available (RA) – set to 1 by the server to indicate that the server supports recursion for future requests.

4.9.4 Name Resolution with DNS

DNS name resolution involves multiple steps, and an example is illustrated in Figure 4.17. Resolution of a simple FQDN consists of the following steps:

- An application program calls an operating system function to resolve the name. An example is a user typing a URL into the browser search bar, with the URL containing a FQDN like www.google.com.
- The recursive resolver first checks its cache to see if it has a recent record of the FQDN. If no record is found or the record is out of date, it begins the query process.
- The recursive resolver queries the root server for the top-level domain server, in this case, the server for the `.com` domain.
- The root server looks up the address record and returns a response to the recursive resolver.
- The recursive resolver queries the top-level domain for the domain's authoritative name server, in this case, the server for the `.google` domain.
- The authoritative name server looks up the address record for the domain's authoritative name server and returns a response to the recursive resolver.
- The recursive resolver queries the authoritative name server for the address of the host/service, in this case, www.
- The authoritative name server looks up the address record for the server and returns a response to the recursive resolver.
- The recursive resolver returns the address to the application.
- The application contacts the server using an appropriate application and transport protocol, and the two begin communication.

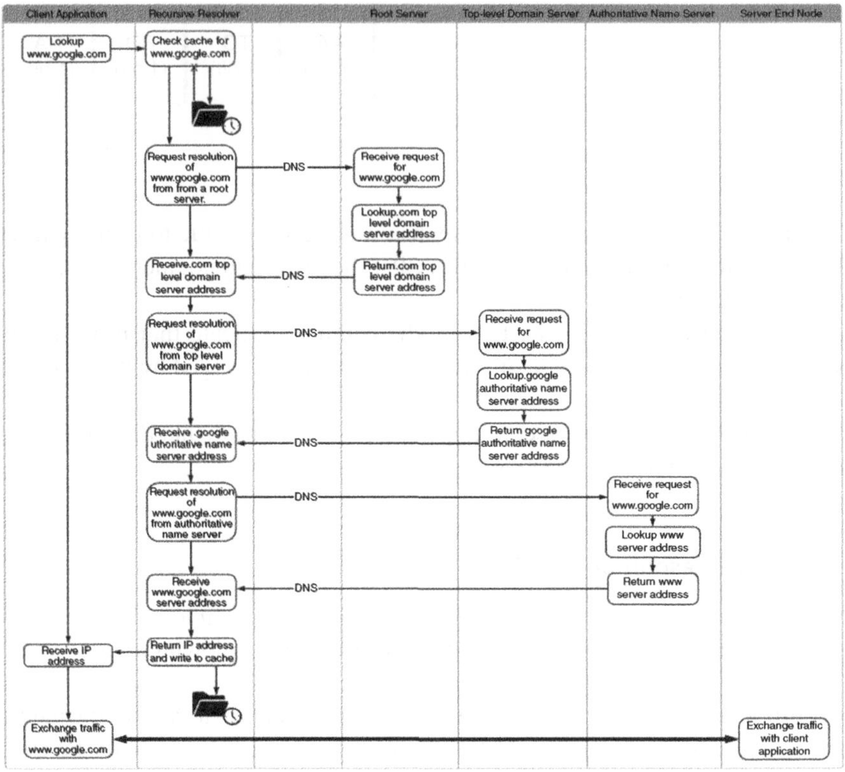

Figure 4.17 DNS Name Resolution

4.10 DHCP

DHCP [61] is an application layer protocol that allows nodes to obtain an IP address and other configuration information automatically from a DHCP server. DHCP was developed in the 1990s as a replacement for the Bootstrap Protocol (BOOTP) [70], an earlier protocol that was used for node configuration in the ARPANET, and many DHCP servers implement BOOTP functionality as well since the DHCP packet format is backward compatible with BOOTP. The primary purpose of DHCP is to automate and simplify the process of configuring network devices, eliminating the need for manual configuration. It enables nodes to request and obtain IP addresses, subnet masks, default routers, DNS servers, and other configuration parameters from a central DHCP server. DHCP can also help improve network security by ensuring that only authorized devices are able to obtain IP addresses, and help improve network performance by ensuring that all nodes on a link are using a consistent set of IP addresses with the same subnet prefix and the same subnet mask. DHCP is a critical part of IP network infrastructure, and it is used on networks of all sizes, since it radically simplifies connecting a node to an IP network without requiring the node to be manually configured by a network administrator.

DHCP is based on a client–server architecture, with the incoming node being the client and the DHCP server being the server. The DHCP server is responsible for managing a pool of IP addresses and other configuration information, and DHCP clients are responsible for requesting and receiving this information from the DHCP server. IPv6 networks have a version of DHCP called DHCPv6, but this section primarily covers DHCP for IPv4 since IPv6 has a rich collection of well-designed autoconfiguration protocols, whereas configuration of a new node entering a subnet in IPv4 networks is accomplished almost exclusively with DHCP. IPv4 nodes typically perform configuration using DHCP just after they boot, then periodically renew their configuration prior to a timeout if necessary. Clients use broadcast to locate a DHCP server, and if the server is located on a different link, a DHCP relay is necessary to convey the request across the subnet router since broadcast messages are not forwarded across routers.

Network administrators have three options for setting up a DHCP server to configure incoming nodes [71]: dynamic allocation, automatic allocation, and static allocation. If the network administrator chooses to use dynamic allocation, they reserve a range of IP addresses for allocation to incoming nodes. Each node entering the network is preconfigured to request an IP address from the DHCP server when the node boots. The address is granted with a lease which requires the node to periodically renew the address, allowing the DHCP server to reclaim and reallocate IP addresses that are not renewed.

In automatic allocation, the DHCP server assigns an IP address to a client device for an unlimited lease time, and the server retains the association between the client's identifier, typically the link-layer address of the client's network interface card, and the IP address. However, the client does not necessarily receive the same IP address every time it boots. The goal of automatic allocation is to provide a permanent IP address to the client, but with no guarantee that the client will always get the same address. However, other tools, such as network monitoring tools, need to be used to detect when the node goes offline. Because automatic allocation can deplete the address pool, it is not widely used.

Static allocation works like automatic allocation, but the client gets the same IP address every time it boots. To set up static DHCP allocation, the network administrator additionally configures the DHCP server with a mapping between the client's link-layer address and the IP address to be reserved for the client. Once the reservation is in place, the DHCP server will always offer the reserved IP address to the client with the matching link-layer address, ensuring that the client gets the same IP address every time it boots, and the address will not be offered to any other client. Static allocation requires that the network administrator manage the mappings between link-layer addresses and IP addresses, removing mappings when nodes are retired.

4.10.1 DHCP Message Types

The DHCP protocol defines several message types for initial node configuration and node configuration renewal. The following list contains a short description of the primary DHCP message types:

- DHCPDISCOVER – sent by the client as a broadcast message to locate available DHCP servers. This message initiates the configuration process and indicates that the client is searching for an IP address and other network configuration parameters,
- DHCPOFFER – sent by the DHCP server(s) in response to a DHCPDISCOVER message. It contains an available IP address and other configuration parameters for the client. A network with multiple DHCP servers may result in the client receiving multiple DHCPOFFER messages,
- DHCPREQUEST – sent by the client to accept an offer from a specific DHCP server, requesting the offered IP address and configuration parameters. This message is also a binding confirmation, indicating that the client agrees to use the offered IP address for the lease duration,
- DHCPACK – sent by the DHCP server to the client to confirm the lease of the IP address and other configuration information. This message finalizes the node configuration process, and the client can now use the assigned IP address and other parameters,
- DHCPNAK – sent by the DHCP server if the requested IP address in a DHCPREQUEST message sent by the client is invalid or unavailable, or if the client's configuration information is otherwise incorrect. The client must then restart the DHCP process with a new DHCPDISCOVER message,
- DHCPDECLINE – sent by the client if it detects that the offered IP address is already in use on the network, typically through an ARP check, and therefore cannot be used. This informs the server that it should update its pool of available IP addresses accordingly. The client then restarts the DHCP process with a new DHCPDISCOVER message,
- DHCPRELEASE – sent by the client when it no longer needs the assigned IP address, typically during a graceful shutdown or when switching to a different network. This message informs the DHCP server that the IP address can be returned to the available pool, allowing the address to be allocated to other nodes,

- DHCPINFORM – sent by the client to request additional local configuration parameters, such as time servers or proxy settings, from a DHCP server without requesting an IP address. The server responds with a DHCPACK message containing the requested information.

4.10.2 DHCP Packet Format

A DHCP packet, shown in Figure 4.18, consists of a header and a variable-length data payload containing options. The header has a fixed size of 236 bytes (1,888 bits) and includes several fields that provide information on the participants in the DHCP transaction as well as on the transaction itself. The data payload consists of options that provide configuration information, information on the type of the DHCP message, and other information to both the client and server. Note that the header was designed to be backward compatible with BOOTP which is why the OpCode field lists the BOOTP message types and the actual DHCP message types are communicated in the options.

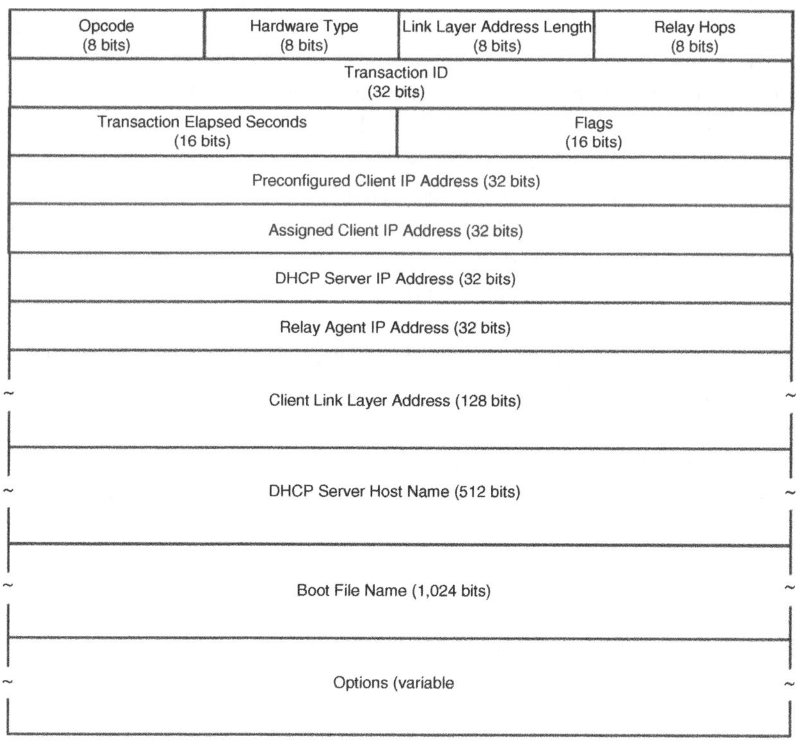

Figure 4.18 DHCP Packet Format

The header consists of the following fields:

- Opcode – the basic message type, either BOOTREQUEST (1) for a client's request or BOOTREPLY (2) for a server's response. Note that the actual DHCP message type is transmitted in an option. This field is from BOOTP,
- Hardware Type – a physical layer hardware type code indicating the kind of link-layer address used in the client's network. The original hardware type codes are defined in RFC 2132 [72] but have been updated over the years,
- Hardware Address Length – the length of the link-layer address, in bytes. For an EUI-48 Ethernet address, this is 6,
- Hops – the number of relay agent hops a packet has traversed. Initially set to 0 by the client,
- Transaction ID – random number generated by the client to uniquely identify a DHCP transaction,
- Seconds – the number of seconds since the client began the address acquisition or renewal process, set to 0 for the initial DHCPDISCOVER message,
- Flags – a field for flags. Only one is currently allocated: the first (leftmost) bit in the field is the BROADCAST flag, set to 0 if the client expects a unicast reply from the server or 1 if a broadcast reply is expected. The remaining bits are reserved for future use,
- Client IP Address – if the client has an existing, valid IP address, it is placed here; otherwise, it is set to the unspecified address, 0.0.0.0,
- Assigned IP Address – the server places the offered or assigned IP address for the client in this field when sending the DHCPOFFER,
- Server IP Address – if the client knows the DHCP server's IP address, it is placed here; otherwise, it is set to 0.0.0.0,
- Gateway IP Address – relay agents specify the IP address of the DHCP server or another relay agent to which the packet should be forwarded in this field,
- Client Link-Layer Address – the link-layer address of the client's network interface. For Ethernet, this is 6 bytes, with the remaining 10 bytes padded with zeros,
- Server Hostname – an optional null-terminated string that can be used to specify the server hostname. If not used, it is filled with zeros,
- Boot Filename – an optional null-terminated string that can contain the name of a boot file, typically for network booting purposes. The client needs to fetch the file from another server using FTP. If not used, it is filled with zeros.

Following the header, a variable-length payload contains DHCP options. The options provide additional configuration parameters and information relevant to the DHCP configuration transaction. The options field starts with a 4 byte "magic cookie" having the value 99.130.83.99. The options are encoded in TLV format. Each option contains a 1-byte option type, a 1-byte option length, and a variable option value. Each option has a unique identifier, the option type. The length is the length of the option value in bytes. The option value contains the contents of the option. The end of the options payload is marked with an End option having type 255.

The options included in the options payload depend on the message type and the information being exchanged. Some common options and their types, with type codes in parenthesis, used in DHCP messages are:

- DHCP Message Type (53) – DHCP type of the message, such as DHCPDISCOVER, DHCPOFFER, DHCPREQUEST, or DHCPACK,
- Subnet Mask (1) – the subnet mask of the offered or requested IP address,
- Subnet Router Address (3) – the IP address of the default subnet router for the offered or requested address,
- Domain Name Server Addresses (6) – the IP addresses of one or more DNS servers,
- IP address Lease Time (51) – the duration of the lease for the offered or requested IP address,
- Requested IP Address (50) – in DHCPREQUEST messages, the client uses this option to request a specific IP address,
- Parameter Request List (55) – a list of option codes for additional configuration parameters that the client requests from the server,
- Client Identifier (61) – a unique identifier for the client, usually its link-layer address,
- Renewal Time Value (58) and Rebinding Time Value (59) – these options specify the time intervals in the lease renewal process,
- Hostname (12) – the hostname to the DHCP server. Optional, sent by the client,
- Vendor Class Identifier (60) – the type of hardware or software the client is using. This option allows the DHCP server to customize the configuration based on the client's hardware and software,
- Vendor-Specific Information (43) – returned by the server in response to the vendor class identifier option. Contains custom configurations for the hardware and software specified in the client option.

These are the most commonly exchanged options. There are many others that are less commonly used.

4.10.3 Node Configuration Using DHCP

DHCP uses UDP for transport, and DHCP servers listen on port 67 while DHCP clients listen on port 68. Note that the client does not have a valid, allocated IP address that it can use until the entire process is complete. The basic DHCP process of configuring a newly booted host with parameters needed to connect to the subnet and successfully communicate is shown in Figure 4.19 and consists of the following steps:

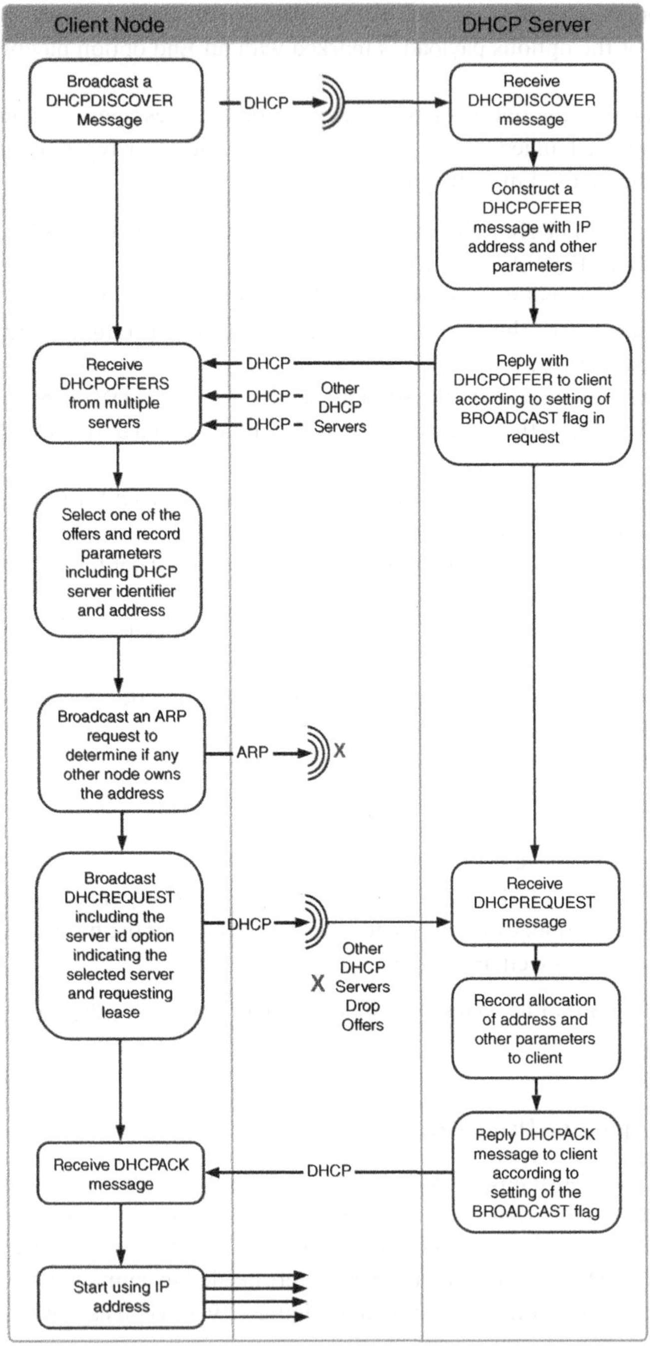

Figure 4.19 IP Node Configuration with DHCP

- The client node broadcasts a DHCPDISCOVER message to locate available DHCP servers on the network. The client sets the destination IP address to either the IP broadcast address (255.255.255.255) or the subnet broadcast address (subnet prefix + 255 for the node identifier), and the source IP address to the unspecified IP address (0.0.0.0). In addition, the client sets the BROADCAST flag in the DHCP header to indicate whether it can accept unicast replies (BROADCAST flag = 0) or requires broadcast replies (BROADCAST flag = 1) without having a valid IP address configured. If the BROADCAST flag is not set, the server replies using the source IP address from the request packet but unicast directly to the link-layer address in the link-layer frame.
- DHCP servers receiving the DHCPDISCOVER message respond with a DHCPOFFER, offering a lease on an IP address. The DHCPOFFER message contains the client's id, typically the link-layer address of the client, the offered IP address, the subnet mask, the lease duration, and the IP address of the DHCP server making the offer.
- The client selects an offer and broadcasts a DHCPREQUEST message requesting a lease on the offered IP address and the other configuration parameters. The client includes a server identifier option indicating the server that made the offer. When other servers hear this message, they drop their offers and return the offered IP addresses to their address pools.
- The DHCP server sends a DHCPACK message confirming the lease duration and includes the other configuration requested by the client. The client can now configure its IP stack and use the assigned IP address and configuration for the duration of the lease.

4.10.3.1 Lease Renewal

Clients must periodically renew their lease on an IP address. To renew a DHCP lease on an address, the client unicasts a DHCPREQUEST message to the server again after roughly 50% of the lease duration has expired. The message includes the client's current IP address and indicates that the client wants to renew the lease. If the server approves the lease renewal, it responds with a DHCPACK message containing the updated lease duration and any changes to the configuration parameters. The client then updates its renewal timer based on the new lease duration.

If the client fails to update its lease in the initial update, perhaps due to the DHCP server not responding, it continues to use the address until roughly 87.5% of the lease duration has expired. At this point, the client enters the rebinding phase and starts sending broadcast DHCPREQUEST messages, seeking any available DHCP server to renew the lease. If the client fails to renew the lease, the lease expires and the client must discontinue using the IP address. The client then restarts the DHCP configuration process with a broadcast DHCPDISCOVER message to obtain a new IP address.

4.11 Open-Source Tools for Monitoring BEMS IP Networks

One advantage of using IP networks is the wide selection of open-source tools that are available for building and monitoring them. In many cases, these tools are the equal of or superior to proprietary tools. This section focuses on Wireshark, an open-source tool for visualizing IP packets for monitoring an IP network.

4.11.1 Packet Tracing with Wireshark

Wireshark [73] is an open-source network protocol analyzer[4] that allows users to capture and examine network traffic in real-time. It is available for various operating systems, including Windows, macOS, and Linux, and supports over 2,000 different network protocols, including many involved in deploying networks for a BEMS. With Wireshark, network administrators can visualize packet flows on operating system network interfaces using a graphical user interface, allowing network issues, security vulnerabilities, and performance problems to be identified. In addition to capturing and analyzing network traffic, Wireshark offers a wide range of advanced features, including the ability to filter network traffic based on various criteria, create custom profiles, conduct performance tests of complex client/server systems, and generate detailed network statistics and reports.

Figure 4.20 shows a screen shot of a Wireshark session tracing a DNS lookup for www.google.com from the Firefox browser. At the very top of the window just under the menu bar, a packet filter has been set up to look for packets with TCP or UDP port 53 (`udp.port == 53 || tcp.port == 53`). Below the filter window is a window showing captured packets that match the filter. Each line corresponds to a captured packet. The source and destination addresses, protocol from the last header in the header stack, length of the packet, and other information for each packet captured are shown. The row with the selected packet is highlighted. For the selected packet, a recursive resolver with a private address of 192.168.0.1 is being used, and the information column shows that the DNS request is for an A (IPv4 address) resource record for the FQDN www.google.com.

Below the captured packets window is a window providing more detail on the selected packet. Here, the IP and UDP headers are displayed in an easy-to-read format. In the UDP header section, the source port is the ephemeral port 37748. The DNS section shows the transaction id is 0x17eb, with a single query indicating the client has requested single question. In the Queries section, the request is shown. The query also has a request for Additional Information shown after the

4 Wireshark source code is available at https://github.com/wireshark/wireshark.

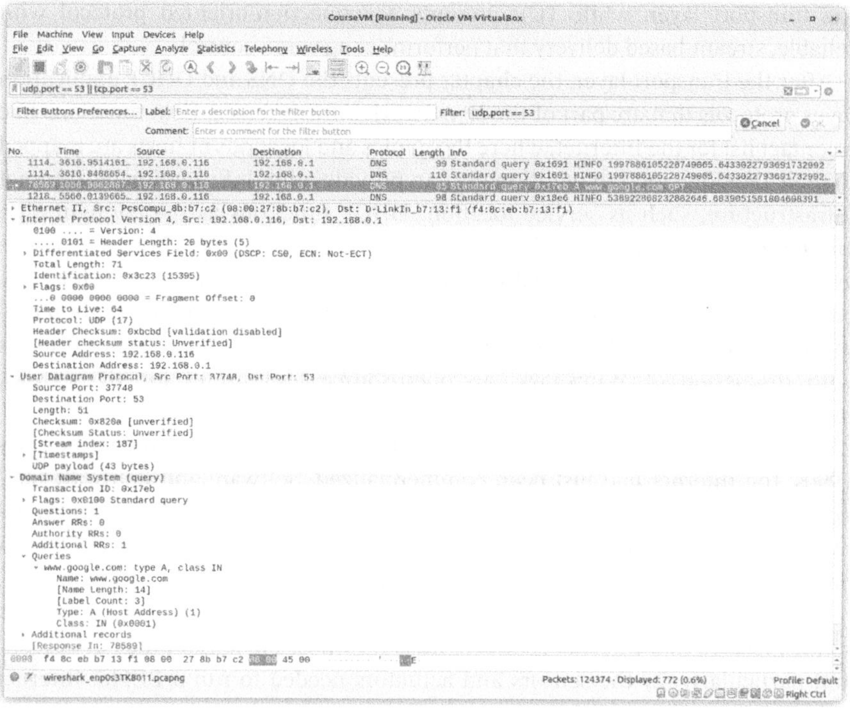

Figure 4.20 Wireshark Protocol Trace for a DNS Query

Queries section. This is a DNS option added according to the DNS extension mechanism, which is specified in RFC 6891 [74].

4.12 Summary

This chapter provides an in-depth review of IP networking and its significance in the BEMS. It begins with a brief introduction to the IP as the backbone of data communication and the Internet. The chapter then delves into the core elements of IP networking, including addressing, subnetting, and routing for both IPv4 and IPv6. The 6LoWPAN protocol is then explored, since it is an important extension to the IPv6 protocol for wireless, low-power personal area network link layers like 802.15.4 which are widely used in devices that act as sensors and actuators in a BEMS. Next, the chapter discusses the transport layer protocols, UDP and TCP. These protocols play an important role in application development and delivery and have very different properties by design. UDP is designed for connectionless delivery of relatively short payloads and models the IP network layer properties at the transport layer, while TCP presents a connection-oriented protocol with reliable, stream-based delivery that performs congestion control.

After the transport layer, the chapter presents the DNS and DHCP application layer protocols that are part of the IP network infrastructure. DNS functions like a contact list for the Internet, where the names and locations of hosts are mapped to their IP addresses, but additionally can provide other information on Internet infrastructure, such as service location. DHCP functions as the configuration engine for an IP node when it first connects to the network. Though DHCP is available and occasionally used in IPv6 networks, it is primarily used for node configuration in IPv4 networks because IPv4 lacks a node autoconfiguration protocol such as the IPv6 Neighbor Discovery protocol. Finally, the chapter finishes with discussion of Wireshark, an open-source tool useful for IP network monitoring.

When standardized BMS systems were first designed and deployed in the early 1990s, the Internet had just been commercialized. Software and hardware supporting IP stack development and deployment were very expensive or unavailable. In contrast, today the Internet engineering community has had over 30 years of experience with developing and deploying widespread, broadly available commercial services. The software and hardware necessary to run an IP stack are now so inexpensive that Internet connectivity is now possible for all kinds of devices, and particularly for the sensors and actuators needed to run a BEMS. Internet connectivity is also available in most buildings, either through a wired ISP connection or through a wireless cellular provider. The IP stack will only continue to grow in importance for building energy management as BEMS device, software, and system vendors become more familiar with it and begin deploying it in their products and open-source projects.

References

1 Wikipedia. Internet [Internet]. [cited 2023 Feb 2]. Available from: https://en.wikipedia.org/wiki/Internet.

2 IETF. IETF [Internet]. [cited 2023 Feb 15]. Available from: https://www.ietf.org.

3 ICANN. ICANN | ASO [Internet]. [cited 2023 Feb 15]. Available from: https://aso.icann.org/about/aso-and-nro/rirs.

4 Wikipedia. Border Gateway Protocol [Internet]. [cited 2023 Feb 28]. Available from: https://en.wikipedia.org/wiki/Border_Gateway_Protocol.

5 Wikipedia. Distance-Vector Routing Protocol [Internet]. [cited 2023 Feb 28]. Available from: https://en.wikipedia.org/wiki/Distance-vector_routing_protocol.

6 AlgoWiki. Bellman-Ford Algorithm [Internet]. [cited 2023 Mar 1]. Available from: https://algowiki-project.org/en/Bellman-Ford_algorithm.

7 Wikipedia. Open Shortest Path First [Internet]. [cited 2023 Mar 1]. Available from: https://en.wikipedia.org/wiki/Open_Shortest_Path_First.

8 Wikipedia. Link-State Routing Protocol [Internet]. [cited 2023 Mar 1]. Available from: https://en.wikipedia.org/wiki/Link-state_routing_protocol.

9 Postel J. Internet Control Message Protocol, RFC 792 [Internet]. [cited 2023 Feb 2]. Available from: https://www.rfc-editor.org/info/rfc792.

10 Conta A., Deering S., and Gupta M. Internet Control Message Protocol (ICMPv6) for the Internet Protocol Version 6 (IPv6) Specification, RFC 4890 [Internet]. [cited 2023 Feb 2]. Available from: https://www.rfc-editor.org/info/std89.

11 Hornig C. A Standard for the Transmission of IP Datagrams over Ethernet Networks, RFC 894 [Internet]. [cited 2023 Feb 2]. Available from: https://www.ietf.org/rfc/rfc894.txt.

12 University of Aberdeen. Address Resolution Protocol (arp) [Internet]. [cited 2023 Feb 2]. Available from: https://erg.abdn.ac.uk/users/gorry/course/inet-pages/arp.html.

13 Narten T., Nordmark E., Simpson W., and Soliman H. Neighbor Discovery for IP version 6 (IPv6), RFC 4861 [Internet]. [cited 2023 Feb 2]. Available from: https://www.rfc-editor.org/rfc/rfc4861.html#section-7.2.

14 Postel J. Internet Protocol, RFC 791 [Internet]. [cited 2022 Feb 13]. Available from: https://www.rfc-editor.org/rfc/rfc791.

15 Wikipedia. Internet Protocol Version 4 [Internet]. [cited 2023 Feb 7]. Available from: https://en.wikipedia.org/wiki/Internet_Protocol_version_4.

16 Nichols K., Blake S., Baker F., and Black D. Definition of the Differentiated Services Field (DS Field) in the IPv4 and IPv6 Headers, RFC 2474 [Internet]. [cited 2023 Feb 7]. Available from: https://www.rfc-editor.org/rfc/rfc2474.

17 Ramakrishnan K., Floyd S., and Black D. The Addition of Explicit Congestion Notification (ECN) to IP, RFC 3168 [Internet]. [cited 2023 Feb 7]. Available from: https://www.rfc-editor.org/rfc/rfc3168.

18 Touch J. Updated Specification of the IPv4 ID Field, RFC 6864 [Internet]. [cited 2023 Feb 7]. Available from: https://www.rfc-editor.org/rfc/rfc6864.

19 Bhardwaj R. Fragment Offset [Internet]. [cited 2023 Feb 7]. Available from: https://ipwithease.com/fragmentation-offset.

20 IANA. Protocol Numbers [Internet]. [cited 2023 Feb 16]. Available from: https://www.iana.org/assignments/protocol-numbers.

21 Wikipedia. Classless Inter-Domain Routing [Internet]. [cited 2023 Feb 12]. Available from: https://en.wikipedia.org/wiki/Classless_Inter-Domain_Routing.

22 Cotton M., Vegoda L., and Meyer D. IANA Guidelines for IPv4 Multicast Address Assignments, RFC 5771 [Internet]. [cited 2023 Feb 12]. Available from: https://datatracker.ietf.org/doc/html/rfc5771.

23 Cheshire S., Aboba B., and Guttman E. Dynamic Configuration of IPv4 Link-Local Addresses, RFC 3927 [Internet]. [cited 2023 Feb 12]. Available from: https://www.rfc-editor.org/rfc/rfc3927.

24 Wikipedia. IPv4 Address Exhaustion [Internet]. [cited 2023 Feb 12]. Available from: https://en.wikipedia.org/wiki/IPv4_address_exhaustion.

25 Wikipedia. Network Address Translation [Internet]. [cited 2023 Feb 13]. Available from: https://en.wikipedia.org/wiki/Network_address_translation.

26 Deering S. and Hinden R. Internet Protocol, Version 6 (IPv6) Specification, RFC 8200 [Internet]. [cited 2023 Feb 16]. Available from: https://www.rfc-editor.org/rfc/rfc8200.

27 Amante S., Carpenter B., Jiang S., and Rajahalme J. IPv6 Flow Label Specification, RFC 6437 [Internet]. [cited 2023 Feb 16]. Available from: https://www.rfc-editor.org/rfc/rfc6437.

28 McCann J., Deering S., Mogul J., and Hinden R. Path MTU Discovery for IP Version 6, RFC 8201 [Internet]. [cited 2023 Feb 16]. Available from: https://datatracker.ietf.org/doc/html/rfc8201.

29 Hinden R. and Deering S. IP Version 6 Addressing Architecture, RFC 4291 [Internet]. [cited 2022 Feb 17]. Available from: https://www.rfc-editor.org/rfc/rfc4291.

30 IETF. Unique Local IPv6 Unicast Addresses, RFC 4193 [Internet]. [cited 2005 Oct]. Available from: https://www.rfc-editor.org/rfc/rfc4193.

31 Wikipedia. Universally Unique Identifier [Internet]. [cited 2023 Feb 22]. Available from: https://en.wikipedia.org/wiki/Universally_unique_identifier.

32 Thomson S., Narten T., and Jinmei T. IPv6 Stateless Address Autoconfiguration, RFC 4862 [Internet]. [cited 2023 Feb 21]. Available from: https://www.rfc-editor.org/rfc/rfc4862.

33 Jeong J., Park S., Beloeil L., and Madanapalli S. IPv6 Router Advertisement Options for DNS Configuration, RFC 8106 [Internet]. [cited 2023 Feb 21]. Available from: https://www.rfc-editor.org/rfc/rfc8106.

34 Montenegro G., Kushalnagar N., Hui J., and Culler D. Transmission of IPv6 Packets over IEEE 802.15.4 Networks, RFC 4944 [Internet]. [cited 2022 Feb 17]. Available from: https://www.rfc-editor.org/rfc/rfc4944.

35 Shelby Z., Chakrabarti S., Nordmark E., and Borman C. Neighbor Discovery Optimization for IPv6 over Low-Power Wireless Personal Area Networks (6LoWPANs), RFC 6775 [Internet]. [cited 2022 Feb 17]. Available from: https://www.rfc-editor.org/rfc/rfc6775.

36 Hui J. and Thubert P. Compression Format for IPv6 Datagrams over IEEE 802.15.4-Based Networks, RFC 6282 [Internet]. [cited 2022 Feb 17]. Available from: https://www.rfc-editor.org/rfc/rfc6282.

37 Watteyne T. and Thurbet P. On Forwarding 6LoWPAN Fragments over a Multi-Hop IPv6 Network, RFC 8930 [Internet]. [cited 2023 Feb 27]. Available from: https://datatracker.ietf.org/doc/html/rfc8930.

38 Thubert P. IPv6 over Low-Power Wireless Personal Area Network (6LoWPAN) Selective Fragment Recovery, RFC 8931 [Internet]. [cited 2022 Feb 17]. Available from: https://datatracker.ietf.org/doc/html/rfc8931.

39 Nieminen J., Savolainen T., Isomaki M., Patil B., Shelby Z., and Gomez C. IPv6 over BLUETOOTH(R) Low Energy, RFC 7668 [Internet]. [cited 2022 Feb 19]. Available from: https://www.rfc-editor.org/rfc/rfc7668.html.

40 Shelby Z. and Borman C. 6LoWPAN: The Wireless Embedded Internet. Chichester, West Sussex: John Wiley & Sons Ltd; 2009.

41 Winter T., Thubert P., Brandt A, et al. RPL: IPv6 Routing Protocol for Low-Power and Lossy Networks, RFC 6550 [Internet]. [cited 2022 Feb 27]. Available from: https://www.rfc-editor.org/rfc/rfc6550.

42 Vasseur J.P., Ed., Kim M., Ed., Pister K., Ed., Dejean, N., Ed. et al. Routing Metrics Used for Path Calculation in Low-Power and Lossy Networks, RFC 6551 [Internet]. [cited 2023 Mar 4]. Available from: https://www.rfc-editor.org/rfc/rfc6551.

43 Thubert P. Objective Function Zero for the Routing Protocol for Low-Power and Lossy Networks (RPL), RFC 6552 [Internet]. [cited 2023 Mar 2]. Available from: https://www.rfc-editor.org/rfc/rfc6552.

44 Gnawali O. and Levis P. The Minimum Rank with Hysteresis Objective Function, RFC 6719 [Internet]. [cited 2023 Mar 2]. Available from: https://www.rfc-editor.org/rfc/rfc6719.

45 IANA. Service Name and Transport Protocol Port Number Registry [Internet]. [cited 2023 Mar 12]. Available from: https://www.iana.org/assignments/service-names-port-numbers/service-names-port-numbers.xhtml.

46 Postel J. User Datagram Protocol, RFC 768 [Internet]. [cited 2023 Mar 12]. Available from: https://www.rfc-editor.org/rfc/rfc768.html.

47 Braden R., Borman D., and Partridge C. Computing the Internet Checksum, RFC 1071 [Internet]. [cited 2023 Mar 10]. Available from: https://datatracker.ietf.org/doc/html/rfc1071.

48 Wikipedia. User Datagram Protocol [Internet]. [cited 2023 Mar 10]. Available from: https://en.wikipedia.org/wiki/User_Datagram_Protocol.

49 Iyengar J., Ed. and Thompson M., Ed. QUIC: A UDP-Based Multiplexed and Secure Transport [Internet]. [cited 2023 Mar 18]. Available from: https://datatracker.ietf.o/doc/html/rfc9000.

50 Information Sciences Institute, University of Southern California. Transmission Control Protocol, RFC 793 [Internet]. [cited 2023 Mar 12]. Available from: https://www.rfc-editor.org/rfc/rfc793.html.

51 Eddy W., Ed. Transmission Control Protocol (TCP), RFC 9293 [Internet]. [cited 2023 Mar 12]. Available from: https://www.rfc-editor.org/rfc/rfc9293.html.

52 Wikipedia. Transmission Control Protocol [Internet]. [cited 2023 Mar 12]. Available from: https://en.wikipedia.org/wiki/Transmission_Control_Protocol.

53 Allman M., Paxson V., and Blanton E. TCP Congestion Control, RFC 5681 [Internet]. [cited 2023 Mar 13]. Available from: https://www.rfc-editor.org/rfc/rfc5681.html.

54 Postel J. and Reynolds J. File Transfer Protocol, RFC 959 [Internet]. [cited 2023 Mar 17]. Available from: https://www.rfc-editor.org/rfc/rfc959.html.

55 Ylonen T., Ed. and Lonvick C., Ed. The Secure Shell (SSH) Protocol Architecture, RFC 4251 [Internet]. [cited 2022 Mar 17]. Available from: https://www.rfc-editor.org/rfc/rfc4251.html.

56 ASHRAE. BACnet: A Data Communication Protocol for Building Automation and Control Networks. American Society of Heating Refrigerating and Air-Conditioning Engineers, ANSI/ASHRAE Standard 135-2020; 2020.

57 W3C. HTML 5 [Internet]. [cited 2022 Mar 17]. Available from: https://html.spec.whatwg.org.

58 OpenAPI Initiative. OpenAPI Specification v3.1.0 [Internet]. [cited 2023 Nov 10]. Available from: https://spec.openapis.org/oas/v3.1.0.

59 Fielding R., Ed., Nottingham M., Ed., and Reschke J., Ed. HTTP Semantics, RFC 9110 [Internet]. [cited 2023 Mar 17]. Available from: https://www.rfc-editor.org/rfc/rfc9110.html.

60 Wikipedia. Domain Name System [Internet]. [cited 2023 Mar 17]. Available from: https://en.wikipedia.org/wiki/Domain_Name_System.

61 Droms R. Dynamic Host Configuration Protocol, RFC 2131 [Internet]. [cited 2023 Mar 17]. Available from: https://www.rfc-editor.org/rfc/rfc2131.html.

62 Mockapetris P. Domain Names – Implementation and Specification, RFC 1035 [Internet]. [cited 2023 Mar 17]. Available from: https://datatracker.ietf.org/doc/html/rfc1035.

63 Educause. Educause Homepage [Internet]. [cited 2023 Mar 19]. Available from: https://www.educause.edu.

64 Lewis E., Ed. and Hoenes A., Ed. DNS Zone Transfer Protocol (AXFR), RFC 5939 [Internet]. [cited 2023 Mar 21]. Available from: https://www.rfc-editor.org/rfc/rfc5936.

65 Ohta M. Incremental Zone Transfer in DNS, RFC 1995 [Internet]. [cited 2023 Mar 21]. Available from: https://www.rfc-editor.org/rfc/rfc1995.

66 Cheshire S. and Krochmal, M. DNS-Based Service Discovery, RFC 6763 [Internet]. [cited 2024 Jan 16]. Available from: https://datatracker.ietf.org/doc/html/rfc6763.

67 Eidnes H., de Groot G., and Vixie P. Classless IN-ADDR.ARPA delegation, RFC 2317 [Internet]. [cited 2023 Mar 22]. Available from: https://www.ietf.org/rfc/rfc2317.html.

68 Vixie P. A Mechanism for Prompt Notification of Zone Changes, RFC 1996 (DNS NOTIFY) [Internet]. [cited 2023 Mar 22]. Available from: https://www.rfc-editor.org/rfc/rfc1996.

69 Vixie P., Ed., Thomson S., Ed., Rekhter Y., Ed., and Bound J., Ed. Dynamic Updates in the Domain Name System (DNS UPDATE), RFC 2136 [Internet]. [cited 2023 Mar 22]. Available from: https://www.rfc-editor.org/rfc/rfc2136.

70 Croft B. and Gilmore J. Bootstrap Protocol (BOOTP), RFC 951 [Internet]. [cited 2023 Mar 23]. Available from: https://www.rfc-editor.org/rfc/rfc951.html.

71 Wikipedia. Dynamic Host Configuration Protocol [Internet]. [cited 2023 Mar 22]. Available from: https://en.wikipedia.org/wiki/Dynamic_Host_Configuration_Protocol.

72 Alexander S. and Droms R. DHCP Options and BOOTP Vendor Extensions, RFC 2132 [Internet]. [cited 2023 Mar 23]. Available from: https://www.rfc-editor.org/rfc/rfc2132.html.

73 Wireshark Foundation. Wireshark: The World's Most Popular Network Protocol Analyzer [Internet]. Available from: www.wireshark.org (Accessed 2024 Jan 13).

74 Damas J., Graff M., and Vixie P. Extension Mechanisms for DNS (EDNS(0)), RFC 6891 [Internet]. [cited 2023 Mar 25]. Available from: https://www.rfc-editor.org/rfc/rfc6891.html.

5

Link Layer and IP Stack Protocol Security

The security services discussed in Chapter 2 are implemented in network protocols using cryptographic operations based on security technology algorithms. This chapter presents security protocols for the link layers discussed in Chapter 3 and IP stack protocol network and transport layers discussed in Chapter 4. While some older application layer protocols also implement their own security services, most modern application protocols depend on the link layer, network layer, and transport layer for their basic communication security services and only implement authentication and access management (IAM) since including security in a protocol tends to introduce additional complexity and complicate deployment. Modern link layer, network layer, and transport layer security protocols are sufficiently robust, straightforward to deploy, and widely available to meet the basic communication security requirements of application layer protocols. However, applications that are built on top of application protocols need to implement security services for those parts of the application that don't involve communication. An example is a Web dashboard for a BEMS that accesses a database. In a zero-trust architecture, the Web dashboard must control access through user and service IAM, and the database records need to be encrypted.

Security for the link layer protocols presented in Chapter 3 is discussed in Sections 5.2 through 5.6. After a short introduction to link layer security protocols in Section 5.2, the EAP and RADIUS protocols that provide the foundation for link layer identity authentication in Ethernet and identity authentication and key exchange for confidentiality protection in Wi-Fi for enterprise networks are presented in Section 5.3. Ethernet security protocols are discussed in Section 5.4. Most deployments of Ethernet don't need confidentiality protection since physical access to the building is required to compromise a building LAN. A more important problem for Ethernet is authenticating access to specific VLANs.

Automating Building Energy Management for Accelerated Building Decarbonization: System Architecture and the Network Layer, First Edition. James Kempf.

Published 2025 by John Wiley & Sons, Inc.

With Wi-Fi, however, the situation is different. Because Wi-Fi is wireless, connectivity to building networks may be possible for a short distance outside the building. Wi-Fi security protocols are discussed in Section 5.5. The 802.15.4 and BLE protocols are in a similar situation, and in addition, they need to ensure that a node advertising itself as a mesh router in 802.15.4 or a Central in BLE is authorized. Security protocols for 802.15.4 and BLE are discussed in Section 5.6.

IP stack security is founded on public key signatures for integrity protection and non-repudiation and shared key technology for confidentiality protection, with key exchange providing the glue between the two. But as discussed in Chapter 2, without further measures, a digital signature cannot provide identity authentication. A public key and digital signature don't say anything about the identity of the node that generated the signature. To trust the sending node, the recipient needs some kind of trustable identity verification. Public key certificates and the public key infrastructure (PKI) define mechanisms widely used in network protocol security standards for verifying node identity and are discussed in Section 5.7.

Security for the IP stack is discussed in Sections 5.8 through 5.11. Section 5.8 introduces IP stack security. Sections 5.9 and 5.10 discuss the Internet Key Exchange (IKE) protocol and the IP security (IPSec) protocols, respectively, which are the primary protocols implementing IP network layer security. Section 5.11 presents TLS which implements transport layer security. The mechanisms in these sections provide the foundation for securing traffic over the Internet. The network and transport layer protocols serve as components in the design of VPNs, a technology for allowing private communication between networks at different sites, and between a node on the Internet and a secure site like an enterprise network. Section 5.12 discusses VPNs as an application of the network and transport layer security protocols, including three open source toolkits for implementing VPNs. Finally, Section 5.13 summarizes the chapter.

5.1 Threats to Link Layer and IP Stack Protocols

Network protocols constitute a threat boundary for BEMS services since they are used to communicate with outside services and between services within the BEMS. Two basic threats to network protocol security are eavesdropping and unauthorized message modification. Eavesdropping occurs when an unauthorized party is able to overhear and record protocol traffic between two nodes sent as clear text. Unauthorized message modification occurs when a message is modified in transit by an unauthorized party. Confidentiality protection and integrity protection, implemented with encryption and data origin authentication, respectively, are the two fundamental security services required for addressing these threats. Encryption protects the content of protocol traffic between two nodes by scrambling the clear text in messages so that nobody but a recipient having a security association with the sender can decrypt the message. Data origin authentication protects against unauthorized message modification by placing a cryptographically verifiable field in the message which the receiver can use to verify that the message was unaltered.

Unauthorized message modification occurs when an unauthorized party inserts itself between the sender and the receiver, modifies the message, and spoofs the sender's addresses, an attack known as a man-in-the-middle attack. Encryption alone cannot protect against man-in-the-middle attacks, since the attacker can simply replace the encrypted body of the message with its own content, either encrypted or not. A cryptographically generated message identity code is required at minimum to ensure that the contents of the message were not modified in transit. But data origin authentication does not provide non-repudiation nor identity authentication of the message sender. Non-repudiation requires a digital signature and identity authentication requires some form of cryptographically verifiable identity from the sender prior to any protocol communication.

Eavesdropping and man-in-the-middle attacks can also lead to a replay attack. A replay attack occurs when an attacker collects legitimate packets and replays them, in order to gain the information needed to compromise the security of the protocol or otherwise compromise the functioning of the service. Features of the protocol security such as per frame or per packet counters that uniquely identify a message can help mitigate replay attacks. They allow a receiving node to determine whether it has already seen a packet or frame, and drop it if so.

Denial of service attacks occur when an attacker compromises multiple servers and recruits them as bots to direct a massive amount of traffic at an uncompromised network service in order to prevent legitimate users from accessing it. There are some measures protocol security designs can incorporate to mitigate denial of service attacks, such as protecting protocol operations that require extensive processing. But at the level of the BEMS service, denial of service protection requires system modifications such as detection of attack traffic and configuration of routers upstream from the BEMS to drop attack traffic when it is identified. Such measures are needed whenever a BEMS service is deployed in a public or private cloud and are not part of the network protocol security design.

5.2 Link Layer Security Protocols

The need for specific security services in building link layer deployments depends on the characteristics of the physical layer and the probability that an attacker can gain access to the physical layer. For example, in building networks, the primary security services needed for wired Ethernet are identity authentication and access control to ensure that only authorized users and devices can connect to the network, and these are provided as part of the 802.1X protocol [1]. But Ethernet has a minimal need for confidentiality protection in general, since it runs over twisted pair or fiber optic cable. In order to eavesdrop on a conversation at the link level, an attacker would need to gain access to the physical link over which the communication is occurring, and physical access to a building can be controlled by other means (key cards, etc.). In addition, through the deployment of encryption at the network or transport layer, an application can ensure that the communication remains confidential over any link between the two endpoints, not just over the building LAN. There are some Ethernet links that may need confidentiality protection, for example, in utility SCADA applications or long-haul WAN links that traverse areas where the service provider has minimal control of physical access, and the IEEE has developed the 802.1AE standard [2] (also known as MACsec) to address these needs, but it is not widely deployed in building LANs. Likewise, IEEE 802.1AR [3] defines a device identity standard for Ethernet-connected devices, but it is also not widely deployed because device identity is typically managed in other ways.

In contrast, because Wi-Fi, WPAN, and BLE run over wireless links, confidentiality protection is critical for their security, since wireless reception is typically possible outside the building for a short distance where an eavesdropper can overhear them. Security is not enabled by default for these protocols, so they require configuration by an end user or network administrator to enable confidentiality protection, identity authentication, access control, and integrity protection.

While both the LTE protocols and LoRa also run over wireless links, the end device and network usually require minimal to no configuration, because the end device cannot access the network without having security enabled. The LTE IoT protocols are only available as a managed service offered by network providers. The network interface cards and modems needed to access LTE networks don't require much, if any, on-site configuration. The most amount of configuration required is typically to insert a SIM card into the modem and sometimes the modem even comes from the service provider preconfigured with identity management and keys. With LoRa, devices come with a root key preconfigured by the manufacturer. While it is possible to set up a private LoRa network that may require a network administrator to configure the network server with the device keys, LoRa is more commonly also purchased as a managed service from a service provider. As with LTE, the service provider handles link layer security. More details on security for LoRa and the LTE IoT protocols can be found in the websites of the LoRa Alliance [4] and the GSM Operator's Association [5].

5.3 EAP and RADIUS Protocols

Both the Ethernet and Wi-Fi protocols, 802.1/3 and 802.11, use a common protocol, the Extensible Authentication Protocol (EAP) for identity authentication and access management, with the base protocol specified in RFC 3748 [6]. In the case of Wi-Fi, EAP is also used to set up the security association for confidentiality protection. This protocol was originally developed for the Point-to-Point Protocol (PPP), a link layer protocol that was widely deployed in the 1990s for dial-up Internet connections. Today, EAP is deployed for link layer authentication in many types of networks, and there are a total of 65 RFCs for EAP dealing with different authentication methods, applications of EAP to other use cases and link layer types, historical drafts that are now superseded, etc. EAP requires a backend server securely holding the identity records and credentials for users and devices and a protocol to connect the authentication gatekeeper on the link layer network with the server. Although there are a number of possible choices for backend protocols, EAP is often paired with a server supporting the Remote Authentication Dial-in User Service (RADIUS), specified in RFC 2865 [7]. Ethernet and Wi-Fi use EAP in different ways, but the basic protocol is the same. This section discusses the basic EAP and RADIUS protocols, while protocol features specific to Ethernet and Wi-Fi are discussed in the Sections 5.4 and 5.5, respectively.

5.3.1 EAP Network Architecture

Figure 5.1 illustrates the generic EAP network reference architecture. An Unauthenticated Device seeks authorization to enter a network controlled by the Network Access Server. The Unauthenticated Device exchanges EAP payloads with the Authentication Server through a link layer tunnel with the Network Access Server acting as an intermediary. Note that the link layer protocol for tunneling is a client–server (request–response) type protocol but, despite its name, the Network Access Server and Authentication Server act as the client and the Unauthenticated Device acts as the server. The Unauthenticated Device can't send

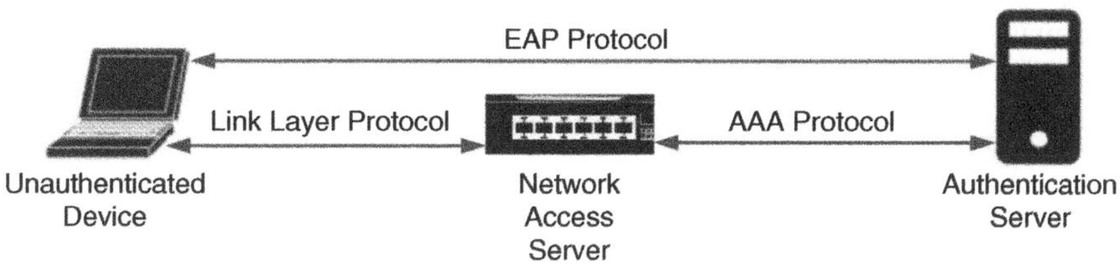

Figure 5.1 Generic EAP Network Reference Architecture

any IP traffic over the link layer until the authentication exchange is successfully completed. The Network Access Server sends the EAP payloads to the Authentication Server using an Authentication, Authorization, and Accounting (AAA) protocol, but the EAP authentication session is end to end between the Unauthenticated Device and Authentication Server. The Authentication Server requests security credentials from the Unauthenticated Device, and if the credentials are valid, it authorizes the Network Access Server to admit the Unauthenticated Device which then is authenticated.

5.3.2 EAP Protocol

EAP provides a standardized interface for various authentication methods. The main advantage of EAP is that it is extensible and can support various authentication methods and device credentials like passwords, token cards, smart cards, and certificates. However, the base EAP protocol doesn't have any mechanism for confidentiality protection, integrity protection, or non-repudiation so it must be run inside of a secure tunnel if these security services are required. Extensions for securing EAP are described in other IETF RFCs, Section 5.11 describes the TLS protocol framework which is the basis of EAP-TLS. TLS provides a secure tunnel for EAP-TLS.

5.3.2.1 EAP Payload Format

Figure 5.2 illustrates the EAP payload format. EAP payloads contain the following fields:

- Code – the type of the EAP message. For basic authentication, the Code can be one of the four values shown in Table 5.1.
- Identifier – a unique number sent in an *EAP-Request* and returned in an *EAP-Response*, and used to match requests with responses.
- Length – the length of the EAP payload including the Code, Identifier, Length, Type, and Type-Data fields. The minimum length is 4 if no Type is present.
- Type – only in *EAP-Request* and *EAP-Response* payloads. It specifies the type of the request or response. Type values are used to identify the Unauthenticated Device and to select a specific authentication mechanism.
- Type-Data – only in *EAP-Request* and *EAP-Response* packets. Its content and format depend on the Type.

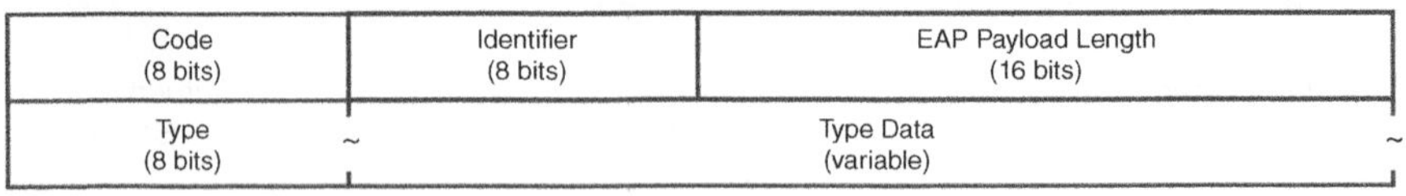

Figure 5.2 EAP Payload Format

Table 5.1 EAP Message Codes

Message	Code	Function
EAP-Request	1	Request from the Network Access Server or Authentication Server for credentials.
EAP-Response	2	Response from the Unauthenticated Device with credentials.
EAP-Success	3	Response from the Authentication Server that the authentication was successful. The Unauthenticated device is now authorized to use the network.
EAP-Failure	4	Response from the Authentication Server that the credentials were not authenticated. The Unauthenticated device is not authorized to use the network.

In addition to the message codes for basic authentication messages shown in Table 5.1, a few additional codes have been added to handle complex reauthentication cases, see RFC 6696 [8] for a description. EAP Code and Type field values are standardized through IANA [9].

The Type and Type-Data fields constitute the message-specific content of the EAP payload. The first EAP message sent from the Network Access Server to the Unauthenticated Device is *EAP-Request/Identity* (type code 1), with an empty Type-Data field. The Unauthenticated Device replies with an *EAP-Response/Identity* message, setting the Type-Data field to a string identifying the user or device. The string is typically a network access identifier having the format "user@domain" where "user" is the username and "domain" is the domain name of the network or service the user is trying to access. But the value in the *EAP-Response/Identity* message Type-Data field can vary depending on the network. For privacy reasons, a device can use an anonymous identity such as "anonymous@domain" in the *EAP-Response/Identity* message if the network supports an EAP method with a protected identity exchange, like EAP-TTLS, since the true identity of the user is revealed when the encrypted tunnel established by the EAP method is set up. In certain Wi-Fi networks, the device's MAC address can be used as the identity. Finally, the identity can also be an alphanumeric string that is not in the form of a network access identifier.

The other common EAP message exchange between the Authentication Server and Unauthenticated Device is *EAP-Request/Authentication* or *EAP-Challenge/Authentication* and *EAP-Response/Authentication*. Of the 56 EAP methods with Type codes listed in the IANA registry, EAP-MD5 (type code 4) is one of three authentication methods specified in the base EAP RFC. EAP-MD5 specifies an authentication method for the Unauthenticated Device, constructing a one-way authenticator by combining a username and a password and then hashing with the MD5 hashing algorithm. It is not considered secure for wireless networks because it does not support key generation or mutual authentication, and in addition, MD5 is considered obsolete due to weak collision resistance and other problems. EAP-TLS (type code 13) is specified in RFC 5216 [10, 11], and it uses the general TLS framework to offer strong security with mutual authentication, key generation, and encryption, but it requires certificates to be installed on the end devices as well as on the Authentication Server. EAP-TTLS (type code 21) is an extension of EAP-TLS that allows the end devices to authenticate using a username and password but only requires the Authentication Server to have a certificate, like the protocol used by Web browsers to authenticate Web servers over the Internet. The initial handshake creates a secure tunnel, and then, the end node credentials are sent inside this tunnel. EAP-TTLS is specified in RFC 5281 [12].

5.3.3 RADIUS Protocol

The RADIUS protocol is used to communicate between a Network Access Server and the Authentication Server. RADIUS runs as an application layer protocol over an IP network through UDP transport at ports 1645 or 1812 for authentication, with the latter more commonly used. Figure 5.3 illustrates the RADIUS application protocol header. The header includes three fields:

- Code – the type of RADIUS message,
- Identifier – a numeric identifier code used to match requests with responses and to avoid responding to duplicate requests received due to packet duplication or retransmission,
- Length – the length of the application layer packet including the RADIUS header, in bytes,

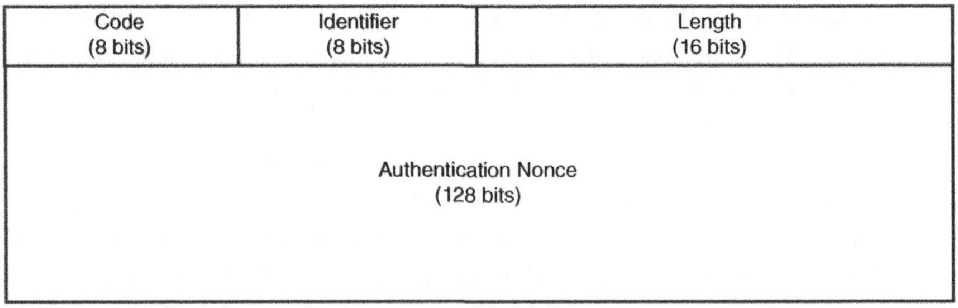

Figure 5.3 RADIUS Header

Table 5.2 Radius Message Type Codes for Use with EAP

Message	Code	Function
Access-Request	1	Request to the RADIUS server for the Unauthenticated Device to access the network based on the contained credentials.
Access-Accept	2	Response from the RADIUS server granting access.
Access-Reject	3	Response from the RADIUS server denying access.
Access-Challenge	11	Response from the RADIUS server containing a challenge. The Network Access Server should present the challenge to the Unauthenticated Device and return the reply in a new *Access-Request*.

- Authentication Nonce – a nonce used by the Network Access Server to verify the response from the Client, see below for more details.[1]

RADIUS has a total of 37 message type codes standardized by IANA [13]. Of these, only a few are relevant for basic EAP authentication and authorization. They are shown in Table 5.2, together with their functions.

The Authentication Nonce has a different value depending on the RADIUS message within which it is embedded, and generation of the nonce assumes that the Network Access Server, acting as a client in the RADIUS client–server protocol, shares a secret such as a password with the Authentication Server, acting as the server. In the *Access-Request* message, the Authentication Nonce is a randomly generated value that must be statistically unique and should not be reused. In the *Access-Accept*, *Access-Reject*, and *Access-Challenge* messages, the Authentication Nonce is calculated by concatenating the header Code, Identifier, and Length fields from the reply, the Authentication Nonce field from the request, the attribute-value pairs from the reply, and the shared secret then taking the MD5 hash of the result. The 16 bytes of the MD5 hash are then inserted into the Authentication Nonce field. Because the MD5 hashing algorithm is considered obsolete, the security of the RADIUS exchange should not depend on the Authentication Nonce field. RADIUS should always be deployed with UDP over Datagram Transport Layer Security (DTLS) or with a VPN to ensure confidentiality, integrity, and non-repudiation. TLS, upon which DTLS is based, is discussed in Section 5.11, and VPNs are discussed in Section 5.12.

1 Note that RFC 2865 calls this field the "Authenticator" but it has been renamed here to avoid confusion with the Network Access Server for 802.1X which is also called the Authenticator.

The header is followed by a variable length payload containing a collection of attribute-value pairs containing the requests from the Network Access Server or Authentication Server or responses from the Unauthenticated Device. The attribute-value pairs contain the authentication exchange information. RFC 2865 specifies a collection of attribute-value pairs as part of the base RADIUS protocol, but most use cases define attribute-value pairs in a separate RFC. Attribute-value pairs for EAP are specified in RFC 3579 [14]. Each attribute-value pair consists of a one-byte attribute type code field, a 1-byte value length field, and a variable length value field containing the value in zero or more bytes, up to the length specified in the length field. The attribute type codes are also standardized by IANA [15].

The RADIUS messages for EAP have two common attribute-value pair types that all messages must carry: the EAP-Message and Message-Authenticator attribute-value pairs. The EAP-Message (attribute type 79) attribute value contains the EAP payload. If multiple EAP-Message attributes are included in a single RADIUS packet, they must be concatenated in order to obtain the complete EAP message. The Message-Authenticator (attribute type 80) attribute value contains a Message Authentication Code (MAC) calculated over the entire RADIUS message, with the value of the Message-Authenticator attribute value itself set to zero for the calculation. The bytes are input into the HMAC-MD5 keyed MAC algorithm together with the shared secret, which should be different from the shared secret used to calculate the Authentication Nonce on the packet header. HMAC is a keyed MAC algorithm described in RFC 2104 [16] that can be used with any hashing algorithm. The security of HMAC combined with MD5 is considered stronger than MD5 alone because HMAC does not rely on the collision resistance of the underlying hash function. The 16 bytes of the hash are inserted into the Message-Authenticator attribute value field.

5.3.4 Generic EAP/RADIUS Authentication Protocol

A generic EAP/RADIUS authentication exchange that results in a successful authentication is shown in Figure 5.4. The exchange between the Unauthenticated Device and the Network Access Server typically runs in a link layer-specific tunnel, while the exchange between the Network Access Server and the Authentication Server runs over an IP network internal to the organization providing the link layer access. The link layer-specific tunneling protocol may have a message for the Unauthenticated Device to initiate the authentication protocol which it sends to the Network Access Server prior to starting the EAP exchange, or the Network Access Server may start the exchange as shown in Figure 5.4. The figure also assumes the Authentication Server is running RADIUS. Here are the steps in the process:

- The Network Access Server sends an *EAP-Request/Identity* message to the Unauthenticated Device.
- The Unauthenticated Device looks up its identity and sends an *EAP-Response/Identity* message to the Network Access Server with its identity, or if the network

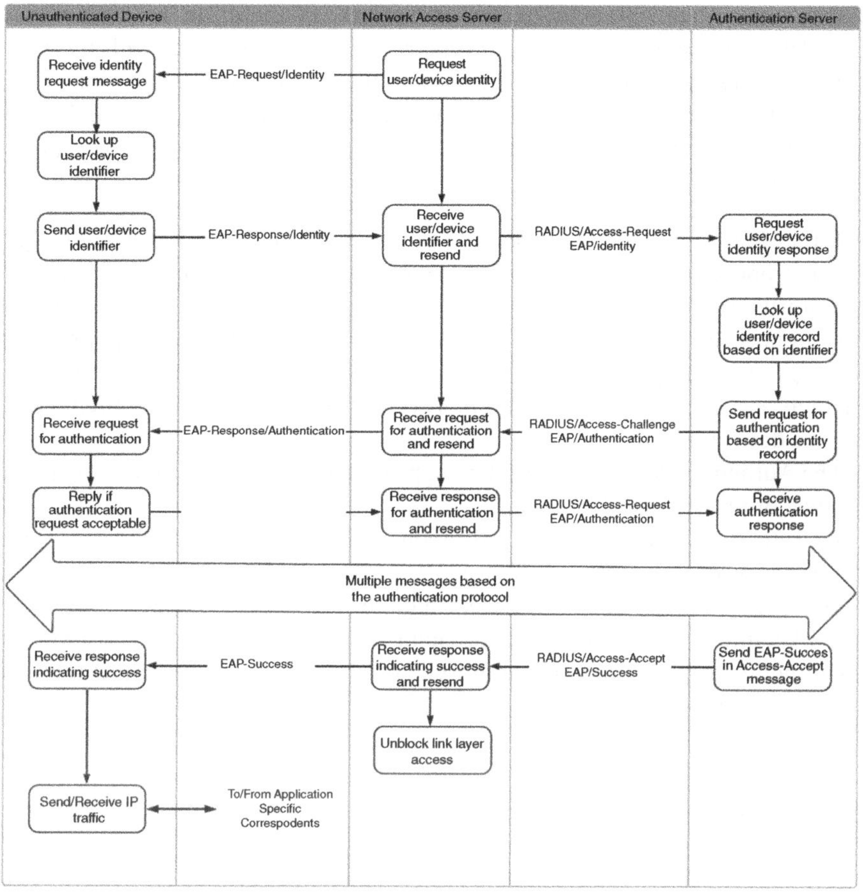

Figure 5.4 Generic EAP/RADIUS Authentication Protocol

uses an EAP method that establishes an encrypted tunnel with the Authentication Server, an anonymous identity.

- The Network Access Server forwards the identity to the Authentication Server in a RADIUS *Access-Request* message.
- The Authentication Server responds with a RADIUS *Access-Challenge* message including an EAP message proposing an authentication method.
- The Network Access Server forwards the message to the Unauthenticated Device. If the Unauthenticated Device is prepared to use the proposed authentication method, the Unauthenticated Device and the Authentication Server then engage in a series of *EAP-Request* and *EAP-Response* messages based on the authentication protocol, with the Network Access Server as the intermediary. The Network Access Server encapsulates the EAP messages in RADIUS *Access-Request* and *Access-Challenge* messages until authentication is complete.

- Once the authentication process is complete, the Authentication Server sends a RADIUS *Access-Accept* message to the Network Access Server with the *EAP-Success* message, which then forwards it to the Unauthenticated Device.
- With the successful authentication, the Unauthenticated Device is now authorized and can exchange application-specific traffic with correspondents over the link.

If the Unauthenticated Device is not prepared to use the authentication algorithm proposed by the Authentication Server, the two sides undergo a negotiation until a mutually acceptable method is agreed upon. If the supplicant does not support the proposed method, it will respond with an *EAP-Response/NAK* message type (Negative Acknowledgment, type code 3 or 254), which can include a suggestion for a different EAP method that the Unauthenticated Device does support.

The Authentication Server can then approve the proposal to use the suggested method or propose a different one.

If a method cannot be agreed upon, the authentication fails and the Authentication Server sends an *Access-Reject* message to the Network Access Server with the *EAP-Failure* message, which the Network Access Server forwards to the Unauthenticated Device. The exchange can also fail if the credentials of the Unauthenticated Device are not approved or if the exchange takes longer than the timeout set on the Network Access Server.

5.4 Ethernet Security

IEEE 802.1X-2020 [1] is the standard protocol for authenticating access to specific Ethernet switch ports. 802.1X defines a link layer tunnel through which end devices can prove identity and obtain access to controlled Ethernet ports, or port-based access control. While 802.1X is currently deployed primarily in service provider networks and some enterprise networks, its adoption in building automation networks has been more limited. However, as the convergence of IT and OT systems continues to grow, there is an increasing need for implementing higher levels of security for building automation networks. In some cases, organizations may choose to deploy 802.1X to provide secure access control for users and devices connected to their building automation systems, particularly if these systems share physical infrastructure with the organization's primary IT network.

The 802.1X-2020 standard includes specifications for two different protocols: the Port Access Control Protocol (PACP), and the MACsec Key Agreement (MKA) protocol. MKA is used to set up security associations for the IEEE 802.1AE link encryption protocol. In this section, only PACP is discussed.

MKA is primarily oriented toward providing confidentiality protection on a single link or a collection of links in the same administrative domain. As discussed, where physical access to the building can be controlled, the risk of an attacker eavesdropping on a wired connection is minimal. Confidentiality protection in building automation systems is, therefore, more useful if it is deployed at the IP and transport layers, since it will additionally protect traffic over links that are not within the building administration domain. This includes WAN links and links in cloud networks where the BEMS central components reside. However, MKA can be deployed for additional link layer security in areas where physical access to the building is harder to control, for example, in municipal buildings such as schools.

5.4.1 802.1X Network Architecture

The 802.1X network reference architecture is shown in Figure 5.5. 802.1X calls the Unauthenticated Device from the generic EAP reference architecture the Supplicant and the Network Access Server the Authenticator. In contemporary 802.1X deployments, the Authentication Server often doesn't hold the client records directly, but rather backends into an IAM service, either in a corporate network or a cloud service, where the actual records are held. The Authentication Server terminates the EAP session and communicates with the backend IAM service using the protocol or API supported by the service.

The Supplicant and the Authenticator exchange Extensible Authentication Protocol over LAN (EAPOL) frames, an Ethernet protocol for tunneling EAP payloads between the Supplicant and the Authenticator. The Authenticator then transfers the EAP payloads into the AAA protocol packet which it sends over a secure IP network to the Authentication Server. A similar process occurs in the reverse direction, from the Authentication Server to the Authenticator, and finally to the Supplicant. One of two AAA protocols is typically deployed with 802.1X: the RADIUS protocol or the Diameter protocol. Diameter is a considerably more complex protocol that is commonly used for access control and accounting in 3G and 4G mobile networks and is specified in RFC 6733 [17].

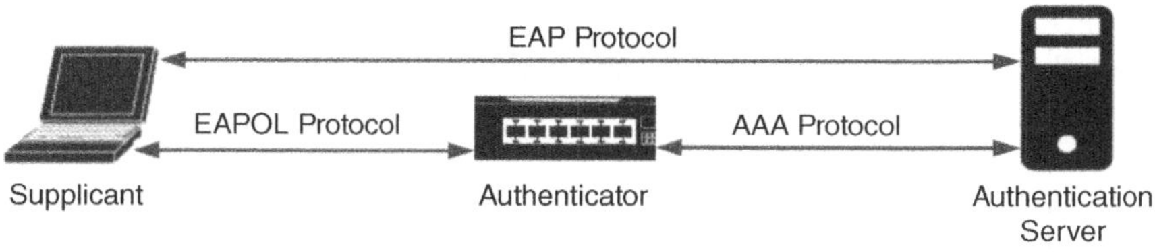

Figure 5.5 802.1X Network Reference Architecture

In the 802.1X PACP model access control is based on a controlled port. A port is an abstraction of the link layer connection between the two entities communicating over an Ethernet link. To send traffic on a controlled port, the Supplicant must authenticate with the Authentication Server through the Authenticator. The Supplicant and the Authenticator each contain a functional entity called the Port Access Controller (PAC) that controls access to a common port from an uncontrolled (unauthenticated) port and a controlled (authenticated) port. The Supplicant's IP layer only has access to the uncontrolled port until it is authenticated. The Supplicant and Authenticator can only exchange EAPOL frames across the uncontrolled port, all other frames are dropped by the PAC. Once the Supplicant has been authenticated, the Authorization Server tells the PAC on the Authenticator to open the controlled port and the Supplicant can send IP traffic. Clause 6.4 of the standard [18] contains a formal model of the PAC.

5.4.2 EAPOL Protocol

The EAPOL protocol is specified in Clause 11 of IEEE 802.1X-2020 [1] and has Ethernet frame type code 0x888e. The EAPOL application protocol header is shown in Figure 5.6, and the fields are the following:

- Protocol Version – The version number for the 802.1X protocol. For 802.1X-2020, the version number is 0x3,
- EAPOL Message Type – The type of the EAPOL message, see Section 5.4.2.1,
- EAPOL Payload Length – The length of the EAPOL payload in bytes, with a value of 0 indicating that no payload is present. The calculated Payload Length field value does not include the header.

The EAPOL application protocol payload, which is typically an EAP message for the *EAPOL-EAP* message type, follows in the frame.

5.4.2.1 EAPOL Message Types for PACP

The EAPOL protocol has a total of eight different message types, but only three are relevant for the PACP. The rest are involved in implementing the MKA protocol and other functions such as announcements and alerts. The message types and their type codes are shown in Table 5.3.

EAPOL-Start can have an optional payload if the least significant (rightmost) bit of the first byte in the payload is set to one, but both *EAPOL-Start* and

Protocol Version (8 bits)	EAPOL Message Type (8 bits)	EAPOL Payload Length (16 bits)

Figure 5.6 EAPOL Header

Table 5.3 EAPOL Message Types used in PACP

Message	Code	Function
EAPOL-EAP	0	Encapsulates an EAP payload.
EAPOL-Start	1	Sent from the Supplicant to the Authenticator to request the PACP to start an authentication session.
EAPOL-Logoff	2	Sent from the Supplicant to the Authenticator to request the PACP to terminate its access to the controlled port or to terminate an ongoing authentication exchange.

EAPOL-Logoff typically have no payload in an enterprise/building LAN deployment. More information on the *EAP-Start* message key-value pairs can be found in Clause 11.6 [19] of the standard.

Either the Supplicant or the Authenticator can start an authentication exchange at any time when the Supplicant is unauthenticated. If the Supplicant starts, it uses the *EAPOL-Start* message to solicit an *EAPOL-EAP* message encapsulating the *EAP-Request/Identity* EAP message from the Authenticator, to start the authentication. If the Authenticator starts, it sends an unsolicited *EAPOL-EAP/EAP-Request/Identity* message to which the Supplicant responds. If the Supplicant wishes to restart the exchange, it sends an *EAPOL-Start*, causing the PACP layer on the Authenticator to reinitialize the EAP layer and restart the authentication.

The network administrator can set a time limit on the authenticated access. If a time limit is set, the Authenticator's PACP layer asks the EAP layer when the limit times out to request reauthentication from the Supplicant. The Supplicant remains authenticated while the reauthentication is in progress. If the Supplicant wishes to release access to the controlled port, it can send an *EAPOL-Logoff*. *EAPOL-Logoff* is typically sent if the credentials used to authenticate the Supplicant are based on user identity and the user logs out of the device. The bulk of the PACP protocol is conducted by the exchange of *EAPOL-EAP* messages between the Supplicant and the Authenticator. The messages contain the EAP payloads that the Authenticator forwards between the Supplicant and the Authentication Server.

5.4.2.2 EAPOL Addressing

Messages in the EAPOL protocol are sent to different addresses depending on the message type. In building LAN deployments, the *EAPOL-EAP* message would most commonly be sent from an individual (unicast) address to an individual address; for example, from the Authenticator's link layer address to the Supplicant's link layer address. The *EAP-Start* and *EAP-Logoff* messages are sent from the Supplicant to the group (multicast) address 01-80-c2-00-00-03 which is monitored by the Authenticator. For other more complex Ethernet deployments such as in service provider networks, other addressing schemes are used, see Table 11-1 in the standard for more information [20].

5.5 Wi-Fi Security

Wi-Fi requires security services beyond those required for Ethernet. In addition to identity authentication and access control, Wi-Fi needs confidentiality protection and integrity protection to prevent unauthorized access to data transmitted over the air, since wireless transmissions can often be overheard for a short distance outside the building in which the network is deployed. The initial IEEE 802.11-1997 standard [21] specified an algorithm and protocol called Wired Equivalent Privacy (WEP) [22] for identity authentication and access control, confidentiality protection, and integrity protection. WEP was based on a static key shared between the access point and end nodes, but the algorithm was quickly found to have many fatal flaws.

In 2003, the Wi-Fi Alliance introduced Wi-Fi Protected Access (WPA) [23] as an interim solution to replace WEP. In order to maintain backward compatibility with the hardware, WPA kept the basic encryption algorithm from WEP but changed the integrity protection algorithm and replaced the static key with an algorithm called the Temporal Key Integrity Protocol (TKIP) in which a new session key was generated from the static root key on each frame transmitted between the access point and end node. After flaws were found in WPA, the Alliance introduced the substantially redesigned WPA2 protocol in 2004, and WPA2 was required for all Wi-Fi Alliance-certified devices as of 2006. WPA2 features an algorithm based on AES for confidentiality protection and integrity protection, and an extension of 802.1X for identity authentication and access control. The Alliance released an improved version of WPA2 in 2018 called WPA3, which became mandatory for all Wi-Fi Alliance-certified devices in 2020. WPA3 additionally included support for a strengthened key management algorithm to prevent password guessing and to implement forward secrecy, so that the compromise of a single session key will not affect any data other than that exchanged in the session encrypted with the compromised key.

Note that WPA, WPA2, and WPA3 are industry standards developed and promoted by the Wi-Fi Alliance rather than technical standards developed by IEEE. Copies of the standard specifications are only available to members of the alliance, but the basic protocols and algorithms required for implementing them are documented in the latest version of the 802.11 specification, IEEE 802.11-2020 [24]. However, WEP and TKIP have not been removed from the current IEEE specification, although WEP is listed as obsolete, TKIP is listed as deprecated, and both are designated as unsuitable for use. In practice, WPA3 should be used in all BEMS deployments unless the access points and devices in the network don't support it, then WPA2 should be used. If the access points and devices in the network don't support WPA2 at minimum, upgrading the network hardware or firmware to include WPA2 support is strongly encouraged.

5.5.1 The Robust Security Network Architecture

WPA2 and WPA3 define an architecture for strong security called the robust security network architecture. The robust security network architecture specifies a framework with stronger security mechanisms than previous protocols such as WEP and WPA. It was originally defined in the IEEE 802.11I amendment and is now included in the 802.11-2020 standard [24]. The architecture mandates 802.1X/EAP with an Authentication Server for identity authentication and network access control in commercial networks, an extension of 802.1X for key provisioning in all types of networks, and robust cryptosystems based on AES for encryption and integrity protection. A four-way handshake is used to derive the session keys for encryption and integrity protection on both the end node and access point, and group and management frame key provisioning is also included. In addition, hashing algorithmic support was strengthened. WPA2 specified the use of the SHA1 algorithm for hashing, WPA3 added the optional use of SHA256 algorithm since SHA1 was found vulnerable to brute force guessing attacks. Mechanisms to protect management frames and group (multicast) transmission from attack are included in the framework. The architecture also contains enhancements to reduce the amount of time needed to set up the security association when an end node moves between access points.

802.1x over Wi-Fi

The link above provides a detailed explanation of how 802.1X works over Wi-Fi. The basic operation is the same as for Ethernet but there are a few details that are slightly different, such as the four-way handshake, which is necessary for encrypting Wi-Fi traffic because wireless links are subject to snooping.

If 802.1X is not deployed with an Authentication Server, WPA2 specified the Pre-shared Key (PSK) protocol for identity authentication and deriving the master session key from a PSK that starts with a password manually entered into the access point and end devices. An algorithm is used to derive the 256-bit master session key from the password, and a pairwise temporal key is derived from the PSK, with SHA1 specified as the hashing algorithm. The pairwise temporal key is used to encrypt traffic over the air. However, the SHA1 algorithm is susceptible to brute force attacks in which the attacker can try all combinations of passwords until the correct one is found. While using a long password helps mitigate such attacks, many users tend to use simple passwords that are vulnerable. In addition, the four-way handshake for confirming the security association was found to be vulnerable to a kind of attack called the key reinstallation attack [25].

WPA3 replaced the key derivation protocol in WPA2 with a protocol called Simultaneous Authentication of Equals (SAE). The SAE protocol uses the Dragonfly key exchange algorithm, which is described in RFC 7664 [26]. Dragonfly specifies a Password-Authenticated Key Exchange (PAKE) protocol that uses the same asymmetric key cryptosystem as for the Diffie–Hellman key exchange algorithm. Diffie–Hellman is the key provisioning protocol that is part of the IKE protocol described in Section 5.9. In addition, SAE uses SHA256 for hashing instead of SHA1 for better security. SAE protects against passive attacks, active attacks, and dictionary attacks.

5.5.2 Overview of the Security Association Setup Process

In a robust security network, the process of setting up a security association goes through the following steps:

- The access point includes a Capability Information field in the *Beacon*, *Probe-Response*, and *Association-Response* Frame Payload field with the Privacy bit set to 1, indicating to the end node that confidentiality is required on the link. Clause 9.4.1.4 [27] in the standard describes the Capability Information field in detail.
- During the association message exchange between the end node and the access point, the two sides exchange information on their available authentication and key management suites and the end node selects an algorithm shared with the access point. The authentication algorithms in the standard have encryption and integrity protection algorithms associated with them.
- The end node engages in the selected authentication algorithm to verify its identity to the access point. For 802.1X/EAP, this process occurs after the association process while for PSK and SAE, identity authentication happens during the *Authentication-Request*/*Authentication-Response* exchange. 802.1X/EAP also exercises control over whether the end node is allowed to enter the network. In addition, both sides derive a master key for use in key derivation after the two sides have authenticated each other.
- The end node and the access point engage in the four-way handshake to derive a temporal key from the master key which will be used to encrypt traffic and

verify its integrity, and two other keys used in further key provisioning. This is done after the association message exchange, using an extension to 802.1X.

- The access point can optionally configure the end node with a group (multicast) message key and a key for encrypting management frames. These keys are encrypted when sent over the air by a key derived in the previous step.

In the second step, a 256-bit pairwise master key is derived which the key derivation algorithm uses in the third step to derive further keys on both the access point and the end node. The pairwise master key is derived in a different way for each of the authentication algorithms:

- 802.1X/EAP – The Authentication Server derives a master session key and the first 256 bits of the master session key are then provisioned to the end node and access point as the pairwise master key.
- PSK – The pairwise master key is generated by applying a key derivation function to a password pre-provisioned on each device and other material.
- SAE – A public key cryptographic algorithm for PAKE based on the Diffie–Hellman key exchange algorithm is used to derive the pairwise master key from a shared passphrase.

In addition, if a group key and a key for protecting management frames are required, these are derived on the access point and provisioned in an encrypted field to the end node after the four-way handshake when the end node has confirmed keys available.

5.5.3 Authentication Algorithm Selection and Parameter Configuration

The robust security network architecture requires the specification of an algorithm to use for the security services and the selection of cryptosystems and parameters for implementing these services. The end node and access point must agree on the algorithm, cryptosystems, and parameters in order to successfully authenticate and share encrypted traffic. The protocols between the access point and end node depend on the selected authentication algorithm:

- If 802.1X is deployed with an Authentication Server, basic 802.1X is used for identity authentication and access control, and the key management extension of 802.1X is used for key provisioning.
- If PSK is used, the key management extension for 802.1X is used for identity authentication and key provisioning.
- For WPA3, if SAE is used, additional *Authentication-Request/Authentication-Response* messages are exchanged between the two sides with public keys and algorithm parameters during which identity authentication is performed, and the key management extension for 802.1X is used for key provisioning.

The actual authentication algorithm determination, cryptosystems selection, and parameter configuration are accomplished through information shared in the Frame Payload fields of messages exchanged between the end node and access point during the initial association protocol which was discussed in Chapter 3. If 802.1X or PSK are used for identity authentication and pairwise master key generation, the end node requests the authentication algorithm in the *Authentication-Request* message. The access point either confirms or declines the request in the *Authentication-Response* message. The end node then requests specific cryptosystems with configuration parameters in the *Association-Request* message and the access point either confirms or declines the request in the *Association-Response* message. If SAE is used for identity authentication and pairwise master key generation, the *Beacon* and *Probe-Response* messages contain the cryptosystem and configuration parameters. The first exchange of *Authentication-Request/Authentication Response* messages confirms the selection of SAE as the authentication algorithm and provides additional configuration parameters for the public key cryptosystem.

5.5.3.1 Determining the Authentication Algorithm

For 802.1X and PSK, the end node and the access point determine the authentication algorithm to be used on the link in the *Authentication-Request/Authentication-Response* exchange of messages. The end node sends a proposed authentication algorithm code in the Frame Payload field of the *Authentication-Request* message and the access point replies either confirming or denying the requested authentication protocol in the *Authentication-Response* Frame Payload field. The format of the Frame Payload field for both messages is shown in Figure 5.7. Additional parameters in the Frame Payload differ depending on the authentication algorithm and are included in the Additional Information Elements field, which can be up to 2,304 bytes long, and includes the information elements identified by a type code. A list of additional information elements can be found in Clause 9.3.3.11 [28] of IEEE 802.11-2020. An example is the Robust Network Security Element discussed Section 5.5.3.2. For SAE, the access point confirms the use of SAE during the authentication exchange since the end node receives notification of the access point's use of SAE in the *Beacon* and *Probe-Response* messages.

The standard defines 7 code values for Authentication Algorithm Code (0-6), plus a vendor-specific value (65535), with all other values reserved. The primary codes of interest for BEMS deployments are Open System – code 0 and SAE – code 3.

Authentication Algorithm Code (16 bits)	Sequence Number (16 bits)	Status Code (16 bits)	~ Additional Information Elements (variable) ~

Figure 5.7 Frame Payload for the *Authentication-Request/Authentication-Response*

If the Open System code is used, then the selection requires further refinement since either 802.1X or PSK could be used for identity authentication, access control, and key provisioning. The SAE code confirms the use of the SAE protocol from the end node side, which the end node learns from the *Beacon* and *Probe-Response* messages.

The Sequence Number field indicates the current state of progress through a multistep transaction. This field is set to 1 for the *Authentication-Request* message and 2 for the *Authentication-Response* message in a two-message authentication exchange for 802.1X/EAP and PSK. If SAE is the selected authentication and key management algorithm, four authentication messages are required so the Sequence Number field on the first exchange is 1 on both messages and 2 on the second exchange. The Status Code field is 0 if the exchange is successful and the access point accepts the authentication algorithm, or an error code if not. The authentication algorithm and status codes are defined in Clause 9.4.1.1 [29], and their use in the Frame Payload field is defined in Clause 9.3.3.11 [28] of IEEE 802.11-2020.

5.5.3.2 Robust Security Network Element

An example of an information element used to specify further cryptosystem parameters is the Robust Security Network Element (RSNE) with element type code 48, which is specified in Clause 9.4.2.24 of 802.11-2020 [30]. Using the RSNE, the access point advertises available cryptosystems and parameters in the *Beacon* and *Probe-Response* messages and the end node requests specific cryptosystems and parameters in the *Association-Request* message for Open System authentication, or the *Authentication-Request* message, for SAE. The RSNE contains information such as the cryptosystem cipher suites for authentication, encryption, and key management that both sides must support to set up a robust network security association.

One of the fields in the RSNE contains the codes for the authentication and key management suites supported by the access point or end device. The code consists of a 3-byte organizationally unique identifier, similar to the identifier used in an Ethernet or Wi-Fi link layer address. The organizationally unique identifier allows vendors to provide custom cipher suite codes. The 3-byte code is followed by a 1-byte cipher suite code. The codes listed in the standard all have the IEEE organizationally unique identifier, 0x000fac, and the codes for algorithms discussed in this chapter that are relevant to BEMS deployments are the following:

- 802.1X with SHA1– 0x01,
- PSK with SHA1 – 0x02,
- 802.1X with SHA256 – 0x05,
- PSK with SHA256 – 0x06,
- SAE – 0x08.

A node can advertise up to four different authentication and key management suites, and Table 9-151 in 802.11-2020 [31] lists a total of 15 different codes.

5.5.4 Identity Authentication, Access Control, and Pairwise Master Key Derivation

The three algorithms for identity authentication and access control take different approaches to authenticating the end node and deriving the pairwise master key.

5.5.4.1 Identity Authentication, Access Control, and Pairwise Master Key Derivation with 802.1X/EAP

If the end node and access point use 802.1X/EAP for authentication and key provisioning with an Authentication Server deployed, the network reference architecture for Wi-Fi is exactly the same as for Ethernet as shown in Figure 5.5, with the Supplicant being the end node and the Authenticator being the access point. 802.11 uses EAP for carrying out the authentication between the Supplicant and the Authentication Server, typically with EAP-TLS [11]. The PACP protocol is used for port-based access control on both the Supplicant and the Authenticator as described in Section 5.4. Clause 4.10.3.2 [32] in 802.11-2020 describes the protocol for identity authentication and access control using 802.1X/EAP.

The access point and the end node agree on a specific version of the 802.1X algorithm during the Wi-Fi access point association exchange. The access point advertises support for 802.1X with the hashing algorithms using the codes described in Section 5.5.3.2 for a RSNE in the Frame Payload field of the *Beacon* and *Probe-Response* fields. The end node and access point exchange the Authentication Algorithm Code for Open System (code 0) in the *Authentication-Request*/*Authentication-Response* Frame Payload field shown in Figure 5.7. The end node requests 802.1X with a particular hashing algorithm in the *Association-Request* Frame Payload field by including one of the 802.1X authentication and key management suite codes from Section 5.5.3.2 in a RSNE. After the access point association exchange is complete, the two sides move to the identity authentication, access control, key generation, and key provisioning stage using 802.1X/EAP, where they agree on an EAP method. The pairwise master key is provisioned by the Authentication Server in this stage.

There are a few minor differences between 802.1X/EAP on Ethernet and Wi-Fi. As discussed in Chapter 3, Wi-Fi has its own association and disassociation protocol controlling end-node association with an access point, so it does not use the *EAPOL-Logoff* message. The access point discards any *EAPOL-Logoff* messages received from the end node. The end node also does not use a group link layer address for *EAPOL-Start*, but rather the access point's BSSID, which is its link layer address. The pairwise master key is derived from a master session key generated on the

Authentication Server and provisioned to the access point and end node in the last step of the 802.1X identity authentication and access control exchange. In addition, any EAPOL frames sent prior to the completion of key provisioning on both sides are not encrypted. The two sides may end up completing key management at slightly different time, and so, one or the other may receive an encrypted frame that it is unable to decrypt, in which case the frame is simply discarded.

5.5.4.2 Identity Authentication and Pairwise Master Key Derivation with PSK

With PSK, both the end node and the access point must be preconfigured with a password. The password is the same for every end node that authenticates with the access point. The *Authentication-Request/Authentication-Response* message for PSK has the same Frame Payload field as for 802.1X/EAP with the Open System algorithm code (code 0). Also similar to 802.1X/EAP, the end node includes an RSNE in the *Association-Request* Frame Body field, with authentication and key management suite code set to one of the two codes from Section 5.5.3.2 identifying which hashing algorithm to use. After the access point association phase is complete, the process moves on to identity authentication, key generation, and key provisioning. The end node and access point use *EAPOL-Key* frames to communicate during this phase, though they do not use the full 802.1X/EAP protocol.

The password together with the network SSID is used to derive the pairwise master key, from which the keys for securing traffic between the end node and access point are derived. If a device can prove it knows the password during the four-way handshake process by correctly deriving the shared keys, then it is authenticated. However, PSK supports no method for fine-grained access control, since all end nodes need to be preconfigured with the same password. Either an end node can join the network because it can prove it knows the password, by generating the pairwise master key, or it can't. PSK identity authentication is described in Clause 4.10.3.4 of 802.11-2020 [33].

5.5.4.3 Identity Authentication and Pairwise Master Key Derivation with SAE

SAE extends the basic two-message Wi-Fi authentication exchange with an additional *Authentication-Request/Authentication-Response* exchange to implement identity authentication and pairwise master key derivation using PAKE. The first exchange is called the Commit exchange, while the second is called the Confirm exchange. The Frame Body fields of these messages contain cryptographically generated identifiers used for the pairwise master key derivation. Like PSK, SAE uses a pre-shared secret, called a passphrase, to seed the algorithm, and it also can't do fine-grained access control because all devices that associate with the access point are required to share the same passphrase. As with both of the other algorithms, the *EAPOL-Key* frame type is used for the four-way handshake to derive the temporal key and other keys.

Figure 5.8 illustrates an example of a successful SAE authentication exchange. The exchange consists of the following steps:

- Prior to initiating the exchange both sides generate a Commit Scalar and Commit Element from their shared passphrase and other cryptographic parameters. The Commit Scalar and Commit Element are very large unsigned integers, with the size depending on the cryptosystem used by the two sides.[2]
- The end node initiates the authentication by sending an *Authentication-Request:Commit* message to the access point with its two values in the Frame Payload field, and sets the Sequence Number to 1.
- The access point verifies the Commit Scalar and Commit Element using an algorithm dependent on the cryptosystem, then sends an *Authentication-Response:Commit* message back to the end node with its Commit Scalar and Commit Element values in the body of the message and the Sequence Number again set to 1.
- The end node similarly verifies the access point's Commit Scalar and Commit Element. If the verification succeeds on both sides, the two are authenticated to each other and the process moves to the key derivation phase.
- In the key derivation phase, the end node and the access point first generate a pairwise master key and key confirmation key. The end node generates a Confirm value by applying a confirmation function with inputs being the key confirmation key, its Commit Scalar and Commit Element, the access point's Commit Scalar and Commit Element, and a 2 byte counter called the send-confirm counter used to prevent replay attacks. The Confirm value is also a large unsigned integer.
- The end node inserts the Confirm value in the Frame Payload field of the *Authentication-Request:Confirm* message with the send-confirm counter, sets the Sequence Number to 2, and sends the message to the access point.
- Upon receiving the message, the access point extracts the Confirm value and the send-confirm counter from the message. It uses the confirmation function with the same inputs as used by the end node, namely the key confirmation key, the end node's Commit Scalar and Commit Element, its own Commit Scalar and Commit Element, and the send-confirm counter to construct a Verifier value. If the verifier and the Confirm value from the end node match, the authentication is confirmed.

---

2 In Wireshark screenshots, the values which the standard calls the Commit Scalar and Commit Element are labelled as "Scalar" and "Finite Field Element" respectively.

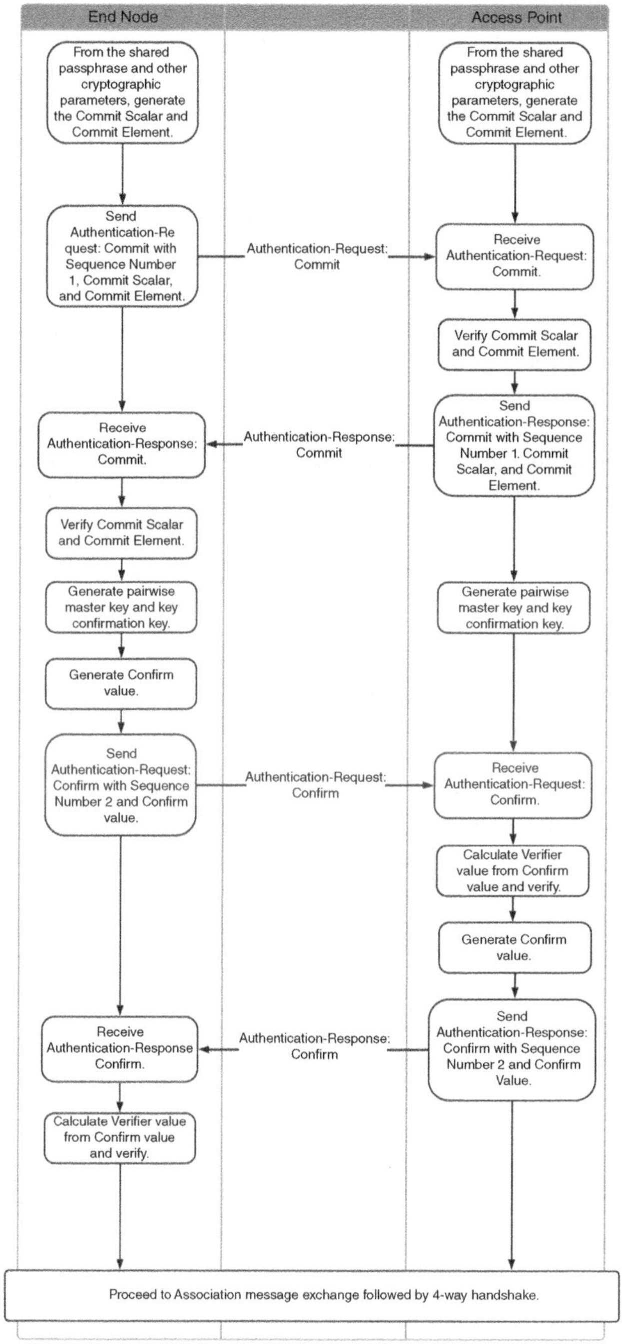

Figure 5.8 SAE Identity Authentication and Pairwise Master Key Derivation

- The access point replies to the end node with an *Authentication-Reply:Confirm* message including a Confirm value calculated in exactly the same way as for the end node, the same send-confirm counter as received from the end node, and the Sequence Number set to 2. The end node constructs a Verifier value in exactly the same way as the access point and compares it to the received Confirm value. If the two match, the authentication transaction has succeeded and the end node can move on to the association message exchange and the 4-way handshake.

The SAE protocol also contains an optional measure called anti-clogging to prevent denial of service attacks on the access point, since the cryptographic computations involved in the protocol are extensive and consume more processing time than for PSK or 802.1X/EAP authentication. An overview of the SAE protocol and how it fits together with 802.1X is available in Clause 4.10.3.3 [34] and the complete protocol including cryptographic algorithms is specified in Clause 12.4.5 of 802.11-2020 [35]. In addition, reference [36] has extensive Wireshark screenshots of the authentication exchange for SAE and details on the cryptographic algorithms, while reference [37] also has screenshots and an overview of the protocol.

5.5.5 The Four-Way Handshake

For all five authentication and key management suites under discussion here in both WPA2 and WPA3, an 802.1X key management extension message *EAPOL-Key* handles the four-way handshake for key provisioning. Table 5.4 contains the message code for *EAPOL-Key* and a short description of its function. Note that since the *EAPOL-Key* message is only exchanged between the access point and the end node, support for the *EAPOL-Key* message does not require the use of an Authentication Server. This enables *EAPOL-Key* to be used with the PSK and SAE four-way handshake protocols, as well as for the 802.1X/EAP.

The Frame Payload field of the *EAPOL-Key* frame, shown in Figure 5.9, contains a 1-byte Descriptor Type field followed by a variable length Descriptor Body field containing the key descriptor. The 802.1X standard [1] gives little guidance on the format for the Key Descriptor field. It states only that the Key Descriptor should start with the Subtype and Version fields but does not specify any format for them.

Table 5.4 *EAPOL-Key* Message Type

Message	Code	Function
EAPOL-Key	3	Four-way handshake and group key provisioning for WPA2 and WPA3 key provisioning.

Figure 5.9 *EAPOL-Key* Frame Payload Format

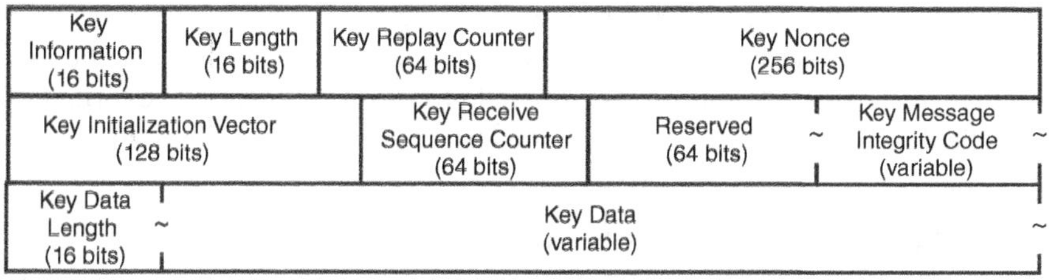

Figure 5.10 Key Descriptor Field for the 802.11 *EAPOL-Key* Message

The Descriptor Type code for 802.11 is 2 and is specified directly in the 802.1X standard document. The contents of the Descriptor Body field format for 802.11 are defined in the 802.11-2020 standard in Clause 12.7.2 [38], and are shown in Figure 5.10.

The fields have the following functions:

- Key Information – a control bitfield that contains information about the key exchange message. Fields of interest for this discussion are:
  - Install – a one-bit flag which the access point uses to instruct the end node to install the derived temporal key,
  - Secure – a one-bit flag which is set once the security association between the access point and the end node is in place,
  - Encrypted – a one-bit flag indicating that the keys in the Key Data field are encrypted,
  - Group – a one bit flag indicating that the keys in the Key Data field are for securing group and management frame traffic.
- Key Length – the length of the pairwise temporal key. The length depends on the encryption and integrity protection algorithms. Lengths for different algorithms are listed in Table 12-7 in 802.11-2020 [39].
- Key Replay Counter – initialized to zero when the pairwise master key is established and is incremented by 1 by the access point on each *EAPOL-Key* message exchange. The end node copies the value into its reply. This field is used to prevent replay attacks.
- Key Nonce – used during the four-way handshake to calculate the temporal key, see Section 5.5.5.1.
- Key Initialization Vector – part of the key derivation protocol for TKIP in WPA, included for backward compatibility only. This field is set to 0 in WPA2 and WPA3,

- Key Receive Sequence Counter – the current receive sequence counter, used to detect replay attacks during the optional group key provisioning which follows the four-way handshake,
- Key Message Integrity Code – a keyed cryptographic hash used by both the access point and the end node to authenticate the *EAPOL-Key* messages,
- Key Data Length – the length of the Key Data field,
- Key Data – a variable length field for additional data needed for the key exchange, particularly during the optional group key exchange. The format of this field is shown in Figure 12-35 of the 802.11-2020 specification [40].

5.5.5.1 The Four-Way Handshake Protocol

In all five Wi-Fi authentication algorithms discussed in this chapter, the end node and access point exchange a series of four *EAPOL-Key* messages in the four-way handshake to establish a security association for encryption and integrity protection. The four-way handshake starts with the end node and access point sharing a pairwise master key and ends with both sides sharing a pairwise key confirmation key, a key encryption key, and a temporal key. The following is an overview of the message exchange:

- After the end node is authenticated and associated, the access point generates an Authenticator nonce and sends an *EAPOL-Key* message to the end node containing the nonce.
- The end node receives the Authenticator nonce and generates its own Supplicant nonce. It uses the Authenticator and Supplicant nonces, along with the pairwise master key, to compute the pairwise transient key. From the pairwise temporal key, the end node derives the following keys:
  - Pairwise Key Confirmation Key – to calculate the message integrity code on *EAPOL-Key* frames,
  - Key Encryption Key – used by the access point to encrypt keys during the group key provisioning which optionally follows the four-way handshake,
  - Temporal Key – used for encryption and integrity protection on general data traffic between the access point and end node.

  The sizes of these keys depend on whether WPA2 or WPA3 is in use. Note that in SAE, the pairwise key confirmation key generated during this step is different from the key confirmation key generated during the Commit phase of the SAE *Authentication-Request/Authentication-Response* exchanges.
- The end node then sends an *EAPOL-Key* message back to the access point containing the Supplicant nonce and a message integrity code. The keyed hash algorithm used to calculate the message integrity code again depends on whether WPA2 or WPA3 is in use. The inputs to the algorithm are the pairwise key confirmation key and all the fields in the *EAPOL-Key* frame starting from

and including the EAPOL protocol version field up to and including the Key Data field, calculated with the Key Message Integrity Code field set to 0.

- The access point validates the message integrity code and similarly computes the same three keys as the end node. The access point then sends a third *EAPOL-Key* frame back to the end node with a message integrity code generated with the same inputs as the end node. This frame also sets the Install bit in the Key Information field and other bits, indicating that the temporal key has been installed in the access point and should now be used for encryption and integrity protection.
- The end node verifies the message integrity code and, if it is correct, installs the three keys. The end node then sends a final *EAPOL-Key* frame back to the access point to confirm the key installation.

Note that no keys are transmitted over the air during the four-way handshake. Both the access point and the end node independently calculate the temporal key, pairwise key confirmation key, and key encryption key and then use the message integrity code values to confirm that they both have the same key. Even though the four-way handshake must be conducted over an unencrypted link, the derived keys can't be compromised by an eavesdropper.

5.5.6 Group Key Provisioning

After the completion of the four-way handshake, another two-message *EAPOL-Key* exchange optionally follows for group and management frame key provisioning. This exchange helps to establish a group temporal key and an integrity group temporal key for protecting broadcast/multicast data frames and management frames, respectively. Since the traffic which the group temporal key and integrity group temporal key protect is shared among multiple end nodes on the access point, the access point derives the group key and management frame key unilaterally.

In the first message, the access point sends an *EAPOL-Key* frame to the end node. The Key Descriptor includes the group temporal key and integrity group temporal key in the Key Data field, encrypted with the key encryption key derived during the four-way handshake. In the Key Information field, the Secure, Encrypted, and Group bits are set in addition to other bits. The Replay Counter is incremented, and the message integrity code over the frame is calculated using the pairwise key confirmation key derived during the four-way handshake. The message integrity code is calculated with the encrypted keys in the Key Data field.

In the second message, the end node sends an EAPOL-Key frame back to the access point acknowledging receipt of the keys. In the Key Information field, the Secure bit is set in addition to other bits, and the Replay Counter matches the one received. The message integrity code is calculated over the frame using the key encryption key. Full details of the group key exchange can be found in Clause 12.7.7 of 802.11-2020 [41].

5.5.7 Commissioning Security Overview

Before a Wi-Fi device can connect with the BEMS services, it needs to be commissioned. The commissioning process involves provisioning the Wi-Fi device with credentials to connect with the BEMS services to which the device reports data and from which it receives commands. Commissioning requires solving the following problems:

- Making sure the device and the BEMS have bidirectional connectivity,
- Performing cryptographic and other operations to establish trust between the device and the BEMS,
- Downloading security credentials and configuration data into the device.

FDO Protocol Deep Dive

The video at the link above contains a technical deep dive on the FDO Alliance onboarding protocol by Geoffrey Cooper from Intel. The video provides details about the four FDO protocols: DI for device initialization, TO and T1 for device rendezvous, and T2 for device onboarding, and also illustrates the role of the ownership credential and how it is formed.

Commissioning is typically done by an installer who may be from a separate company or otherwise less trusted. Two commissioning techniques have been developed for Wi-Fi: the Wi-Fi Alliance Easy Connect method for untrusted installers and the Firmware Device Onboard (FDO, pronounced "Fido") Alliance protocol for trusted installers

The Wi-Fi Alliance Easy Connect onboarding protocol is designed to simplify device onboarding and is suitable for use with an untrusted installer. It streamlines the integration of IoT devices into Wi-Fi networks, through the use of a QR code or numerical code printed on the box in which the device is packaged. The installer connects the device by scanning the QR code or entering the numerical code into a smartphone or tablet, removing the need for a password. Public key cryptography is used to authenticate the device to the BEMS and secure the connection between the device and the network. Easy Connect is designed to simplify the setup of devices with minimal or no display interfaces.

The FDO Alliance was established in 2013 by manufacturers of IoT devices and IoT software products and services to develop a standardized, secure onboarding protocol for IoT devices. The FDO onboarding protocol provides the same basic services as Easy Connect through a framework for device security that uses public key cryptography to identify and authenticate the device being onboarded without exposing sensitive information. In addition to bidirectional authentication and encryption, the protocol introduces a security credential called an ownership voucher. An ownership voucher is a digitally signed certificate that cryptographically certifies each step in the transfer of ownership from the manufacturer to the final installation. It serves as a secure token that allows the new owner to claim and configure the device without direct physical access. The ownership voucher links a device to its owner through cryptographic evidence and ensures no compromise of the device's credentials occurs during the transfer of the device from the manufacturer to the final owner.

The video link to the left provides more information on the FDO Alliance onboarding protocol, and the blog post by Geoffrey Cooper [42] compares Easy Connect with the FDO Alliance protocol.

5.6 802.15.4 WPAN Security

In contrast with Wi-Fi, 802.15.4 WPAN does not provide a comprehensive set of security services but instead specifies basic confidentiality and integrity protection with a collection of cryptographic operations and supporting infrastructure that higher layers such as Zigbee, 6LoWPAN, and Thread can build upon. The IEEE 802.15.4-2020 standard [43] specifies algorithms for encryption, data origin authentication, and a mechanism for key management, but does not define mechanisms for key exchange/provisioning, security association setup, non-repudiation, end node identity authentication, or end node access control. The encryption and data origin authentication algorithms that implement confidentiality and integrity protection are based on the 128-bit AES algorithm, though a recent amendment, IEEE 802.15.4y-2021, specifies an option for the 256-bit AES algorithm for better security [44].

The basic model for 802.15.4 security is that a node shares security associations with other nodes with which it is communicating. The security associations consist of a set of indexed keys and identifiers representing the nodes. The key management mechanism allows flexible representation of the identifiers. When the sending node communicates with a receiving node, it selects the appropriate key from the key management store and uses that key to encrypt the packet and/or calculate a message integrity code. The node includes information in the Auxiliary Security Header field in the Frame Header of the transmitted frame about the security association used to secure the packet, so the receiving node can perform the reverse operation. How the security associations are set up is not discussed in the standard, as that is specified by higher layers.

5.6.1 Security Fields in the 802.15.4 Packet

The 802.15.4 protocol uses the 1-bit Security Enabled flag in the Frame Control field discussed in Chapter 3 to indicate that the packet is secured. If the bit is set, then the packet contains an Auxiliary Security Header in the Frame Header. The Auxiliary Security Header provides information on the security association used to secure the packet. Depending on the contents of the Auxiliary Security Header, the Frame Payload contains either a message integrity code alone or a message integrity code and the payload contents are encrypted.

5.6.2 Auxiliary Security Header

Figure 5.11 illustrates the Auxiliary Security Header field. The subfields include:

- Security Control – information about the security association, see below for details,
- Frame Counter – a unique frame counter value different for each frame. The Frame Counter field is checked by the receiving node to prevent replay attacks,
- Key Identifier – a variable length field, the size of which depends on the Key Identifier Mode field in the Security Control field, that identifies to the receiving node the security association to be used to process the packet.

The Security Control field, shown in Figure 5.12 contains information about the cryptographic algorithms and operations performed on the packet, so the receiving node can successfully extract the plain text from cyphertext if the packet is encrypted and/or verify the packet's integrity if a message integrity code is incorporated.

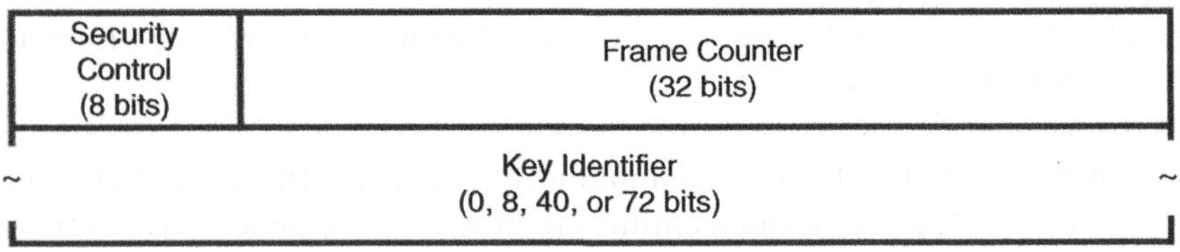

Figure 5.11 Auxiliary Security Header Field

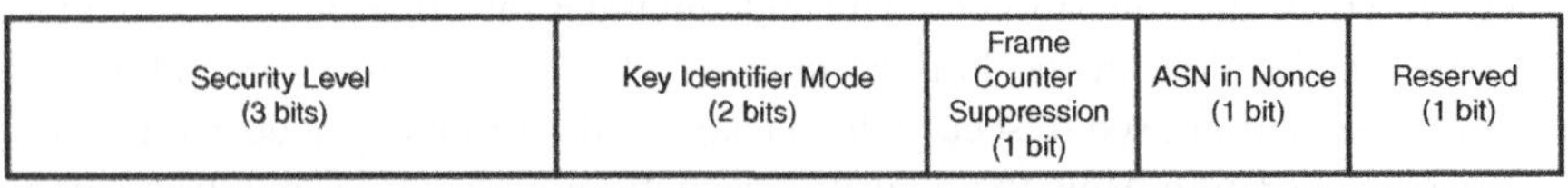

Figure 5.12 Security Control Field in the Auxiliary Security Header

The Security Level bitfield encodes what protection has been applied to the frame. The bitfield values correspond to the following protection levels:

- 0x00 – No protection,
- 0x01 – 32 bit message integrity code only,
- 0x02 – 64 bit message integrity code only,
- 0x03 – 128 bit message integrity code only,
- 0x04 – reserved,
- 0x05 – 32 bit message integrity code and encryption,
- 0x06 – 64 bit message integrity code and encryption,
- 0x07 – 128 bit message integrity code and encryption.

The previous version of the standard, 802.15.4-2015, included the option of encryption but no data origin authentication. This was deprecated as it is not considered secure, since the message is susceptible to man-in-the-middle attacks. An attacker could potentially intercept communications between two nodes, modify the data, and then forward it to the original recipient. Even if the data is encrypted, the attacker could replace it entirely with their own encrypted message.

Different-sized message integrity codes are provided for end nodes with different power and processing constraints. A reduced-function device that has a small battery or is powered by energy harvesting can use a 32-bit message integrity code and thereby reduce the amount of energy needed to construct it. 802.15.4-2020 specifies the AES-CCM* shared key cryptosystem [45] with a 128-bit key as the algorithm. AES-CCM* is a combined encryption and authentication block cipher with some additional properties that make it easier to use in WPANs, and it is compatible with the older AES-CCM used for encryption and authentication in 802.15.4-2003. It can be used to perform authentication, authentication with encryption, or only encryption, but in 802.15.4 it is used for the former two.

The Key Identifier Mode field in the Security Control field determines the form and size of the Key Identifier in the main Auxiliary Security Header. The bitfield values indicate the following Key Identifier field properties:

- 00 –indicates that the Key Identifier field is not present. The key used for security processing is determined implicitly based on the current security context. This is typically used in networks where a single key is shared across all nodes and does not need to be explicitly identified.
- 01 –indicates that the Key Identifier field consists of a single-byte Key Index, and the size of the field is 8 bits. This is used in environments where multiple keys may be in use, but they can be uniquely identified by a simple index.
- 10 –indicates that the Key Identifier field consists of a 4-byte Key Source and a 1-byte Key Index, and the size of the field is 40 bits. This mode allows for the differentiation of keys from multiple sources.

- 11 –indicates that the Key Identifier field consists of an 8-byte Key Source and a 1-byte Key Index, and the size of the field is 72 bits. This mode allows for an even greater number of key sources and is typically used in large networks with multiple key providers.

The Key Source could be a unique identifier for a node on the network like its 64-bit extended link layer address, or it could identify the PAN coordinator or some other entity responsible for key distribution. The Key Index field is used in conjunction with the Key Source to identify the specific key that should be used for security processing. This is necessary in scenarios where a single source provides multiple keys, or keys are rotated over time for enhanced security. The Key Index can range from 1 to 255, with 0 being reserved. The Key Index and Key Source allow a node to switch between keys on a frame-by-frame basis based on the information in the Auxiliary Security Header of the incoming frame.

The Frame Counter Suppression bit and the Absolute Slot Number (ASN) in Nonce bit are used to indicate how the nonce is formed on end nodes using frequency hopping as opposed to end nodes not using frequency hopping. The nonce is part of the cryptographic calculations and its formation is described in Section 5.6.3. ASN is a counter that increments for each timeslot since the start of the network. It provides a unique identifier for each timeslot in the network's operation. So, for example, ASN = 0 corresponds to the first timeslot when the network was started, ASN = 1 to the second timeslot, etc. In networks using frequency hopping, the ASN can be used instead of the frame counter to form the nonce since it will be unique per frame. If the Frame Counter Suppression bit and the ASN in Nonce bit are set to 1, there is no Frame Counter field in the Auxiliary Security Header and the ASN is used to form the nonce instead.

5.6.3 Calculation of the Frame Payload Field in Secured Frames

The encryption and message authentication algorithms in 802.15.4 use a nonce in their calculations. The nonce is formed by concatenating the source address, the frame counter from the Frame Counter field in the Auxiliary Security Header or the ASN for a frequency hopping network, and the security level from the Security Level field. The nonce is used only once, and it helps to ensure that even if the same data is sent multiple times, the encrypted payload and message integrity code will be different each time, increasing security.

Calculation of the message integrity code and encryption use the AES-CCM* symmetric key cryptosystem with 128 bit keys. The algorithm is a block cipher so it operates on blocks of data 128 bits long. The message integrity code is formed by applying the AES-CCM* message authenticator calculation algorithm with input of the key, the nonce, the frame header, Auxiliary Security Header, and the unencrypted contents of the Frame Payload field, sequencing the data into 128 bit chunks. The output is the message integrity code with 32, 64, or 128 bits depending on the Security Level field. After the message integrity code is formed, it is encrypted before being inserted into the frame. The AES-CCM* encryption algorithm also uses the key and nonce together with a counter which is incremented on each block of payload data. The message integrity code is calculated on the unencrypted payload. To verify the frame, the receiver must unencrypt the payload and the message integrity code first, then calculate the message integrity code from the key, nonce, frame header, Auxiliary Security Header, and clear text payload, then compare the two. If they match, the frame is verified.

5.6.4 Bluetooth Low Energy Security

In contrast with 802.15.4, the BLE 5.3 specification [46] includes procedures for identity authentication, access management, anonymity, and privacy, as well as mechanisms for key provisioning and security association construction. In addition, BLE 5.3 supports confidentiality protection and integrity protection, as does 802.15.4. Unlike Wi-Fi, however, BLE does not support forward secrecy, nor does it have any mechanisms for DDoS protection. As shown in Figure 3.18, the BLE architecture defines the Security Manager functional entity to handle the implementation of security services. The Security Manager supports a protocol between the Central and the Peripheral for setting up the security association. The BLE security architecture calls the Central the Initiator because it initiates the security association protocol exchange and the Peripheral the Responder because it responds to the request for the protocol exchange. BLE calls the process of establishing a temporary security association pairing, and the process of storing the security association for future communications bonding. As a consequence of BLE's evolution from the original Bluetooth which has a long history as a protocol for communicating between mobile devices and peripherals, BLE security has many options and suboptions. Some of the options have insecure features because they were released prior to the development of strong security for Internet-connected devices in the early 2000s. This section only briefly touches on the options with security issues and focuses primarily on the options with the strongest security.

5.6.5 BLE Security Modes and Levels

The BLE specification categorizes the security association between the two sides into three security modes for nodes that exchange data frames, each providing different levels of security services. Security Mode 1 is focused on establishing an encrypted link between the two sides, while Security Mode 2 establishes a security association to implement integrity protection. Security Mode 1 supports two identity

Introduction to BLE Security

Mohammad Afaneh provides a detailed introduction to BLE security in this video. The video discusses security services, which he calls security concerns, an overview of BLE security including paring and bonding, and some misconceptions and myths surrounding BLE security.

authentication and key provisioning methods, Legacy Pairing, and Secure Connections. Legacy Pairing supports a key provisioning method not based on any cryptographic algorithm, while Secure Connections supports a key provisioning method based on the elliptic curve Diffie–Hellman key exchange algorithm.

The levels associated with Security Mode 1 are:

- Level 1 – No security services implemented on the link,
- Level 2 – Pairing with encryption but no identity authentication. In this level, devices are paired using the Legacy Pairing method and a security association is established without authenticating the Initiator and Responder to each other, leading to an encrypted link,
- Level 3 – Pairing with encryption and identity authentication. After identity authentication, devices are paired using the Legacy Pairing method,
- Level 4 – Pairing with encryption and identity authentication using Diffie–Hellman key exchange. This level uses the BLE Secure Connections method, which provides strong protection against passive eavesdropping and man in the middle attacks, by using the Diffie–Hellman key exchange method to derive the encryption keys.

Security Mode 2 has the following levels:

- Level 1 – Pairing with integrity protection but no identity authentication. This level adds a message integrity code to a message, providing a way to ensure the data was not modified in transit, but provides no identity authentication,
- Level 2 – Pairing with integrity protection and identity authentication. This level uses a pairing method with identity authentication along with integrity protection.

The selection of security modes and levels depends on the requirements of the application and the capabilities of the devices. Note that the security modes are not mutually exclusive. A security association can be set up for either Security Mode 1 for confidentiality protection or Security Mode 2 for integrity protection, or both Security Mode 1 and 2 if both confidentiality protection and integrity protection are desired.

BLE 5.3 also supports Security Mode 3 specifically for Broadcaster-type nodes that don't exchange data frames with other nodes. Security Mode 3 provides encryption and identity authentication on Broadcaster transmissions and is described in Volume 3, Part C, Section 10.2.5 of the Bluetooth Core specification [47].

5.6.6 Pairing and Bonding Procedure

As shown in Figure 5.13, BLE begins the process of setting up a security association by having the Initiator and Responder execute a *Pairing-Request/Pairing-Response* exchange. The messages contain a capabilities and security association descriptor which provides information about the i/o capabilities of the two devices, whether Legacy Pairing or Secure Connections is used to set up the security association, which of the three security modes and levels they would like to set up, and the keys they would like to exchange during the bonding phase. This exchange governs the rest of the protocol, and the exact messages exchanged depend on the contents of the capabilities and security association descriptors exchanged between the two sides during this phase.

After the initial pairing capabilities and security association descriptors exchange, the two sides undergo three steps. The exact messages and algorithms executed depend on the security association descriptors exchanged during the *Pairing-Request/Pairing-Response* exchange. The three steps are:

- Mutual authentication using one of four authentication methods. During this step, cryptographic material for the construction of a temporary key, short-term key, and long-term key may be exchanged,
- Generation of a short-term key for Legacy Pairing or long-term key for Secure Connections to encrypt further key exchanges,
- Exchange of keys for other security services, such as integrity protection over a channel encrypted with the short-term key or long-term key. The *Central-Information* message is shown in the figure as a dashed arrow because it can only be sent for the Legacy Pairing identity authentication and key provisioning method.

The payloads sent in the pairing and bonding procedure protocol frames are formatted as BLE Attribute Protocol frames. The BLE Attribute Protocol is described briefly in Chapter 7.

The i/o capabilities exchange during the *Pairing-Request/Pairing-Response* exchange is used primarily to determine how to perform identity authentication on the two nodes. Reflecting its legacy as a protocol for connecting mobile phones

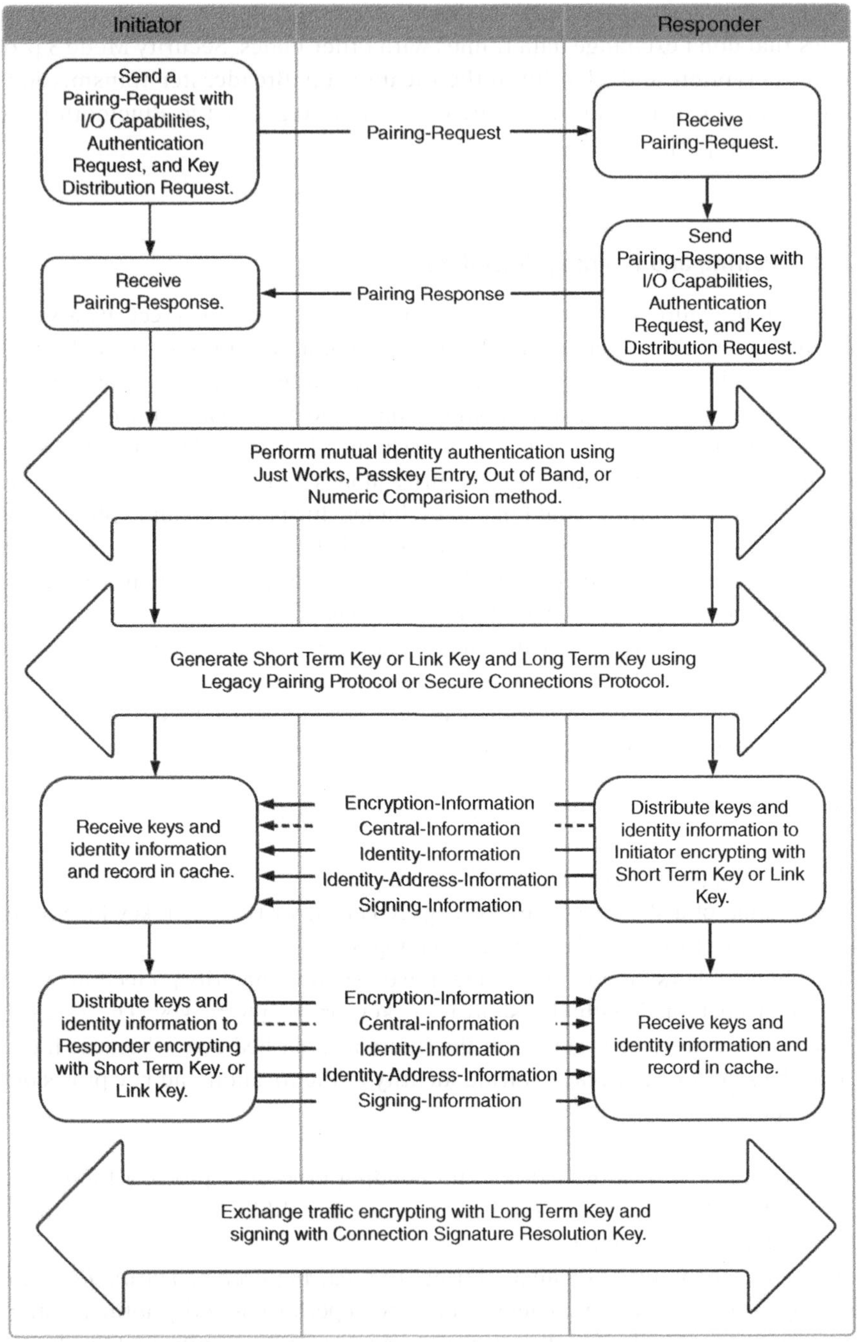

Figure 5.13 Generic Pairing and Bonding Procedure for BLE

to peripherals, some of the authentication methods require human intervention to confirm the pairing. There are four different authentication methods:

- Just Works – This method is used when neither node has sufficient i/o capabilities to support the other methods. It provides no protection against passive eavesdropping or man-in-the-middle attacks, since it doesn't authenticate the nodes,
- Passkey Entry – One node generates a six-digit passkey and displays it on an output display. A person then enters the key either on the other node or on both nodes depending on whether one or both nodes have an entry device. This method provides protection against passive eavesdropping and man-in-the-middle attacks, and authenticates both nodes to each other,
- Out-of-Band – In this method, the nodes exchange identity authentication information using an out-of-band method, such as near-field communication. The level of security provided depends on the security of the out-of-band communication channel used, but requires no i/o capability on either node except for the out-of-band communication channel.
- Numeric Comparison – Both nodes independently compute and display a 6-digit passkey. A person must check that these values match and confirm the match on both nodes. This method provides protection against passive eavesdropping and man-in-the-middle attacks and authenticates both nodes to each other.

5.6.7 Legacy Pairing Method

The Legacy Pairing method only supports the Just Works, Passkey Entry, and Out-of-Band methods for identity authentication and the security of the keys derived using the Legacy Pairing method depends on which of the three authentication methods is used. In the Legacy Pairing method, Just Works and Passkey Entry involve exchanging private cryptographic material for constructing the short-term key over an unencrypted connection. The short-term key is used to encrypt messages in the key exchange step. This exposes the key exchange step to a potential attacker eavesdropping on the channel during the initial exchange. The Out-of-Band method is more secure if the out-of-band link is encrypted, since the two nodes exchange the temporary key over the out-of-band link and only exchange confirmations that they have the same temporary key over the unencrypted BLE link. However, these security issues make the Legacy Pairing method unsuitable for deployment in a BEMS based on a zero-trust security architecture, so the rest of the discussion in this section is focused on the Secure Connections method.

5.6.8 Pairing Exchange Payload Format

Figure 5.14 shows the general format for the *Pairing-Request/Pairing-Response* message payload. The fields have the following functions:

- Message Type Code – for the *Pairing-Request* message, the code is 0x01. For the *Pairing-Response* message, the code is 0x02,
- I/O Capabilities – the type of i/o devices available for identity authentication, see Table 5.5,
- Out-of-Band Flag – set to 0x00 if Out-of-Band identity authentication should not be used or 0x01 is Out-of-Band identity authentication is available,
- AuthReq – a bitfield containing further configuration information about the pairing and bonding process, see Section 5.6.8.1 below,
- Max Encryption Key Size – the maximum size of the encryption key in bytes, between 7 to 16 bytes,
- Initiator Key Distribution and Responder Key Distribution – these are bitfields indicating what keys the Initiator and Responder should generate and distribute during the key distribution phase after identity authentication and short-term key or long-term key generation is complete, see Section 5.6.8.2.

Message Type Code (8 bits)	I/O Capabilities (8 bits)	Out-of-Band Flag (8 bits)	AuthReq (8 bits)	Max Encryption Key Size (8 bits)	Initiator Key Distribution (8 bits)	Responder Key Distribution (8 bits)

Figure 5.14 *Pairing-Request/Pairing-Response* Message Payload Format

Table 5.5 I/O Capabilities Field Encoding

Value	Description
0x00	The node has a numeric display but no input device.
0x01	The node has a numeric display and a button or some other device for a human to indicate "yes" or "no" in response to the display of the passkey.
0x02	The node has a numeric keyboard that can input the numbers 0 to 9, and at least two buttons that can be easily mapped to "yes" and "no" or some other mechanism by which the user can respond either "yes" or "no."
0x03	The node has no device allowing a "yes" or "no" response and no device capable of displaying a six-digit passkey.
0x04	The node has a device capable of displaying a six-digit passkey and both a keyboard allowing input of the numbers 0 to 9, and at least two buttons that can be mapped to "yes" and "no" or some other mechanism by which the user can respond either "yes" or "no."
0x05-0xff	Reserved for future use.

Table 5.5 shows the encoding of the I/O Capabilities field and its mapping to device hardware. These device types determine which of the four authentication methods can be used for identity authentication.

The pairing method is selected according to the following rules:

- Just Works – Used when at least one of the nodes has no input and no output capability (code 0x03) since no human interaction is required,
- Passkey Entry – Used when one node has a keyboard and the other can display a passkey. The node with the display generates a six-digit passkey and a human enters the passkey on the node with the keyboard,
- Numeric Comparison – Used when both nodes have a numeric display and a yes/no capability and Secure Connections will be used for key derivation and provisioning. Both nodes display a six-digit number and a human confirms that they match.

Out-of-Band requires no input devices since the authentication is performed through a side channel.

5.6.8.1 AuthReq Bitfield

Figure 5.15 illustrates the AuthReq bitfield format. The Bonding Flags bits can take one of two values:

- 00 – no bonding,
- 01 – bonding.

The other values of the field are reserved for future use.

If the Man in the Middle Protection bit is set, the node requires protection against man-in-the-middle attacks. This requires that the pairing process include an authentication step that can't be replicated by an attacker. In Legacy Pairing, if the Man in the Middle Protection bit is set, the short-term key, used to encrypt further keys during the key exchange phase, must be generated using a pairing method that provides man-in-the-middle protection, one of Passkey Entry or Out of Band, rather than the Just Works method which does not provide any protection. In Secure Connections, if the Man in the Middle Protection bit is set, the long-term key, used to encrypt data traffic between the two nodes after pairing and bonding, must be generated using a pairing method that provides man in the middle protection, one of Numeric Comparison, Passkey Entry, or Out of Band, rather than the Just Works method.

Bonding Flags (2 bits)	Man in the Middle Protection (1 bit)	Secure Connections (1 bit)	Keypress (1 bit)	CT2 (1 bit)	Reserved (2 bits

Figure 5.15 AuthReq Bitfield Format

If the Secure Connections bit is set on both the *Pairing-Request* and *Paring-Response*, then the Secure Connections method must be used. Otherwise, the Legacy Pairing method is used.

If the Keypress bit is set, it indicates that the node can support sending keypress notifications during Passkey Entry. In Passkey Entry, one node displays a six-digit number and a human enters this number on the other node. The Keypress bit signals that the node where the number is being entered can send a notification to the other node every time a key is pressed, released, or cleared, allowing feedback to the human on the node showing the passkey. For example, the node can display asterisks (*) to indicate each digit entered. This feature can help improve user experience, but it doesn't affect the security of the pairing process and is also optional.

Finally, the CT2 bit indicates that a specific hash function should be used when converting a key from BLE to regular Bluetooth and back. This procedure is described in Volume 3, Part H, Sections 2.4.2.4–2.4.2.5 of the Bluetooth Core specification [48].

5.6.8.2 Initiator and Responder Key Distribution Bitfields

The Initiator and Responder Key Distribution bitfields allow a node to indicate what keys it needs from the other node during the key distribution phase of the protocol. Figure 5.16 illustrates the Initiator and Responder Key Distribution bitfield format. For the Secure Connections method, the Encryption Key bit is ignored because both sides have independently and in secret derived the same long-term key using the Diffie–Hellman key exchange algorithm. This bitfield is only used in the Legacy Pairing method. The Link Key bit indicates that the sending node would like to derive the receiving node's legacy Bluetooth link key from the BLE long-term key.

The Identity Key bit indicates that the sending node needs the receiving node's identity resolving key. The identity resolving key is used for resolving random resolvable addresses. As discussed in Chapter 3, every device has a globally unique, EUA-64 address, assigned by the manufacturer. A random resolvable private address is a 48-bit node address that is randomly generated. Nodes can utilize two types of random resolvable addresses: static and dynamic. Static random resolvable addresses are generated when a node boots and are only valid until the node shuts down. Dynamic random resolvable addresses change periodically and

Encryption Key (1 bit)	Identity Key (1 bit)	Message Signing Key (1 bit)	Link Key (1 bit)	Reserved (4 bits)

Figure 5.16 Initiator and Responder Key Distribution Bitfield Format

can only be recognized and resolved by nodes that have the corresponding identity resolving key. The identity resolving key allows nodes to recognize each other's address even when one or both are using changing, private device addresses for privacy protection, protecting their permanent addresses from potential eavesdroppers.

The Message Signing Key bit is used to indicate that the sending node would like the receiving node's connection signature resolving key. The connecting signature resolving key is used to verify a message integrity code calculated on the frame, to ensure message integrity.

5.6.9 Confidentiality and Integrity Protection

BLE specifies one algorithm that provides both confidentiality and integrity protection and one that only provides integrity protection. BLE uses the same confidentiality and integrity protection as legacy Bluetooth, AES-CCM. AES-CCM is specified in RFC 3610 [49]. The key size for the Secure Connections AES-CCM long-term key is 128 bits. The encryption process is described in Volume 2, Part H, Section 9.0 of the Bluetooth Core specification [50].

For jurisdictions in which encryption is prohibited or for other cases where only integrity protection is desired, BLE specifies AES-CMAC. AES-CMAC is specified in RFC 4493 [51]. The key size for the AES-CMAC Secure Connections connection signature resolving key is 128 bits. The message integrity code size is 64 bits and is calculated over the message data to be signed, the connection signature resolving key, and a 32-bit counter initialized to zero when the connection signature resolving key is provisioned. Calculation of the message integrity code is described in Volume 3, Part H, Section 2.4.5 of the Bluetooth Core specification [52].

5.6.10 Secure Connections Method

Secure Connections supports all four methods of identity authentication and uses the elliptic curve Diffie–Hellman algorithm to generate a shared key on both sides. No private cryptographic material is sent over any channel in any of the four methods. This ensures that keys derived and exchanged through Secure Connections are safe from eavesdroppers and attackers. Figure 5.17 gives an overview of the protocol. The following steps are involved:

- As discussed in Section 5.6.6, the Initiator sends a *Pairing-Request* message to the Responder, setting the Secure Connections bit in the AuthReq field and providing information about its i/o capabilities in the I/O Capabilities field The Responder replies in a similar manner.
- Both sides calculate a public/private key pair using the elliptic curve Diffie–Hellman algorithm and exchange their public keys.

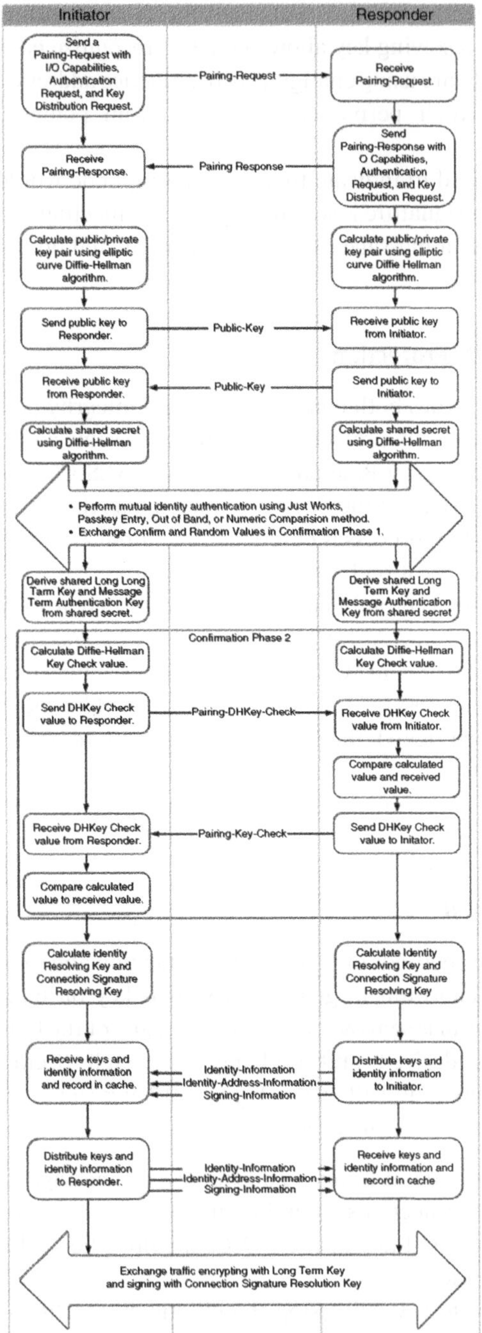

Figure 5.17 Secure Connections Pairing and Bonding Protocol

- With the opposite party's public key in hand, both nodes use the elliptic curve Diffie–Hellman algorithm to calculate the shared secret without exchanging any further cryptographic material over any link.
- The two sides utilize an identity authentication method based on the i/o capabilities they advertised in the I/O Capabilities field, as discussed in Section 5.6.8, to authenticate each other. The identity authentication is confirmed with Confirmation Phase 1, in which the two sides calculate and exchange Confirm and Random values. The number and type exchanged depend on the identity authentication method. The Confirm and Random value exchange is discussed in more detail in Section 5.6.10.1.
- Both nodes use their shared secret to calculate a long-term key for encrypting the key exchange and for encrypting the link when data is exchanged. They also calculate a message authentication key for use in the next step, Confirmation Phase 2.
- The two sides execute Confirmation Phase 2. Both nodes calculate a Diffie–Hellman Pairing Key Check value based on the message authentication key, the two Random values exchanged during Authentication Phase 1, the addresses of both devices used during pairing, and other inputs. They then exchange their calculated values and compare the received and calculated values. If the two match, the validity of the Diffie–Hellman key generation has been confirmed. From this point on, the link is encrypted with the long-term key.
- Depending on the keys requested in the Initiator and Responder Key Distribution bitfields from the initial pairing exchange, the nodes generate an identity resolving key and a connection signature resolving key. The identity resolving key is used for masking a node's address and provides privacy and access control. The connection signature resolving key is used for calculating and resolving a message integrity code and provides integrity protection on messages. Key management is discussed in Volume 3, Part H, Appendix B of the Bluetooth Core specification [53].
- The nodes exchange the requested keys. The *Signing-Information* message contains the connection signature resolving key, for calculating a message integrity code on data frames. The *Identity-Information* message contains the identity resolving key, and the *Identity-Address-Information* key contains the node's address that will be used for masking. Address masking is briefly discussed in Section 5.6.11.
- The nodes can now exchange application data frames encrypted with the long-term key for confidentiality protection and protect the messages with message integrity code calculated using the connection signature resolving key for integrity protection.

The above protocol description holds if the pairing and bonding are successful and details of the identity authentication protocol are not shown. Section C.2.2 in the BLE 5.3 specification contains complete message sequence figures for Secure Connections, including all four authentication protocols and the message sequences if errors occur [54].

5.6.10.1 Confirm and Random Values

The nodes conduct an exchange of Confirm values during and after the identity authentication as a means of foiling any eavesdropping attackers that may attempt a replay attack. Calculation of the Confirm value involves generating a Random value prior to beginning the identity authentication step. The Random value is hashed together with the two public keys and other material, which differs depending on the authentication method used, on one or both nodes to calculate the Confirm values after the identity authentication method has been completed. For example, the Passkey method also requires the entered passkey to generate the Confirm value. The Confirm value is sent in Confirmation Phase 1 after the identity authentication phase in *Pairing-Confirm* messages exchanged over the BLE link for the Just Works, Passkey Entry, and Numeric Comparison methods. For the Out-of-Band method, the Confirm value is exchanged over the out-of-band link. Calculation of the Confirm value is described in Volume 3, Part H, Sections 2.3.5.6.2 to 2.3.5.6.4 of the Bluetooth Core specification [55].

The Random value also is sent between the Initiator and Responder nodes during the identity authentication step in an exchange of *Pairing-Random* messages because the Random value is used during Confirmation Phase 2. The DHKey Check value is calculated as a hash of the message authentication key, the two Random values, the addresses of both devices, and other material. Calculation of the DHKey Check value is described in Volume 3, Part H, Section 2.3.5.6.5 of the Bluetooth Core specification [56].

5.6.11 Access Control and Privacy

Access control in BLE is implemented primarily through device filtering. The Central can be configured with a filter accept the list of the device identity addresses from nodes that are permitted to enter the WPAN. The device identity address is one of the public addresses, the static random address, or the result of using the identity resolving key to resolve a dynamic random address. Different device filtering policies can be configured for when the Central is advertising, scanning, initiating, or periodically establishing sync but all policies use the same device filter list. The device filter list includes a special, anonymous record that matches all advertisements sent with no address. An example of an application for messages without any address is a Broadcaster that advertises sensor data but prefers not to establish a connection to save energy. Link layer device filtering is described in more detail in Volume 6, Part B, Section 4.3 of the Bluetooth Core specification [57] and a figure showing the device filtering list is in Volume 1, Part A, Section 5.4.5 [58].

BLE access control is tightly linked with privacy. Dynamic random addresses are privacy-protecting mechanisms, since they make tracking a particular node more difficult. The pairing and bonding process is another. A node can reject a request to pair and bond with another node if it does not recognize the address, or if the other node declines to make its identity resolving key available. Volume 3, Part C, Section 10.7 of the Bluetooth Core specification [59] has a detailed description of the privacy mechanisms that different BLE node types (Broadcaster, Observer, Central, Peripheral) can utilize.

In general, BLE supports two privacy protection modes:

- Device Privacy Mode – a node in device privacy mode is only concerned about its own privacy. It accepts advertising frames from peer devices that contain their public and static random addresses as well as their dynamic random addresses, even if the peer node has distributed its identity-resolving key.
- Network Privacy Mode – a node in network privacy mode does not accept advertising packets containing the public and static random address of peer nodes that have distributed their identity resolving key.

In device privacy mode, while a node can be recognized by peers that have its identity resolving key, it is less likely to be tracked by untrusted devices. This is because the device can use dynamic random addresses that change periodically, making it harder for untrusted devices to consistently identify it. Because the node can still respond to advertisements with public and static random addresses, the node can communicate more freely with peers.

Network privacy mode enforces stricter privacy guarantees. When a node shares its identity-resolving key, the node essentially grants its peer the ability to recognize it even though it uses a resolvable private address. The peer cannot, however, pin down the node's device identity through a public or static random address. If the peer then sends advertising packets with its own public or static random address, the sender may be trying to trick the receiver into revealing its identity. This could happen, for example, if the advertising packets are intended to prompt a response from the receiving device. To prevent this from happening, a device in network privacy mode simply ignores advertising packets with public or static random addresses.

5.7 Public Key Certificates and the Public Key Infrastructure

One of the security services that has not been addressed by the link layer security protocols is non-repudiation. Non-repudiation requires ensuring that both parties cannot deny their involvement in the communication or transaction. Using a shared key to calculate a MAC cannot establish non-repudiation because both sides hold the key, so a message cannot be traced to one side or the other. In addition, with the exception of EAP-TLS, the methods of identity authentication described so far do not tie a node's identity to some offline entity, like a company or nonprofit organization. In many cases, identity authentication requires verifying that a service, node, or user has membership in some organization. Such a verification can extend to the service, node, or user's role, and be used for access management as well.

X.509 certificates and the PKI are designed to support non-repudiation, identity authentication, and access management. While public keys provide non-repudiation without relying on pre-shared secrets, a naked public key without any tie to identifying information about the owner says very little about whether the communication can be trusted. X.509 certificates provide a standardized document format for verifying the identity of individuals or entities by associating their public keys with a digital document signed by a trusted certificate authority (CA). The PKI establishes the infrastructure, processes, and protocols to manage the generation, distribution, validation, and revocation of these certificates, enabling non-repudiation, identity authentication, and access management for entities in BEMS networks.

5.7.1 X.509 Certificates

X.509 certificates are digital documents used to verify the identity of a public key holder. They are in a standardized format defined by the International Telecommunication Union Telecommunication Standardization Committee (ITU-T) [60] X.509 certificates are used in the context of PKI to establish secure and trustworthy communication channels by verifying the identity of services, entities, or individuals. An X.509 certificate is encoded in the Abstract Syntax 1 format (ASN.1, ISO standard 8824 [61]) and contains the following information:

- Subject – The subject field identifies the entity to which the certificate is issued, such as a service, device, person, or organization. It typically includes information like the common name (CN), organizational unit (OU), organization (O), locality (L), state (ST), and country (C).
- Issuer Name – The issuer field identifies the entity that has digitally signed and issued the certificate. The issuer is usually a trusted CA responsible for verifying the identity of the subject and that the subject does hold the private key corresponding to the public key in the certificate. The CA indicates its seal of approval by signing the certificate with its private key.
- Public Key Info – The subject's public key is included in the certificate, in addition to the key size and public key algorithm to use when verifying a digital signature created with the public key.
- Validity – X.509 certificates have a defined period during which they are considered valid. The validity period includes a start date and time and an

expiration date and time. Certificates need to be renewed or reissued before they expire or they will become invalid and will be rejected in protocol communications.

- Subject Alt Names – The certificate can optionally include a list of alternative DNS names for the subject.
- Miscellaneous – A collection of miscellaneous information can be included, such as the serial number, version, etc.
- Other fields – Certificates can include other fields that allow the subject to provide verified information to recipients.

The CA issuing the certificate digitally signs it using the CA's private key. This signature ensures the integrity and authenticity of the certificate. By verifying the signature using the issuing CA's public key, the recipient can be confident that the certificate has not been tampered with and that the issuer indeed vouches for the subject's identity.

5.7.2 Public Key Infrastructure (PKI)

PKI is a framework of hardware, software, policies, and procedures used to manage the generation, distribution, validation, and revocation of X.509 certificates. PKI enables secure and trustworthy communication and supports various security services, including identity authentication, data integrity, and non-repudiation.

The PKI supports a hierarchical trust model, with multiple levels of CAs forming a chain of trust. The CA is the trusted entity responsible for issuing and managing X.509 certificates. CAs verify the identity of the certificate subjects and digitally sign the certificates they issue. CAs are organized into a hierarchy, with the subject certificate's signing CA at the bottom and intermediate CAs between the subject certificate's signing CA and the root CA. The root CA acts as a trust anchor at the top of the hierarchy and signs its own certificate. The certificates of intermediate CAs between the root and the subject certificate are signed by the CA immediately above them in the chain. Finally, the subject certificate is signed by the second to last CA in the chain. A certificate is typically distributed along with a certificate chain, including a certificate containing

PKI Explained

This video provides a detailed explanation of how PKI works. The video includes protocol flow diagrams showing how a certificate is issued and how certificate revocation works.

the signing CA's public key and a hierarchical list of CAs up to the root CA. By verifying the signatures from the subject certificate up to the root, a receiver can ensure that the identity of the subject and the other CAs in the chain are vouched for by their signing CAs.

When the private key associated with the public key in the certificate is compromised or the certificate expires, the certificate may need to be revoked. Certificate revocation lists (CRLs) or certificate revocation status checking mechanisms like Online Certificate Status Protocol (OCSP) are used to communicate the revocation status of certificates. OSCP is specified in RFC 6960 [62]. Prior to the specification of OCSP, the receiver of a certificate had to download the CRL and check whether the certificate had been revoked before verifying the signature. With OCSP, the client can send an OCSP request to the corresponding CA's OCSP service. The service checks its records and responds with an OCSP response containing the revocation status information. OCSP allows clients to obtain up-to-date revocation information without the need to download and process the entire CRL, which could run into megabytes.

5.8 IP Stack Security Protocols

Security for the IP stack has evolved considerably since the original ARPAnet research project in the 1980s. In the early days when the ARPAnet was strictly an academic research project, the focus was more on establishing connectivity and ensuring functionality, not on securing communication. The idea of a global network of computers was still in its infancy and the potential scale of security threats was not yet fully understood. As the Internet began to expand in the late 1980s, and especially 1990s as it was commercialized, the need for security became clear. In response to increasing threats, new security protocols were developed and added to the IP stack. These included protocols like Secure Sockets Layer (SSL) and its successor TLS to provide encrypted and authenticated connections at the transport layer. TLS is the foundation of server identity authentication for the web, without which trust in Internet commerce and the enormous growth in online shopping would have been impossible. The IKE protocol and IPSec protocol were developed to provide security at the network layer through encrypted and authenticated IP packets. IKE and IPSec are key technologies for implementing network layer VPNs, which allow employees to securely connect with an enterprise network and enable a BEMS network with a private address space local to a building to securely connect to the control system in the cloud. TLS is also deployed for transport layer VPNs. In this section, security protocols for the IP stack are discussed.

5.9 IKE

A basic step in establishing security between two nodes at the IP network layer is to establish a security association with a shared key for symmetric key encryption, the cryptographic algorithms to be used in processing traffic, and an indication of to what traffic the security association applies. While encryption can also be implemented with public key cryptography, IP stack protocols overwhelmingly use symmetric key encryption because the performance is superior. Similarly, once two nodes have authenticated each other's identity, keyed message authentication algorithms implement integrity protection through message integrity codes. Establishing a shared key and security association between two nodes in an automated fashion over the Internet requires that the keys be kept secret and no cryptographically significant information be exchanged over the Internet in clear text that could lead an attacker to reconstruct the key if they could eavesdrop on the key provisioning exchange.

IKE is the key exchange solution standardized by the IETF that allows two parties to compute a shared secret using the Diffie-Hellman algorithm primarily to set up security associations for IPSec. IKE is defined in a series of RFCs by the IETF, with the two versions being IKEv1 in RFC 2409 [63] and IKEv2 in RFC 7296 [64]. IKEv1 was designed with a companion protocol called the Internet Security Association and Key Management Protocol (ISAKMP) defined in RFC 2408 [65] for setting up an IPSec security association. The two were kept separate because other key provisioning protocols had been developed, and the designers of IPSec wanted to allow flexibility in how key provisioning was implemented. However, over the years, the other key protocols lost implementation momentum, and in 2014, RFC 7296 folded the security association configuration for IPSec into the IKEv2 protocol.

5.9.1 Overview of the IKEv2 Protocol

The IKEv2 protocol consists of three phases:

- Initialization – the Initiator and Responder exchange public Diffie–Hellman values and propose and accept security association configurations. After this phase is complete, shared keys are derived on both sides and all further messages are encrypted. Both sides have a security association within which to continue further IKE communication.
- Authentication – the Initiator and Responder engage in one of several possible identity authentication protocols. Available protocols include a simple certificate exchange with or without a CA certificate chain and EAP, among others.

- Create Child Security Association – if additional security associations are required for IPSec, the Initiator and Responder create children of the root security association. Child security associations can provide perfect forward secrecy if a new set of keys is configured for each child security association, since compromise of one child's keys does not compromise any others.

The various IKEv2 type codes and their functions are maintained by IANA [66].

5.9.2 IKEv2 Header

IKE uses UDP ports 500 and 4500 with one IKE message per UDP packet. When sent on UDP port 500, IKE messages begin immediately following the UDP header, but when sent on UDP port 4500, 4 bytes of zeros are prepended before the IKE message to differentiate IKE packets from IPSec packets during NAT traversal. The 4 bytes of zeros are not part of the IKE message and are not included in any of the length fields or checksums defined by IKE.

Figure 5.18 illustrates the IKEv2 header. The fields are the following:

- Initiator's and Responder's Security Association Security Parameter Index (SPI) – in the Initialization message from the Initiator, the Initiator's Security Association SPI contains a unique value to identity the IKE security association for the key exchange the two sides are setting up, while the Responder's Security Association SPI is zero. The Responder replies by echoing the Initiator's value and inserting a unique value of its own into the Responder's Security Association SPI field,
- Next Payload – a code for the type of the payload immediately following the header,
- MjV – major version number. For IKEv2, this is 2,
- MnV – minor version number. For IKEv2, this is 0,
- Exchange Type – a message type code indicating the IKE message type. The message type codes from RFC 7296 and their function are described in Table 5.6,

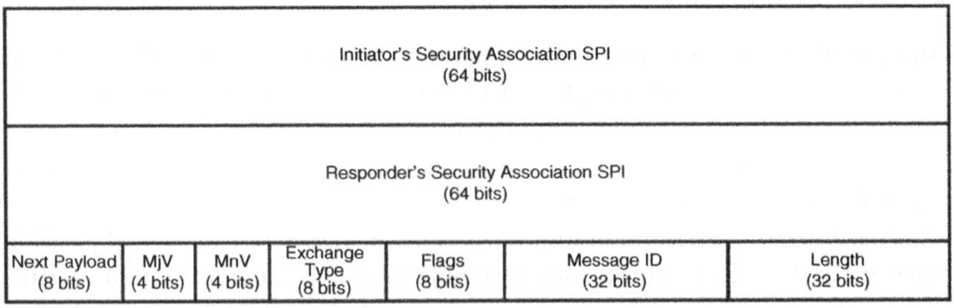

Figure 5.18 IKEv2 Header Format

Table 5.6 IKEv2 Message Type Codes

Message	Code	Function
IKE_SA_INIT	34	Initialize the security association for the IKE key exchange.
IKE_AUTH	35	Establish mutual identity authentication between the Initiator and Responder.
CREATE_CHILD_SA	36	Create a child security association of the main key exchange security association for a specific IPSec purpose.
INFORMATIONAL	37	Convey control messages between peers regarding errors or notification of other events.

- Flags – options for the message. Of the eight bits, only bits 2-4 have defined functions:
  - Bit 2 – indicates the message is a response to a message having the same Message ID,
  - Bit 3 – indicates that the sender supports a higher version than in the major version number field, but this field must not be set for IKEv2,
  - Bit 4 – set in a message sent by the Initiator and cleared in messages sent by the Responder.
- Message ID – a message identifier used to control retransmission of lost packets and matching of requests and responses. The Message ID also helps to prevent replay attacks,
- Length – the length of the message including the header and payload, in bytes.

5.9.3 IKEv2 Payload Elements

After the header, the IKEv2 message contains zero or more payload elements, including a header and body. The generic payload element header is shown in Figure 5.19. Each header contains a type code in the Next Payload field indicating the type of the next payload element. Table 5.7 shows the payload element type codes defined in RFC 7296 and a brief description of their contents. If the current payload element is the last, the Next Payload field contains 0. Type codes from 1 through 32 are reserved to avoid overlap with IKEv1. The Payload Length field contains the length of the current payload element in bytes, including the payload element header.

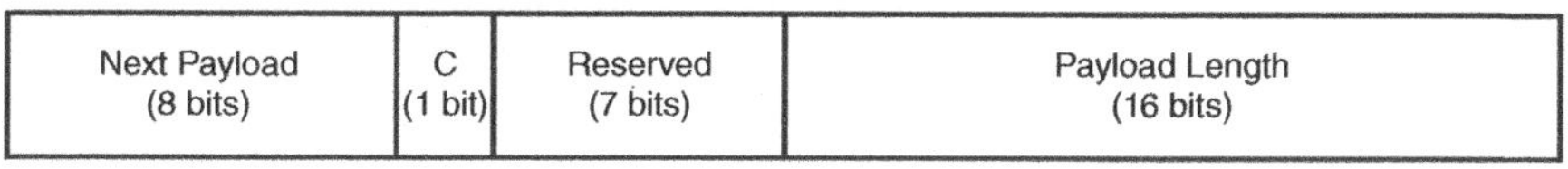

Figure 5.19 IKEv2 Generic Payload Element Header

Table 5.7 IKEv2 Payload Element Type Codes and Functions

Message	Code	Function
No Next Payload	0	The current payload is the last.
Reserved	1–32	These codes are reserved to avoid overlap with IKEv1.
Security Association	33	One or more proposals for an IKE or IPSec security association.
Key Exchange	34	The Diffie–Hellman public key and group number parameter for the key exchange.
Identification-Initiator	35	The Initiator's identity and a code indicating the identity type.
Identification-Responder	36	The Responder's identifier and a code indicating the identifier type.
Certificate	37	The sender's certificate and code indicating the certificate format.
Certificate Request	38	Request a certificate in a preferred format from a preferred CA.
Authentication	39	Authentication data and a code indicating the authentication method type.
Nonce	40	A randomly generated value between 16 and 256 bytes.
Notify	41	Information on error conditions or state transitions in a response.
Delete	42	Delete an IKE or IPSec security association, indicated by the SPI.
Vendor ID	43	A vendor-defined constant is used to recognize a vendor's equipment.
Traffic Selector-Initiator	44	A specification identifying what packet flows should be processed by IPSec services on the Initiator side.
Traffic Selector-Responder	45	A specification identifying what packet flows should be processed by IPSec services on the Responder side.
Encrypted and Authenticated	46	A wrapper for other payloads in encrypted form.
Configuration	47	IP remote access client request for an internal IP address and IP remote access server to provide DHCP configuration information.
EAP	48	Wrapper for an EAP message.

In addition to the next element's type and the length, the header contains a Critical, or C, bit followed by 7 bits that are reserved and must be set to zero. The C bit indicates what the recipient should do if it does not have code to process the payload element. The C bit must be set to zero when the sender wants the recipient to skip this payload element if it cannot process the element based on the element's type code from the previous element header's Next Payload field. If the sender instead wants the recipient to reject the entire message, the C bit must be set to 1. The C bit must be ignored if the recipient can handle the payload element type, and must not be set for any payload type codes in RFC 7296, since the payload type codes in the base specification define the basic IKEv2 protocol.

Of the payload element types in Table 5.7, the Security Association type (code 33) is the most important for the current discussion. A Security Association element encapsulates one or more security association proposals for the IKE key exchange or for one of the IPSec protocols. The Initiator and the Responder exchange Security Association elements during all three phases of the IKEv2 protocol. During the Initialization phase, Security Association elements are exchanged containing proposals for the IKEv2 key provisioning process itself. During the Authentication phase, Security Association elements are exchanged for both the IKE authentication and the Create Child Security Association phases. During the Create Child Security Association, Security Association elements are only exchanged for the IPSec child security associations. A node may send multiple proposals within a single Security Association element, but the most preferred proposal must be listed first.

A security association proposal contains the following fields, among others:

- A type code indicating to which protocol the proposal applies: 0 for IKE, 1 for IPSec Authentication Header (AH), and 2 for IPSec Encapsulating Security Payload (ESP). The IPSec protocols are discussed in Section 5.10,
- The sending entity's SPI,
- One or more transform elements, describing the cryptographic algorithm and parameters to be used in the security association.

Each transform element contains a number of proposed transforms applicable to the security association. Transforms can be of five different types having code numbers 1 to 5 respectively: encryption method, pseudorandom function, integrity protection method, key exchange method, and extended sequence numbers. Transform elements for encryption, integrity protection, and key exchange contain a specification for the cryptographic algorithm along with public parameters for the algorithm. For example, one side can propose the elliptic curve Diffie–Hellman algorithm for key exchange, along with the proposer's public key and a random nonce which are the public parameters for the algorithm. The pseudorandom function transform proposes one of several keyed MAC functions to use during the key exchange, and the extended sequence numbers transform simply proposes whether or not extended (64 bit) sequence numbers rather than regular (32 bit) sequence numbers should be used for the IPSec security associations. The format of the transform proposal is described in Section 3.2.2 of RFC 7296 [64].

The key exchange method and pseudorandom function transforms are negotiated in the Initialization phase of the protocol since they apply to key provisioning. The encryption and integrity protection transforms are negotiated in all three phases of the protocol. In the Initialization phase, the encryption and integrity protection transforms apply to the IKE messages themselves while those in the Authorization and Create Child Security Association phase apply to the child security associations for IPSec. The extended sequence numbers transform is only negotiated during the Create Child Security Association phase of the protocol since extended sequence numbers only apply to IPSec.

5.9.4 Example IKEv2 Protocol Exchange

The basic IKEv2 protocol defined in RFC 7296 specifies only the 4 message types listed in Table 5.6 with an additional 6 added since the original protocol was specified to address more specialized use cases, as recorded in the IANA IKEv2 registry. While the basic protocol is fairly simple, there are many different possible combinations of the 16 payload elements from Table 5.7, which contribute to constructing security associations, that nodes can use to address specific IPSec security association needs. Figure 5.20 illustrates one example of a successful IKEv2 protocol exchange for the Initialization and Authentication phases, in which the two sides authenticate each other by exchanging certificates. In this exchange, the Initiator and Responder agree on security associations and traffic selectors for the first IPSec child security association as part of the Authentication phase. Traffic selectors are included in a payload element and consist of the source IP address, destination IP address, protocol, source port, and destination port of packets protected by the security association. If only one IPSec security association is required, the IKEv2 exchange is finished after the Authentication phase completes. The example exchange includes the following steps:

- The Initiator sends an *IKE_SA_INIT* message to the Responder with a collection of proposed Security Associations for the key exchange security association. The Initiator also includes their SPI, the Diffie–Hellman public value (public key), and a Nonce.
- The Responder receives the message and selects one of the proposed Security Associations, then stores the parameters from the Initiator's message and replies to the Initiator with an *IKE_SA_INIT* message. The response includes both SPIs, the Security Association accepted for the key exchange, and their Diffie–Hellman key and Nonce.

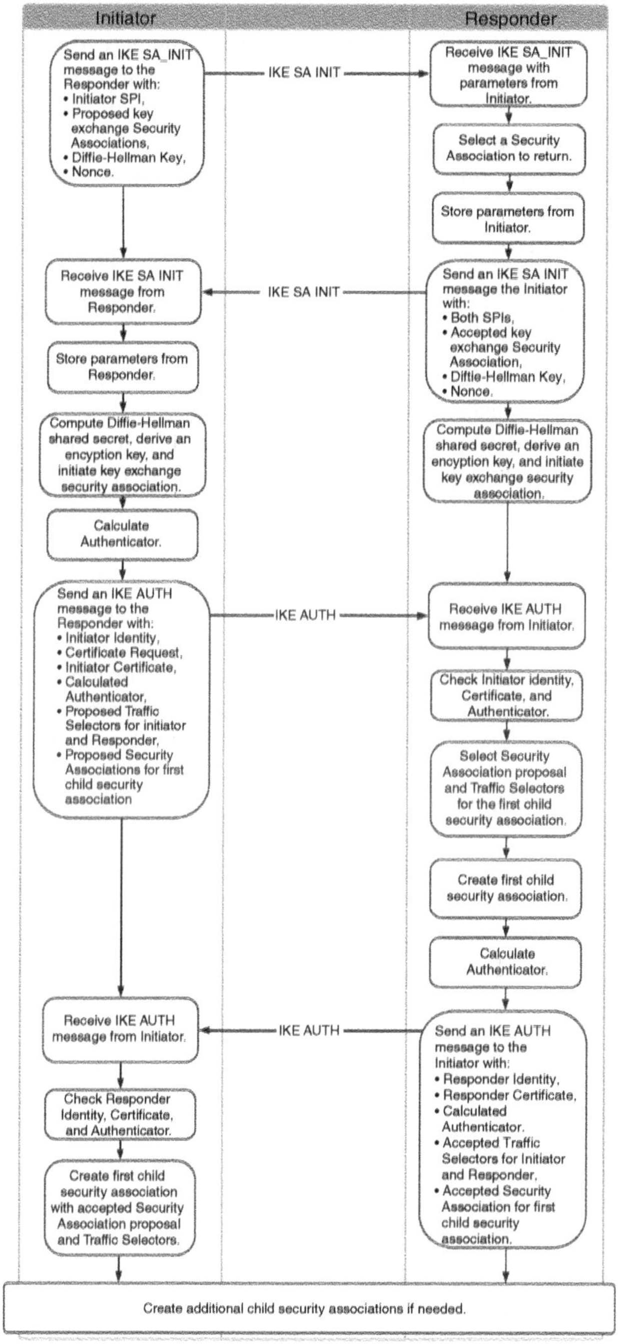

Figure 5.20 Example IKEv2 Initialization and Authentication Exchange

- The Initiator stores the parameters from the Responder.
- Both sides compute the Diffie–Hellman shared secret using the agreed-upon algorithm, derive the encryption key, and set up the security association for the key exchange. From this point on, all messages are encrypted with the key exchange security association encryption key.
- The Initiator calculates an Authenticator consisting of a portion of their original *IKE_SA_INIT* message combined with the Responder's nonce and some other values and signs the Authenticator with the private key corresponding to the public key in their certificate. It is essential for the security of the protocol that the Initiator and Responder incorporate the other side's nonce in their Authenticator calculations.
- The Initiator sends an *IKE_AUTH* to the Responder with their Identity, a Certificate Request, their Certificate, the calculated Authenticator, proposed Initiator and Responder traffic selectors elements for the first IPSec child security association, and proposed Security Associations for the first IPSec child security association.
- The Responder receives the *IKE_AUTH* message and checks the Initiator's Identity, Certificate, and Authenticator. If the validation succeeds, the Responder chooses Traffic Selectors and a Security Association for the first IPSec child security association and creates the security association.
- The Responder calculates its Authenticator, which consists of a portion of its original IKE_SA_INIT message, the Initiator's nonce, and some other values, and signs it with the private key corresponding to the public key in their certificate.
- The Responder replies with an *IKE_AUTH* message to the Initiator with their Identity, a Certificate Request, their Certificate, the calculated Authenticator, accepted Initiator and Responder Traffic selectors, and accepted Security Association.
- The Initiator receives the *IKE_AUTH* message and checks the Responder's Identity, Certificate, and Authenticator. If the validation succeeds, the Initiator creates the first IPSec child security association with the accepted Traffic Selectors and Security Association.
- If additional security associations are required, the two sides move to the Create Child Security Association phase and exchange *CREATE_CHILD_SA* messages.

Since certain steps require a significant amount of computation, IKEv2 includes a DDoS protection provision in case the Responder comes under volumetric traffic attack. After a certain number of *IKE_SA_INIT* initialization requests from different IP addresses, the Responder can include a cookie in its response and refuse to accept further requests until the Initiator replies with the cookie. The Responder can then demand a certificate with a Certificate Request in the

IKE_SA_INIT response. The DDoS protection provision is described in Section 2.6 of RFC 7296 [67]. RFC 7296 also includes protocol exchanges for many common cases, including error cases, throughout the document.

5.10 IPSec

Internet Protocol Security (IPSec) is the standard protocol suite for security at the network layer in the IP stack. The development of IPSec dates back to the 1990s when enterprise organizations needed a way to connect branch offices or remote employees into the corporate network. Previously, organizations relied on dedicated private networks, such as leased lines, to establish secure connections with their branch offices or to connect remote employees. However, these dedicated networks were expensive and often limited in scalability. IPSec emerged as a solution to provide secure and cost-effective communication channels over the Internet, enabling the creation of VPNs.

While IPSec defines the mechanisms for securing IP packets, IKE handles the negotiation and establishment of the security associations required by IPSec. As discussed in Section 5.9, IKE enables the secure provisioning and management of cryptographic keys and the establishment of shared security associations between participating devices for IPSec. The security association includes the cryptographic algorithms and parameters for confidentiality protection and/or integrity protection. A security association is identified by an SPI value included in the IKEv2 protocol packets. By combining IPSec and IKE, secure communication channels can be established between nodes at the network layer.

IPSec provides several security services essential for securing network layer traffic. Identity authentication ensures that the participating parties in a communication are mutually authenticated to each other. The most common method of mutual authentication is to exchange public key certificates that have been issued by a trusted CA but other methods are also supported. Access control is enforced by the distribution of cryptographic keys and traffic selectors for the security associations through IKE. IPsec supports encryption of payloads and optionally headers to ensure confidentiality, and calculation and inclusion of message integrity codes to ensure integrity of the traffic. To protect against replay attacks, IPsec packets incorporate sequence numbers and timestamps, and any packets that are duplicate or out-of-order are discarded.

The security services for IPSec are implemented through two different protocols:

- Authentication Header (AH) – AH provides data origin authentication, integrity protection, and protection against replay attacks. It computes a keyed hash-based MAC (HMAC) using a shared key between the sender and receiver to ensure that the data has not been tampered with during transit and to verify that the data did originate from the sender, and uses sequence numbers to protect against replay attacks,

- Encapsulating Security Payload (ESP) – ESP provides data origin authentication, integrity protection, confidentiality protection, and protection against replay attacks. Data origin authentication, integrity protection, and protection against replay attacks are implemented similarly to AH, while confidentiality protection is implemented with encryption using a shared key between the sender and receiver.

IPSec is specified in a collection of RFCs which have been revised several times over the years since the protocols were originally designed. RFC 4301 [68] provides an overview of the IPSec architecture and is paired with RFC 4835 [69] which describes the cryptographic requirements for AH and ESP. RFC 4302 [70] describes the AH protocol while ESP is specified in RFC 4303 [71]. In addition to the IPsec specifications, a thorough knowledge of IKEv2, which was discussed in Section 5.9, is necessary to understand how IPsec security associations are constructed and torn down. IKEv2 is specified in RFC 7296 [64].

5.10.1 IPSec Reference Architecture

IPSec deployments fall into one of three different scenarios:

- Site-to-site VPN,
- Machine-to-site VPN,
- Machine-to-machine VPN.

In the site-to-site VPN deployment scenario, IPSec is used to establish secure communication channels between two or more geographically separated networks. This scenario is commonly employed by organizations that have private networks in multiple branch offices or remote sites interconnected over the Internet or for interconnecting a private network with resources in a cloud network. Each site has a VPN gateway node that serves as the endpoint for secure communication with the other sites. The VPN gateway peers authenticate each other, establish a security association, and encrypt and encapsulate the traffic between the sites.

In the machine-to-site VPN deployment scenario, IPsec is utilized to establish secure communication between an individual machine and a central network. This scenario is often employed by remote workers or mobile devices that need secure access to resources hosted within an organization's private enterprise network or in a cloud network. The machines act as VPN clients and establish a secure connection with a VPN gateway or concentrator at the organization's site. IPsec is used to authenticate the client, establish a security association, and encrypt the traffic between the machine and the site. This enables remote workers or mobile devices to securely access resources, such as files, applications, or internal services, as if they were directly connected to the organization's network.

In a machine-to-machine VPN deployment scenario, IPsec is utilized to establish secure communication channels between individual machines or devices. This scenario is particularly relevant in situations where specific devices, such as servers, IoT devices, or remote sensors, need to communicate securely with one another over public or untrusted networks. Each machine involved in the VPN deployment acts as an IPSec endpoint that handles authentication, security association establishment, and encryption of packets. Machine-to-machine VPNs provide a secure and isolated communication environment for the devices, protecting sensitive data, ensuring data integrity, and enabling secure interactions between machines regardless of their physical location.

The specifications for AH and ESP in RFC 4302 and 4303 specify two modes for handling these deployment scenarios. Transport mode is used for the machine-to-machine scenario while tunnel mode is used for the site-to-site and machine-to-site scenarios. In transport mode, AH authentication is calculated on that part of the IP header that does not change during routing and on the payload, and ESP only encrypts the payload. In tunnel mode, an authenticated and/or encrypted tunnel is created between the two IPSec peers, with the entire packet being authenticated and/or encrypted, including all the header fields. A separate header is added for routing over the Internet between the peers. After IPSec processing, a transport mode packet is sent up the receiving peer node's IP stack for further processing by an application running on the host. If the packet is sent in tunnel mode and the receiving peer is an end node, the packet is handled similarly to a transport node packet. If the receiving peer is an IPSec gateway, the packet is routed out of the gateway's interface through the protected network and to the destination host. The two IPsec peers specify whether the security association is to run in transport mode or tunnel mode through a parameter in the Security Association payload element of the original IKE exchange that set up the security association.

In Figure 5.21 illustrates the architectural elements for an IPSec Gateway in a site-to-site VPN. For a machine-to-machine deployment or a machine-to-site deployment on the machine side, the architectural elements in the node would be similar, except packets would be routed up the IP stack rather than out to other nodes through the Protected Network Interface. The left side shows a building network connected through the VPN over the Internet to the BEMS in a cloud deployment network on the right. The IPSec Gateway typically runs on a virtual or physical machine.

The Security Policy Database (SPD) is responsible for storing security policies that determine how IP packets should be processed and protected. It contains a set

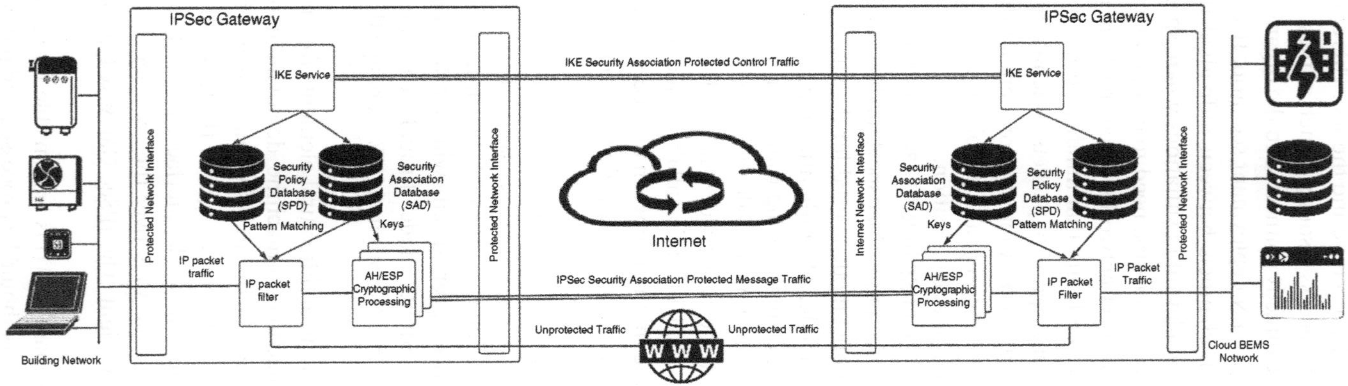

Figure 5.21 IPSec Reference Architecture

of rules that define the traffic selectors and the associated security services (AH, ESP, or both) to be applied to the packets. The Security Association Database (SAD) maintains the security associations negotiated through IKE between IPSec peers. IPSec defines a security association as a unidirectional relationship between two endpoints, containing security parameters such as encryption and authentication algorithms, keys, SPI, and other security attributes. For any endpoint, there are two security associations: one for outgoing and one for incoming traffic. The SAD keeps track of active security associations for packet processing. Policies and security associations are managed by the IKE Service, which negotiates these with peers over the Internet through a connection protected by the IKE SA. The SPD and SAD may also be accessible through a control application running on the IPSec Gateway node that allows network management engineers to examine and change the stored policies and security associations.

IP traffic to and from the protected network exits and enters the IPSec Gateway through the protected network interface, from which it is directed through the IP Packet Filter module. The IP Packet Filter applies matching rules to packet headers with patterns obtained from the SPD and SAD. If a packet's header does not match any rules, it is sent out through the Internet network interface unprotected. Incoming packets that do not match any rules may be dropped. If a packet's header matches an IPSec rule, the packet is sent to the AH/ESP Cryptographic Processing module which obtains keys for applying the appropriate security association from the SAD. The packets protected with AH, ESP, or both are then sent out to the Internet through the gateway's Internet network interface, from which they traverse the Internet until they reach the peer gateway. At the peer gateway, the process is reversed, decrypting and validating the packets before they are sent to protected network nodes.

5.10.2 Authentication Header (AH)

Figure 5.22 shows the IPSec header format for AH. The fields in the header are the following:

- Next Header – code number of the next header after the AH header. These are chosen from the IANA IP protocol codes [72], for example, 6 if the next header is for TCP.
- Header Length – length of the AH header in 4-byte words minus 2. In IPv6, this must be a multiple of 8 bytes, so the Integrity Check Value needs to be padded with zeros if the Header Length field value is shorter.
- Security Parameters Index (SPI) – the 8-byte index used by IKE to designate the applicable security association from the SAD for cryptographic processing.
- Sequence Number – a 32-bit unsigned integer that is incremented for each packet sent under the security association designated by the SPI. If the two sides

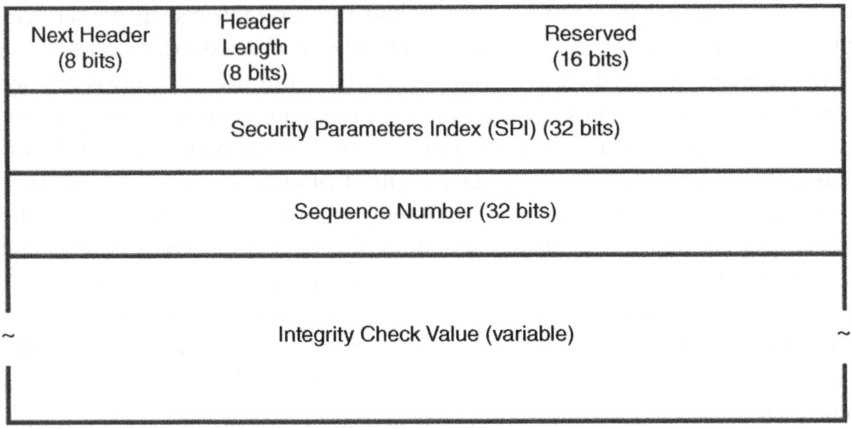

Figure 5.22 IPSec AH Header Format

have negotiated a 64-bit extended sequence number for the security association using IKE, then only the low-order 32 bits are transmitted. The high-order bits are maintained by the sender and receiver as part of the security association but are not transmitted to save space in the header.

- Integrity Check Value – a variable-sized message integrity code calculated over the packet using a keyed cryptographic hash algorithm that is specified in the security association.

If AH is in use, the IPv4 or IPv6 header Next Header field should include the type code 51 for AH when the AH header is positioned immediately after the network header.

The contents of the packet that are included in the message integrity code calculation depend on whether tunnel mode or transport mode processing is in effect. If tunnel mode processing is being used, the entire IP packet including all fields in the header is included in the message integrity code calculation. For outgoing packets, the message integrity code is written into the Integrity Check Value field of the AH header and the outer, unauthenticated IP header and the AH header are prepended to the inner packet, with the IP header leading. For incoming packets, the outer header and AH header are removed before the message integrity code is calculated on the inner IP packet, and then the calculated message integrity code is compared with the Integrity Check Value in the AH header.

If AH processing is running in transport mode, the entire payload is included in the message integrity code calculation, but parts of the AH header and IP header are left out of the calculation. Any fields in the AH and IP header that are left out of the calculation are filled with zeros for the message integrity code calculation. For the AH header, the Integrity Check Value field is naturally not included in the calculation because it is the value being calculated. In addition, if extended sequence numbers are in effect, the high order 32 bits of the sequence number are included in message integrity code calculation even though they are not included in the Sequence Number field.

For the IP header, only fields that are immutable or predictable at the receiver are included in the calculation. Fields that are left out in IPv4 include the Differentiated Services Code Point field, the Explicit Congestion Notification field, the Flags field, the Fragment Offset field, the Time to Live field, and the Header Checksum field. Excluded fields in IPv6 include the Differentiated Services Code Point field, the Explicit Congestion Notification field, the Flow Label field, and the Hop Limit field. These fields can all be changed in transit, and so, a message integrity code cannot be calculated if they are. In addition, option headers in IPv4 or extension headers in IPv6 that are mutable and can't be predicted at the receiving end are left out of the message integrity code calculation. Appendix A of RFC 4302 [73] contains a list of the IPv4 and IPv6 headers that are immutable, mutable in transit but predictable, or mutable.

5.10.3 Encapsulating Security Payload (ESP)

The format of a packet protected by ESP is shown in Figure 5.23. The SPI and Sequence Number constitute the ESP header, while the Integrity Check Value constitutes the ESP trailer and the other fields are part of the payload. The SPI and

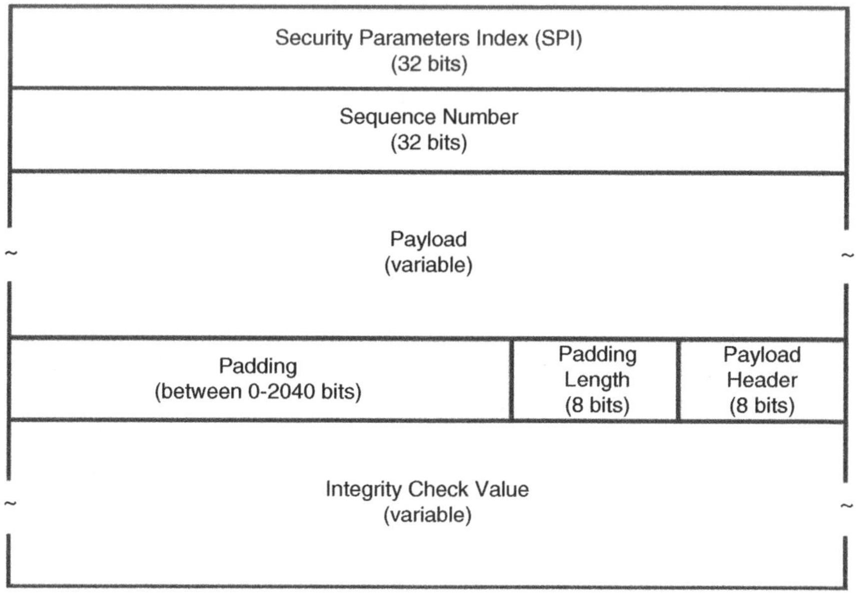

Figure 5.23 IPSec ESP Packet Format

Sequence Number perform the same function as in the AH protocol. The Payload may contain cryptographic synchronization data for the receiver. In addition, extended (64-bit) sequence numbers can be used in ESP as well, and the same rules hold (only put the lower 32 bits into the packet header). The Integrity Check Value is also the same as in AH. The size of the Integrity Check Value is determined by the message integrity code algorithm that is part of the security association. Integrity protection is provided to the fields from the SPI through the Payload Header field, while confidentiality protection is only provided for the Payload, Padding, Padding Length, and Payload Header fields.

The Payload contains the encrypted contents from the original packet, including any headers, and is variable in size but an integral number of bytes in length. Following the Payload is the Padding field containing 0–255 bytes of padding. The padding in the Padding field is required by some encryption algorithms on input, particularly block ciphers, to align the cleartext data to a certain block size boundary before processing by the encryption algorithm. The Padding Length field provides the length of the padding bytes. The Payload Header[3] field is the IP protocol number for the first header in the encrypted payload from the IANA Protocol Numbers registry [72], for example, 6 if a TCP payload is protected or 4 if an entire IPv4 packet is encapsulated.

For processing of a packet with padding, the decryption process is applied to the entire payload, from the start of the encrypted data to the end of the Payload field, including the padding and trailer. After decryption, the receiver can read the Pad Length field in the decrypted trailer to determine how much of the decrypted data is padding.

The security associations setup for ESP can specify one of five different types of cryptosystem algorithms depending on the level and type of protection required:

- encryption only,
- authentication only,
- encryption and authentication,
- combined mode.

If encryption only or authentication only is specified in the security association, the other algorithm must be null in the security association. If authentication is not used, the Integrity Check Value trailer is not included in the packet, while if encryption is not used, the Payload Data, Padding, Pad Length, and Next Header fields are included in plaintext. A major difference between AH and ESP authentication is that AH protects parts of the IP header while ESP does not. If protection of the IP header is desired in transport mode, ESP can be used for

3 RFC 4303 calls this field Next Header.

encryption with AH for authentication rather than or in addition to using the native authentication capability, in which case, security associations for both AH and ESP are required.

Depending on the encryption algorithm specified in the SA, the algorithm may require an initialization vector or other parameters to synchronize the encryption between the sender and receiver. If the encryption algorithm requires an initialization vector as a parameter, the RFC specifying how to use the algorithm with ESP must include it and any other such explicit, per-packet synchronization parameters – including the length, any structure for such parameters, and the location of the parameters – at the start of the Payload field. Synchronization parameters included in the Payload field are not encrypted. In addition, the header at the beginning of the encrypted Payload field must be aligned to 4 bytes in IPv4 and 6 bytes in IPv6. Similarly for integrity protection algorithms, the RFC specifying how to use the algorithm with ESP needs to specify the amount of padding if the algorithm requires padding to align the message on a specific byte boundary before generating the message integrity code.

Combined mode algorithms include encryption and message authentication in one operation. As with encryption, the RFC describing how to use a combined mode algorithm with ESP is required to specify the Payload field substructure used to send the message integrity code. Since the SPI and Sequence Number are required to be integrity protected, the RFC must also specify how the algorithm provides protection for these fields, as they are not encrypted. If the combined mode algorithm only provides integrity protection for encrypted data, the SPI and Sequence Number must be replicated and encrypted with the payload. Since the integrity check code is included in the encrypted Payload field for combined mode algorithms, the Integrity Check Value field can be omitted.

When processing an IPv4 packet in transport mode, the ESP packet payload from Figure 5.23 is inserted directly after the IP header and any option headers, with the ESP Payload field encapsulating the transport and application layer protocol payloads. If AH is also applied to a packet, it is applied to the ESP header fields, Payload field, and ESP trailer fields if present. For IPv6, ESP is considered to be an end-to-end payload. Therefore, it should appear after hop-by-hop, routing, and fragmentation extension headers. Other destination options extension headers if any should be placed after the ESP header to ensure that they derive the benefit of ESP encryption and authentication. In tunnel mode, the entire IPv4 or IPv6 packet is placed within the ESP Payload field, and a new header, including any header options, is affixed to the packet. The inner IP packet is tunneled to the receiver address in the outer header, where it is decapsulated and routed further to the ultimate receiver address on the inner packet.

5.11 TLS

TLS is the standard IP stack security protocol for the transport layer. TLS evolved from the SSL protocol which was initially developed by Netscape in the mid-1990s to secure web traffic. SSL version 1.0 was never publicly released due to severe security flaws. SSL versions 2.0 and 3.0 were released in 1995 and 1996 respectively, each improving on the security of the previous version. By 1998, however, it was clear that SSL needed a complete overhaul to address various security vulnerabilities. The IETF released a redesigned transport layer security protocol named TLS 1.0 in early 1999, which quickly achieved widespread deployment momentum. Since then, TLS has gone through multiple versions with TLS 1.3 [74], released in 2018, being the latest.

TLS has been combined with a broad variety of transport-level protocols. The most common case is TLS running over TCP. Protocols like HTTP for secure web browsing (forming the Hypertext Transfer Protocol Secure (HTTPS) when combined with TLS), SMTP for secure email transfer, FTP for secure file transfers, and XMPP for secure chat and messaging services are all protocols that use TCP for their transport layer and can leverage TLS to secure their communication. DTLS, specified in RFC 9147 [75], is a version of TLS adapted for use with connectionless protocols like UDP. DTLS provides the security benefits of TLS but is suitable for an environment where the transport model is based on a connectionless, packet-based channel rather than a connection-based stream. EAP-TLS [11] provides mutual authentication at the link layer and is used in Ethernet and Wi-Fi networks as discussed in Section 5.3.2.

Security services provided by TLS include identity authentication, confidentiality protection, integrity protection, data origin authentication, and forward secrecy. TLS uses PKI and digital certificates to authenticate the server and optionally the client. This authentication ensures that the communicating parties are who they claim to be. Two parties to the communication can utilize the public keys in their certificates to perform a Diffie-Hellman key exchange. Through data encryption using a shared key, TLS ensures that the data being transferred remains confidential. Using message integrity codes, TLS provides a method to ensure the data has not been altered in transit. The process of authentication in TLS also ensures that the data received has originated from the expected source, providing data origin authentication. By employing ephemeral Diffie–Hellman key exchange, modern versions of TLS (1.2 with certain cipher suites, and 1.3) support forward secrecy, ensuring that even if a node's private key is compromised, past communications are secure and cannot be decrypted.

TLS was originally developed for client/server communications, specifically on the World Wide Web but also for other TCP-based client/server application protocols. In the early days of the World Wide Web, there was no way for a client to authenticate the identity of an ecommerce server it contacted, making any financial transaction subject to possible fraud and theft. With SSL and later TLS, the server presents the client with a certificate that the client verifies, and an encrypted and authenticated exchange then can proceed between the client and server. However, TLS as used in web browsers does not require the client to present a certificate. With the advent of the zero trust architecture for cloud and microservices in the late 2010s, mutual authentication is required between both sides. TLS can also be used for mutual authentication; in which case it is sometimes called mTLS for mutual TLS.

5.11.1 TLS Overview

Unlike IPsec and IKE, TLS incorporates the key exchange and security association creation into a subprotocol called the Handshake protocol. With the Handshake protocol, the client and server agree on the protocol version, select cryptographic algorithms, exchange parameters, authenticate each other, and establish a shared secret for derivation of the encryption and authentication keys. Once the handshake is successfully completed, the client and server can start exchanging protected application messages during data exchange. The application data is encrypted and authenticated using keys derived from the shared secret established with the Handshake protocol. In addition to these protocols, TLS 1.3 has built-in mechanisms for session resumption and retrying the negotiation if the server and client can't immediately agree on the parameters.

The Handshake protocol can be broken into three phases:

- Key Exchange – The client and server select cryptographic algorithms and parameters and derive a shared secret and keys for authentication and encryption. All messages exchanged after this phase are encrypted,
- Server Parameters – Other handshake parameters such as whether the client's identity is authenticated, application layer support, etc., are established. Because messages in this phase are encrypted, they can be used to send extensions and server information that must be kept confidential,
- Authentication – The server, and optionally the client, provide proof of identity and both sides provide key confirmation and prove handshake integrity.

The Handshake protocol and the protected messages exchanged during data exchange are encapsulated in packets of another protocol, the Record protocol. Record protocol packets can be either unencrypted, such as for the Key Exchange phase of the Handshake protocol, or encrypted and authenticated such as during the Server Parameters and Authentication phases of the Handshake protocol or during data exchange. Once the client and the server have established a security association for the channel, the encrypted Record protocol messages are used to complete the Handshake protocol and transfer application data between the two sides.

TLS 1.3 supports three different methods of generating a shared secret for key derivation:

- The Diffie–Hellman key exchange, either regular Diffie–Hellmanor elliptic curve Diffie–Hellman. The client and server must exchange large random numbers and their public keys during the Key Exchange phase,
- Preshared keys, in which the client and server are preconfigured through some mechanism outside the TLS protocol with a shared secret,
- Preshared keys combined with regular Diffie–Hellman or elliptic curve Diffie–Hellman.

For authentication, TLS 1.3 uses either a preshared key or a certificate, but not both at the same time.

5.11.2 TLS Record Protocol Messages

RFC 8446 does not use character-based block diagrams to define the header and message body element formats like most other IETF protocol specifications. Instead, it uses a bespoke specification language similar to ASN.1. Because the TLS 1.3 specification is extremely complex and requires several accommodations for backward compatibility with TLS 1.2, the authors needed more flexibility in defining the message formats.

Figure 5.24 illustrates the TLS Record protocol message format. The Content Type field contains a code for the type of the subprotocol message in the Message Body. The subprotocol message types are listed in Table 5.8. The Version field contains the protocol version. For TLS 1.3, this must be set to 0x0303 which is actually the version number for TLS 1.2. The reason is to ensure that middleboxes like firewalls and web proxies can pass TLS 1.3 without having to be updated. The actual version of the protocol, which is 0x0304, is negotiated in the supported versions extension during the Handshake protocol. The Length field specifies the length of the Message Body field following the header.

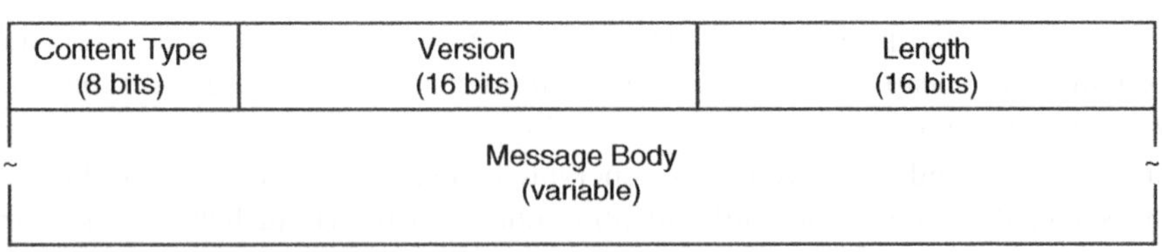

Figure 5.24 TLS Record Protocol Packet Format

Table 5.8 TLS Record Protocol Message Types

Message	Code	Function
Invalid	0	Not a valid Record protocol message type code.
ChangeCypherSpec	20	Single byte message adopted from TLS 1.2 and included for middlebox compatibility purposes only.
Alert	21	Error or warning message from the peer node.
Handshake	22	Message is one of the Handshake protocol messages.
ApplicationData	23	Message includes application data.
Heartbeat	24	Used as a keep-alive message and for discovering the maximum packet size over a link.

The *Handshake* message type is used to signal encapsulation of Handshake protocol messages. *Alert* messages can be either one of two types: closure alerts or error alerts. Closure alerts indicate that the sender wants to close the connection. After a closure alert, the sender does not send or receive any further messages under the security association. Error alerts indicate some problem with the message exchange, which can be either a warning or fatal. If the alert indicates a fatal condition, the connection is closed without sending any further data. The Record protocol uses *ApplicationData* messages during application data exchange. Messages from the application layer are fragmented into blocks, the blocks are encrypted, a message integrity code is added, and the secured blocks are transmitted to the other side in *ApplicationData* messages. The *Heartbeat* message type is not widely used since the discovery of the HeartBleed bug in 2014 [76].

5.11.3 TLS Handshake Protocol Messages

The TLS message format for Handshake protocol messages is shown in Figure 5.25. In the header, the Message Type field contains the message code for the Handshake message and the Payload Length field contains the length of the Handshake Message Elements field. The Handshake Message Elements

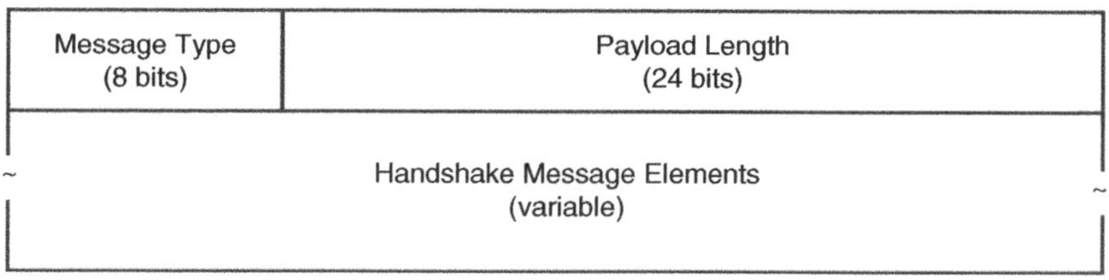

Figure 5.25 TLS Handshake Protocol Packet Format

Table 5.9 TLS Handshake Message Type Codes

Message	Code	Phase	Function
HelloRequest	0	N.A.	Reserved, not used in TLS 1.3
ClientHello	1	Key Exchange	Client sends version, preferred cryptosystem types, client parameters, and any extensions to server.
ServerHello	2	Key Exchange	Server replies with selected cryptosystem, server parameters, and accepted extensions.
HelloRetryRequest	2	Key Exchange	Server requests the client to retry the *ClientHello* with different parameters. It uses the same message number as *ServerHello* except the Random message element field is set to the SHA-256 hash of "HelloRetryRequest."
EncryptedExtensions	8	Server Parameters	Server sends extensions to the client after the encryption keys have been derived, ensuring the confidentiality of the message contents.
Certificate	11	Authentication	Server and optionally the client send the certificate to the other party.
CertificateRequest	13	Authentication	Client and optionally the server request that the other party send their certificate.
CertificateVerify	15	Authentication	Server and optionally the client prove that they possess the private key corresponding to the public key in the certificate.
Finished	20	Authentication	Sent by both the client and the server to verify that the key exchange and authentication processes were successful and to signal the completion of the Handshake protocol.

field holds the various message body elements the two sides are exchanging, which depend on the message type.

Table 5.9 contains a list of the message types, the phase in which they are used, and a brief description of their functions in the core Handshake protocol. RFC 8446 contains a few additional messages not shown in the table that are used for enabling a client to restart a previous session and to update keys. These messages make the overall protocol more efficient by not requiring a security association to be negotiated if some of the security association agreements can be reused from the old one.

5.11.4 Example TLS 1.3 Handshake Protocol Exchange

Figure 5.26 shows an example TLS 1.3 Handshake protocol exchange in the case where the client and server are using Diffie–Hellman key exchange for generating a shared secret and the server is supplying a certificate for authentication. The optional mutual TLS steps where the client also authenticates with a certificate are shown with dotted arrows. Other exchanges are possible, for example, with a preshared key, and are documented in RFC 8446 [74].

The steps in the protocol correspond to the following:

- The client sends a *ClientHello* message to the server with its Diffie–Hellman public key, a random nonce, and a list of proposed cryptosystem algorithms for key derivation, encryption, and message integrity code calculation.
- The server selects one of each type of cryptosystem algorithm, stores the parameters from the client, and replies with a *ServerHello*, including its public key, random nonce, and the selected algorithms. If the server can't find an algorithm it supports, it can request the client to retry.
- The client stores the parameters from the server.
- Both sides derive their Diffie–Hellman shared secret, and shared keys for encryption and message integrity code calculation, and initiate the security association. From this point on, all messages sent between the client and server are encrypted.
- If mutual TLS is desired, the server sends a *CertificateRequest* message to the client.
- The server sends a *Certificate* message to the client with its public key certificate, followed by a *CertificateVerify*. The *CertificateVerify* message contains a digital signature of all the handshake exchange messages up to this point, signed with the private key corresponding to the public key in the certificate.
- The client receives the *Certificate* message and stores the server's certificate, then receives the *CertificateVerify* message and verifies the signature with the public key from the certificate.
- The server sends a *Finished* message, including a message integrity code calculated over all the handshake exchange messages up to this point, using the shared message integrity code calculated during the key exchange.
- If mutual TLS is desired and the server requests a certificate from the client, the client sends its certificate in a *Certificate* message and *CertificateVerify* message, with a digital signature over all the handshake messages up to this point.
- If mutual TLS is desired, the server receives the certificate and stores it, then receives the *CertificateVerify* message and verifies the digital signature.

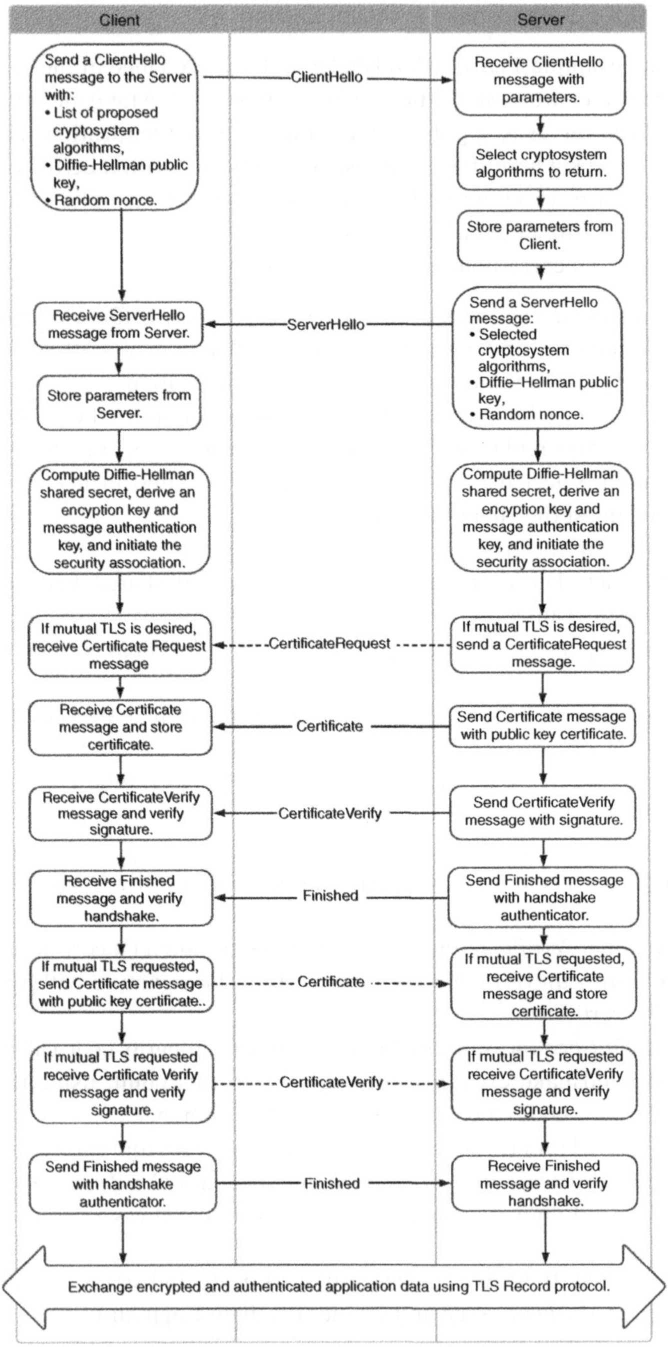

Figure 5.26 Example TLS 1.3 Handshake Protocol Exchange

- The client sends a *Finished* message, including a message integrity code calculated over all the handshake messages up to this point. The server receives the *Finished* message and verifies the message integrity code, concluding the Handshake protocol.
- The two sides can now exchange encrypted and authenticated application data using the TLS Record protocol.

5.12 Virtual Private Networks

An example use case of how security protocols are used for BEMS deployments is in routing traffic over the public Internet between geographically dispersed network sites with private address spaces. For such use cases, VPNs [77] provide security protection for packets in route over the Internet. VPNs are commonly deployed between buildings or from buildings on remote sites to cloud or private data center deployments where the BEMS central components are located. With VPNs, each building site along with the central cloud deployment can be assigned a subnet in the private address space so nodes within the geographically dispersed sites under management can address each other transparently.

An example is shown in Figure 5.27. The EV charger with IP address 192.168.7.5 in the Commercial Building Network is sending a charging update to the cloud BEMS control center with IP address 192.168.1.42. The packet traverses the VPN Gateway at the Commercial Building Border Router, where it is encrypted, authenticated, and encapsulated in a packet with a destination address of the VPN Gateway in the Cloud Data Center, 178.42.42.76. The source and the destination address on the inner packet are not changed. At the Cloud Data Center VPN Gateway, the packet is decapsulated, decrypted, and verified and then sent to the destination address of the BEMS control center on the inner packet, 192.168.1.42.

Tunneling from the building sites happens through a VPN gateway which is often integrated into the BEMS gateway. A packet is encrypted, authenticated, and encapsulated into a new IP packet with globally routable source and destination IP addresses. The packet is then tunneled across the Internet to the destination, the contents being secured from eavesdropping and attackers. The IP header on the inner packet keeps the private addresses and is decapsulated from the tunnel at the VPN gateway in the destination network. In addition to encapsulation, VPNs add authentication and encryption to avoid man-in-the-middle attacks and snooping. For applications accessing public websites and other addresses on the Internet, a VPN is not needed. Traffic is routed through a firewall without being encrypted, although if the website uses HTTPS, it will be encrypted end-to-end between the client in the private address space and the server on the Internet.

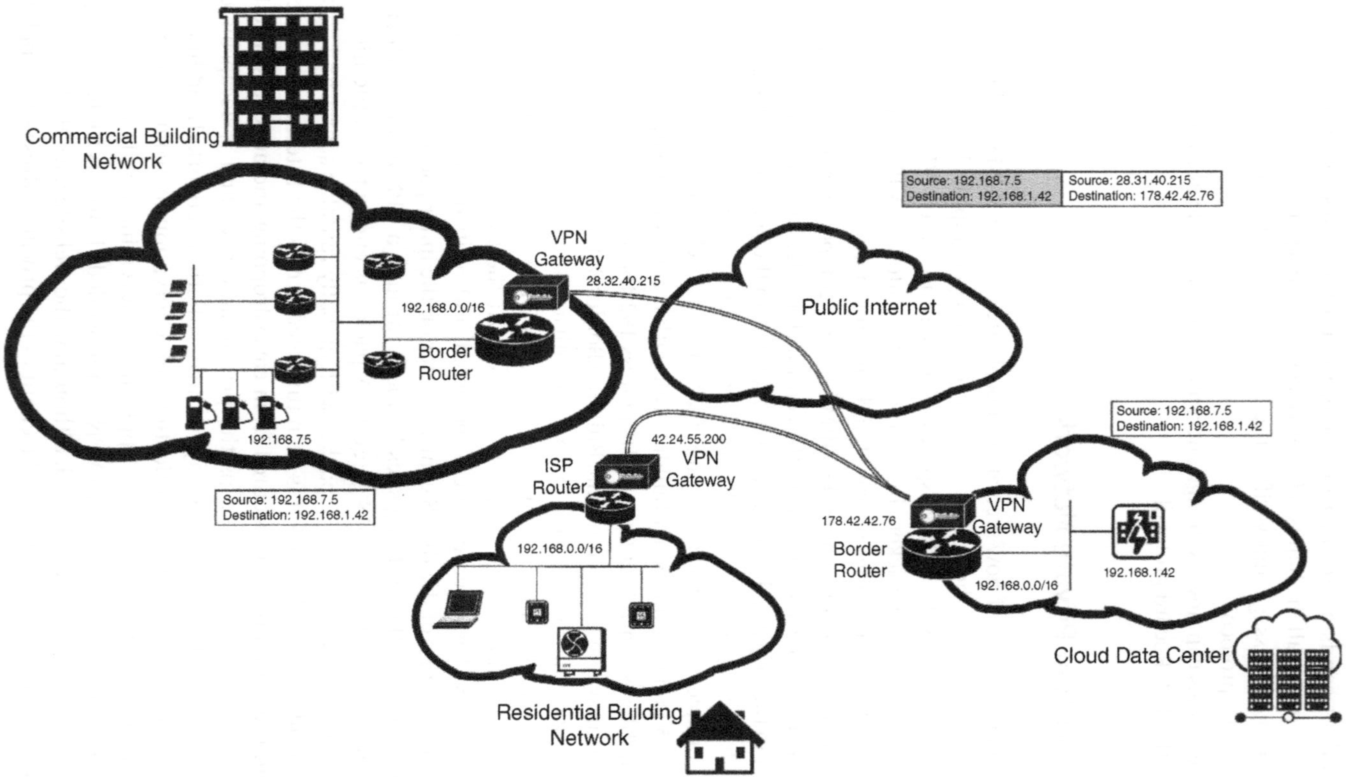

Figure 5.27 VPN Tunnel Deployment for Remote Site to Cloud Communication

5.12.1 Open Source Tools for Deploying VPNs

For BEMS deployments, open source VPNs are widely available and typically reliable and secure. Open source VPNs come in two basic types:

- IPSec VPNs use IKE to establish a security association and IPSec ESP to tunnel traffic,
- Layer 3 VPNs use some other protocol to establish the security association and an application layer protocol on top of UDP or sometimes TCP with the contents encrypted and authenticated to tunnel traffic.

Comparing the two types of VPNs, both offer comparably secure network connections. However, they differ somewhat in ease of deployment and feature set. Deployment of IPSec VPNs can be complex, especially in heterogeneous environments. They require careful configuration and management of security policies, keys, and certificates, which can be time-consuming and error-prone, especially when deployment conditions change. Layer 3 VPNs are typically easier to deploy, configure, and manage compared to IPSec VPNs. With respect to feature set, IPSec VPNs offer a wide range of features, including support for multiple encryption algorithms, key exchange protocols, authentication methods, and more. Layer 3 VPNs offer a similar feature set, including robust encryption, key exchange, and authentication options.

While IPSec VPNs are more likely to be used for site-to-site VPNs, it is possible to use Layer 3 VPNs for VPN gateways in a site-to-site deployment. Using a Layer 3 network for a site-to-site deployment requires configuring one node in the site network as the VPN gateway and setting up the routing so that the addresses in the private address space on the remote site are recognized and routed into the VPN interface. At the VPN interface, the packets are encapsulated, authenticated, and sent to the other site gateway. Setting up a site-to-site VPN with a Layer 3 solution requires a more complex configuration than for a simple node-to-node VPN.

5.12.2 Open Source IPsec VPNs

There are a number of open-source IPsec VPNs available, but OpenSwan [78] is one of the most widely deployed. OpenSwan is an open-source implementation of IPsec and IKE for Linux. It was initially released in 2003 as a fork of the FreeS/WAN project, a now-defunct pioneer in Linux-based IPsec implementations, and has, itself, been forked into various other child projects such as strongSwan and LibreSwan. OpenSwan provides a secure VPN for Linux, offering both node-to-node and network-to-network connections, and supports all the security services supported by IPsec and IKE.

OpenSwan supports both IPsec transport and tunnel modes, allowing encrypted and secure communication between two endpoints or encapsulating an entire

IP packet for transmission across the Internet. For establishing a security association, OpenSwan uses the IKE protocol. It supports both IKEv1 and IKEv2 and a variety of encryption algorithms and hash functions. Authentication with OpenSwan can be done through PSKs or X.509 certificates. The use of X.509 certificates is more secure and scalable, especially in larger deployments, as it provides each user or device with its own unique credentials, and a CA can centrally manage these credentials.

5.12.3 Open-Source Layer 3 VPNs

Layer 3 VPNs are based on a common architectural pattern of implementing the tunneling protocol as an application layer protocol on top of a standard transport layer, usually UDP. Figure 5.28 illustrates the generic Layer 3 VPN packet format. The application layer payload consists of the encrypted packet and any encryption parameters needed for synchronizing the decryption. The message integrity code is provided either in the Layer 3 VPN Header or appended at the end of the packet after the encrypted packet payload in the Layer 3 Trailer. The source and destination address fields of the outer IP packet header are the source and destination VPN gateways, while the source and destination address fields of the encrypted inner packet are the source node and destination node addresses in the networks behind the VPN gateways. The transport layer header destination port number in the outer packet transport header is the port on which the Layer 3 VPN application

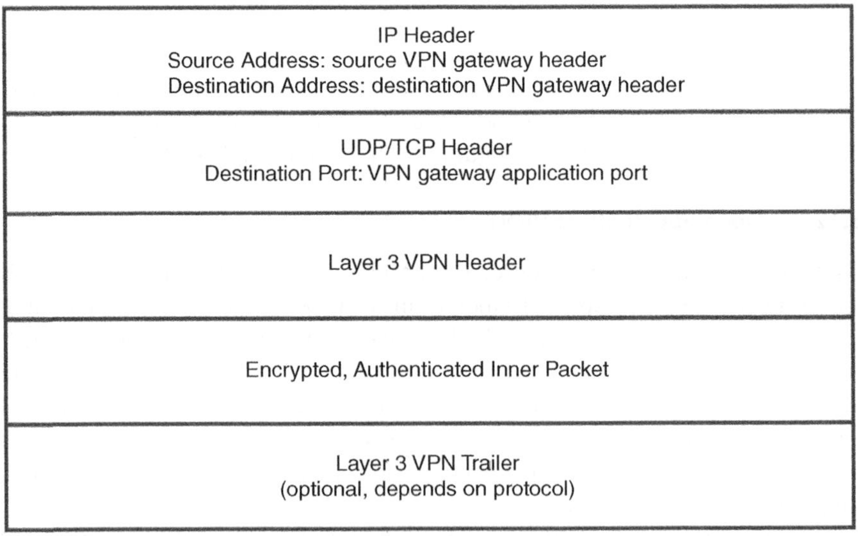

Figure 5.28 Generic Layer 3 VPN Packet Format

is expecting to receive packets. When a packet comes in, the Layer 3 VPN application verifies and decrypts it, and forwards the packet to the destination node address on the inner packet header.

5.12.3.1 OpenVPN

OpenVPN is an open-source VPN protocol that is designed to be highly configurable and extensible with many advanced features and a wealth of configuration options [79]. It supports the following security services: identity authentication, access control, confidentiality protection, integrity protection, forward secrecy, and data origin authentication. OpenVPN uses the OpenSSL package for cryptosystem support with the TLS 1.3 Handshake protocol to establish the security associations between gateways; however, it does not use the TLS 1.3 Record protocol for tunneling packets. Instead, it uses an application-specific protocol with its own application protocol header. The message integrity code is calculated over the concatenation of any parameters needed for the encryption algorithm and the encrypted packet being tunneled, then it is inserted into the packet header along with other fields. The concatenated encryption parameters and encrypted packet are then inserted into the application payload field. OpenVPN can run over either the TCP or UDP transport layers, though UDP typically provides more reliable performance. The OpenVPN application runs in user space on port 1194 for UDP and port 443 for TCP.

5.12.3.2 Wireguard

Wireguard is an alternative to OpenVPN that is considerably more lightweight [80, 81]. It supports the same list of security services as OpenVPN. Wireguard does not use any of the standard IPSec protocols. Instead, it utilizes the Noise protocol [82] for establishing a security association and a collection of modern cryptographic algorithms for authentication and encryption. In addition, the shared key is changed on a regular and automatic basis to provide forward secrecy. Wireguard is based on a principle called cryptokey routing. Cryptokey routing associates a public key with an IP address that is allowed inside the tunnel. This enables the receiving

Wireguard Details

This longer video is a talk by Jason Donenfeld, the original implementor of Wireguard, from the 2018 Black Hat security conference in Las Vegas. He describes some of the design considerations that went into Wireguard, why he chose to cut across the IP stack layers rather than work within them the way IPsec does, and how cryptokey routing works, among other topics.

Wireguard node to perform access control on incoming packets. Wireguard uses UDP for the transport protocol and the port number for the first tunnel is 51820. Additional tunnels must use a different port.

The Wireguard is implemented in a kernel module for speed, and a node running Wireguard can be configured as a VPN gateway. However, it is designed to be lightweight and simple to use, with fewer features than OpenVPN. Performance-wise, it is faster because of its streamlined cryptographic operations. In addition, Wireguard utilizes a key management algorithm that is designed to minimize the risk of key compromise.

5.13 Summary

As the Internet has become more deeply embedded in the way individuals and commercial, nonprofit, and government organizations transact business, it has become a larger target for attackers seeking to gain advantage at the expense of others. Since a BEMS depends on the Internet to communicate sensor data and actuator commands from the BEMS control system in a private or public cloud data center to and from building gateways at remote sites, protection of the traffic from eavesdroppers and attackers is vital. While protocol security is not the only security measure required to protect the BEMS, it is nevertheless an important component of an overall security architecture.

This chapter discussed the details of link layer and IP stack security protocols. For each protocol, the security services provided by the protocol are outlined, along with protocol packet or header formats and, if relevant, a representative protocol exchange between the two sides seeking to establish a security association and exchange encrypted and authenticated data. Security architecture and design for the link layer protocols important in BEMS deployments, including Ethernet, Wi-Fi, 802.15.4, and BLE, are described in detail in Sections 5.2 through 5.6.4. Included in this discussion are 802.1X, Radius, and EAP, protocols that play a major role in authentication for wired and wireless networks. X.509 public key certificates and the PKI, which form the basis of identity authentication for some of the link layer protocols and for the IP stack, are presented in Section 5.7. For the IP stack, the IKE protocol and IPsec on the network layer, and TLS on the transport layer are discussed in Sections 5.8 through 5.11. Finally, the use of security protocols in VPNs is described in Section 5.12, as an example of an important security system component for BEMS network deployments. By understanding the important features of the link layer and IP stack security protocols and what security services they provide, BEMS engineers can gain an understanding of the role network protocol security plays in a comprehensive BEMS security architecture design and apply the understanding to BEMS deployments.

References

1 IEEE 802.1X-2020. IEEE Standard for Local and Metropolitan Area Networks – Port-Based Network Access Control. In IEEE Std 802.1X-2020 (Revision of IEEE Std 802.1X-2010 Incorporating IEEE Std 802.1Xbx-2014 and IEEE Std 802.1Xck-2018), pp. 1–289, 28 Feb. 2020, doi: 10.1109/IEEESTD.2020.9018454.

2 IEEE 802.1AE-2006. IEEE Standard for Local and Metropolitan Area Networks-Media Access Control (MAC) Security. In IEEE Std 802.1AE-2018 (Revision of IEEE Std 802.1AE-2006), pp. 1–239, 26 Dec. 2018, doi: 10.1109/IEEESTD.2018.8585421.

3 IEEE 802.1AR-2009. IEEE Standard for Local and Metropolitan Area Networks – Secure Device Identity. In IEEE Std 802.1AR-2018 (Revision of IEEE Std 802.1AR-2009), pp. 1–73, 2 Aug. 2018, doi: 10.1109/IEEESTD.2018.8423794.

4 LoRa Alliance. LoRaWAN® Is Secure (but Implementation Matters) [Internet]. [cited 2023 May 3]. Available from: https://lora-alliance.org/resource_hub/lorawan-is-secure-but-implementation-matters/.

5 GSMA. Security Features of LTE-M and NB-IoT Networks [Internet]. [cited 2023 May 3]. Available from: https://www.gsma.com/iot/wp-content/uploads/2019/09/Security-Features-of-LTE-M-and-NB-IoT-Networks.pdf.

6 Aboba B., Blunk L., Vollbrecht J., Carlson J., and Levkowetz H., Ed. Extensible Authentication Protocol (EAP), RF 3748 [Internet]. [cited 2023 May 7]. Available from: https://datatracker.ietf.org/doc/html/rfc3748.html.

7 Rigney C., Willens S., Rubens A., and Simpson W. Remote Authentication Dial In User Service (RADIUS), RFC 2865 [Internet]. [cited 2023 May 8]. Available from: https://datatracker.ietf.org/doc/html/rfc2865.

8 Cao Z., He B., Shi Y., Wu Q., Ed., and Zorn G., Ed. EAP Extensions for the EAP Re-authentication Protocol (ERP), RFC 6696 [Internet]. [cited 2023 May 12]. Available from: https://www.rfc-editor.org/rfc/rfc6696.html.

9 IANA. Extensible Authentication Protocol (EAP) Registry [Internet]. [cited 2023 May 12]. Available from: https://www.iana.org/assignments/eap-numbers/eap-numbers.xhtml.

10 Simon D., Aboba B., and Hurst R. The EAP-TLS Authentication Protocol, RFC 5216 [Internet]. [cited 2023 May 12]. Available from: https://datatracker.ietf.org/doc/html/rfc5216.

11 Mattsson J., Preuß, and Sethi M. EAP-TLS 1.3: Using the Extensible Authentication Protocol with TLS 1.3, RFC 9190 [Internet]. [cited 2022 May 12]. Available from: https://datatracker.ietf.org/doc/html/rfc9190.

12 Funk P. and Blake-Wilson S. Extensible Authentication Protocol Tunneled Transport Layer Security Authenticated Protocol Version 0 (EAP-TTLSv0), RFC 5281 [Internet]. [cited 2023 May 12]. Available from: https://datatracker.ietf.org/doc/html/rfc5281.

13 IANA. RADIUS Packet Type Codes [Internet]. [cited 2023 May 9]. Available from: https://www.iana.org/assignments/radius-types/radius-types.xhtml#radius-packet-types.

14 Aboba B. and Calhoun P. RADIUS (Remote Authentication Dial In User Service) Support For Extensible Authentication Protocol (EAP), RFC 3579 [Internet]. [cited 2023 May 9]. Available from: https://www.rfc-editor.org/rfc/rfc3579.html.

15 IANA. RADIUS Attribute Types [Internet]. [cited 2023 May 9]. Available from: https://www.iana.org/assignments/radius-types/radius-types.xhtml.

16 Krawczyk H., Bellare M., and Canetti R. HMAC: Keyed-Hashing for Message Authentication, RFC 2104 [Internet]. [cited 2022 May 12]. Available from: https://datatracker.ietf.org/doc/html/rfc2104.

17 Fajardo V., Ed., Arkko J., Loughney J., and Zorn G., Ed. Diameter Base Protocol, RFC 6733 [Internet]. [cited 2023 May 8]. Available from: https://datatracker.ietf.org/doc/html/rfc6733.

18 Clause 6.4, IEEE 802.1X-2020, pp. 46–48.

19 Clause 11.6, IEEE 802.1X-2020, p. 118.

20 Table 11-1, IEEE 802.1X-2020, p. 113.

21 IEEE. IEEE Standard for Wireless LAN Medium Access Control (MAC) and Physical Layer (PHY) specifications. IEEE Std 802.11-1997, pp. 1–445, 1997.

22 Wikipedia. Wired Equivalent Privacy [Internet]. [cited 2023 May 16]. Available from: https://en.wikipedia.org/wiki/Wired_Equivalent_Privacy.

23 Wikipedia. Wi-Fi Protected Access:WPA [Internet]. [cited 2023 May 16]. Available from: https://en.wikipedia.org/wiki/Wi-Fi_Protected_Access.

24 IEEE. IEEE Standard for Information Technology – Telecommunications and Information Exchange Between Systems Local and Metropolitan Area Networks – Specific Requirements – Part 11: Wireless LAN Medium Access Control (MAC) and Physical Layer (PHY) Specifications. IEEE Std 802.11-2020 (Revision of IEEE Std 802.11-2016), pp. 1–4379, 2020.

25 Vanhoef M. Key Reinstallation Attacks: Breaking WPA2 by Forcing Nonce Reuse [Internet]. [cited 2023 May 25]. Available from: https://www.krackattacks.com/.

26 Harkins D., Ed. Dragonfly Key Exchange, RFC 7664 [Internet]. [cited 2023 Apr 19]. Available from: https://datatracker.ietf.org/doc/html/rfc7664.html.

27 Clause 9.4.1.4, IEEE 802.11-2020, pp. 867–869.

28 Clause 9.3.3.11, IEEE 802.11-2020, pp. 849–854.

29 Clause 9.4.1.1, IEEE 802.11-2020, p. 866.

30 Clause 9.4.2.24, IEEE 802.11-2020, pp. 1050–1059.

31 Table 9-151, IEEE 802.11-2020, pp. 1054–1056.

32 Clause 4.10.3.2, IEEE 802.11-2020, pp. 283–286.

33 Clause 4.10.3.4, IEEE 802.11-2020, p. 287.

34 Clause 4.10.3.3, IEEE 802.11-2020, pp. 286–287.

35 Clause 12.4.5, IEEE 802.11-2020, pp. 2461–2464.

36 Mtroi. WPA3 – Improving Your WLAN Security [Internet]. [cited 2023 May 25]. Available from: https://wlan1nde.wordpress.com/2018/09/14/wpa3-improving-your-wlan-security/.

37 Nayanajith R. WPA3-SAE Mode [Internet]. [cited 2023 May 25]. Available from: https://mrncciew.com/2019/11/29/wpa3-sae-mode/.

38 Clause 12.7.2, IEEE 802.11-2020, pp. 2546–2554.

39 Table 12-7, IEEE 802.11-2020, p. 2749.

40 Figure 12-35, IEEE 802.11-2020, p. 2550.

41 Clause 12.7.7, IEEE 802.11-2020, pp. 2567–2571.

42 Cooper G. Device Onboarding Using FDO and the Untrusted Installer Model [Internet]. [cited 2024 April 3]. Available from https://cacm.acm.org/practice/device-onboarding-using-fdo-and-the-untrusted-installer-model/.

43 IEEE. IEEE Standard for Low-Rate Wireless Networks. In IEEE Std 802.15.4-2020 (Revision of IEEE Std 802.15.4-2015), pp. 1–800, 23 July 2020.

44 IEEE. IEEE Standard for Low-Rate Wireless Networks Amendment 3: Advanced Encryption Standard (AES)-256 Encryption and Security Extensions. In IEEE Std 802.15.4y-2021 (Amendment to IEEE Std 802.15.4-2020 as amended by IEEE Std 802.15.4z-2020 and IEEE Std 802.15.4w-2020), pp. 1–23, 31 May 2021.

45 Struik R. Formal Specification of the CCM* Mode of Operation [Internet]. [cited 2023 May 29]. Available from: https://mentor.ieee.org/802.15/dcn/04/15-04-0537-02-004b-formal-specification-ccm-star-mode-operation.pdf.

46 Bluetooth-SIG. Bluetooth Core Specification. Bluetooth Core Specification, Version 5.4, January, 2023.

47 Volume 3, Part C, Section 10.2.5, Bluetooth Core Specification, Version 5.4, p. 1335.

48 Volume 3, Part H, Sections 2.4.2.4–2.4.2.5, Bluetooth Core Specification, Version 5.4, pp. 1580–1581.

49 Whiting D., Housley R., and Ferguson N. Counter with CBC-MAC (CCM), RFC 3610 [Internet]. [cited 2023 Jun 8]. Available from: https://www.ietf.org/rfc/rfc3610.html.

50 Volume 2, Part H, Section 9.0, Bluetooth Core Specification, Version 5.4, pp. 1002–1006.

51 Song J. H., Poovendran R., Lee J., and Iwata T. The AES-CMAC Algorithm, RFC 4493 [Internet]. [cited 2023 Jun 8]. Available from: https://www.ietf.org/rfc/rfc4493.html.

52 Volume 3, Part H, Section 2.4.5, Bluetooth Core Specification, Version 5.4, pp. 1584–1585.

53 Volume 3, Part H, Appendix B, Bluetooth Core Specification, Version 5.4, pp. 1612–1615.

54 Volume 3, Part H, Section C.2.2, Bluetooth Core Specification, Version 5.4, pp. 1630–1642.

55 Volume 3, Part H, Sections 2.3.5.6.2–2.3.5.6.4, Bluetooth Core Specification, Version 5.4, pp. 1570–1576.

56 Volume 3, Part H, Section 2.3.5.6.5, Bluetooth Core Specification, Version 5.4, pp. 1576–1578.

57 Volume 6, Part B, Section 4.3, Bluetooth Core Specification, Version 5.4, pp. 2750–2756.

58 Volume 1, Part A, Section 5.4.5, Bluetooth Core Specification, Version 5.4, pp. 275.

59 Volume 3, Part C, Section 10.7, Bluetooth Core Specification, Version 5.4, pp. 1346–1350.

60 ITU-T. ITU-T Recommendations: ITU-T X.509 (10/2019) [Internet]. [cited 2023 Jun 8]. Available from: https://www.itu.int/ITU-T/recommendations/rec.aspx?rec=X.509.

61 ISO/IEC 8824 (1990). Information technology - Open Systems Interconnection – Specification of Abstract Syntax Notation One (ASN.1) [Internet]. [cited 2023 Apr 19]. Available from: https://www.iso.org/standard/16288.html.

62 Santesson S., Myers N., Ankney R., et al. X.509 Internet Public Key Infrastructure: Online Certificate Status Protocol – OCSP, RFC 6960 [Internet]. [cited 2023 Jun 9]. Available from: https://www.rfc-editor.org/rfc/rfc6960.html.

63 Harkins D. and Carrel D. The Internet Key Exchange (IKE), RFC 2409 [Internet]. [cited 2023 Jun 10]. Available from: https://datatracker.ietf.org/doc/html/rfc2409.html.

64 Kaufman C., Hoffman P., Nir Y., et al. Internet Key Exchange Protocol Version 2 (IKEv2), RFC 7296 [Internet]. [cited 2023 Jun 10]. Available from: https://www.rfc-editor.org/rfc/rfc7296.html.

65 Maughan D., Schertler M., Schneider M., and Turner J. Internet Security Association and Key Management Protocol (ISAKMP), RFC 2408 [Internet]. [cited 2023 Jun 11]. Available from: https://www.rfc-editor.org/rfc/rfc2408.html.

66 IANA. Internet Key Exchange Version 2 (IKEv2) Parameters [Internet]. [cited 2023 Jun 12]. Available from: https://www.iana.org/assignments/ikev2-parameters/ikev2-parameters.xhtml.

67 Section 2.6, RFC 7296 [Internet]. [cited 2023 Jun 16]. Available from: https://www.rfc-editor.org/rfc/rfc7296.html#page-32.

68 Kent S. and Seo K. Security Architecture for the Internet Protocol, RFC 4301 [Internet]. [cited 2023 Jun 16]. Available from: http://www.rfc-editor.org/info/rfc4301.html.

69 Manral V. Cryptographic Algorithm Implementation Requirements for Encapsulating Security Payload (ESP) and Authentication Header (AH), RFC 4835 [Internet]. [cited 2023 Jun 16]. Available from: https://www.rfc-editor.org/rfc/rfc4835.html.

70 Kent S. IP Authentication Header, RFC 4302 [Internet]. [cited 2023 Jun 16]. Available from: http://www.rfc-editor.org/info/rfc4302.html.

71 Kent S. IP Encapsulating Security Payload (ESP), RFC 4303 [Internet]. [cited 2023 Jun 16]. Available from: http://www.rfc-editor.org/info/rfc4303.html.

72 IANA. Protocol Numbers [Internet]. [cited 2023 Jun 18]. Available from: https://www.iana.org/assignments/protocol-numbers/protocol-numbers.xhtml.

73 Appendix A, RFC 4302 [Internet]. [cited 2023 Jun 18]. Available from: https://www.rfc-editor.org/rfc/rfc4302.html#page-25.

74 Rescorla E. The Transport Layer Security (TLS) Protocol Version 1.3, RFC 8446 [Internet]. [cited 2023 Jun 21]. Available from: https://www.rfc-editor.org/rfc/rfc8446.html.

75 Rescorla E., Tschofenig H., and Modadugu N. The Datagram Transport Layer Security (DTLS) Protocol Version 1.3, RFC 9147 [Internet]. [cited 2023 Jun 23]. Available from: https://www.rfc-editor.org/rfc/rfc9147.html.

76 Synopsis. The Heartbleed Bug [Internet]. [cited 2023 Jun 24]. Available from: https://heartbleed.com/.

77 Wikipedia. Virtual Private Network [Internet]. [cited 2023 Feb 13]. Available from: https://en.wikipedia.org/wiki/Virtual_private_network.

78 OpenSwan. Welcome to Openswan! [Internet]. [cited 2023 Jun 25]. Available from: https://openswan.org/.

79 OpenVPN. Secure Access and Network Connectivity Reimagined [Internet]. [cited 2023 Jun 25]. Available from: https://openvpn.net.

80 Wireguard. Wireguard: Fast, Modern, Secure VPN Tunnel [Internet]. [cited 2023 Jun 25]. Available from: https://www.wireguard.com.

81 Dodenfeld J. WireGuard: Next Generation Kernel Network Tunnel [Internet]. [cited 2023 Jun 25]. Available from: https://www.wireguard.com/papers/wireguard.pdf.

82 Perrin T. The Noise Protocol Framework [Internet]. [cited 2023 Jun 25]. Available from: http://noiseprotocol.org/noise.pdf.

6

Building Automation Stack for Large Commercial and Public Buildings

Large commercial and public buildings consume around 18% of the US primary energy according to the DOE [1]. Reducing large building energy use overall and controlling it to reduce energy use during times of high grid energy carbon intensity is a critical step in decarbonizing the built environment. Controlling energy use requires conveying sensor information from the building to the BEMS and actuator control from the BEMS to the building's energy-consuming equipment. Building automation protocols are designed to accomplish this task. These protocols run on top of the link layer and the IP stack within the building, and connect energy-intensive equipment such as HVAC systems, lighting, and other building-related energy appliances and systems through the building gateway to the BEMS supervisory control in a public or private cloud.

Building automation protocols facilitate the optimization of energy usage, reduction of dependence on non-renewable energy, and curtailment of the carbon emissions from large buildings. As such, building automation protocols form the backbone of sustainable and smart building operations. In this chapter, an introduction to BEMS's use of building automation stacks for large commercial and public buildings is presented. Since BACnet is estimated to constitute over 60% of the market for commercial BASs worldwide [2], an understanding of BACnet architecture and capabilities is essential for designing and implementing a BEMS, and the bulk of the chapter is devoted to a detailed examination of the BACnet architecture and stack.

Section 6.1 introduces BACnet, while Section 6.2 presents an overview of the BACnet stack architecture. In Section 6.3, the design of the BACnet link layer is presented. Sections 6.4 and 6.5 discuss two of the most widely deployed link layers, the serial line-based MS/TP link layer and the UDP/IPv4-based "virtual" link layer. Section 6.6 presents BACnet network layer. The BACnet network layer

Automating Building Energy Management for Accelerated Building Decarbonization: System Architecture and the Network Layer, First Edition. James Kempf.

Published 2025 by John Wiley & Sons, Inc.

allows multiple floors in a building or multiple buildings in a campus to be efficiently grouped into a single control domain.

Section 6.7 presents the BACnet application layer, which is where building automation devices model their sensor and actuator functionality for external measurement and control. The discussion in this section primarily emphasizes application layer objects and messages suitable for implementing a BEMS oriented toward decarbonization; however, BACnet in general has a wider scope, including such building automation functions as access control and preventive maintenance. Section 6.8 discusses a few approaches to BACnet security, in the absence of any security mechanisms built into the protocol itself that are universally applicable to all link layers. The detailed examination of the BACnet protocol concludes in Section 6.9, where interoperability specifications are presented. These mechanisms are defined in the standard based on functional architecture building blocks.

BACnet is not the only building automation protocol in use for large commercial and public buildings, though it is by far the most widely deployed. Section 6.10 contains brief descriptions of a few other building automation and control protocols that have some amount of legacy deployment: KNX, Digital Addressable Lighting Interface (DALI), and LonWorks. Finally, Section 6.11 summarizes the chapter.

A Short Introduction to BACnet

This video presents a short, 10-minute introduction to BACnet. The video covers the motivation for the development of BACnet, BACnet objects and services, and discusses the BACnet stack layers from the application layer down. More details are presented on the BACnet link layer adaptations, like BACnet over Ethernet, which are not covered in this chapter.

6.1 Introduction to the BACnet Protocol

The BACnet protocol is an open, standarized communication protocol designed for building automation and control systems in commercial buildings. It was developed by the ASHRAE in the late 1980s to ensure communication interoperability between building automation devices and systems from different manufacturers. ASHRAE has published a series of BACnet standards, including the original BACnet standard (ASHRAE 135) in 1995, and subsequent revisions and extensions to the standard over the years. The latest version of the BACnet standard is ANSI/ASHRAE Standard 135-2020 [3], which was published in 2020. BACnet has also been approved by the International Organization for Standardization (ISO) and was published in ISO 16484-5 [4] as a global standard for interoperable building automation and control systems. The ISO standard includes the BACnet protocol as one of its approved communication profiles.

BACnet is designed to provide a common language that allows devices and management systems from different manufacturers to communicate interoperability. The protocol is based on a complete network stack from the application layer down to the data link layer with a standardized set of messages and data formats. Interoperability allows building automation design engineers, building owners, and facility managers to mix and match devices and applications from different manufacturers to obtain the best combination of energy efficiency, reduced costs, and improved occupant comfort and safety. BACnet has continued to evolve over the years, with new features and capabilities being added to support the latest building automation technologies. Today, BACnet is widely deployed in a variety of buildings, including commercial buildings, hospitals, schools, and government facilities.

6.2 The BACnet Stack Architecture

Figure 6.1 contains an overview of the BACnet stack architecture. Like the IP stack discussed in Chapter 2, BACnet omits the session and presentation layer but it also omits the transport layer. The transport layer is not needed in the BACnet stack because the stack only supports one application protocol so there is no need for the transport layer to multiplex between different applications. The application layer protocol is based on a collection of formally defined service primitives

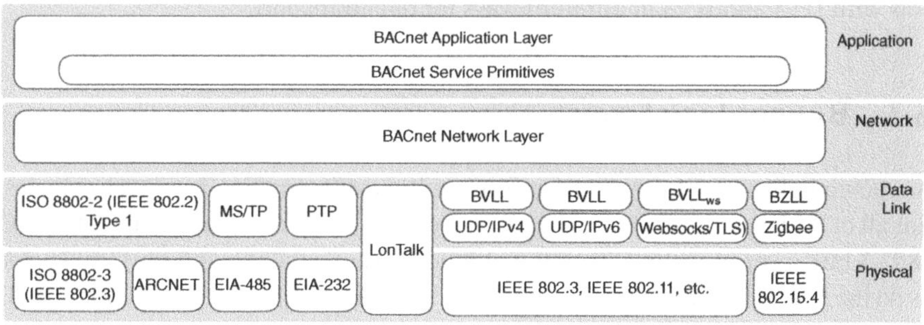

Figure 6.1 BACnet Stack Architecture[1]

1 Note that the BACnet standard denotes the virtual link layer for Websockets as BVLL also without the subscript, but the subscript has been added here because the packet header format is different from the UDP/IPv4 and UDP/IPv6 virtual link layers.

which are discussed in more detail in Section 6.7.7. In addition, BACnet supports only a single communication abstraction, a connectionless, packet-oriented protocol like UDP. Guaranteed end-to-end delivery and error recovery are provided in the application layer, as are message segmentation and reassembly and end-to-end flow control.

The most notable feature of the stack is the presence of a custom network layer rather than IP as the network layer. This structure is primarily a matter of how the standard developed historically. Prior to the early 2000s, networks based on the Internet protocol stack were expensive and difficult to deploy especially for operational technology networks. They were viewed as boutique networks suitable only for information technology use cases in large enterprise companies. Since standardization of BACnet began in the late 1980s before the commercialization of the Internet protocol stack, it was defined on top of a custom network stack, including its own network layer. The stack was designed to be implementable on devices powered by microcontrollers using a variety of different physical layer technologies including inexpensive serial lines, making it affordable and straightforward to deploy, as well as on various other types of data link layers in common use among OT networks at the time.

In the early to mid-2000s, the hardware supporting LANs and software based on the IP stack became cheaper and the software became easier to deploy and configure. Ethernet-based LANs and the IP stack became ubiquitously available in buildings around the world. Rather than simply revise the version number from 1 to 2 and port the BACnet application layer directly onto an IP stack transport layer as an IP stack application layer, ASHRAE chose to instead design a collection of "virtual" link layers that utilize different IP stack transport or application protocols with IPv4 and IPv6 in different ways for data transport.

6.3 BACnet Link Layers

As shown in Figure 6.1, BACnet supports multiple different link layer technologies, not all of which run over mutually exclusive physical layers. For example, the ISO 8802-3 link layer runs directly on top of the Ethernet data link and physical layers, as do the UDP/IPv4, UDP/IPv6, and Websockets/TLS link layers. The standard calls the latter "virtual" link layers and defines different adaptation layers called BACnet Virtual Link Layers (BVLLs), similar to the 6LoWPAN adaptation layer discussed in Chapter 4, for implementing them. The UDP/IPv4 and UDP/IPv6 BVLLs share the same basic adaptation layer packet header format and their signaling protocols are similar. Zigbee and Websockets/TLS have different packet header formats and signaling in their adaptation layers. This chapter focuses on the most commonly deployed link layers, namely, MS/TP and UDP/IPv4.

MS/TP stands for Master Slave[2]/Token Passing and the name describes at a high level how the link layer functions. MS/TP is discussed in Section 6.4. MS/TP devices that are allowed to autonomously send messages without receiving a message in response are called "masters," while those that can only send messages after receiving a message are called "slaves." Token passing is the technology used for medium access control and is described in more detail in Section 6.4.1. MS/TP is widely utilized in BACnet deployments because it is inexpensive and simple to install and configure, though somewhat limited in scope.

UDP/IPv4 was the first virtual link layer to utilize IP for data transport. It was released by ASHRAE in 2004 [5]. Prior to the release of the UDP/IPv4 BVLL, the ASHRAE standard contained a way to tunnel frames over IP from devices running the BACnet ISO 8802 2 Ethernet link layer between links [6]. As the name implies, the UDP/IPv4 BVLL utilizes the UDP transport over IPv4 to implement the virtual link layer. UDP as a transport is consistent with BACnet's basic connectionless, packet-oriented nature. As with all BVLLs, the UDP/IPv4 BVLL defines basic attributes of the link, including addressing based on the underlying transport and network layer protocol addressing, routing, and the handling of broadcast. With UDP/IPv4 and the proper building network configuration, BACnet communication can be sent over the Internet. The UDP/IPv4 virtual link layer is discussed in Section 6.5.

6.4 MS/TP

The BACnet MS/TP link layer runs on top of the EIA-425/RS-425 [7] serial line standard physical layer. The maximum length of the RS-425 serial bus is 1,200 m and MS/TP supports data transfer rates of 9,600, 19,200, 38,400, and 76,800 bits/sec. Nodes on the BACnet MS/TP link layer are divided into two types [3] :

- Masters that are allowed to send and receive data if the link layer on the node has access to the physical medium through the medium access control algorithm.
- Slaves such as HVAC controllers and thermostats that are only allowed to send data after receiving a message from a master.

Since the RS-425 is a serial bus, only one master can have access at a time, so the MS/TP link layer requires that the masters on a link cooperate using a medium access control algorithm to limit access to the bus.

2 The author apologizes to anyone who finds this archaic terminology offensive, but, unfortunately, ASHRAE has yet to clean up its terminology in this area, unlike many other technical standards organizations and open-source projects.

The original design thinking behind dividing BACnet MS/TP devices into one fully functional device type (the master) and one limited device type (the slave) was that the slave device would be simple and cheap to implement and therefore would appeal to manufacturers that wanted to keep their equipment economical. However, most manufacturers have concluded that the incremental cost to build a fully functional device is minimal given that they already have the software to build a fully functional device, so very few slave devices have been implemented [8].

6.4.1 MS/TP Medium Access Control

The MS/TP network uses token passing to grant a master access to send over the link. The BACnet masters on the link pass a token from one to the other to control which station has authorization to send. When a station receives a token, it has control over the bus and is allowed to send messages. Only a BACnet master can receive a token and initiate a data exchange. When there are two or more masters on an MS/TP network, the master with the token must forward it to one of the other master stations after a fixed time period. In Sections 6.4.4 and 6.4.5, discuss the frame types involved in the token passing protocol and an example of how the protocol is used to bootstrap an MS/TP link.

The MS/TP masters and slaves run state machines to determine what actions to take. Transitions are controlled by a series of conditions described in the standard, followed by actions that are taken if the specific conditions are met. The last action in the series results in the transition of the state machine into the next state, which may be the same as the previous state. The BACnet protocol specification contains 19 parameters on a bus that can be configured to control the state machines. The specification assumes that the master and slave do not receive frames that they themselves have transmitted. More information and state machine diagrams can be found in Clauses 9.5.6 through 9.5.7 in ANSI/ASHRAE Standard 135-2020 [9].

6.4.2 MS/TP Frame Format

The MS/TP frame format is shown in Figure 6.2. The fields are as follows:

- Preamble – the preamble consisting of the two bytes 0x55 and 0xFF,
- Frame Type – the frame type, see Section 6.4.4 for a discussion of MS/TP frame types,
- Destination Address – the BACnet MS/TP address of the receiving node on the link,
- Source Address – the BACnet MS/TP address of the sending node on the link,

Preamble (16 bits)	Frame Type (8 bits)	Destination Address (8 bits)	Source Address (8 bits)	Length (16 bits)
Header CRC (8 bits)	Payload (variable)			Data CRC (16 bits)

Figure 6.2 BACnet MS/TP Frame Format

- Length – the length of the Payload field, in bytes. The length value has the most significant byte first, that is, it is in big-endian format,
- Header CRC – a cyclic redundancy check on the header calculated according to a polynomial defined in the standard. The CRC allows the receiving node to detect if any of the header bits were changed due to a problem on the physical layer,
- Payload – the data contents of the frame that the sending node is transmitting to the receiving node. This field is only present if the length field is nonzero,
- Data CRC – A cyclic redundancy check on the data calculated according to a polynomial defined in the standard. As with the Header CRC, the Data CRC ensures that the payload was not modified in transit due to a problem on the physical layer. This field is also only present if the length field is nonzero.

6.4.3 MS/TP Addressing

As Figure 6.2 indicates, MS/TP addresses are 8 bits (1 byte) in size. MS/TP restricts which addresses can be assigned to masters and slaves. Addresses 0–127 are only valid for master nodes, while addresses 0–254 are valid for both master and slave nodes. Address 255 is reserved for broadcast to all nodes on the link. The link-layer address of masters and slaves is configurable within the specified ranges when the network is installed and must be unique on a single link, that is, no two nodes are allowed to have the same address on the same link.

6.4.4 MS/TP Frame Types

The BACnet standard specifies 10 frame types (0–7, 32–33). Table 6.1 contains a complete list of the BACnet MS/TP frame type codes and their functions as specified in the standard. Types 8–31 and 34–127 are reserved by ASHRAE, while types 128–255 are available to manufacturers for proprietary extensions.

The first three frame types are used to implement the token-passing protocol for medium access control and have no Payload field. Frame type 00 is sent when a master passes the token to another master. Frame types 01 and 02 allow masters to poll and receive a response from other masters on the link. Masters also use these messages to determine the next master in the sequence to receive the token.

Table 6.1 MS/TP Frame Types and Functions

Number	Function
00	Send token
01	Poll for master
02	Reply to poll for master
03	Test request
04	Test response
05	BACnet data – expecting reply
06	BACnet data – not expecting reply
07	Reply postponed
32	BACnet extended data – expecting reply
33	BACnet extended data -not expecting reply

Both masters and slaves can receive frame type 01 since it is typically broadcast to all nodes on the link, but slaves must ignore it.

Frame types 03 and 04 are used to complete a loopback test of the MS/TP network. A test request (type 03) message initiates a loopback test to a destination address, and the test response (type 04) message is returned from the destination address to the source address to complete it. The length of the Payload field ranges from 0 to 501 bytes, and the test response message must include the same Payload field as in the triggering test request. These frame types allow a master to determine if there is a path to a particular destination address on the link and are handled by both masters and slaves.

Frame types 05 through 07 and 32 and 33 implement sending network and application layer data between MS/TP nodes on the link. Types 05 and 06 are used to send between 0 and 501 bytes of data, while types 32 and 33 are used to send between 502 and 1497 bytes. The extended data types require the sending node to compress the data in the Payload field using a technique called consistent overhead byte stuffing (COBS). For frames with COBS-encoded payloads, the payload includes both the encoded data subfield and an additional encoded CRC subfield covering the encoded payload. The Length field for the frame includes the combined length of both these subfields −2[3]. For more information on COBS, see Section 9.10 of the ASHRAE BACnet standard [3].

3 The length is reduced by two to maintain backward compatibility with earlier versions of the protocol.

The general pattern for communicating data on the MS/TP link layer is for the master to send one of the expecting reply message types (types 5 or 32) to the recipient address, with the Payload field containing a message from the upper layers for the requested data. The replying nodes, in the case of a broadcast message, return the data using the corresponding not expecting reply types (types 6 or 33) to the requesting master. These messages can be sent between masters or between masters and slaves. Some BACnet upper-layer messages do not require a reply to the sender and are sent to the receivers over broadcast or unicast to inform the receivers of the sender's status. These messages use the frame types 6 or 33 with the status information (if any) in the Payload.

Frame type 7 (Reply Postponed) is sent by a master to defer sending a reply to a received frame expecting a reply (types 3, 5, and 32). If the processing required to calculate and/or fetch the reply data exceeds 250 msec, the MS/TP receive state machine needs to move on to the IDLE state or the link layer access control state machine on the sending node or it will hang. In addition, the master uses this message type if it receives a request expecting a reply on the broadcast address and it does not hold the token. The master replies with Reply Postponed and then waits until it receives the next periodic Token (type 0) frame before sending the reply.

6.4.5 Bootstrapping the MS/TP Link Layer

An MS/TP node is bootstrapped by the master node having the smallest address on the link. Figure 6.3 shows bootstrapping for a network and provides an illustration of how the MS/TP protocol works. In the figure, there are two masters, one with address 1 and a second with address 3 [10]. Master 1 enters the INITIALIZE state and stays in the state for a maximum of 500 msec with no incoming traffic before beginning to initialize the network. As the master with the lowest address on the network, it exits the state and broadcasts a poll for master (frame type 1) to address 2. Receiving no reply after 20 msec, it polls for Master 3. Master 3 replies and Master 1 passes the token to Master 3. Master 3 then proceeds to poll for further masters, up to the maximum configured for the link and, receiving no replies, it rolls over to Master 1 and polls for it. Master 1 answers, Master 3 passes the token to Master 1, and since Master 1 has seen the token before, it takes control of the link. The two masters periodically pass the token between themselves according to a configurable timeout until Master 1 has had the token 50 times. Master 1 then restarts the master discovery phase to ensure that no new masters have joined the network.

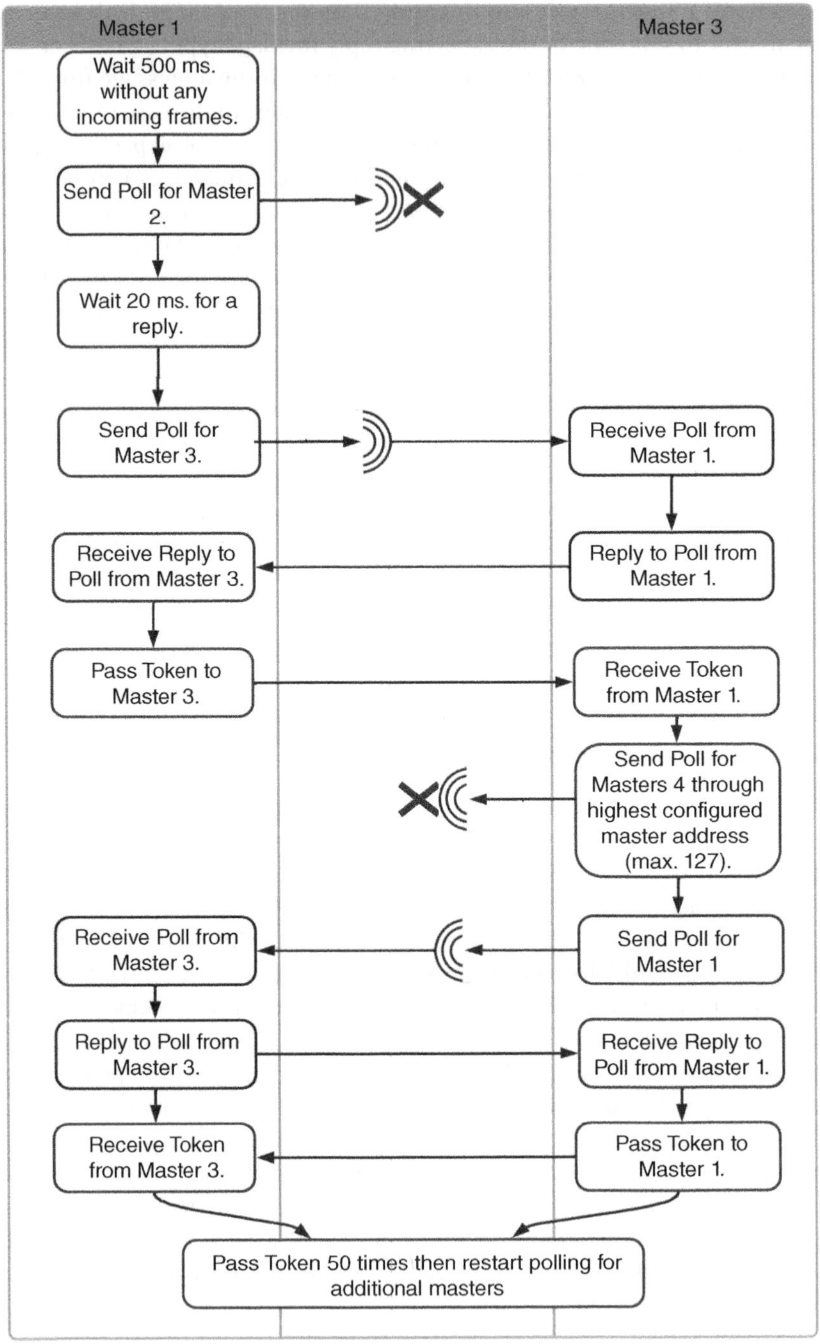

Figure 6.3 MS/TP Network Bootstrap Process

6.5 UDP/IPv4 Virtual Link Layer

The UDP/IPv4 BVLL encapsulates a BACnet network layer and application layer message into a UDP datagram and transmits the datagram over IPv4. The BVLL adaptation layer maps the BACnet unicast and broadcast services onto a combination of the IPv4 unicast and broadcast. The UDP/IPv4 virtual link layer architecture distinguishes between four different types of nodes:

- Regular BACnet devices are nodes that send and receive unicast and broadcast BACnet packets through the UDP/IPv4 link layer. Unlike MS/TP, UDP/IPv4 does not distinguish between full-function and limited-function.
- BACnet routers are nodes that function like BACnet device nodes with respect to the UDP/IPv4 link layer but additionally route packets between BACnet networks. BACnet networks and routers are described in more detail in Section 6.6.
- A foreign device is a BACnet device that has an IP subnet prefix different from the BACnet/IP subnet prefix of the BACnet network the foreign device wants to join. Foreign devices allow a BACnet device to be located on a different floor of a building, or in a different building altogether from the BACnet devices and networks in the main deployment but still participate in BACnet networks.
- BACnet Broadcast Management Devices (BBMD) proxy BACnet broadcasts off the local IP subnet. They forward BACnet broadcasts across IP routers and forward broadcast messages to and from foreign devices.

Unlike MS/TP, IPv4 already has a subnet structure so BACnet networks need to be mapped onto IPv4 subnets. The mapping between BACnet networks and IPv4 subnets can be one-to-one, one-to-many, where a single BACnet network spans multiple IP subnets, or many-to-one, where multiple BACnet networks map to a single IPv4 subnet. In the one-to-one case where a single BACnet network maps to a single IP subnet, both unicast and broadcast are straightforward between nodes on the local BACnet network. In the many-to-one case where multiple BACnet networks map to a single IP subnet, the BACnet routers will route between the devices with different BACnet network addresses mapped to the local subnet. However, for the one-to-many case where a single BACnet network maps to multiple IP subnets, and for multiple BACnet networks on different subnets all connected into a local building or campus BACnet deployment (the BACnet "internet"), there is a problem with broadcast. Since most IP network administrators configure their IP routers to not route broadcast packets off the local subnet, the BBMD provides a proxy service for broadcast between subnets.

The UDP/IPv4 BVLL adaptation layer implements a message and control protocol called the BACnet Virtual Link Control (BVLC) protocol for sending messages, controlling BBMD broadcast distribution, and foreign device registration. This protocol is used to identify application and network data messages, in addition to creating, maintaining, and deleting information from the tables used by the BBMD to manage foreign devices and BACnet broadcast across IP routers.

The UDP/IPv4 link layer was the first BVLL developed. Since it was introduced in 2004, ASHRAE has also specified a virtual link layer for UDP/IPv6 with the same BVLL header format, and two different virtual link layers, for ZigBee and Websockets/TLS. The virtual link layer architecture has proven to be a flexible solution allowing adaptation of the BACnet stack to different network and transport layers without having to change the BACnet network and application layers, although at the expense of considerable complexity (e.g. the latest BACnet standard specification [3] is over 1,400 pages long).

6.5.1 UDP/IPv4 BVLL "Frame" Format

The virtual link "frame" header is shown in Figure 6.4, and consists of three fields:

- BVLL Type – code for the BVLL type. For UDP/IPv4, the code is 0x81,
- BVLC Function Code –function code for the type of BVLC message. Function codes for UDP/IPv4 range between 0x00 and 0x0b. Tables 6.2 and 6.3 contain a list of function codes with information about what they do,
- BVLC Length –length of the virtual "frame" in bytes, including the BVLC virtual frame header and the payload.

After the BVLC header, depending on the BVLC message type, the payload contains either a BACnet network layer header followed by a network layer message, an application layer message, or a BVLC message.

6.5.2 UDP/IPv4 BVLL Addressing

BACnet nodes on a UDP/IPv4 BVLL network have 6-byte virtual BACnet link-layer addresses. The 6-byte address for unicast transmission is formed by combining the 2-byte UDP port number and the 4-byte IPv4 node address of the target node, including both the IPv4 subnet prefix and node identifier fields. The default BACnet UDP port number is 0xbac0 (47808) and the ranges 47808–47832 and 49152–65535 have also been designated for BACnet nodes. An example of a BACnet address in a private building network is: 192.168.42.24:47808. Note that from an IP networking standpoint, this addressing architecture allows multiple BACnet devices to reside on a single IP node if the port number is different. Each BACnet device will have a separate BACnet address.

BVLC Type (8 bits): 0x81	BVLC Function Code (8 bits): 0x00-0x0c	BVLC Length (16 bits)

Figure 6.4 BVLC Protocol Virtual "Frame" Header for UDP/IPv4 Virtual Link Layer

Table 6.2 UDP/IPv4 BVLC Message Distribution Types

Type Code	Name	Transmission Mode	Payload Contents and Return	Function
0x0a	*Original-Unicast-Network-Protocol-Data-Unit*	Unicast by BACnet device or router node.	Network or application layer message.	Send a BACnet network or application message to another BACnet device or router node.
0x0b	*Original-Broadcast-Network-Protocol-Data-Unit*	Broadcast by BACnet device or router node.	Network or application layer message.	Send a BACnet network or application message to all BACnet nodes in the local IP subnet.
0x04	*Forwarded-Network-Protocol-Data-Unit*	Unicast or broadcast from a BBMD node to BBMD peers, unicast to foreign devices, and broadcast on the receiving BBMD's local link.	6-byte address of the original sending BACnet node followed by network or application layer message.	Unicast or broadcast, depending on the broadcast distribution table record, a broadcast message broadcast on the BBMD's local subnet to the BBMD's peers for rebroadcast on the receiving BBMDs' IP subnets, and to foreign devices in the foreign device table.
0x09	*Distribute-Broadcast-To-Network*	Unicast from a foreign device to a BBMD with which it is registered.	Network or application layer message. Returns a *BVLC-Result* with *Distribute-Broadcast-To-Network-NAK* if the broadcast fails. See Table 6.3 for definitions of *BVLC-Result* and *Distribute-Broadcast-To-Network-NAK*.	Unicast a message from a foreign device to a BBMD with which it has registered, for broadcast on the BBMD's IP subnet link and relay to other foreign devices and peer BBMDs for rebroadcast on their IP subnet links.

Table 6.3 UDP/IPv4 BVLC BBMD Table Management Message Types

Type Code	Name	Transmission Mode	Payload Contents and Return	Function
0x00	*BVLC-Result*	Unicast to sending node.	One of the following 2-byte status codes: • Success – 0x00 • Write-Broadcast-Distribution-Table-NAK – 0x01 • Read-Broadcast-Distribution-Table-NAK – 0x02 • Register-Foreign-Device-NAK – 0x03 • Read-Foreign-Device-Table-NAK – 0x04 • Delete-Foreign-Device-Table-Entry-NAK – 0x05 • Distribute-Broadcast-To-Network-NAK – 0x06.	Return status for those messages that require a status return.
0x01	*Write-Broadcast-Distribution-Table*	Unicast to a BBMD.	The list of 10-byte broadcast distribution table records. Returns a *BVLC-Result* with status of request.	Insert new records or update existing records in the BBMD's broadcast distribution table.
0x02	*Read-Broadcast-Distribution-Table*	Unicast to a BBMD.	None. Returns either a *Read-Broadcast-Distribution-Table-ACK* with contents of the broadcast distribution table or a *BVLC-Result* with Read-Broadcast-Distribution-Table-NAK if the request fails.	Request a *Read-Broadcast-Distribution-Table-ACK* message including a copy of the receiving BBMD's broadcast distribution table.

0x03	*Read-Broadcast-Distribution-Table-ACK*	Unicast from a BBMD.	Returns zero or more 10-byte records containing the broadcast distribution table contents. An empty list indicates the table is empty.	Returned in response to a *Read-Broadcast-Distribution-Table* with the contents of the sending BBMD's broadcast distribution table.
0x05	*Register-Foreign-Device*	Unicast from a foreign device to a BBMD.	A 2-byte positive integer giving the duration, in seconds, of the foreign device registration should be maintained by the BBMD. Returns a *BVLC-Result* with the status of the request.	Register a foreign device with the receiving BBMD in the BBMD's foreign device table. The foreign device BACnet address is taken from the UDP/IPv4 address and port number.
0x06	*Read-Foreign-Device-Table*	Unicast to a BBMD.	None. Returns either a *Read-Foreign-Device-Table-ACK* with the contents of the foreign device table or *BVLC-Result* with *Read*-Foreign-Device-Table-NAK if the request fails.	Request a *Read-Foreign-Device-Table-ACK* message including a copy of the receiving BBMD's foreign device table.
0x07	*Read-Foreign-Device-Table-ACK*	Unicast from a BBMD.	Returns zero or more 10-byte records containing the foreign device table contents. An empty list indicates the table is empty.	Returned in response to a *Read-Foreign-Device-Table* with the contents of the sending BBMD's foreign device table.
0x08	*Delete-Foreign-Device-Table-Entry*	Unicast to a BBMD.	The 6-byte BACnet UDP/IPv4 address of the record for the foreign device to be deleted. Returns a *BVLC-Result* with the status of the request.	Delete a foreign device table record.

For broadcast on BACnet networks, the BACnet UDP/IPv4 virtual link layer uses the IPv4 subnet broadcast address (e.g. 192.168.42.255) and the UDP port, typically 47808, for both directed and global broadcast. From the standpoint of IPv4, broadcasts to this address do not propagate off the local IPv4 subnet. BACnet does not use the global IPv4 broadcast address (255.255.255.255) since the IPv4 global broadcast address may be blocked by some routers in the local subnet. BBMDs provide more targeted control over BACnet broadcast forwarding.

6.5.3 BBMD Internal Data Structures for Broadcast Control and Foreign Device Registration

BBMDs were introduced into the UDP/IPv4 BVLL architecture because BACnet makes much heavier use of broadcast than a typical IP stack application protocol. BBMDs allow a collection of BACnet networks covering multiple IP subnets to nevertheless exchange broadcast traffic between themselves regardless of whether the IP network administrators have configured the IP routers to route broadcast traffic. In addition, the BBMDs act as proxies for foreign devices that are not part of any BACnet network but still need to send and receive BACnet broadcast traffic. To perform these functions, the BBMDs maintain two data structures: the broadcast distribution table and the foreign device table.

6.5.3.1 BBMD Broadcast Distribution Table

The broadcast distribution table consists of a series of records containing information allowing a BBMD to forward broadcasts from the local subnet to peer BBMDs in other subnets. As shown in Figure 6.5, a broadcast distribution table record contains the BACnet node address of the peer BBMD on another subnet, which consists of the IPv4 address and UDP port, and a broadcast distribution mask. The broadcast distribution mask determines whether the BBMD forwards a broadcast by broadcasting it on the subnet broadcast address of the peer BBMD's subnet or unicasting the message to the peer BBMD's IPv4 address, allowing the peer BBMD to then rebroadcast it on the subnet broadcast address. If the broadcast distribution mask has all 1s in the subnet prefix portion and all 0s in the node identifier portion, then the message is broadcast. If the broadcast distribution mask is all 1s, then the message is unicast. Distribution by rebroadcast on the peer BBMD's subnet without the peer's involvement is called "one hop distribution" while rebroadcast by the peer BBMD is called "two hop distribution." However, note

IPv4 Address (32 bits)	UDP Port (16 bits)	Broadcast Distribution Mask (32 bits)

Figure 6.5 BBMD Broadcast Distribution Table Record

that broadcast routing needs to be configured on the IP routers between the peer BBMD's subnet and the sending BBMD's subnet if one hop distribution is used; otherwise, the packet will be dropped.

When a broadcast arrives, the BBMD performs a logical NOT on the broadcast distribution mask, inverting the mask (all 1s become 0s and vice versa). It then performs a logical OR with the IP address of BBMD, so the result is 1 if either operand is 1, otherwise 0. For example, suppose the peer BBMD's IPv4 address is 192.168.42.24/24 and the broadcast distribution mask is 255.255.255.000/24, meaning the BBMD should rebroadcast the message. Inverting the broadcast distribution mask gives 000.000.000.255/24 and logically ORing it with the peer's IPv4 address gives 192.168.42.255, which is the subnet broadcast address for the peer's subnet. On the contrary, if the broadcast distribution mask is all 1s (255.255.255.255/24), inverting it results in a mask that is all 0s, and ORing it with 192.168.42.24 returns the peer's BBMD address again.

6.5.3.2 BBMD Foreign Device Table

The foreign device table contains the records of registered foreign devices that are not registered with a BACnet address but nevertheless must send and receive broadcast BACnet traffic on the BACnet network where the BBMD is located. As shown in Figure 6.6, a foreign device table record contains the IPv4 address and UDP port number, which together form the BACnet node address, of the foreign device, along with the amount of time, in seconds, the foreign device would like the registration to stay active and the remaining amount of time in the registration. These are provided when the foreign device registers with the BBMD. The BBMD adds the time remaining for the registration, which is initialized to the time to live value plus 30 seconds, giving the foreign device a 30-second grace period to renew its registration. If the foreign device does not reregister by the end of the time to live period plus 30 seconds, its record is purged from the foreign device table.

6.5.4 UDP/IPv4 BVLC Message Types

The BVLC is the link layer protocol for communicating between UDP/IPv4 link layer nodes. It contains two general classes of message types: message types for communicating between UDP/IPv4 link layer nodes and message types for controlling the contents of the BBMD broadcast distribution and foreign device registration tables [11].

IPv4 Address (32 bits)	UDP Port (16 bits)	Time to Live (16 bits)	Time Remaining (16 bits)

Figure 6.6 BBMD Foreign Device Table Record

6.5.4.1 UDP/IPv4 BVLC Message Types for Communicating Between Nodes

Table 6.2 contains the BVLC message types for distributing BACnet applications and network messages. *The Original-Unicast-Network-Protocol-Data-Unit* and *Original-Broadcast-Protocol-Data-Unit* message types are used when a BACnet device or router node wants to transmit a message to one or more recipients over the UDP/IPv4 virtual link layer. The payload for these messages is the BACnet application or network message for distribution. In addition to device nodes and BACnet routers on the local link, the *Original-Broadcast-Protocol-Data-Unit* is received by any BBMDs in the local subnet for rebroadcast on peer BBMDs' subnets, and unicast to any foreign devices that have registered with it.

The BBMD uses the BVLC message type *Forwarded-Network-Protocol-Data-Unit* for relaying a broadcast message to peers and foreign devices. The payload in the *Forwarded-Network-Protocol-Data-Unit* is the 6-byte UDP/IPv4 virtual link-layer address of the node that originated the broadcast followed by the network or application layer message that the node broadcast. The source BBMD uses broadcast or unicast to its peers depending on the contents of the peer's record in the BBMD's broadcast distribution table. Receiving BBMDs then rebroadcast the original message as a *Forwarded-Network-Protocol-Data-Unit* BVLC message on their local IP subnets. They do not use the *Original-Broadcast-Network-Protocol-Data-Unit* BVLC message to rebroadcast. The source BBMD also unicasts the message to foreign devices registered in its foreign device table.

The *Distribute-Broadcast-To-Network* BVLC message type is unicast by a foreign device to a BBMD with which it is registered to retransmit a message via broadcast as a *Forwarded-Network-Protocol-Data-Unit* message type. The BBMD then broadcasts the message on its local link and distributes the message to peer BBMDs for rebroadcast on their local links. The BBMD also sends the message to other registered foreign devices. The payload is the message the foreign device would like to have broadcast. If the broadcast fails, the BBMD returns a *BVLC-Result* message, described in Table 6.3, with the *Distribute-Broadcast-To-Network-NAK* code. The standard does not say that a *BVLC-Result* message should be returned with the *Success* error code if the broadcast is successful, but as a practical matter, some stack implementations may choose to do this to simplify the control flow.

6.5.4.2 UDP/IPv4 BVLC Table Management Message Types

Table 6.3 contains the BVLC messages used by the BBMDs to manage the broadcast distribution and foreign device tables. The *BVLC-Result* message type is a return message for those BVLC message types that need a status return. The payload is a 2-byte status code for the operation. Not all BVLC messages receive *BVLC-Result* responses, and some only receive a response if the operation fails.

Two message types enable management of the BBMD broadcast distribution table:

- The *Write-Broadcast-Distribution-Table* message is unicast to a BBMD to either insert new broadcast distribution table records or update existing records. The BBMD returns a *BVLC-Result* message with the status code for the request.
- The *Read-Broadcast-Distribution-Table* message is unicast to a BBMD to read its broadcast distribution table. The BBMD returns either a *Read-Broadcast-Distribution-Table-ACK* with a copy of the BBMD's table or a *BVLC-Result* with the Read-Broadcast-Distribution-Table-NAK code indicating the operation failed.

Three message types allow foreign devices to manage their registrations in the BBMD foreign device table:

- The *Register-Foreign-Device* message is unicast by a foreign device to a BBMD to register for receipt of broadcast messages. The BBMD returns a *BVLC-Result* message with the request status.
- The *Read-Foreign-Device-Table* message is unicast to a BBMD. The BBMD returns either a *Read-Foreign-Device-Table-ACK* message with the contents of the foreign device table or *BVLC-Result* with a Read-Foreign-Device-Table-NAK code indicating the operation failed.
- The *Delete-Foreign-Device-Table-Entry* is unicast to the BBMD to remove a foreign device registration. The BBMD returns a *BVLC-Result* message with the request status.

6.5.5 UDP/IPv4 BVLL Broadcast Example

Figure 6.7 contains a protocol flow diagram with an example UDP/IPv4 BVLL deployment showing how a broadcast message is propagated. In this deployment, we have the following IP subnets and BACnet nodes:

- Subnet A contains a broadcasting BACnet device node and the Subnet A BBMD.
- Subnet B contains the Subnet B BBMD plus some BACnet device nodes (not shown).
- Subnet C likewise contains the Subnet C BBMD plus some BACnet device nodes (not shown).
- Foreign Device A.

The Subnet A BBMD broadcast distribution table contains the following entries:

- Subnet B's BBMD address (IPv4 address + UDP port number) and the broadcast mask 255.255.255.000/24.
- Subnet C's BBMD address (IPv4 address + UDP port number) and the broadcast mask 255.255.255.255/24.

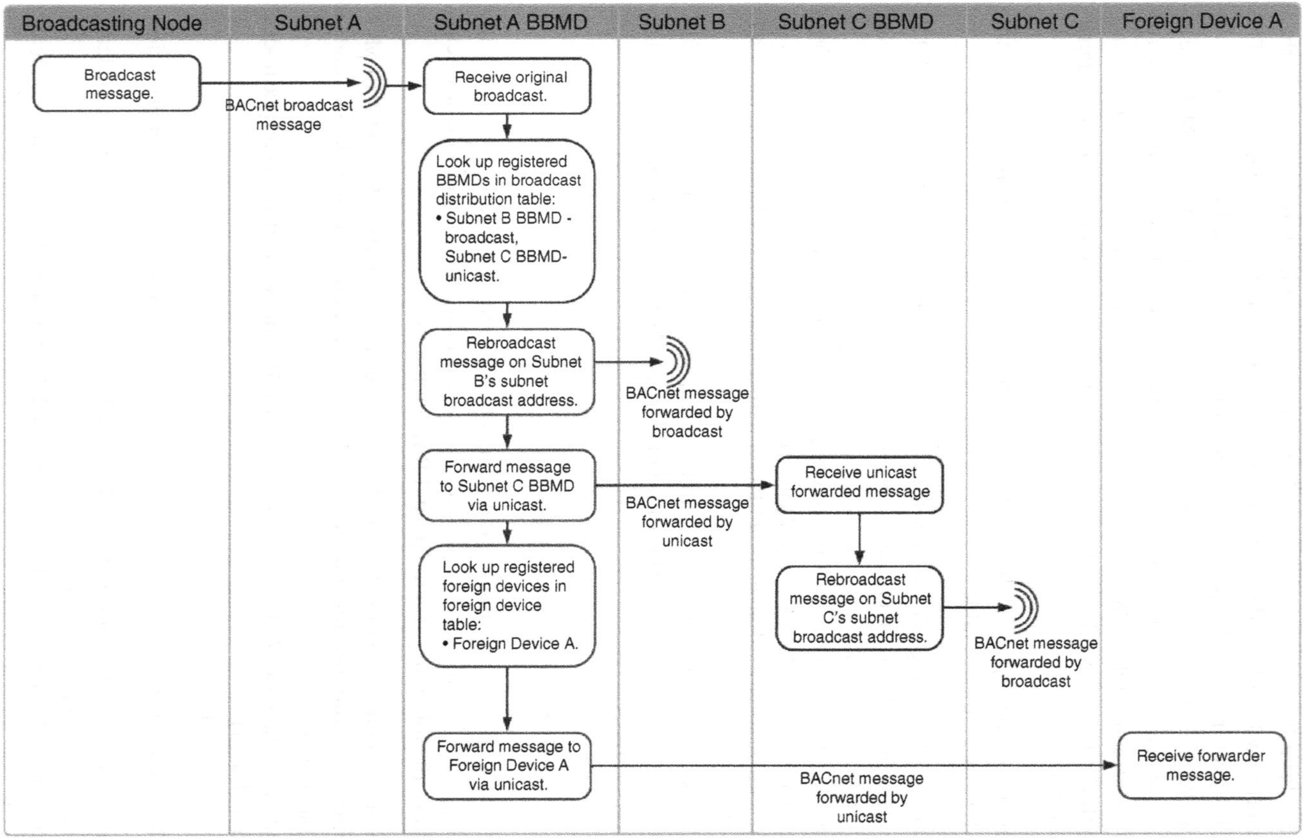

Figure 6.7 Protocol Flow for Example UDP/IPv4 Broadcast

In addition, the Subnet A BBMD foreign device table contains an entry for Foreign Device A.

As shown in the figure, the BACnet device node on Subnet A broadcasts a message out to Subnet A, and the message is received by the Subnet A BBMD. Since Subnet B can receive broadcasts directly on the subnet broadcast address, the Subnet A BBMD uses one hop distribution for Subnet B. Since Subnet C requires the message to be rebroadcast, the Subnet A BBMD uses two-hop distribution for Subnet C and the Subnet C BBMD receives a unicast that it rebroadcasts. Foreign Device A receives the message via unicast since Subnet A's foreign device table contains an entry for it.

6.6 BACnet Network Layer

The BACnet network layer is responsible for managing the communication between devices, including device discovery, addressing, routing, and message forwarding. The BACnet network layer includes support for both broadcast and unicast messaging, and for specific link layers that also support multicast. A BACnet network layer packet encapsulates the application layer message payload and is in turn encapsulated by the link layer frame or virtual link layer "frame" before being sent over the physical medium. Additionally, the BACnet network layer provides mechanisms for managing network congestion and handling errors and retries.

6.6.1 BACnet Network Layer Header Format

Figure 6.8 shows the full BACnet network layer header format. The fields correspond to the following:

- Version – version number of the BACnet protocol, currently "1,"
- Control –bitfield that indicates which further header fields are present and which are absent. Section 6.6.3 provides more details,
- Destination Network Address– network address of the destination node's BACnet network,

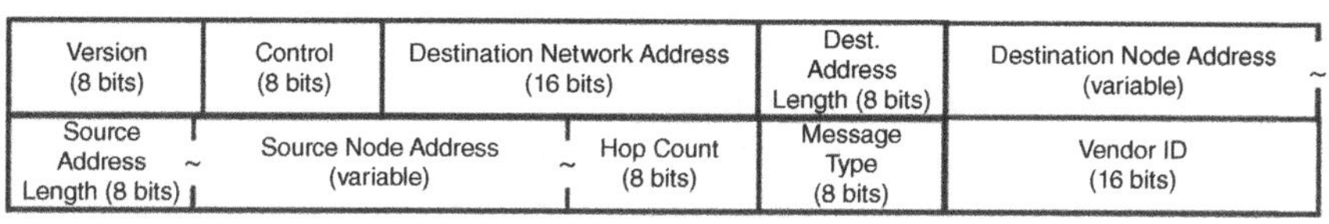

Figure 6.8 BACnet Network Layer Header Format

- Dest. Address Length –length of the Destination Node Address field, in bytes,
- Destination Node Address – BACnet link-layer address of the destination node. This field is variably sized because different link layers have different-sized addresses,
- Source Network Address –address of the source node's BACnet network,
- Source Address Length –length of the Source Node Address field, in bytes. This field is also variably sized because different link layers have different-sized addresses,
- Source Node Address –BACnet link-layer address of the source node,
- Hop Count – set to 0xff initially and decremented every time the packet traverses a BACnet router. When the hop count reaches zero, the packet is discarded. This field prevents packets from being routed indefinitely on misconfigured networks,
- Message Type – if the message is a network management message, this field contains the type,
- Vendor Id – if bit 7 in the Control field is set and the Message Type field is in the range of 0x80–0xff, then the message is a proprietary message and the Vendor Id field contains a numeric identifier for the vendor.

The header is followed by a variably sized payload.

Fields in the BACnet network layer header that encode BACnet-specific information like the Control field and the Destination Network Address field are encoded in big-endian format (most significant bit first), while if the Source and Destination Node Address fields contain addresses from the underlying link layer directly, they use the encoding of the link layer to which the address applies.

6.6.2 BACnet Network Layer Addressing and Routing

BACnet uses a 2-byte network address to designate separate links, with 65,535 as the maximum number of networks on a BACnet "internet," and the network address 0xffff reserved for global broadcast, that is, broadcast to all nodes on all links in the BACnet "internet." On the BACnet network layer, a BACnet node is identified uniquely by the node's network address and the node's link-layer address. How the BACnet network layer forms the link-layer address for the Destination Node Address and Source Node Address fields depends on the link layer type. Sections 6.4.3 and 6.5.2 describe the unicast and link-layer addressing schemes for MS/TP and UDP/IPv4, respectively.

BACnet network links are connected by BACnet routers. A BACnet router is a BACnet system application connected to two or more BACnet networks. Each connection is called a port, but note that because the mapping between BACnet networks and the underlying BACnet link layer networks may not be one-to-one, there may also not be a one-to-one mapping between the underlying link layer ports on the node and the BACnet router ports. The algorithm used in BACnet routing is very simple but requires that there be only one path between any two nodes on the wider BACnet network, unlike IP routing. This eliminates the need to prioritize paths during route computation and avoids routing loops if the network is properly configured. Note that if the BACnet network layer is run on top of a UDP/IPv4 virtual link layer, the BACnet routers are separate from the IP routers, and the BACnet virtual link-layer address is different from a node's IPv4 address. BACnet link-layer addressing is built independently on top of the UDP/IPv4 virtual link layer, as discussed in Section 6.5.

BACnet makes much heavier use of broadcast than most IP application layer protocols. The BACnet network layer supports three different types of broadcast transmission: local link, designated remote link, and global. For local broadcast, the network layer Destination Node Address field value is different depending on the underlying link layer and is discussed for MS/TP and UDP/IPv4 in Sections 6.4.3 and 6.5.2, where addressing for these link-layer types is discussed. Broadcast to a designated remote link is initiated by setting the Destination Network Address field to the network address of the designated remote link in the network layer header, the Dest. Address Length field to zero, and omitting the Destination Node Address field. Global broadcast throughout the entire network of connected BACnet links to all devices is supported by setting the Destination Network Address field to 0xffff, the Dest. Address Length field to zero, and omitting the Destination Node Address field.

BACnet routers handle local broadcast and designated link broadcast by broadcasting the packet on the link using the local link broadcast address and the protocol for the link layer type in question. For global broadcast, a BACnet node initiates a message by setting the network layer and link layer headers on the broadcast packet as follows:

- The network layer header Destination Network Address field is set to 0xffff.
- The network layer header the Dest. Address Length field is set to zero.
- The network layer header Destination Node Address field is omitted.
- The link layer header destination address is set to the broadcast address for the current link layer type.

This ensures that all BACnet routers and nodes on the link receive the message via broadcast. Upon receipt of a global broadcast message, a BACnet router decrements the hop count by one. If the hop count is greater than zero, the router forwards the packet to all directly connected links on the local link broadcast address. If the hop count is zero, the message is discarded.

6.6.3 Header Control field

Unlike the IP header, the BACnet network layer header format varies depending on whether the source and destination nodes are on the same link or on different links. The Control field provides information about which network layer fields are present and constrains the content of some fields. An example of how the Control field works is shown in Figure 6.9. The Control field indicates a packet with bitfield 100 or 0x04, that is, a Normal Priority packet with the source and destination nodes on the same link. Consequently, no further network headers are required since the destination node link-layer address is available from the link layer header.

Table 6.4 is a complete list of the control bits and their functions. The Priority bits (bits 0 and 1) are used by BACnet routers to decide in which queue to place an incoming packet, called quality of service or class of service in IP routing. A packet having priority 0x3 (11) is given the highest priority in routing since the message is of critical life safety importance. Conversely, a packet having priority 0x0 (00) is given the lowest priority and must wait for higher-priority packets to be sent before it is sent. If the Application Field Type bit (bit 7) is set to zero, the data in the Payload field of the network layer packet is from the BACnet application layer, and there is no message type field. The BACnet router can simply forward the message based on the Destination Network Address field and the Destination Node Address field and need not be concerned with interpreting the message contents. If set to 1, the data in the Payload field is a network control message that must be examined by the BACnet router to see if any information would require updating its routing tables or replying.

6.6.4 Network Layer Management Message Types

If bit 7 is set in the Control field, the message is a BACnet network management message and the Message Type field is present. Table 6.5 contains a complete list of the BACnet network management message codes, their names, transmission modes, Payload field contents, and a short description of their functions. The *I-Could-Be-Router-To-Network*, *Establish-Connection-To-Network*, and *Disconnect-Connection-To-Network* messages are all associated with routers that have at least one port connected to a point-to-point (PTP) link layers and are not discussed further here. More information on them can be found in the BACnet specification [3].

Version: 0x1	Control: 0x4	Payload: Application Layer Data

Figure 6.9 Example BACnet Network Layer Packet for the Source and Destination Nodes on the Same Link

Table 6.4 Control Bits in the BACnet Network Layer Control Field

Bit #	Function	Value Range
0 and 1	Priority	Network priority: • 11 – life safety message. • 10 – critical equipment message. • 01 – urgent message. • 00 – normal message
2	Expecting Reply	Indicates whether a reply is expected: • 0 – message does not expect a reply. • 1 – message expects a reply.
3	Source Addresses Present	Indicates which source address information is present: • 0 – Source Network Address, Source Node Address Length, and Source Node Address absent. • 1 – Source Network Address, Source Node Address Length, and Source Node Address present. Source Node Address Length must be nonzero.
4	Reserved	Set to 0
5	Destination Addresses Present	Indicates which destination address information is present: • 0 – Destination Network Address, Dest. Address Length, Destination Node Address, and Hop Count absent. • 1 – Destination Network Address, Dest. Address Length, Destination Node Address, and Hop Count present. A Dest. Address Length of zero indicates broadcast and the Destination Node Address is absent.
6	Reserved	Set to 0
7	Application Field Type.	Indicates the type of the Payload field: • 1 – Payload field contains network management information; Message Type field is present. • 0 – Payload field contains BACnet application layer information; Message Type field is absent.

Table 6.5 BACnet Network Layer Management Message Types

Type Code	Name	Transmission Mode	Payload Contents	Function
0x00	*Who-Is-Router-To-Network*	Broadcast and unicast by non-routing nodes to a router.	Optional 2-byte network address. If omitted, receiving router returns all reachable network addresses.	When broadcast, to ascertain the next router to a specific network or as an aid for routers in building their routing tables. When unicast, to discover routing tables.
0x01	*I-Am-Router-To-Network*	Broadcast by router in response to *Who-Is-Router-To-Network*.	One or more 2-byte network addresses.	Indicates networks accessible through a router.
0x02	*I-Could-Be-Router-To-Network*	Unicast response to *Who-Is-Router-To-Network*.	2-byte network address followed by a 1-byte performance index.	Sent by router connected to a point-to-point (PTP) link if the link needs to be brought up.
0x03	*Reject-Message-To-Network*	Unicast to a node originating a message to a network that the router temporarily cannot route.	1-byte code indicating reason for rejection followed by 2-byte network address. See text for rejection codes.	Indicates that the router is currently unable to route to a network.
0x04	*Router-Busy-To-Network*	Broadcast.	Optional list of 2-byte network addresses. If no list, indicates all networks reachable through the router.	Indicates on which networks the router is currently experiencing congestion and cannot accept new packets.
0x05	*Router-Available-To-Network*	Broadcast.	Optional list of two byte network address. If no list, indicates all networks reachable through the router.	Indicate congestion is cleared and routing to listed networks is again available.
0x06	*Initialize-Routing-Table*	Unicast to a router.	1 byte number of ports field followed by optional routing table. See the text for more details on routing table format.	See above for the description of this message's function.
0x07	*Initialize-Routing-Table-Ack*	Unicast acknowledgment of an *Initialize-Routing-Table* message.	Optional copy of the router's routing table. See above for more details on routing table format.	If the *Initialize-Routing-Table* message's number of ports field was zero, returns a complete copy of the routing table in the Payload field. If the number of ports was nonzero, the Payload field is empty.

0x08	*Establish-Connection-To-Network*	Unicast.	2-byte network address and 1-byte termination time, in seconds, link should remain active if no packets are transmitted. Value 0 means permanent.	Instructs a router connected to a PTP link which is currently inactive to activate a connection to the link.
0x09	*Disconnect-Connection-To-Network*	Unicast.	None.	Instructs a router connected to an active PTP link to deactivate the link.
0x12	*What-Is-Network-Number*	Broadcast or unicast on local link only (i.e. not routed)	Empty.	Sent to discover the local network number. Ignored by the recipient if the Source Network and Source Node Address fields or Destination Network and Destination Node Address fields are present or if the recipient does not know the local network address.
0x13	*Network-Number-Is*	Broadcast on local link only (i.e. not routed). If query received as broadcast, non-routing node can wait for 10 seconds to see if another device responds before responding or not respond if a different device responds.	2-byte local network address followed by one byte flag. The flag is 0 if network address was learned through a *What-Is-Network-Number* message or 1 if the network address was configured.	Returned in response to *What-Is-Network-Number*. The recipient caches the network address unless it has already been configured. Ignored by the recipient if the Source Network and Source Node Address fields or Destination Network and Destination Node Address fields are present.

For most of the messages, the Payload format is either empty or a list of 2-byte network addresses, but several have structured payload fields. The *Reject-Message-To-Network* payload consists of a 1-byte code indicating the reason for rejecting the message followed by the 2-byte network address for which the message was rejected. The following is the list of rejection codes:

- 0 – Other error (unspecified).
- 1 – Router is not directly connected to Destination Network Address or cannot find a next hop router that can route to the network on any directly connected network discovered using *Who-Is-Router-To-Network*.
- 2 – Router is busy and cannot accept a message to Destination Network Address.
- 3 – The Message Type field is for an unknown network management message type.
- 4 – The message is too long to be routed to the Destination Network Address.
- 5 – Deprecated.
- 6 – The message was rejected due to errors in the Source Node Address or Destination Node Address or the length of the Source Node Address or Destination Node Address is invalid.

In addition to the network management message type codes, the following Message Type code ranges are relevant:

- 0x0A-0x11 – obsolete and removed;
- 0x14-0x7f – reserved for ASHRAE;
- 0x80-0xff – reserved for vendor-proprietary messages and the Vendor Id field is also present in the network layer header to indicate the vendor that specified the proprietary message.

The *Initialize-Routing-Table* and *Initialize-Routing-Table-Ack* messages also have structured payloads and can both send a routing table configuration to the receiving BACnet router in the Payload field. The payload starts with the number of ports field, which gives the number of port records to follow. A port record consists of the following fields:

- The 2-byte address of the BACnet network,
- The 1-byte port identifier of the port through which the designated network address is accessible.
- A 1-byte port information length field, giving the length, N, in bytes, of the following port information field, N bytes of port information for this port. If N is zero, this field is absent.

An example of the kind of information that may be in the port information field is a telephone number and access control information for a dial-up, point-to-point link.

Like IP routers, BACnet routers maintain a routing table internally that guides the routing algorithm on the reachability of BACnet networks. The BACnet router internal routing table consists of the following entries, one for each connection to a BACnet network, called a port in the standard:

- The BACnet link-layer address of the port's connection to its BACnet network.
- The 2-byte BACnet network address of the directly connected network corresponding to the port.
- A list of BACnet network addresses reachable through the port.
- The link-layer address of the next router on the path to the network address.
- An implementation-dependent reachability status indicator with, at minimum, status values indicating permanent failure and temporary unreachability due to congestion.

If the port identifier for a BACnet network address is zero in a BACnet network layer management message, the router should purge all entries from its routing table about that network address. If the port identifier is nonzero, the router should replace entries for the network address in its routing table with the information in the message, or, if there are no pre-existing entries, it should add a new record to the table with the network address.

A router receiving an *Initialize-Routing-Table* message with the number of ports field set to zero returns a complete copy of its routing table in an *Initialize-Routing-Table-Ack* message. If the routing table is too large to fit into one message, it returns the routing table in multiple messages. If the number of ports field is nonzero in the *Initialize-Routing-Table* message, the router responds with an *Initialize-Routing-Table-Ack* message having an empty Payload field.

6.6.5 Transmitting and Receiving Packets on BACnet Networks

In general, the BACnet stack is not as strict in separation of concerns and keeping information compartmentalized within specific layers as the IP stack. Unlike the IP stack, the BACnet stack does not use end-to-end addressing at the network layer. For packets sent to a destination address on the local link, no address information appears in the network layer header, as shown in Figure 6.9. Address management is handled by the link layer. For packets sent to remote links, BACnet routers rewrite the network layer header at the first hop router and last hop router. In contrast, IP routers do not touch the IP network layer header except to update the hop count and explicit congestion notification bits, but they do rewrite the link-layer address with the address of the next hop router or end node. BACnet routing therefore requires careful coordination between the link layer and the network layer. The next sections present a case-by-case discussion of packet transmission on BACnet networks.

6.6.5.1 Sending and Receiving Packets by a Non-routing Node

For a non-routing node to send a packet to another node on the local link, the link-layer address of the destination must be known. The Source and Destination Network Address fields in the network layer header are omitted because the packet will not be routed off the local link. The Source and Destination Node Address fields are also unnecessary because the link layer knows the source address of the sending node and the link-layer address of the destination is passed directly to the link layer by the network layer rather than through the network layer header. The network layer packet looks like Figure 6.9, that is, no address information appears in the network layer header.

The network layer on a non-routing node performs the following steps to send a packet to another node on the local link:

- Sets the Control field bits indicating the priority, whether the message is a network layer message or application layer message, and whether the message is expecting a reply.
- If the message is a network layer message, sets the Message Type field in the header.
- Sets the Payload field to the application layer message or to the network layer message depending on message type.
- Passes the packet to the link layer along with the destination link-layer address.

The link layer then performs the following steps:

- Sets the Source Address field in the link layer header to the sending node's link-layer address.
- Sets the Destination Address field in the link layer header to the destination node's link-layer address.
- Inserts the application or network layer payload into the Payload field.
- Places the frame onto the wire.

When a non-routing node receives a packet with the Destination Network Address field in the network layer header, it discards the packet unless the value of the field is the global broadcast address (0xfff) because the Destination Network Address field should be stripped off by the last hop router. The network layer on the receiving node, either a router or a non-routing node, checks the Control byte for whether the packet contains an application layer message or a network layer message. If the former, the network layer passes the Payload to the application layer for processing. If the latter, the network layer processes the message. If the node is a router and it can not interpret the message, the router sends a *Reject-Message-To-Network* message back to the sender.

To send a packet to a node on a remote link, a non-routing node must know the Destination Network Address and the Destination Node Address. If the node wants to perform directed broadcast to a specific remote link, the Destination

Network Address is the address of the link to which the broadcast is sent and the Destination Node Address is zero. If instead the node wants to broadcast the packet to all remote links, the Destination Network Address is the global broadcast address (0xffff) and the Destination Node Address is zero.

A non-routing node performs the following steps to send a packet to a remote link:

- Sets the Control field bits indicating the message priority and whether the packet contains an application layer message or a network layer message.
- Sets the Control field bits indicating that the network layer header contains the Destination Network Address field, and, if not a broadcast packet, the Dest. Address Length field and Destination Node Address field.
- Sets the Control field bits indicating the Source Network or Source Node Address fields are not present in the network layer header.
- If the packet is to be unicast:
  - Sets the Destination Network Address field to the address of the destination network.
  - Sets the Dest. Address Length field to the length in bytes of the destination node's address and Destination Node Address field to the destination node's address.
- If the packet is to be broadcast:
  - If the packet is to be delivered by directed broadcast, sets the Destination Network Address field to the address of the destination network.
  - If the packet is to be delivered by global broadcast, sets the Destination Network Address field to 0xffff.
  - Sets the Dest. Address Length field to zero and omits the Destination Node Address field.
- If the packet is a network layer message, sets the Message Type field.
- Sets the Hop Count field to 0xff (255).
- Sets the Payload field to the application layer message or network message depending on message type.
- Passes the packet to the link layer with the address of a router that has a route to the destination network.

An example of a network layer header for an application data packet expecting a reply that is to be unicast to a node on a remote link is shown in Figure 6.10.

Version: 0x1	Control: 0x24	Destination Network Address: 0xXXXX	Dest. Addr. Length: 0x06	Destination Node Address: 0xXXXXXXXXXXXX
Hop Count: 0xff	Payload: Application Layer Data			

Figure 6.10 BACnet Network Layer Packet for an Application Data Packet Unicast to a Node on a Remote Link

In this case, the size of the destination node address is 6 bytes, which means it could, for example, be on a UDP/IPv4 BVLL network.

The link layer then performs the following steps:

- Sets the Source Address field in the link layer header to the node's source link-layer address.
- Depending on whether the frame is to be unicast or broadcast or if the node does not have the address of a router to the remote link:
  - If the frame is to be unicast or directed broadcast, sets the Destination Address field in the link layer header to the link-layer address of a router that has a route to the destination network passed down from the network layer.
  - If the packet is to be global broadcast or the sending node does not have the address of a router that routes to the destination network, sets the Destination Address field to the local link broadcast address.
- Inserts the payload received from the network layer into the Payload field.
- Places the frame on the wire.

See Section 6.6.6 for how a non-routing node discovers a router on the same link as the sending node but with information about how to route to the remote link.

6.6.5.2 BACnet Router Unicast Forwarding Actions

If a router receives a unicast message with the Destination Network Address field set, it checks whether the Source Network Address and Source Node address are present or absent. If they are absent, the packet arrives from a node on a local link directly connected to a router port and must be routed to a remote link. The router adds the Source Network Address, Source Node Address Length, and Source Node Address (which it receives from the link layer) of the sending node to the network layer header and enqueues the packet into a routing queue for the destination network. The packet will be forwarded in a first-in first-out manner within a priority class.

If the packet has the Source Network Address, Source Address Length, and Source Node Address fields, the packet came from another router. If the Destination Network Address field refers to a network that is directly connected to a router port, then the packet is destined to a node on a local link. The router takes the following actions:

- Removes the Destination Network Address field, Dest. Address Length field, and Destination Node Address field from the network layer header. The router saves the Destination Node Address field for the link layer.
- Adjusts the Control byte to indicate that the destination node and network address information was removed.
- Removes the Hop Count field from the network layer header.

- Looks up the link layer port for the Destination Network Address in its routing table.
- Hands the packet off to the link layer along with the destination address and the link layer port for the local link through which the packet should be forwarded.

The link layer takes the following actions:

- Sets the Source Address field in the link layer header to the router's link-layer address,
- Sets the Destination Address field in the link layer header to the local node's destination address passed down from the network layer,
- Inserts the payload received from the network layer into the Payload field,
- Sends the frame out the port for the correct local link passed down from the network layer.

If the Destination Network Address does not refer to a directly connected link, the packet arrives from another router and must be routed on the next hop to its destination. The router takes the following actions:

- Decrements the Hop Count field and discards the message if the result is zero.
- If the router has an entry in its routing table for the next hop router:
  - The network layer hands the packet off to the link layer with the port for the next hop router and the next hop router's link-layer address.
- If the router does not have an entry in its routing table for the next hop router, it must perform router discovery:
  - The router broadcasts the *Who-Is-Router-To-Network* message on its local links.
  - If another router returns an *I-Am-Router-To-Network* with the network address for the destination link, the router:
    - Updates its routing table with the new routing information.
    - Hands the packet off to the link layer with the newly discovered router's address and the port on which to forward the packet.
  - If the router discovery fails, the router sends a *Reject-Message-To-Network* message back to the sender with rejection code 1: router is not directly connected to Destination Network Address or cannot find a next hop router that can route to the network on any directly connected network.

The link layer performs the following actions:

- Sets the Source Address field in the link layer header to this router's link-layer address.
- Sets the Destination Address field in the link layer header to the next hop router's link-layer address.
- Inserts the payload received from the network layer into the Payload field.
- Sends the frame out through the port corresponding to the next hop router's link.

6.6.5.3 BACnet Router Broadcast Forwarding Actions

If the router receives a directed broadcast message – the Dest. Address Length Field is zero and the Destination Node Address field is absent but the Destination Network Address field is present and does not equal the global broadcast address in the network layer header – the router performs the following actions:

- If the Destination Network Address does not correspond to one of the networks on a local link, the router proceeds to forward the packet as if it were a unicast packet received from another router destined to a remote link, as described in Section 6.6.5.2.
- If the Destination Network Address corresponds to one of the networks on a local link the router performs the following actions:
  - Removes the Destination Network Address, Dest. Address Length, and Hop Count fields from the network layer header.
  - Passes the packet to the link layer for the port corresponding to the network address with the link-layer destination address set to the local link broadcast address rather than a specific node address.

If the Destination Network address corresponds to the global broadcast (0xffff), the router performs the following actions:

- Decrements the Hop Count field value and discards the packet if it is zero.
- Passes the packet to all link layers for all ports with the appropriate link-layer broadcast address for the link as the destination address.

The link layer then performs the following actions:

- Copies the router's address into the Source Address field of the link layer header.
- Copies the link's broadcast address into the Destination Address field of the link layer header.
- Inserts the payload received from the network layer into the Payload field.
- Places the frame on the wire.

6.6.6 Discovering a Router to a Remote Link

When sending packets to a remote link, a non-routing node does not need to know the address of its own link because the first hop router takes care of inserting the network address into the network layer header. However, a non-routing node does need to know the address of a router with a route to the remote link because the node's network layer must supply such an address when sending a packet to a node on a remote link. The standard suggests the following approaches for how a node can discover a router with a route to a remote link [3]:

- Configuration during or after commissioning.
- If the node knows the address of a node on the remote link, it can send a *Who-Is* message and note the link layer source address of the local router

that returns the response. *Who-Is* is an application services message that is discussed further in Section 6.7.

- Broadcast a *Who-Is-Router-To-Network* message with the remote link's network address in the payload contents section.
- Note the source link-layer address on the frame from a previous request that originated in the destination network, as it will be link-layer address of a local router with a route to the remote link.
- Instead of unicasting a request to a remote device, set the local broadcast address on the frame but put the remote node address in the network layer as the destination address. Then note the source address of a response from the device, as this will be the source address of a local router with a route to the remote link.

The last approach does not follow the standard procedure for sending broadcast to remote links and the second approach mixes concerns from the application layer with the network layer, but as mentioned, the BACnet stack is not as strict as the IP stack in the separation of concerns between various network layers and transmission procedures.

6.7 BACnet Application Layer

The application layer sits at the top of the stack and is where device sensors and actuators interface with the building's control network and control applications. The BACnet application layer has two parts: BACnet objects and BACnet services. BACnet object types provide a schema covering many different types of sensors and actuators, as well as other data types such as audit logs, types for performing operations like staging and looping, etc. BACnet services are a collection of protocol messages that are exchanged over the network between applications running the BACnet stack. These messages allow applications to read and write to BACnet objects, subscribe to updates on the objects' properties, and perform other operations.

6.7.1 BACnet Objects and Object Types

ASHRAE Standard 135-2020 [3] defines 64 different types of BACnet objects. Object types represent the functions of a BACnet device in a way that is visible over the network and standardized. Instances of object types model various components of a BAS, such as sensors, actuators, alarms, and other system elements. Each BACnet object type has a collection of properties that define the type. The properties are assigned values of either a primitive or a sequence datatype. Primitive datatypes have no property structure while sequence datatypes have a key-value structure and are like data classes in Python and *struct* datatypes in C. These types are discussed in more detail in Section 6.7.3. BACnet objects are the basic building blocks of a BACnet deployment, representing different components within the building automation and control network. They serve as a standardized way to model and manipulate a building's automation system over the network.

BACnet uses a formal schema definition language called Abstract Syntax Notation 1 (ASN.1) to describe primitive datatypes, properties, sequence types, and object types [12]. However, it does not use the ASN.1 specification for data serialization to transmit the data over the network. Instead, BACnet defines its own scheme in Clause 20.2 of the standard [13]. A BACnet object type is designated by its common name, which is a string of words strung together, for example, *Analog Value*. The identifier also corresponds to an ASN.1 enumerated type identifier consisting of the words in all lower case separated by a hyphen ("–") rather than a space to conform with ASN.1 syntax, for example: *analog-value*. The formal, ASN.1 name for object types corresponds to a numeric type code in Clause 21.6 of the standard [14]. In this section, the names of primitive datatypes, sequence types, properties, and object types are additionally set in italics to distinguish them from informal descriptions, for example, of architectural components.

The common name for properties is formed by a string of words in "snake case"[4], that is separated by underscores instead of spaces or hyphens, with the individual words capitalized. The ASN.1 name uses the same naming convention as for types. An example of a common name identifier for a property is *Object_Identifier*, the corresponding formal, ASN.1 name is *object-identifier*, and Clause 21.6 also contains a list of numeric type codes for all the properties in the standard [15]. The BACnet services use the numeric type code when referring to primitive datatypes, object types, and properties in network communications with devices. As with type names, property names are set in italics as well.

BACnet uses the code 4194303 to indicate that the property value is currently uninitialized, disabled, or unused. Here we use the type identifier *BACnetUndefined* to correspond to 4194303.

6.7.2 Properties

A property comes with three attributes:

- Property Identifier – the common name, ASN.1 name, and corresponding numeric type code for the property.

4 The term "snake case" comes the use of these types of identifiers in the Python programming language.

- Property Datatype – a primitive datatype or a sequence datatype of the property, see Section 6.7.3 for more on these datatypes. Values assigned to the property are of this type.
- Conformance Code – whether the property is readable (*R*), readable/writable (*W*), or optional (*O*). Some properties are writable only under certain conditions.

Four properties are required for all object types: *Object_Identifier*, *Object_Name*, *Object_Type*, and *Property_List*. All object types have other required and optional properties, with the *Device* type having the most (73).

The *Object_Identifier* property is a unique identifier for the object within the BACnet device. The identifier is a 32-bit value that is made up of two parts: the object type's numeric code, encoded in the upper 10 bits of the number, and the object's instance number, encoded in the remaining 22 bits. The object instance number is a number that identifies the specific instance of the object unique within the device. For example, an *Analog Input* object has an object type code of 0 and if the object instance number is 1, it would identify the first *Analog Input* object instance in the device. The *Object_Identifier* property can have a value of *BACnetUndefined* if object is uninitialized. *Device* type objects must have an instance identifier that is globally unique within the BACnet "internet" since they have global visibility. The ASN.1 identifier for this property is *object-identifier*, the numeric type code is 75, and the property conformance code is readable.

The *Object_Name* property is a string containing a human-readable, descriptive name for the object. The *Object_Name* property should be unique within the device to make it easier for users to differentiate between objects. The minimum length of the string is one character, and the maximum length is specified in the device's Protocol Implementation Conformance Statement (PICS) (see Section 6.7.8). The ASN.1 identifier for this property is *object-name*, the numeric type code is 77, and the property conformance code is readable.

Object_Type is the integer type code for the BACnet object's type, such as *Analog Input* (code 0), *Binary_Output* (code 4), or *Device* (code 8). Codes 0–127 are reserved for ASHRAE, and codes 128–1023 allow manufacturers to define their own types. These codes correspond to the ASN.1 type code for the object type. The ASN.1 identifier for this property is *object-type*, the numeric type code is 79, and the property conformance code is readable.

Property_List is a *BACnetArray* collection datatype of unsigned integer numeric code property identifiers for the properties within the object type, with the required properties (*Object_Identifier*, *Object_Name*, *Object_Type*, and *Property_List*) not included. The ASN.1 identifier for this property is *object-type*, the numeric type code is 371, and the property conformance code is readable.

The *Object_Identifier* and *Object_Name* properties can be configured by either the installer or the manufacturer, depending on the specific device and its configuration options. In some cases, the manufacturer pre-configures the properties for BACnet objects within a device, providing default identifiers and descriptive names, avoiding confusion and potential conflicts between objects assigned the same identifier and name. In other cases, the installer can change the values during the installation and commissioning process to match the building naming conventions or requirements providing meaningful context for managing the system. These two properties need to be descriptive and unique within the device context to help building managers easily identify and manage BACnet objects.

6.7.2.1 *Profile_Name* and *Profile_Location* Properties

Profile_Name and *Profile_Location* are optional properties in all standard BACnet object types. *Profile_Name* contains the name of a BACnet object profile describing the object while *Profile_Location* contains a URL with the location of a machine-readable file containing the object profile. The object profile contains more information about the optional properties, the types of their values, the value ranges, whether the properties persist after the device reboots, etc. Object profiles provide an extra level of configurability to BACnet objects. Because an object profile is defined per object and is not tied to the object type, implementors can support devices with configurable objects. These objects can be reconfigured by an application that writes to a proprietary property and changes the property set and behavior of the object. The application can then update the *Profile_Name* and *Profile_Location* with the new definition.

The *Profile_Name* property is a character string that contains the profile name. The format of the name consists of the vendor's organization identifier followed by a "–" followed by a human-readable name that is under the control of the organization. For example, an *Analog_Value* type object that supports a temperature sensor profile from Vendor 42 might have a *Profile_Name* property value of "42-TS" and might contain additional properties, for example, properties to record the high, low, and average temperature over a certain period. Another *Analog_Value* object from Vendor 73 might have a *Profile_Name* property value of "73-TSenseModelY" for their temperature sensor and might support different optional properties. The numeric code "0" is reserved for ASHRAE. Object profiles are written using a formal schema definition language called CSML, for Control Systems Markup Language, an extension of XML, defined in Annex Q of the standard [16].

The *Profile_Location* property points to the location of the machine-readable file with the definition of the object profile. Object profiles are packaged as zip format compressed files called extended data definition (`xdd`) files. The location is a URL pointing to the file location, for example, https://bems.example.com/xd/dev4253.xdd. ASHRAE 135-2020 recommends that all nonstandard and proprietary objects have a *Profile_Name* property and that the *Device* object for the device in which the objects are located have a *Profile_Location* property covering all objects in the device. More about object profiles, the format of the `xdd` definitions, and some examples can be found in Annex X of the standard [17].

6.7.2.2 "Commandable" Properties

In some cases, an object property can be manipulated by different controlling entities, and some method is required to decide which entity has precedence at any given time. Such properties are defined to be "commandable." A commandable property is associated with a priority array. The priority array elements correspond to different controlling entities and are arranged in a hierarchical fashion, with the controlling entities assigned different priority levels. The value of the property is the priority array element controlled by the highest priority controlling entity. This feature is used to enable a hierarchical control structure within a BACnet device, ensuring that commands from higher-priority control entities take precedence over lower-priority ones.

The *Priority_Array* property holds the priority array data structure for a commandable object. The data structure is an array of 16 elements having the same datatype as the commandable property. Each element represents a different priority level, with element 1 being the highest priority and element 16 being the lowest. The elements hold a value for the commandable property, or a *BACnetNull* value if the priority level is not being used. The device determines the active value of a commandable property by evaluating the *Priority_Array*. The element with the lowest index in the array that has a non-NULL value becomes the active value for the commandable property. If all the array elements are null, then the value of a third property, *Relinquish_Default*, is assigned to the commandable property. The *Relinquish_Default* also has the same datatype as the commandable property. Finally, the *Current_Command_Priority* property contains the array index in the priority array from which the *Present_Value* property was taken. A more detailed explanation of how commandable properties work, including a few special cases for specific object types, can be found in ASHRAE 135-2020 Clause 19.2 [18].

As an example, the *Present_Value* property in the *Analog Value* object is specified as commandable. Suppose an *Analog Value* object is used to represent the setpoint temperature for an HVAC system, and that there are three possible control entities: a wall-mounted thermostat that room occupants can set, the centralized building automation control system that is controlled by the building manager, and an automated emergency shutdown system. Each of these components can control the temperature setpoint at different priority levels. The thermostat has a priority of 8, the building automation control system has a priority of 5, and the emergency shutdown system has a priority of 1.

When the system is operating normally, the building automation control sets the temperature setpoint based on a predefined schedule or occupancy data by writing a setpoint temperature into the *Priority_Array* array at element 5. Element 1 is where the emergency shutdown system writes a value if there is an emergency overriding all other values in the array, but normally it has value *BACnetNull* because no emergency is in progress. If a room occupant adjusts the temperature using the thermostat, the new setpoint value is written to the *Priority_Array* array at element 8, but since the building automation control has a higher priority (5), the room occupant's change will not take effect. However, if the building automation control writes a *BACnetNull* value at element 5, the room occupant's command at priority 8 will take effect.

6.7.3 Primitive, Collection, Enumeration, Sequence, and Choice Datatypes

The primitive, collection, and enumeration BACnet datatypes are listed in Table 6.6. These types are similar to nonobject datatypes in an object-oriented programming language like `int` and `float` in Python. They have no property structure like objects do. The table contains a descriptive name for the datatype, along with the ASN.1 name, and tag number used in the serialization encoding. The ASN.1 name and tag number are specified in Clause 20.2.1.4 [19].

For some of the primitive datatypes, BACnet uses well-established standards. Signed integers are represented in 2's complement format. *BACnetFloat* and *BACnetDouble* are defined according to the IEEE 754-2008 standard [20] for single-precision and double-precision floating numbers. Each element in a *BACnetCharacterString* represents a single character from a supported character set, for example, ISO 8859-1 (Latin-1) [21] or ISO 10646 (UCS-4)/Unicode [22]. The size of an element in the *BACnetCharacterString* can range from 1 byte (8 bits) for single-byte character sets like ISO 8859-1 to 4 bytes (32 bits) for multi-byte character sets like Unicode (UCS-4). The actual character sets supported by a BACnet device are listed in its PICS. In addition, the *BACnetOctetString*, *BACnetCharacterString*, and *BACnetBitString* all have absolute maximum lengths determined by the 16-bit integer representing their length, but their actual maximum lengths in devices may differ depending on the manufacturer and are specified in the PICS.

In addition to the primitive datatypes, BACnet supports two collection datatypes that can be assigned to properties, *BACnetArray* and *BACnetList*. A *BACnetArray* is an ordered sequence of elements, each having the same datatype. Array elements are accessed using an unsigned integer index, beginning at 1. An array index of 0 returns the length of the array as a *BACnetUnsigned*. Arrays that are part of a proprietary extension can be resized but not arrays defined in the standard. Similarly,

Table 6.6 BACnet Primitive Datatypes

Data Type	ASN.1 Datatype Name	ASN.1 Numeric Code	Length	Description
BACnetNull	NULL	0	8 bits	A null datatype with no value.
BACnetBoolean	BOOLEAN	1	8 bits	A binary value, either TRUE or FALSE.
BACnetUnsigned	Unsigned Integer	2	8–32 bits	An unsigned integer value.
BACnetSigned	INTEGER	3	8–32 bits	A variable-length signed integer that uses two's complement representation.
BACnetReal	REAL	4	32 bits	A single-precision floating-point number that conforms to the IEEE 754-2008 standard for floating-point arithmetic.
BACnetDouble	DOUBLE	5	64 bits	A double-precision floating-point number that conforms to the IEEE 754-2008 standard for floating-point arithmetic.
BACnetOctetString	OCTET STRING	6	8-bit bytes, maximum length 65,535	A variable-length sequence of bytes.
BACnetCharacterString	Character String	7	Between 8 and 32 bits per element depending on character set, maximum length 65,535	A variable-length sequence of characters, with support for multiple character encodings.
BACnetBitString	BIT STRING	8	Maximum length 524,280 bits	A variable-length sequence of bits, used to represent bitmasks or bit arrays.
BACnetEnumerated	ENUMERATED	9	8–32 bits per value	An enumerated value, represented by an unsigned integer.
BACnetDate	Date	10	32 bits	A calendar date, represented by year, month, day, and day of the week.
BACnetTime	Time	11	32 bits	A time of day, represented by hour, minute, second, and hundredths of a second.
BACnetObjectIdentifier	BACnetObjectIdentifier		32 bits	A unique identifier for a BACnet object, consisting of the object type and instance number (value of the object's *Object_Identifier* property).

a *BACnetList* is a collection of elements, each having the same datatype, but the difference between a *BACnetList* and a *BACnetArray* is that elements in a *BACnetArray* can be accessed using a numeric index while elements in a *BACnetList* can only be accessed using the *ReadRange* service. In addition, it is not possible to obtain the total number of elements in a *BACnetList* except by reading the elements and counting them.

A *BACnetEnumerated* is an enumeration type similar to an `enum` in C. Their values are drawn from a set of numeric type codes assigned to identifiers in the ASN.1 definition. Enumeration types are defined using the ASN.1 *ENUMERATED* production. For example, the enumeration type *BACnetDeviceStatus* has the following ASN.1 definition:

```
BACnetDeviceStatus ::= ENUMERATED {
        operational (0),
        operational-read-only (1),
        download-required (2),
        download-in-progress (3),
        non-operational (4),
        backup-in-progress (5),
}
```

The common names for the value tags of the enumeration are formed by capitalizing all the words in the string and converting to "snake case," for example, *OPERATIONAL_READ_ONLY*.

The sequence datatype consists of a sequence of named and typed members assigned values, essentially key-value pairs. Sequences are defined using the ASN.1 SEQUENCE production. An example of a sequence datatype is *BACnetAddress* (in ASN.1):

```
BACnetAddress ::= SEQUENCE {
        network-number : Unsigned16,
        mac-address : OctetString
}
```

which is used to represent the BACnet network address of a device or application. Note that sequences are not object types even though they have a similar key-value structure as object types, but are similar to `struct` types in C and C++ and data classes in Python. They are used for grouping related values under a single type name and have no network visibility outside of their being assigned to a property on an object, unlike instances of object types.

Finally, the choice datatype consists of a collection of typed keys, only one of which is present in an instance. Choice datatypes are defined with the ASN.1 CHOICE production, and allows a data value to have one of a number of different datatypes. The choice datatype corresponds to the `union` type in C or C++. For example, the following ASN.1 definition is a choice type that specifies the different ways to designate the amount of load that should be shed by the BACnet *Load Control* object:

```
BACnetShedLevel ::= CHOICE {
        percent : Unsigned,
        level : Unsigned,
        amount : REAL
}
```

6.7.4 Basic BACnet Object Types

Some BACnet object types are found in every BACnet device and practically all BACnet applications. This section reviews these basic object types.

6.7.4.1 Device

The *Device* object is one of two top-level objects within a device that are directly accessible over the BACnet network. Every BACnet device is required to support a *Device* object. The *Device* object type represents a top-level access point for all network-visible properties of a BACnet device, except for network connectivity configuration information in the *Network Port* object. It defines a naming scope within which other BACnet objects are implemented and acts as a container for other objects. There can only be one *Device* object per device and any software or hardware device supporting BACnet needs to implement a *Device* type object.

Within the scope of a *Device* object, objects of the same type need to have unique *Object_Identifier* and *Object_Name* property values. The *Device* object itself must have a unique *Object_Name* and *Object_Identifier* property within the BACnet "internet." Devices are uniquely identified over the network within the BACnet "internet" by their identifier, and objects are uniquely defined within their containing *Device* object by a combination of their device identifier and object identifier, allowing all objects to be addressed over the network.

With 72 properties, the Device object has far too many to describe in complete detail here, but the following may be relevant for BEMS applications:

- *Local_Time*, *Local_Date*, *UTC_Offset*, *Daylight_Savings_Status* – allow an application to determine how the device computes time and the device's current time setting,
- *Object_List* – a *BACnetArray* of *Object_Identifiers* for each object contained within the device,

- *Device_Address_Binding* – a list of address bindings containing information about other BACnet devices on the network. Each entry in the list is an instance of the *BACnetAddressBinding* sequence type, having two members: the BACnet object identifier for the *Device* object and a *BACnetAddress* sequence datatype value with the BACnet network address and link-layer address of the device. This property allows building managers and installers to debug possible network misconfigurations,
- *System_Status* – a value of *BACnetDeviceStatus* which is a *BACnetEnumerated* datatype value indicating the device status. The possible values are as follows:
  - *OPERATIONAL* – the device is functioning normally and is available for communication,
  - *OPERATIONAL_READ_ONLY* – the device is functioning normally but is in a read-only mode. It can be queried for information, but its properties cannot be modified,
  - *DOWNLOAD_REQUIRED* – the device is not fully operational and requires additional configuration or programming before it can function normally. This status is typically used when a device is first installed or after a firmware upgrade,
  - *DOWNLOAD_IN_PROGRESS* – the device is in the process of receiving additional configuration. During this state, the device might not be fully operational or available for communication,
  - *NON_OPERATIONAL* – the device is not operational due to a fault or other issue and is not available for communication.

Further *Device* properties can be found in Clause 12.11 of ASHRAE 135-2020 [23].

6.7.4.2 *Network* Port

The *Network Port* object type is the other top-level object directly accessible from outside the device. A *Network Port* object acts as a container for the device's network configuration and properties. All BACnet devices contain at least one *Network Port* object per physical port on a BACnet network. The *Network Port* object provides the configuration of the BACnet network connection on the physical port. *Network Port* objects can optionally be arranged in a hierarchical structure, with a *Network Port* object per stack layer, having an object representing the BACnet application layer at the top and an object representing the physical layer at the bottom, with *Network Port* objects representing the network and link layers in between. A device's *Network Port* objects are accessed from outside the device by using the *Network Port's* object identifier, which is assigned by the manufacturer and is readable.

Some representative properties of the *Networt_Port* object include the following:

- *Network_Type* – a *BACnetEnumerated* datatype with values matching the different BACnet link layers. The value for MS/TP is *MSTP* and the value for UDP/IPv4 is *IPV4*,

- *Protocol_Level* – a *BACnetEnumerated* datatype with values indicating the stack layer corresponding to the configuration parameters in the *Network Port* object. The values are as follows:
  - *PHYSICAL* – the object represents the physical layer,
  - *PROTOCOL* – the object represents an intermediate layer (network or link layer),
  - *BACNET_APPLICATION* – the object represents the BACnet application layer,
  - *NON_BACNET_APPLICATION* – the object represents a nonBACnet application layer.
- *Reference_Port* – if this object is part of a hierarchy, the *Reference_Port* property contains the object identifier of the next *Network Port* object below it in the hierarchy. If there are no further ports in the hierarchy or the network configuration is not hierarchically defined, the *Reference_Port* property is set to *BACnetUndefined*,
- *Network_Number* – the BACnet network address of the port,
- *MAC_Address* – the BACnet link-layer address of the port on the BACnet network (not the underlying link-layer address),
- *BACnet_IP_Mode* – a *BACnetEnumerated* datatype with values indicating the architectural role of the device in the IP network, if the link layer is UDP/IPv4 or UDP/IPv6. The values are as follows:
  - *NORMAL* – the device is not operating as a BBMD or foreign device,
  - *BBMD* – the device is operating as a BBMD,
  - *FOREIGN* – the device is operating as a foreign device.

Other properties provide access to the IP configuration for the UDP/IPv4 link layer including the IPv4 address, UDP port, DHCP server address (if any), DNS server address, and default IP router address as well as BACnet configuration parameters including the routing table if the device is acting as a BACnet router, and BBMD broadcast distribution table and foreign device table if the device is acting as a BBMD. Further information on *Network Port* properties can be found in Clause 12.56 of ASHRAE 135-2020 [24].

6.7.4.3 *Analog Input*, *Analog Output*, and *Analog Value*

The BACnet *Analog Input*, *Analog Output*, and *Analog Value* object types are used to represent control, and monitoring analog variables like temperature, humidity, pressure, or current. The *Analog Input* objects are used by the device to represent analog measurements from sensors. The *Analog Output* objects are read and set by control entities to directly control actuators or to otherwise adjust a specific device setting. The *Analog Value* objects are used to indirectly control actuators or a device setting, like a thermostat setpoint. The distinction between the *Analog Output* and *Analog Value* object types is that *Analog Output* objects directly control physical variables while *Analog Value* objects control internal or "virtual" variables.

Some of the common properties between these object types are:

- *Present_Value* – the current value of the object (e.g. the measured input value, output value, or internal value),
- *Units* – the units of measurement for the *Present_Value* property (e.g. degrees Celsius, percent relative humidity, or amperes current).

The *Present_Value* property in *Analog Output* and *Analog_Value* is also commandable, so changes from different controlling entities can be prioritized.

In addition, these object types support change of value (COV) notification and event notification, see Section 6.7.6. More information on the *Analog Input*, *Analog Output*, and *Analog Value* types can be found in ASHRAE 135-2020 Clauses 12.2, 12.3, and 12.4 [25].

6.7.4.4 *Binary Input*, *Binary Output*, and *Binary Value*

The BACnet *Binary_Input*, *Binary_Output*, and *Binary_Value* object types are used to represent values that can be in one of two states; for example, a switch that can either be on or off. As with the analog types, *Binary_Input* represents the network visible state of a physical device and can only be read, *Binary_Output* also represents the state of a physical device, can be read or written and is commandable, and *Binary_Value* represents the state of a software control variable represented in the device memory and is also commandable. All three properties have a *Present_Value* property which can have one of two *BACnetBinaryPV* type values: *ACTIVE* or *INACTIVE*. *BACnetBinaryPV* has datatype *BACnetEnumerated*.

Binary_Input, and *Binary_Output* have two additional properties that control the *Present_Value* property:

- *Polarity* – a value of *BACnetEnumerated* datatype *BACnetPolarity* that determines the relationship between the property value and the physical state of the device the object represents. The *BACnetEnumerated* datatype has two possible values:
  - *NORMAL* – an *ACTIVE* value in the *Present_Value* property indicates that the physical state of the input is on, an *INACTIVE* value indicates that the physical state of the input is off,
  - *REVERSE* – an *ACTIVE* value in the *Present_Value* property indicates that the physical state is off, an *INACTIVE* value indicates that the physical state is on.
- *Out_Of_Service* – a *BACnetBoolean* value indicating whether the *Present_Value* property is controllable by software local to the device. Although the *Present_Value* property is normally read-only, when the *Out_Of_Service* Boolean property is *TRUE*, the property becomes writable and changes in the *Polarity* property value do not affect the *Present_Value* property value. If the *Out_Of_Service* property is *FALSE*, changes in the Polarity property do affect the *Present_Value* property and the property is read-only.

These types also support event notification, see Clauses 12.6, 12.7, and 12.8 of ASHRAE 135-2020 for full definition of the types [26].

6.7.4.5 *Multi-state Input, Multi-state Output,* and *Multi-state Value*

The *Multi-state Input, Multi-state Output,* and *Multi-state Value* object types expand the idea behind the binary types to include devices that may have discrete states. For example, in commercial HVAC systems, a variable air volume (VAV) box is an air-handling device that controls the amount of air flow into a room. The VAV box typically has multiple states, which can be represented by a *Multi-state Input* object:

- 0% flow – the air flow into the room is completely shut off,
- Minimum flow – the minimum amount of air flow required for ventilation is provided,
- Occupied mode – the air flow is adjusted based on the occupancy of the room,
- Unoccupied mode – the air flow is reduced when the room is unoccupied,
- Maximum flow – the maximum amount of air flow for peak demand.

Each state of the VAV box can be assigned a numeric value, and the *Present_Value* property of the *Multi-state Input* object can be used to indicate the current state of the VAV box. By monitoring the state of the VAV box, the BEMS can adjust the air flow to optimize energy conservation while maintaining comfortable conditions for the building occupants.

The *Present_Value* property value of the *Multi-state Input* and *Multi-state Output* types is a *BACnetUnsigned* representing the current state of a physical device, while the value for the *Multi-state Value* type represents a control parameter in the device memory. These properties support the following additional properties:

- *Number_Of_States* – a *BACnetUnsigned* giving the number of states. The property is writable but if changed, the size of the *State_Text* array needs to change to match the new value,
- *State_Text* – a *BACnetArray* of *BACnetCharacterString* elements with descriptive names for the states. The size of the array matches with *Number_Of_States*.

In addition, as with the analog and binary value types, the *Present_Value* properties of the *Multi-state Output* and *Multi-state Value* objects are commandable.

For complete definitions of the *Multi-state Input, Multi-state Output,* and *Multi-state Value* types see Clauses 12.18-12.20.

6.7.5 Object Types Useful for BEMS Implementations

With 64 total object types, the BACnet standard has many more specialized types than are relevant for BEMS applications, though all are useful for the broader application area of a BAS. Object types that support other BAS applications include processes, collections (these are different from the collection primitive datatypes), notifications and events, logging, life and safety applications, access control applications like key card doors, and simple object types that mimic the primitive datatypes. Their specifications can be found in Clause 12 of ASHRAE 130-2023 [27]. In this section, a few specialized object types particularly suited to BEMS deployments are briefly discussed.

6.7.5.1 Schedule

The BACnet *Schedule* object type supports the scheduling of various energy consuming devices in a building according to a predefined time-based schedule. The schedule can be constructed to control energy consumption enabling energy savings and improving the building's performance and carbon emissions profile. The *Schedule* type has provisions for common daily schedules and exceptions, for example, on holidays.

Schedule utilizes the *BACnetTimeValue* sequence type to represent a scheduled value. *BACnetTimeValue* consists of a time of day and an arbitrary value. Here is the ASN.1 definition:

```
BACnetTimeValue ::= SEQUENCE {
        time : Unsigned32,
        value : <any primitive datatype>
}
```

The time of day is represented by a 32-bit unsigned integer that specifies the number of seconds since midnight. The arbitrary value can be of an instance of any primitive datatype. The *Present_Value* property is set to a *BACnetTimeValue* instance when the schedule period changes. The *Schedule* type implements control by writing the value in the *BACnetTimeValue* instance to the scheduled objects' properties at the day and time when the *Present_Value* property changes.

A schedule is divided into two classes of days: normal and exceptional. Daily schedules can be set for either the full day or part of a day. A priority mechanism allows specifying which scheduled event is in control at a particular time if there is a conflict, like the commandable mechanism used on individual properties.

A *Schedule* object has the following properties:

- *Schedule_Default* – the default value to be used when no other scheduled value is in effect. The type is the same as the type of the controlled objects' properties,
- *Effective_Period* – a *BACnetDateRange* specifying the dates during which the schedule is in effect. *BACnetDateRange* is a sequence type with two keys, *start-date* and *end-date*, giving the starting and ending date of the range. The values have the primitive datatype *BACnetDate*. This property provides a way to implement seasonal schedules, by having multiple *Schedule* objects with date ranges that do not overlap,

- *Weekly_Schedule* – a seven-element *BACnetArray* of *BACnetDailySchedule* instances, each representing a day of the week, starting with Monday and ending with Sunday. Each *BACnetDailySchedule* is an unkeyed (i.e. there are no key names) sequence type of *BACnetTimeValue* instances, defining periods during the day with the time when the scheduled value should go into effect and a value to which the controlled objects' properties are set at the start of the period, except when an *Exception_Schedule* is in effect,
- *Exception_Schedule* – a *BACnetArray* of *BACnetSpecialEvents* instances that takes precedence over the normal schedule during periods defined in the instance. Each instance is a 3-tuple consisting of instances with values of the following types: an instance indicating the period the exception is in force, a list of *BACnetTimeValue* datatype values, and an unsigned integer ranging from 1 to 16 indicating the priority similar to a commandable property. Note that the priority is not for writing to the controlled object, but rather for determining which scheduled event has priority if two or more are scheduled for the same or overlapping periods. The period instance can be a *BACnetCalendarEntry* or a reference to a *Calendar* object. If the date matches any of the period criteria for an entry in the *Exception_Schedule* array, the exception schedule is activated,
- *List_Of_Object_Property_References* – a *BACnetList* of object and property references for the objects that are under the *Schedule* object's control. The elements of the list are instances of the *BACnetDeviceObjectPropertyReference* which has the following ASN.1 definition:

```
BACnetDeviceObjectPropertyReference ::=
SEQUENCE {
    object-identifier : BACnetObjectIdentifier,
    property-identifier :
    BACnetPropertyIdentifier,
    property-array-index : Unsigned OPTIONAL,
    device-identifier : BACnetObjectIdentifier
}
```

  If the property is array-valued and the *property-array-index* value is present, it specifies the controlled array element in the property. If the *property-array-index* value is absent, the entire array is controlled. The *device-identifier* value specifies the *Device* object in which the controlled object resides, if it is absent then the controlled object and the *Schedule* object are in the same device. This allows certain devices or applications to function as controllers of multiple other devices,
- *Priority_For_Writing* – the priority level assigned to the *Schedule* object when writing to the controlled objects' properties that are commandable. As with commandable properties, the lower the numerical value, the higher the priority. All properties in the controlled object are written with the same priority.

The algorithm for deciding which schedule value to use to set a controlled object's *Present_Value* property and therefore what is written to the controlled object is the following:

1) Determine the highest priority element (lowest priority level) in the *Exception_Schedule* list for the current day whose value is not *BACnetNull* and set *Present_Value* to that.
2) If Step 1 fails, then find the element in the *Weekly_Schedule* that corresponds to the current day whose value is not *BACnetNull* and set *Present_Value* to that.
3) If Step 2 fails, then set *Present_Value* to the *Schedule_Default*.

The *Schedule* object type is specified in Clause 12.24 of ASHRAE 135-2020 [28] and supports event notification. The *Calendar* object, which can be used to specify dates for the *Exception_Schedule* property list items, is specified in Clause 12.9 [29].

6.7.5.2 Load Control

The *Load Control* type object provides support for the BEMS to manage electrical loads. A *Load Control* object can enable a utility demand response service to reduce building loads when the grid is under stress. It can also allow the BEMS to exercise flexible load control for decarbonization purposes, for example, by increasing the utilization of grid power when grid carbon emissions are low and decreasing it when carbon emissions are high. Note that the energy-utilizing devices may not actually be under the direct control of the BACnet automation system, and BACnet may need to communicate with another controller that connects to the device with a different protocol to actually control the load. *Load Control* objects can also be arranged in a hierarchy, with a top-level controller object communicating with the utility that then assigns load reduction to lower level objects controlling specific buildings or devices.

Figure 6.11 contains a state machine diagram for the *Load Control* object. The ovals indicate states of the load controller, the arcs indicate state transitions. More details on the states and state transitions are given in Table 6.7. The *Load Control* type implements the state machine variables with properties. The properties are written by control entities outside the object as well as by the *Load_Control* object itself, triggering state transitions. The following properties serve as state variables:

- *Start_Time* – the start of the duty window for a load control event. Load shedding may need to begin before this time for the load to be compliant with the request by *Start_Time*. The property datatype is an instance of the *BACnetDateTime* sequence type having the following ASN.1 definition:

```
BACnetDateTime ::= SEQUENCE {
      date : BACnetDate,
      time : BACnetTime
}
```

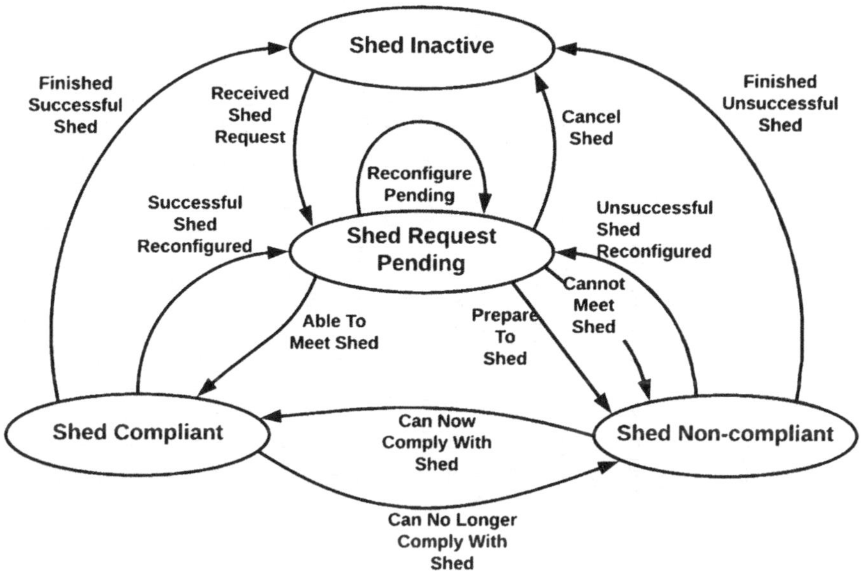

Figure 6.11 State Machine for the Load Control Object Type. *Source:* Adapted from [30]

- *Duty_Window* – a *BACnetUnsigned* value indicating the time interval in minutes within which the load control actions should be applied. The average power consumption over *Duty_Window* must be less than or equal to the requested power reduction for the Load Control object to remain in the Shed Compliant state.
- *Requested_Shed_Level* – the amount of power reduction being requested by the control entity, which can be a utility or another Load Control object,
- *Expected_Shed_Level* – the anticipated amount of electrical load reduction that should be achieved at *Start_Time,*
- *Actual_Shed_Level* – after *Start_Time* + *Duty_Window* time has elapsed, the actual power reduction achieved based on the average value over the previous duty window,
- *Shed_Duration* – a *BACnetUnsigned* value indicating the duration of the load-shedding event in minutes starting at *Start_Time.* If no load reduction command is active or pending or if a command has completed, the value is 0.

The *Requested_Shed_Level, Expected_Shed_Level,* and *Actual_Shed_Level* values are instances of *BACnetShedLevel* which is a choice datatype with three possible variants whose ASN.1 tags are: *percent, level,* or *amount.* A *Load Control* object must support the *level* variant, the others are optional. The three properties should ideally have the same variant type to maintain consistency and facilitate accurate comparisons.

Table 6.7 Description of Load Control Object States

State	Function	Transition	Transitions
Shed Inactive	Wait for a load shed request.	Received Shed Request	The *Start_Time* property has been written. Calculate the value of the *Expected_Shed_Level* and *Actual_Shed_Level* properties and enter the Shed Request Pending state.
Shed Request Pending	Determine from the properties whether the request needs to be fulfilled immediately or at a future time.	Cancel Shed	One of: • The current time is after *Start_Time* plus *Shed_Duration*. The request is for an invalid time. Enter the Shed Inactive state. • The *Requested_Shed_Level* is equal to the default value for the choice, or *Start_Time* is *BACnetUndefined*. Cancel shedding and enter the Shed Inactive state
		Prepare To Shed	The current time is prior to *Start_Time* but the loads require time to decrease usage to the requested level. Command the loads to start shedding prior to *Start_Time*. Calculate *Expected_Shed_Level* and *Actual_Shed_Level* and enter the Shed Non-compliant state.
		Cannot Meet Shed	The current time is after *Start_Time* and the object cannot meet the request immediately. Begin shedding, calculate the *Expected_Shed_Level* and *Actual_Shed_Level*, and enter the Shed Non-compliant state
		Reconfigure Pending	The current time is prior to *Start_Time*, and one of the *Start_Time*, *Requested_Shed_Level*, *Shed_Duration*, or *Duty_Window* properties have been written. The request is being reconfigured. Calculate *Expected_Shed_Level* and *Actual_Shed_Level* and enter the Shed Request Pending state.

		Able To Meet Shed	One of: • The current time is after *Start_Time* and the object can meet the shed request immediately. Shed the loads, calculate *Expected_Shed_Level* and *Actual_Shed_Level*, and enter the Shed Compliant state. • The current time is before *Start_Time*, shedding was started prior to *Start_Time*, and the requested shed level has been achieved. Calculate the *Expected_Shed_Level* and *Actual_Shed_Level* and enter the Shed Compliant state.
Shed Compliant	Continue meeting the shed request until either the object is reconfigured or it can no longer meet the request.	Finished Successful Shed	The current time is after *Start_Time* plus *Shed_Duration* and the request completed successfully. Set *Start_Time* to *BACnetUndefined*, and enter the Shed Inactive state.
		Successful Shed Reconfigured	One of the properties *Start_Time*, *Requested_Shed_Level*, *Shed_Duration*, or *Duty_Window* has been written. Enter the Shed Request Pending state.
		Can No Longer Comply With Shed	Not possible to maintain the *Requested_Shed_Level*. Calculate *Expected_Shed_Level* and *Actual_Shed_Level* and enter the Shed Non-compliant state.
Shed Noncompliant	Attempting to meet the shed request, the object has been reconfigured, or the request completed unsuccessfully.	Finished Unsuccessful Shed	The current time is after *Start_Time* plus *Shed_Duration* and the shed request has completed unsuccessfully. Stop shedding and enter the Shed Inactive state.
		Unsuccessful Shed Reconfigured	One of the properties *Start_Time*, *Requested_Shed_Level*, *Shed_Duration*, or *Duty_Window* was written. Enter the Shed Request Pending state.
		Can Now Comply With Shed	The *Requested_Shed_Level* was achieved. Calculate *Expected_Shed_Level* and *Actual_Shed_Level* and enter the Shed Compliant state

Each variant comes with a default *Requested_Shed_Level* value and power target reduction in kilowatts (kW):

- *percent* – default: 100%
  power target reduction: (current baseline) * *Requested_Shed_Level*/100
- *level* – default: 0
  power target reduction: specified from the *Shed_Levels* property array
- *amount* – default: 0.0
  power target reduction: (current baseline) – *Requested_Shed_Level*

The current baseline value can be either visible in the *Full_Duty_Baseline* property or it can be published by the Load Control object in the *Expected_Shed_Level* property. The *Full_Duty_Baseline* property value has primitive datatype *BACnetReal* and units of kW.

If the properties are represented in *percent*, the target power reduction is a percent of the current baseline. For example, if the current baseline is 10 kW and the *Requested_Shed_Level* is 50%, the Load Control object attempts to reduce the load to 5 kW. If instead the properties are represented as *amount*, a value of 5.0 produces the same target reduction, 5 kW.

The *level* variant is a bit more complicated. The basic idea is that the reduction can be specified by a discrete set of actions corresponding to qualitative tags. These values might match up more closely with the preferences of occupants or the building manager than a strictly numeric Kw or percent target would. The *Shed_Level_Descriptions* property contains a *BACnetArray* instance with *BACnetCharacterStrings* containing the tags. The *Shed_Levels* property is a parallel *BACnetArray* of *BACnetUnsigned* with a numeric code for the level, ordered by increasing size.

If *level* is used, load reduction happens in the following way. When the *Requested_Shed_Level level* value is received, the object scans through the *Shed_Levels* array and finds an array element whose value is less than or equal to the requested value. The Load Control object then invokes an internal algorithm to determine the actual Kw reduction and tries to reach the target. The actual Kw or percent reduction corresponding to the *Shed_Levels* array element values is not available through the standard BACnet protocol, however.

The *Load Control* object has a "main switch" that activates or deactivates the state machine. The *Enable*, property is a *BACnetBoolean* which, when set to *TRUE*, causes the object to take load control requests and follow the state machine in Figure 6.11. When set to *FALSE*, the object no longer takes load control requests. The state machine transitions into the Shed Inactive state and remains in that state until the property is set to *TRUE* again.

In an object as complex as *Load_Control*, deciding what the *Present_Value* property should represent is difficult, but what the ASHRAE committee did was have the

Present_Value property represent the current state of the state machine. The type of the property is a *BACnetEnumerated* datatype with the following ASN.1 definition:

```
BACnetShedState ::= ENUMERATED {
    shed-inactive (0),
    shed-request-pending (1),
    shed-compliant (2),
    shed-non-compliant (3)
}
```

with the tags corresponding to the *Load Control* finite state machine states.

The Load Control object type supports event notification and is specified in Clause 12.28 of ASHRAE 135-2020 [31].

6.7.5.3 Meter Object Types

The BACnet meter object types represent meters for recording the flow of signals measuring some consumable quantity over time. The standard supports two different meter object types: the *Accumulator* type and the *Pulse Converter* type. The *Accumulator* type represents a device that accurately measures total consumption over time but not the current rate of consumption. The *Pulse Converter* type represents a device that accurately measures the current rate of consumption but not the total consumption over time. The consumable quantity can be electricity, gas, water, etc. This section discusses the *Accumulator* object type in detail, the specification for the *Pulse Converter* object type can be found in Clause 12.23 of ASHRAE 135-2020 [32].

The *Accumulator* object type supports utility electrical meters that measure the total amount of energy consumed by a building over time. The accumulated count is recorded as the number of incoming pulses, potentially pre-scaled to convert from the units of the device doing the measuring to the units represented by the *Accumulator* object's *Present_Value* property. An additional property provides a post-scaling value for the monitoring entity to convert from the accumulated count to units specified in *Units* property.

The properties involved in providing the accumulated value are the following:

- *Present_Value* – a *BACnetUnsigned* representing the number of inputs acquired since the value was reset. The incoming pulse signals can be scaled by the *Prescale* property prior to recording in the *Present_Value* property and *Present_Value* wraps around when it exceeds the *Max_Present_Value* property,
- *Units* – indicates the units of quantity that are output by the *Accumulator* after performing post-scaling with the *Scale* property, of the *BACnetEnumerated* datatype *BACnetEngineeringUnits*, defined in Clause 21.6 of the standard [33]. An example is *kilowatt-hours* with numeric code 19,

- *Max_Present_Value* – a *BACnetUnsigned* indicating the maximum value the *Present_Value* property can take on,
- *Scale* – the conversion factor which converts the *Present_Value* property into a value in the units indicated by the *Units* property. The value has choice type *BACnetScale* that can be either *float* or *integer*. The conversion is performed according to the *Scale* type as follows:
  - *float* – value in units is calculated as: $Units = Present\_Value * Scale$,
  - *integer* – value in units is calculated as: $Units = Present\_Value * 10^{Scale}$.

  For example, suppose the meter pulses are in watts (potentially prescaled), *Units* is *kilowatt-hours*, and *Scale* is the *float* choice 0.001. If the *Present_Value* property is showing 2543 then the value in kilowatt-hours is 2543 * 0.001 or 2.543 kW-hours. Note that the conversion must be done by the monitoring entity reading the *Accumulator* object, it is not done by the *Accumulator* object itself.

- *Prescale* – a conversion factor used to convert the actual measured quantity coming from the measured device into *Present_Value*. The conversion factor has sequence type *BACnetPrescale* which consists of two *BACnetUnsigned* type keys: *multiplier* and *modulo-divide*. The *Accumulator* object can either perform the pre-scale calculation itself or the *Prescale* property can simply document the conversion done by the measured device prior to setting the *Present_Value* property.

The algorithm used by the pre-scale process utilizes an internal accumulator variable to calculate the *Present_Value* property using the *multiplier* and *modulo-divide* key values from the *Prescale* property, and consists of the following steps:

For each pulse coming from the measuring device:
    Add the value of the *multiplier* key to the accumulator,
    While the accumulator is greater than or equal to the value of *modulo-divide*,
        Increment *Present_Value* by one,
        Decrement the accumulator by the value of *modulo-divide*

The conversion is designed to accurately convert pulse signals using integer arithmetic so no signals are lost in the conversion. The standard gives the example of an electrical metering application in which the output of voltage and current measuring systems is 7.5 pulses per Kwh, or 2/15 Kwh/pulse. This requires the *Accumulator* object to accumulate 2/15 Kwh/pulse. The algorithm allows the *Accumulator* object to accurately accumulate pulses if the units of the *Present_Value* property (which may not be the same as the *Units* property) are *kilowatt-hours*.

The *Accumulator* also supports a collection of properties allowing the *Present_Value* property to be adjusted or reset, as well as event reporting and logging. The Accumulator type is specified in Clause 12.61 of ASHRAE 135-2020 [34].

6.7.5.4 Lighting Control Object Types

Lighting constitutes around 11% of the energy consumption in commercial buildings in the US, the second highest category after HVAC [35]. Lighting control is a specialized area and the BACnet lighting types were standardized through the Lighting Applications Working Group to ensure they met the needs of lighting applications. Control of lighting involves more than simple on/off commands and can involve ramping up or down, blinking prior to shutting lights off, adjusting lighting to specific luminosities depending on whether daylight harvesting is being used to supplement artificial lighting, etc. From a BEMS standpoint, issues of energy management are of primary concern. BACnet supports three different lighting object types: *Channel*, *Lighting Output*, and *Binary Lighting Output*. This section provides a brief overview of the lighting control object types in BACnet with an emphasis on energy management features.

The *Channel* object type represents an actuator channel used for controlling lighting or other devices. In lighting applications, the *Channel* object is associated with one or more *Lighting Output* objects, which represent actual lighting devices such as lamps, fixtures, or groups of lamps. The *Lighting Output* objects are linked to the *Channel* object through the *List_Of_Object_Property_References* property, which is a *BACnetArray* of *BACnetObjectProperty* references to the controlled objects. To control the linked *Lighting Output* objects, a controlling entity writes to the *Channel* object's *Present_Value* property, which the *Channel* object then propagates to the associated *Lighting Output* objects. The Present_*Value* property has type *BACnetChannelValue*, which is a choice type that can be any one of the primitive datatypes or *BACnetLightingCommand*. *BACnetLightingCommand* is a sequence type with a set of commands that provide a detailed description of what the linked lighting devices should do, and is described in Clause 21.6 of ASHRAE 135-2020 [36]. The *Channel* object is not restricted to controlling lighting and can also be used to control other objects like HVAC equipment through the use of primitive datatypes in the *Present_Value* property instead of a *BACnetLightingCommand* instance. The *Channel* object type supports event notification and is described in Clause 12.53 of ASHRAE 135-2020 [37].

The *Lighting Output* object type represents a lighting device, such as a lamp, fixture, or group of lamps. A *Lighting Output* object is used to monitor and control the state of the lighting device and can be associated with one or more *Channel* objects. In addition to controlling the state of the lighting device, a *Lighting Output* object can also provide status information about the device. The *Lighting_Command* property of type *BACnetLightingCommand* is used to request specific operations on the device, such as switching the light device on or off, fading over a specified time period, and so on. The *Present_Value* property is a *BACnetReal* with the target value, in percent, for the lighting output within the range 0-100%, and is commandable. There are also a few special values that can be written to the *Present_Value* property to trigger warning actions. The *Power* and *Instantaneous_Power* properties are *BACnetReal* values that provide information about power consumption. The *Power* property indicates the power consumption in Kw when the device is at 100% lighting output. The *Instantaneous_Power* property indicates the power consumption given the current lighting level in the *Present_Value* property. These properties can be used by the BEMS exercising load control to decide what level to set the lights at when a utility demand response request is issued. The *Lighting Output* object type supports both COV notification and event notification and is described in Clause 12.54 of ASHRAE 135-2020 [38].

The *Binary Lighting Output* object represents a simple binary lighting device that can be either on or off. Devices controlled by the *Binary Lighting Output* object do not have ramping and dimming control settings like devices controlled by a *Lighting Output* object. *Binary Lighting Output* objects can also be associated with a *Channel* object allowing multiple light fixtures to be controlled through one actuator channel. Writing to the *Present_Value* property of the *Binary Lighting Output* object controls the physical lighting device. The *Present_Value* property value has *BACnetEnumerated* datatype *BACnetBinaryLightingPV* which has five values: *on*, *off*, *warn*, *warn-off*, *warn-relinquish*, and *stop*. The *on* and *off* values turn the controlled lighting on and off, the other values are involved in a blink warning mechanism. In addition, the *Present_Value* property is commandable. The *Power* property holds the power consumption of the lighting device when the lighting device is switched on, in Kw, and can be used by the BAS to implement a utility demand response application. The *Binary Lighting Output* type supports event notification and is described in Clause 12.55 of ASHRAE 135-2020 [39].

6.7.6 BACnet Event and Alarm Handling

Events in BACnet are defined as changes in the values of properties of an object or some internal status change that meets a set of predetermined criteria and therefore needs to be reported to an outside monitoring or control entity. BACnet has two different mechanisms for handling events: COV notification and event notification.

COV notification allows an external monitoring entity to register with a device to receive notifications when the value of a monitored property changes by some predetermined amount such as an increase or decrease in value. This reduces unnecessary network traffic and the load on the device by not requiring applications to continuously poll the device for updates. COV notification is optional and is supported by some of the standard BACnet object types. Proprietary object types may also support COV notification at the implementor's discretion. Section 6.7.6.1 describes some of the objects and properties involved in COV notification.

Event notification allows BACnet objects to notify one or more monitoring or control entities of when an event occurs, or, alternatively, have the monitoring or control entities query the object to receive an event summary from an event summarization service. There are three classes of event notification: intrinsic notification, algorithmic notification, and alert notification. Event notifications are sent to processes within the control or monitoring entity. Section 6.7.6.2 discusses the system architecture for the three different types of event notification.

For a complete specification of how BACnet handles events, see Clause 13 of ASHRAE 135-2020 [40].

6.7.6.1 COV Notification

COV notification can be useful in a BEMS for reporting when electricity use by an HVAC system, lighting system, or other energy-consuming device changes. A stream of changes can be correlated with various other data such as weather, grid carbon intensity, and occupancy level and fed into analytics algorithms to help fine-tune control algorithms, and, when displayed graphically on the BEMS dashboard, to provide the building manager with a visualization of their building's carbon emissions performance. COV notification requires that the *Device* object type and the object types doing COV notification support certain properties.

In the *Device* object type, the *Active_COV_Subscriptions* property contains a readable *BACnetList* datatype value with a list of subscribers to COV event notifications for objects managed by the device. Monitoring entities use the BACnet service messages to subscribe and unsubscribe to COV events, causing an entry to be made into the *Active_COV_Subscriptions* list when a subscription is made, or to be removed when a subscription is removed. The list entries are of *BACnetCOVSubscription* sequence datatype.

The *BACnetCOVSubscription* datatype has the following ASN.1 definition:

```
BACnetCOVSubscription ::= SEQUENCE {
     recipient : BACnetRecipientProcess,
     monitored-property-reference :
     BACnetObjectPropertyReference,
     issue-confirmed-notifications : BOOLEAN,
     time-remaining : Unsigned,
     cov-increment : REAL OPTIONAL
}
```

The values in this sequence datatype correspond to the following:

- *recipient* value – the *BACnetRecipientProcess* sequence type value contains identifiers for the subscribing entity that receives the COV notification,
- *monitored-property-reference* – the *BACnetObjectPropertyReference* sequence type value contains the object identifier and property identifier (and index of the element if the property value is an array) of the object and property that are being monitored,
- *issue-confirmed-notifications* – whether or not the *Device* should issue a notification that needs to be confirmed through a reply by the recipient,
- *time-remaining* – time remaining in the recipient's COV subscription, in seconds,
- *cov-increment* – an optional value that specifies the amount the property value must change before a COV notification is issued. Only present if the property is real-valued.

BACnetRecipientProcess is also a sequence type having the following ASN.1 definition:

```
BACnetRecipientProcess ::= SEQUENCE {
      recipient : BACnetRecipient,
      process-identifier : Unsigned32
}
```

The *recipient* value is either a *BACnetObjectIdentifier* for the *Device* object that signed up for the COV notification or a *BACnetAddress* if the subscribing entity is a foreign device or an application not otherwise associated with a *Device* object. The *process-identifier* is a numeric value allowing differentiation between multiple COV subscriptions from the same entity, and is assigned by the subscribing entity, with a value of 0 reserved for when the entity wants to unsubscribe.

Object types that support issuing COV notifications need to support the *COV_Increment* property. The *COV_Increment* property specifies the minimum change in the monitored property necessary to trigger a COV notification. When the change in the monitored property meets or exceeds the *COV_Increment* value, the object will generate a COV notification. The datatype of the *COV_increment* property corresponds to whatever the datatype of the monitored property is, for example, one of the real types for *Analog_Output*.

For more information on COV notification including a table of object types that support COV notification, BACnet services involved in COV notification, and an explanation of how multiple property COV notification works, see Clause 13.1 in ASHRAE 135-2020 [41].

6.7.6.2 Event Notification

Event notification serves an important function in the BEMS and the broader BAS, in that it allows monitoring and control entities to obtain automated notification when some device has detected a condition that is not normal. In some cases, the condition may point to the need for configuration adjustment or maintenance while in others it may trigger an alarm due to an immediate threat to health and safety like a fire. For the BEMS, off-normal conditions may lead to increased grid energy usage during periods of elevated grid carbon intensity, and therefore require investigation to ensure the building meets its decarbonization goals. In this section, the system architecture for the three different classes of BACnet event notification is presented.

Intrinsic notification is defined as when an object monitors its own properties and reports a transition in the monitored state. BACnet standard object types that support intrinsic notification specify in their type definitions which monitoring algorithms from Clause 13.1 of ASHRAE 135-2020 [40] can be used. They also support optional properties that provide configuration and reporting parameters for event notification and properties for handling and routing of the notifications. Proprietary object types may also support intrinsic notification at the implementor's discretion.

Algorithmic notification is defined as when an object monitors the properties of other objects and reports a transition in the monitored state of the other objects. Algorithmic notification allows the monitoring of objects that do not provide intrinsic notification, or using an algorithm or algorithmic parameters for monitoring that differ from those specified for the monitored object's type. The BACnet standard algorithms for monitoring are defined in Clauses 13.3 and 13.4 of ASHRAE 135-2020 [42], and any of the standard algorithms can be used to define criteria for algorithmic notification. Algorithmic notification differs from intrinsic notification in that an *Event Enrollment* type object is used to determine the event conditions. *Event Enrollment* object types are defined in Clause 12.12 of the standard [43].

Alert notification is defined as when an object sends an event notification that is unrelated to the object's state or the object's intrinsic notification algorithm. When the need for an alert notification is identified by the object, the alert is passed to an *Alert Enrollment* type object for distribution. Alerts differ from intrinsic notification and algorithmic notification in that alerts are not generated by any standard conditions, they are stateless, and they cannot be acknowledged. *Alert Enrollment* object types are defined in Clause 12.52 of the standard [44].

Notifications report transitions in event state. Event state transitions are broadly categorized into one of three possible classes: transitions to a fault condition, transitions to an off-normal condition, and transitions to a normal condition. A fault is a malfunction that nearly always represents a failure of some physical system or software in a device. An off-normal condition is a condition within an object that is not normally expected or is outside the bounds of normal operation. A normal condition is any other condition. Notifications are triggered when a transition from one to another of these classes is detected by an object monitoring its state or the state of another object. Transitions to an off-normal state are TO_OFFNORMAL transitions. Transitions to a fault state are TO_FAULT transitions. All other transitions are TO_NORMAL transitions. These labels are used in the standard to indicate the element in the bit arrays and other reporting datatypes that record event state transition reporting but do not correspond to any specific datatype[5].

Figure 6.12 illustrates the event notification system architecture in BACnet. The figure applies to all three classes of event notification. Objects are shown as squares with rounded corners and processes are shown as circles. Abstract notifications to different processes or objects are shown as arrows with labels next to them. These are not actual BACnet services (protocol messages) but may correspond to one or more of the BACnet services. Five different classes of objects are required for event notification:

- The Event Initiating Device object which the standard calls the "notification-server"[6]. This object acts as a container for the object generating the event notification and other objects involved in event notification, and provides the execution context for the processes that perform event notification,
- The Notified Device object which the standard calls the "notification-client."[7] This object is the recipient of the event notification message and may support instances of the *Event Log* or *Notification Forwarder* type objects. It is also the source of an acknowledgment, if one is required,
- The Event Initiating Object which can be either the object where the event state transition is detected for intrinsic notification and alert notification or a monitoring object for algorithmic notification. In addition to other properties specific to event notification, the object type must support a *Reliability* property, and may optionally support properties for the Alarm Acknowledgment process,
- The *Event Log* type object where events are logged to a data store for later viewing. These may be present in the notification-server and notification-clients,

5 If the naming convention had been followed here, these would have been an enumeration type.

6 This does not correspond to any specific BACnet object type, ASN.1 enumeration type value, or sequence type key.

7 Ditto for this name.

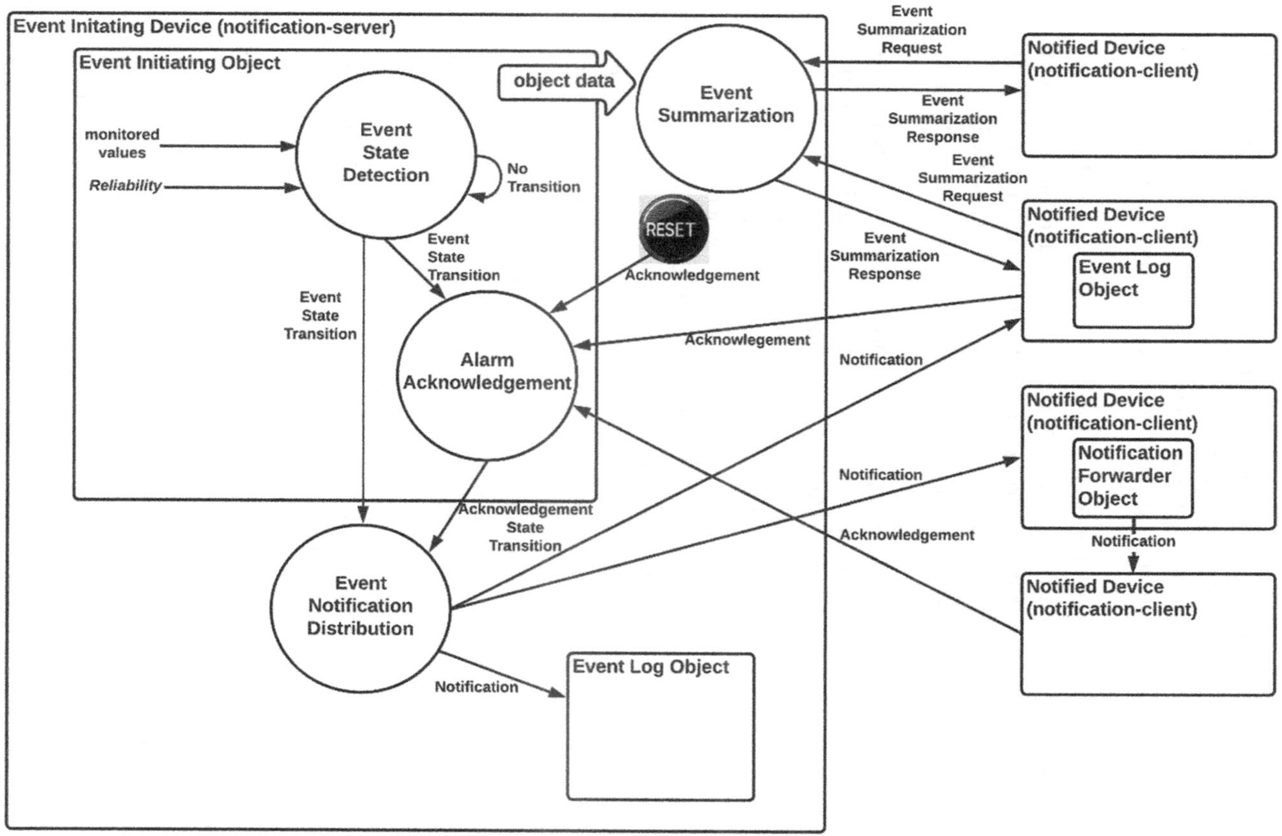

Figure 6.12 Event Notification System Architecture. *Source:* Adapted from [45]

- The *Notification Forwarder* type object which forwards notifications from one notification-client to another that then performs the acknowledgment if required.

Device objects and objects generating event notifications are required to have certain properties which are included in the object type definitions in Clause 12 along with the definitions of the *Event Log* and *Notification Forwarder* object types. Clause 13 provides more context about how the property values and objects are used in event notification.

As shown in the figure, event notification involves four processes: Event State Detection, Event Notification Distribution, Alarm Acknowledgment, and Event Summarization. An object that supports Event State Detection may optionally support Alarm Acknowledgment if a transition requires an acknowledgment. Note that Alarm Acknowledgment supports all event notification classes not just the alert notification class.

The Event State Detection process requires monitoring of one or more properties to evaluate the object's event state, along with the *Reliability* property. Detection of a change in a property value may result in no state transition or may result in a transition from the current state to one of the other states (TO_FAULT, TO_OFFNORMAL, TO_NORMAL). The event algorithm determines the normal or off-normal states and the *Reliability* property determines whether the event state transition is TO_FAULT. When a transition occurs, the Event State Detection process notifies the Event Notification process and the Alarm Acknowledgment process. Reference [46] pulls together a detailed explanation of how the Reliability property is used to determine whether a fault occurred from different parts of ASHRAE 135-2020.

Upon receipt of a state transition notification, the Event State Notification process sends a Notification to notification-clients that have subscribed to the event. It also sends a Notification to the *Event Log* type object in the *Device* object, if the object has one.

The Alarm Acknowledgment process awaits an Acknowledgment corresponding to the state transition from clients that have been notified by the Event State Notification process. When such an acknowledgment has been received, the Alarm Acknowledgment process notifies the Event State Notification process with an Acknowledgment State Transition. For an alert class event, the acknowledgment must come by means local to the device like an alarm reset button.

Event Summarization provides the ability for a device to retrieve event information independently from receiving notifications directly. This allows notification-clients that miss or have not subscribed to notifications to determine what event state transitions have occurred on the device. Event Summarization involves all the object properties and not just the Event Initiating Object. A device is required to support Event Summarization if it supports Event State Detection.

6.7.7 BACnet Services

The application layer protocol is responsible for providing the interface between external objects and applications communication through the BACnet network layer and the BACnet objects and processes in the local application. BACnet defines a set of application layer messages, which the specification calls application services, that allow the objects in BACnet devices to communicate between themselves and with outside applications. The application protocol itself is defined on top of a set of service primitives that implement the application layer protocol, as shown in Figure 6.1. The service primitives are a collection of operations on a state machine that describes how application protocol messages are processed by the application layer, which are translated into different types of messages. The service primitives are defined using a formal specification methodology based on the ISO technical report TR 8509 [47] and are briefly discussed in Section 6.7.7.1.

Altogether, BACnet has 49 different application service messages grouped into the following categories:

- Object Access Services – allow control and monitoring entities to access and manipulate BACnet objects over the network, described in Clause 15 [48],
- Device Management Services – discover, manage, and maintain BACnet devices and objects, described in Clause 16 [49],
- Alarm and Event Services – enable devices to report COV notifications, alarms, and events, described in Clause 13 [50],
- File Access Services – facilitate the transfer and manipulation of files between BACnet devices and applications, described in Clause 14 [51],
- Virtual Terminal Services – enable the interaction between a user of a networked application and a BACnet device using a text-based interface, described in Clause 17 [52].

Error codes for the confirmed services are given as part of the service definition in symbolic form (e.g. *UNKOWN_OBJECT* for an attempt to read a property on a nonexistent object) and the numeric codes for the symbolic tags are defined in Clause 21.4 [53]. A few of the most important services in the object access and device management categories are described in Sections 6.7.7.2 and 6.7.7.3.

Application layer messages need to be passed to the network layer for transmission over the network. Before passing the application layer message to the network layer, application messages are serialized for insertion into the network layer packet. The standard calls the serialized application message an Application Protocol Data Unit (APDU). Clause 21.1 [54] contains the formal ASN.1 APDU definitions while the serialization and deserialization protocol for the APDU messages including objects and base types is defined in Clause 20 of the standard [55].

After insertion into a network layer packet, the application message is sent to a device or application over the network.

6.7.7.1 BACnet Service Primitives

From an application implementation standpoint, the most important point of the service primitives is that they allow sending an application service message request in two different ways: either as a confirmed or an unconfirmed message. The confirmed version results in a response returned to the sender regardless of whether the request was completed successfully or if there was an error; while the unconfirmed version results in no response even if there was an error. Additional service primitives are involved in handling segmented messages, returning responses to confirmed requests if the response completed successfully, or returning responses if the message was rejected due to a syntax error, protocol error, or errors involving the semantics of the request like trying to read an object property that does not exist.

These categories correspond to eight different types of messages:

- *BACnet-Confirmed-Request-PDU* – initiates a confirmed service request, ensuring that the receiving device acknowledges the request,
- *BACnet-Unconfirmed-Request-PDU* – initiates an unconfirmed service request, which does not require an acknowledgment,
- *BACnet-SimpleACK-PDU* – acknowledges a confirmed request that does not require additional data in the response,
- *BACnet-ComplexACK-PDU* – acknowledges a confirmed request with returned data, such as property values,
- *BACnet-SegmentACK-PDU* – acknowledges individual segments of a segmented message,
- *BACnet-Error-PDU* – returned when a device encounters an error while processing a request, providing an error code with details about the nature of the error,
- *BACnet-Reject-PDU* – returned when a device rejects a request due to issues like incorrect formatting or invalid parameters,
- *BACnet-Abort-PDU* – returned when a device terminates an ongoing communication, usually due to a network or system issue.

Clause 5 of ASHRAE 135-2023 provides more information on the service primitives [56].

6.7.7.2 BACnet Services for Remote Object Access

The BACnet object access services are a collection of messages that allow BACnet devices and applications to create and delete objects, read and write properties of objects, and perform other operations over the network on objects and their properties. Altogether, there are 10 different object access messages. In this section, we present an overview of the four most important: *ReadProperty*, *WriteProperty*, *ReadPropertyMultiple*, and *WritePropertyMultiple*.

The *ReadProperty* service allows a BACnet device or application to request the value of a specific property in a BACnet object over the network. For example, a BEMS might use the *ReadProperty* service to query the current temperature from a temperature sensor tracked as the *Present_Value* property of an *Analog_Input* object in the sensor. This message is unicast as a *BACnet-Confirmed-Request-PDU* to the device address where the object is located since a response is required. The device receiving the request responds with the requested property value in a *BACnet-ComplexACK-PDU*. The request contains the following parameters:

- Object identifier – a *BACnetObjectIdentifier* for the object from which the property value is requested,
- Property identifier – a *BACnetPropertyIdentifier* for the property to be read,
- Property array index (optional) – if the property is an array, a *BACnetUnsigned* specifying the array index of the value to be returned. If no array index is included, the entire array is returned.

The return message contains the parameters in the request, in addition to the value of the property. If an error is encountered, a *BACnet-Error-PDU* is sent with an error code. Possible errors result from attempting the following operations:

- Accessing an object or property that does not exist,
- Including a property array index for a property that is not an array or an invalid array index for an array valued property,
- Reading a property that is read protected.

The *WriteProperty* service enables a BACnet device or application to modify the value of a property in a BACnet object on another device. This service is particularly useful for changing the configuration parameters of a device remotely. For instance, a BEMS might use the *WriteProperty* service to adjust the setpoint temperature of a thermostat based on occupancy or time of day in which the setpoint was represented as the *Present_Value* property of an *Analog_Value* object in the thermostat. This message is unicast to the device address where the object is located as a *BACnet-Confirmed-Request-PDU* since a response is required. The request contains the following parameters:

- Object Identifier – a *BACnetObjectIdentifier* for the object to which a property value is written,
- Property Identifier – a *BACnetPropertyIdentifier* for the property to be written,
- Property Array Index (optional) – if the property is an array, a *BACnetUnsigned* specifying the array index of the value to be written. If no array index is included, the entire array is replaced,

- Property Value – the new value for the specified property,
- Priority (Optional) – if the property is commandable, a *BACnetUnsigned* between 1 and 16 specifying the priority level at which the property value should be written.

The return message is either a *BACnet-SimpleACK-PDU* if the response is completed successfully or a *BACnet-Error-PDU* with an error code if the operation was unsuccessful. Possible errors result from attempting the following operations:

- Accessing an object or property that does not exist,
- Including a property array index for a property that is not array valued or an invalid array index for an array valued property,
- Writing a property with a value having the wrong or an unsupported datatype for that property,
- Including a priority parameter for a commandable property and the priority parameter is not within the range 1–16,
- Writing the *Object_Name* or *Object_Identifier* property with a duplicate of an existing object in the device,
- Writing a property that is write protected,
- Running out of memory on the device while writing the property.

The *ReadPropertyMultiple* service is an extension of the *ReadProperty* service, allowing a BACnet device to request multiple properties from one or more BACnet objects in a device with a single network message. For example, a BEMS might use the *ReadPropertyMultiple* service to request temperature, humidity, and occupancy data from a multi-sensor represented as the *Present_Value* properties of *Analog_Value* objects. *ReadPropertyMultiple* reduces the amount of network traffic required to read multiple values from a device. This message is unicast to the device address where the object is located as a *BACnet-Confirmed-Request-PDU* since a response is required.

The message includes a list of read access specifications. Each specification element in the object list consists of the following:

- Object Identifier – a *BACnetObjectIdentifier* for the object in which the properties to be read are located,
- Property Reference List – a list of property references for the properties to be read.

Each property reference in the property reference list consists of the following:

- Property Identifier – a *BACnetPropertyIdentifier* for the property to be read,
- Property Array Index (Optional) – an optional *BACnetUnsigned* if the property is array valued for the array element to be read. If the property is array valued and no array index is included, the entire array is returned.

ReadPropertyMultiple also supports the use of *REQUIRED*, *OPTIONAL*, and *ALL* tags for property identifiers. REQUIRED specifies that the receiving device should return those properties that have *R* and *W* conformance codes, OPTIONAL specifies properties with the *O* conformance code, and ALL specifies all the properties should be returned except for the *Property_List* property and any vendor-specific properties.

A successful response is returned in a *BACnet-ComplexACK-PDU* as a list of read access results. The results list elements mirror the request list, with the exception that the property reference list elements in the result contain the property value and an error code in case the operation on the referenced property fails. The *ReadPropertyMultiple* service can therefore return partial results if the operations on some properties fail. If the entire request fails, a *BACnet-Error-PDU* is returned with an error code. Possible errors for the failure of the entire request are the same as for *ReadProperty*.

The *WritePropertyMultiple* service provides similar extended functionality for the *WriteProperty* service, enabling a BACnet device to modify the values of multiple properties to one or more BACnet objects in a device with a single network request. As with *ReadPropertyMultiple*, the message is unicast to the device address where the object is located as a *BACnet-Confirmed-Request-PDU* and the message parameter is a list of write access specifications with each element containing:

- Object Identifier – a *BACnetObjectIdentifier* for the object in which the properties to be written are located,
- Property Reference List – a list of property references for the properties to be written.

Each property reference in the property reference list consists of the following:

- Property Identifier – a *BACnetPropertyIdentifier* for the property to be written,
- Property array index (optional) – an optional *BACnetUnsigned* if the property is array valued for the array element to be written. If the property is array valued and no array index is included, the entire array is replaced,
- Property Value – the new value for the specified property,
- Priority (Optional) – if the property is commandable, a *BACnetUnsigned* between 1 and 16 specifying the priority level at which the property value should be written.

If the entire request is successful, a *BACnet-SimpleACK-PDU* is returned without any data. Otherwise, the receiving device keeps writing properties until it encounters an error, then it returns a *BACnet-Error-PDU* with the error code and a *BACnetObjectPropertyReference* containing the object identifier and property identifier of the first write attempt that failed. The error codes are the same as for *WriteProperty*, with the addition of an error code indicating if a syntax error in the request parameters occurred after writing one or more properties successfully.

6.7.7.3 BACnet Services for Device Management

The BACnet remote device management services provide a way for discovering, identifying, and managing devices. By utilizing these services, devices can locate and communicate with other BACnet devices on the network. There are a total of 14 device management services. In this section, the four primary BACnet device management services are discussed: *Who-Has*, *I-Have*, *Who-Is*, and *I-Am*. These four services are used by controlling and monitoring entities to perform device and object discovery over the network.

The *Who-Has*/*I-Have* message pair is used to discover which device hosts a specific object. These messages are particularly useful when the requesting device or application does not know the network address of the target device. Both are sent as a *BACnet-Unconfirmed-Request-PDU*.

The *Who-Has* message is broadcast with the following parameters:

- Device Instance Lower Limit and Device Instance Upper Limit (Optional) – unsigned integers between 0 and *BACnetUndefined* that limit the search for the requested object to a specific range of devices whose *Device* object identifiers are within the provided range. Either both limits or neither need to be present. Only devices with identifiers within the range are allowed to respond. If this parameter is omitted, all devices receiving the message are allowed to respond,
- Object Identifier – a *BACnetObjectIdentifier* for the object being sought. Either this parameter or the object name parameter must be present,
- Object Name – a *BACnetCharacterString* with the name of the object being sought. Either this parameter or the object identifier must be present.

A device responds to *Who-Has* if either its identifier falls within the lower and upper limit or there is no upper and lower limit, and the object identifier or object name parameter matches, whichever is included.

The responding device can either unicast the *I-Have* response message directly to the requesting device or broadcast in a way that allows the response to reach the requesting device. The broadcast scope can be either local, directed, or global.

The parameters to the *I-Have* message are as follows:

- Device Identifier – the *BACnetObjectIdentifier* of device sending the response. This is the value of the responding *Device* object's *Object_Identifier* property,
- Object Identifier – the *BACnetObjectIdentifier* of the object being sought. This is the value of the *Object_Identifier* property of the object that was requested in the *Who-Has* message,
- Object Name – a *BACnetCharacterString* with the object's name. This is the value of the *Object_Name* property of the object that was requested in the *Who-Has* message.

The *I-Have* message has the BACnet address of the responding device in the network layer packet which the sender of the *Who-Has* message can then use for further unicast communication.

The *Who-Is/I-Am* message pair is used for device discovery in BACnet. Both messages are sent as a *BACnet-Unconfirmed-Request-PDU*. The *Who-Is* message has an optional device identifier upper and lower limit parameters which work the same as for the *Who-Has* message, but are often omitted. If the parameter is included, only devices with device identifiers within the limits are allowed to respond. *Who-Has* is broadcast.

The *I-Am* message is normally sent as a reply to the *Who-Is* message. As with *I-Have*, the message can be unicast directly to the device or application from which the *Who-Is* message originated, or it can be broadcast within the local, directed, or global scope, in a manner that ensures that the requesting device or application receives a response. Additionally, a device can broadcast an *I-Am* message at any time, and many BACnet devices and applications broadcast one when they first boot up as an announcement of their address and identity, and periodically thereafter.

The parameters to the *I-Am* message are as follows:

- Device Identifier – a *BACnetObjectIdentifier* with the responding device's *Device* object identifier from the *Object_Identifier* property,
- Max APDU Length Accepted – a *BACnetUnsigned* with the maximum application data length that the responding device can accept. Future communications from the requesting device need to ensure that the length of future APDU requests and responses stays within the length limit,
- Segmentation Support – the level of segmentation support provided by the responding device. Segmentation support can be one of the following *BACnetSegmentation* values enumerated datatype values: *segmented-both*, *segmented-transmit*, *segmented-receive*, or *no-segmentation*. *BACnetSegmentation* is defined in Clause 21.6 of the standard [57],
- Vendor ID – a *BACnetUnsigned* with the vendor code of the responding device. The vendor code allows the requesting device to determine if the responding device has any vendor-specific properties and objects.

Upon receiving an *I-Am* message, the recipient can update its internal device database with the information provided including the BACnet network and link-layer address of the device sending the *I-Am* message.

6.7.8 Protocol Implementation Conformance Statement (PICS)

All BACnet devices come with a PICS. The PICS is a document that provides detailed information about a BACnet device's capabilities, features, and conformance to the BACnet standard. Manufacturers of BACnet devices create PICS documents to specify the level of support for various BACnet services, object types, link layers, and other aspects of the BACnet stack that their device implements. System integrators, engineers, and end-users rely on the PICS document to understand a device's capabilities and ensure that it meets the requirements of a specific BACnet deployment. The PICS document also helps ensure interoperability between BACnet devices from different manufacturers by providing a clear and standardized description of each device's BACnet functionality.

The PICS document contains sections providing the following information:

- Device Overview –A general description of the device, its purpose, and its intended use,
- BACnet Standard Version – The version and revision of the BACnet standard the device conforms to,
- Device Profile – The BACnet profile the device complies with, if applicable (e.g. B-ASC, B-AAC, B-BC, etc.),
- Object Types Supported – A list of the BACnet object types supported by the device, along with their properties, optional features, and any proprietary extensions,
- BACnet Services Supported – A list of the BACnet services the device supports, including both standard and proprietary services,
- Data Link Layer Options – The supported data link layers and network options (e.g. BACnet/IP, MS/TP, Ethernet, etc.),
- Character Sets Supported – The supported character sets for text encoding in the device,
- Any other relevant device-specific information, such as hardware requirements, firmware version, and application software version.

PICS is described in Clause 21.1 of ASHRAE 135-2020 [58].

6.8 BACnet Security

During the initial discussions for the BACnet standard in the late 1980s and early 1990s, the thinking around security on the ASHRAE committee was that the only threat would be an insider attack, from a disgruntled operator, so revoking login credentials for control stations and applications would be sufficient to control the threat. Since application account management is strictly a "local matter" (as the specification states), there were initially no plans to put anything on security into the specification. But at the last minute, the ASHRAE committee did add security in Clause 24 [59], including a *Network Security* object and extending the network layer with ten messages to accommodate various security operations, including authentication and encryption. Because it was defined at the BACnet network layer, the specification applied to all BACnet link layers. The security specification was subsequently reviewed and approved by NIST. But no manufacturer implemented the security specification in their stack, since the design was very complex.

In 2016, the committee removed Clause 24 and instead developed BACnet/Secure Connect (BACnet/SC), defined in Annex AB [60]. BACnet/SC defines a new virtual link adaptation layer on top of the IETF Websocket transport protocol running TLS 1.3(RFC 6455) [61]. The Websocket protocol presents a substantially different communication model from the other BACnet link layers, basically of a bidirectional stream of bytes that was designed for real time, streaming communication between Web browser clients and servers. More on Websockets is available from the video link in Chapter 8. BACnet/SC constitutes a new BVLL next to UDP/IPv4, UDP/IPv6, and Zigbee. It also defines a new network "hub and spoke" architecture for BACnet networks.

More on BACnet/SC

This link contains a brief description of BACnet/SC. In addition to providing security directly in the BACnet/SC BVLL, BACnet/SC removes a few other limitations of the IP virtual link layers. It eliminates the use of the BBMD and does not use broadcast. The BACnet/SC network architecture is based on a node and hub concept, where a node can originate and respond to requests and a hub acts as a forwarding agent.

However, this still leaves the problem of how to secure networks that are running the other BACnet link layers. Not all deployments will want to or can switch to BACnet/SC to deploy security, or modify their network architecture. While there is little that can be done to secure the MS/TP link layer, a security for the UDP/IPv4 virtual link layer can be implemented by deploying TLS for UDP. IETF RFC 9147 [62] defines the DTLS protocol, which applies TLS to UDP. DTLS will also work to secure the UDP/IPv6 layer. If all IP nodes running BACnet stacks in a UDP/IPv4 BVLL network are deployed with certificates and configured to do mutual TLS, communication on the UDP/IPv4 layer can be secured. One problem with this approach is that many BACnet devices have closed stacks built into their firmware that do not support DTLS. DTLS only works if the BACnet device or application is sufficiently configurable to allow certificates to be provisioned for the transport layer. With closed stack BACnet devices, end-to-end encryption of BACnet traffic is not possible.

Without DTLS, partial encryption of network traffic can still be achieved if the first hop routers in the building LAN route the traffic to and from the BEMS through a VPN. This approach will also work for MS/TP if the MS/TP to IP router directs the traffic into a VPN. Using a VPN at the first hop router only leaves the traffic between the devices and the first hop routers unencrypted. By deploying the first hop routers sufficiently close to the end devices, the network segment over which the traffic is unencrypted can be limited. Further control over access to the unencrypted device segments can be established by using a dedicated building automation VLAN with 802.1x between the VPN gateway and any devices connecting in from the building LAN side, and deploying ACLs on the first hop routers to not allow any traffic onto the device segments except from the VPN endpoint to the WAN in the building gateway where the traffic from the devices egresses and commands to the devices ingress. If the devices support Wi-Fi, a related approach is to enable WEP encryption on the device and Wi-Fi router, and deploy a VPN on the Wi-Fi router, if the device and Wi-Fi router both support WEP encryption. With this approach, encryption on the device segment is achieved, since the device segment runs over wireless.

Figure 6.13 shows an example architecture illustrating these points. The VPN gateways implement site-to-site or tunnel mode VPNs between the BEMS network and the BACnet device networks. While site-to-site VPNs are typically used for an entire enterprise site network, in this case, they are being used for the device subnets in order to get the traffic into an encrypted tunnel as soon as possible. The VPN tunnel subnet is 172.24.00.00/16. The BACnet UDP/IPv4 devices (heat pump and hot water heater) are configured with IP addresses on the 172.24.176.00/24 subnet and have the VPN gateway configured as their default router. The MS/TP network devices (thermostats) route through an MS/TP to IP router which is configured on the 172.24.178.00/24 subnet with the VPN Gateway as its uplink router. Over the WAN, the building gateway might perform its own VPN tunneling, but it does not have to for the BEMS traffic since the traffic to and from the BACnet devices is already fully encrypted. In the BEMS, the VPN gateway runs the 172.24.180.00/24 network.

Users and services attempting to access BACnet devices need to be authenticated and their access permissions verified. Applications can use standard Web access IAM protocols, to identify users, but a more difficult problem is posed by autonomous services that act without any human intervention. If DTLS is used, in theory, it should be possible to encode access permissions into the DTLS X.509 certificates using the Subject Alternative Name (SAN) extension to the certificate [63, 64], but this might require writing a custom server extension. The SAN extension allows for multiple names to be associated with a certificate, such as hostnames, IP addresses, and URIs.

Another solution to this problem is to add a device management layer above BACnet through which all BAMS services must access BACnet devices. This solution is used by some open-source energy management platforms, such as

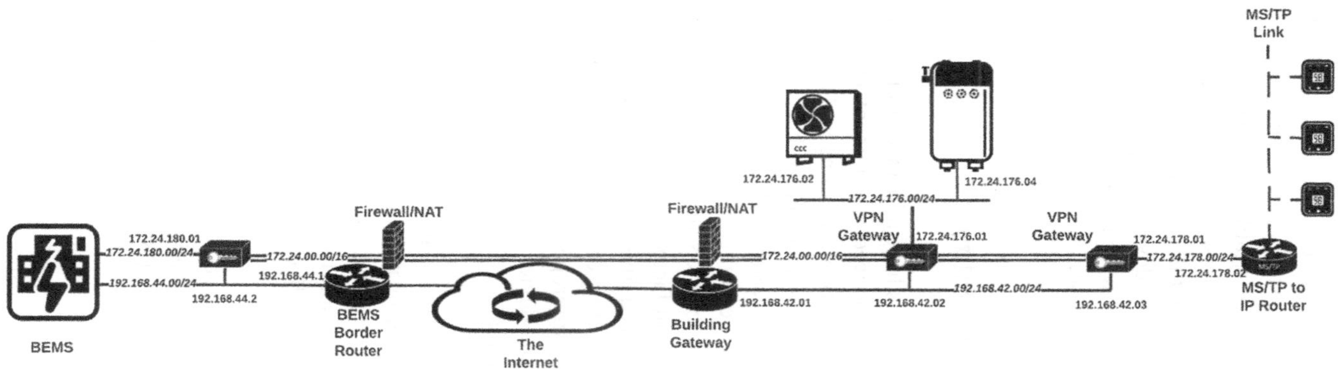

Figure 6.13 Security Architecture for BACnet MS/TP and UDP/IPv4 Network

Volttron [65]. BACnet devices are only accessed through proxies, which can perform IAM on the requests. This allows an IAM system to be incorporated into the device management layer, so the device management services can check whether an autonomous service has authorization to access a BACnet device and, if so, what access permissions the service is allowed. An unauthorized service would then be unable to access the BACnet devices from above. An attempt to connect an unauthorized BACnet device would also fail if the device management layer life cycle management software detected that the device was unauthorized by its identity in the IAM database.

6.9 BACnet Interoperability

As discussed in Chapter 2, BASs have a history of propriety, vertically integrated systems with incompatibility between equipment from different vendors. This increases the cost and complexity of installing and managing a BAS for building construction firms, building owners, and other customers. The point of the BACnet specification, and the reason ASHRAE originally started the standardization effort, is to foster interoperability between BAS vendor equipment. BACnet interoperability is important for the seamless integration of different devices and applications from various manufacturers into a BAS. Interoperability ensures that the BAS and BEMS built with equipment and applications from multiple different vendors will work without any difficulty. Three specifications that facilitate BACnet interoperability are Interoperability Areas (IAs), BACnet Interoperability Building Blocks (BIBBs), and device profiles.

IAs define the specific domains of the building automation and control systems in which interoperability is required. At a high level, they describe the functionality that is important in a practical implementation of a BAS to minimize the amount of additional engineering required by the installation and commissioning engineering teams. The IAs suggest a set of capabilities that require elements of the BACnet device or application to be implemented in a way that will support interoperability. The following five IAs are defined in the Clause 22.2 of the standard [66]:

- Data Sharing – ensures that BACnet devices and applications can seamlessly exchange information. Interoperability permits collecting data between devices and applications from different vendors and coordinating modification of device control parameters,
- Alarm and Event Management – ensures that BACnet devices and applications define the conditions for alarms and events to meet specific criteria and to communicate alarms and events in an interoperable fashion. Interoperability allows

devices and applications from different vendors to share announcing and acknowledging of alarms, displaying alarm data, sharing events for logging or control, modifying alarm limits and routing, and producing summaries of alarms and events,

- Scheduling – ensures that BACnet devices and applications from different vendors can coordinate scheduling activities and operations among themselves,
- Trending – ensures that the collection, storage, and retrieval of historical data from BACnet devices for analysis and optimization purposes is interoperable among devices and applications from different vendors,
- Device and Network Management – ensures that devices and applications from different vendors can connect to the network and communicate together and that they can be managed in an interoperable fashion.

BIBBs are a set of detailed, standardized functional specifications that break down the IAs into the required capabilities and features of BACnet devices and applications to implement interoperability. BIBBs are organized into four categories, each corresponding to one of the IAs. BIBBs define the necessary data models, services, and communication patterns that devices and applications from different vendors must support to interoperate in a BACnet network. Each BIBB consists of two parts: a device's capability to provide a particular service (server-side) and the device's capability to consume or use the service (client-side). Vendors can use BIBBs as an implementation reference in their devices and applications to ensure that they are compatible with BACnet devices and applications from other vendors. BIBBs are defined in Annex K of the standard [67].

A BACnet device profile is a predefined set of requirements and guidelines based on the BIBBs that specify the minimum capabilities and features a BACnet device must support for a particular type of use case. These profiles are defined by ASHRAE and other organizations to ensure interoperability among devices from different manufacturers that serve a similar purpose within the BAS. Note that device profiles are different from object profiles, described in Section 6.7.2.1. Object profiles specify object customizations in a machine-readable way, while device profiles specify the functional architecture for devices to ensure interoperability.

A device that complies with a specific BACnet device profile must support a defined set of BIBBs, as well as meet other performance and conformance criteria outlined in the profile. Device profiles are defined in Annex L of the standard [68]. Some examples of BACnet device profiles include the following:

- B-BC (BACnet Building Controller) – a general-purpose, field-programmable device capable of carrying out a variety of building automation and control tasks,
- B-AAC (BACnet Advanced Application Controller) – a device with limited computational resources relative to the B-BC. It may support specific applications and some programmability,

- B-ASC (BACnet Application Specific Controller) – a device with limited computational resources relative to the B-AAC. It supports a limited set of BACnet object types and services for a specific application,
- B-OWS (BACnet Operator Workstation) – an operator interface used for monitoring and basic control but does not support configuration or advanced troubleshooting capabilities,
- B-OD (BACnet Operator Display) – a device that provides a human-machine interface for monitoring BACnet devices designed for use in hand-held terminals or wall-mounted displays. It is not intended to perform direct digital control.

While there is not a specific device profile solely dedicated to the energy management use case, some profiles have features and capabilities that are applicable to BEMS. For instance, the B-AAC and B-BC profiles include support for various object types and services that can be used to monitor and control energy consumption through a BACnet network.

By adhering to a standardized device profile, manufacturers can ensure that their BACnet devices can interoperate with other devices within a BACnet system, making it easier for system integrators to integrate devices from different vendors. Additionally, system integrators and end-users can use the profile information to evaluate and select devices that meet their specific needs or the requirements of a particular building automation project.

6.10 Other Protocols for Large Commercial and Public Building Automation

While BACnet has by far the largest deployed footprint and product offering of any building automation protocol in the large commercial and public building market, there are three other legacy BAS protocols that manufacturers offer in products that have a deployment footprint: KNX, LonWorks, and DALI. Space does not permit an extensive discussion of these protocols here but this section provides a short one-paragraph description of each protocol. Complete descriptions of the KNX and LonWorks as well as descriptions of available equipment are available in Mertz, Hanseman, and Hübner [69].

The KNX protocol is widely used in Europe and the low-level physical and link layers have been standardized as ISO/IEC 14543-3. Unlike BACnet, KNX is also deployed in medium to high-end residential buildings in Europe. The KNX Association has a variety of tools and specifications for building KNX networks. KNX requires the use of a software tool from the KNX association called ETS (Engineering Tool Software) to design and configure the network. With this tool, integrators can design the system layout, configure devices, and download their configurations to the devices. KNXnet/IP is the standard for KNX communication over IP networks and has been standardized as ISO 22510. It is used to connect KNX installations to IP networks and to connect IP devices to KNX networks. KNX Web Services is a specification for accessing KNX devices using the HTTP and SOAP protocols. KNX Data Secure and KNX IP Secure are recent extensions to the KNX standard. KNX Data Secure provides end-to-end data encryption and authentication at the application layer, while KNX IP Secure provides security at the IP layer. More information on these specifications and software packages is available from the KNX Association [70].

Introduction to KNX

The link above is from the KNX Association and it introduces the basics of the KNX protocol. The video starts with a discussion of KNX devices and then describes the basic KNX line and area topology concepts. It moves on to describe the basic KNX functions and how KNX objects appear in the ETS tool for designing and configuring a KNX network.

LonWorks is a networking platform specifically created for many IoT automation applications, including building, home, and industrial systems. It utilizes a proprietary protocol that has been standardized as ISO/IEC 14908. Developed by Echelon Corporation, the platform uses the LonTalk protocol and the Neuron Chip to facilitate communication among various devices on the network. The neuron chip is a proprietary microprocessor which Echelon licenses to other manufacturers that serves as the heart of any LonWorks device. The Neuron Chip has built-in firmware that runs the LonTalk networking protocol. The LonTalk protocol supports several physical layers, including twisted pair, power line, fiber optics, and wireless. LonWorks Network Services and NodeBuilder are a collection of services for managing LonTalk networks and a software development tool for creating applications for LonWorks devices. OpenLDV is a network interface driver that allows applications running on Ethernet and IP to access LonWorks networks. The LonWorks platform is maintained and promoted by the LonMark Interoperability Association [71], a trade group of manufacturers that offer LonWorks-based devices and software. The association works to ensure interoperability among devices from different manufacturers.

DALI is a protocol specifically designed for the control of lighting in buildings and is not a general building automation protocol. DALI is specified in IEC 62386. It features a bi-directional, digital, and open standard protocol that provides a single interface for all light sources. It enables communication between controllers, such as switches and sensors, software controllers, and gateways that interface with other systems, and lighting devices, such as electronic ballasts, transformers, LED drivers, and other similar devices. These communicate through the DALI protocol, a set of commands and responses that operate at low speed, with a typical message taking 16.7 ms to transmit. Each lighting device has a unique short address (0–63). DALI also supports a kind of multicast in which lighting devices can be assigned to one or more multicast groups for simultaneous control. The latest standard is DALI-2, an update to the original DALI standard, and adds specifications for additional lighting devices, along with improved interoperability and testing procedures. DALI is designed as a single-controller, multiple-device system, where each lighting device is associated with a single controller. A DALI network can have up to 64 lighting devices and 16 controllers. In 2021, the DALI Alliance and the Bluetooth Special Interest group announced the specification of an interoperability gateway between Bluetooth and DALI networks, allowing DALI-controlled lighting to be provisioned, managed, and controlled on a Bluetooth mesh lighting control network. The DALI Alliance [72] is responsible for promoting, maintaining, and developing DALI as an industry standard and for interoperability testing.

These platforms define their own networking stack from the physical layer, in some cases, all the way up to the application layer, making interoperability with existing IP stack-based applications difficult. Commercially available gateways facilitate interoperability with BACnet, allowing the BEMS to communicate with equipment speaking different protocols without having to support the protocols themselves. For instance, a KNX-BACnet gateway translates the KNX protocol to BACnet, enabling seamless integration of KNX devices into a BACnet network. Similarly, DALI lighting systems can be controlled from a BACnet-based BEMS through a DALI-BACnet gateway, streamlining lighting control operations. A LonWorks-BACnet gateway offers the possibility to connect and control LonWorks devices from a BACnet-based BEMS. In essence, these gateways enable the interoperability of manufacturer equipment speaking different protocols through a single, BEMS control system using BACnet to communicate with the buildings' equipment.

6.11 Summary

This chapter contains a detailed review of the BACnet protocol, a standard communication protocol for building automation and control systems developed by the ASHRAE for interoperable BAS and BEMS. While there is more to the BACnet standard than covered in this chapter, the material here provides a basis for facility engineers and BEMS architects and engineers to understand and begin using BACnet in their designs.

Section 6.2 provides an overview of the BACnet architecture. One of the characteristics of the BACnet architecture that distinguishes it from other BAS communication protocols is that it defines a full network stack, from the link layer up to the application layer. ASHRAE started the BACnet standardization effort in the late 1980s, when the IP stack was still in the research phase. Support for running the BACnet stack on different physical, link, network, and transport layers was included in the standard through adaptation layers, which the BACnet standard defined as a "virtual link layer." This solution neatly solved the compatibility problem and allowed the standard to adapt to new underlying networking support as it developed. The MS/TP serial line link layer is reviewed in Section 6.4 and BACnet UDP/IPv4, which defines a virtual link layer on top of the UDP/IPv4 stack layers, is reviewed in Section 6.5.

The BACnet networking layer, described in Section 6.6, provides a means of interconnecting wider deployments of BACnet devices and applications, such as multiple floors in a large building or multiple buildings on a campus. While not as flexible as IP, BACnet routing ensures that packets reach their destinations in multi-link BACnet networks. A distinguishing characteristic of the BACnet network layer is that it makes heavy use of broadcast, in contrast to the IP network layer where broadcast is limited.

The BACnet application layer provides a rich collection of data objects modeling many kinds of devices found in BAS. Basic objects allow a more complex device to be built by combining them into a single *Device* object. The application layer additionally defines a few more complex objects for load controllers, meters, and lighting which are specifically of interest for BEMS applications. The standard also defines messages for remote reading and writing of object properties. The messages are called services and support both confirmed and unconfirmed communication through both broadcast and unicast. The BACnet application layer is discussed in Section 6.7.

Since ASHRAE started developing BACnet before the widespread deployment of the Internet, attention to security was confined to attacks that could only be carried out by the physical presence of the attacker in the building. The initial security design in the standard was quite complex and was removed in 2016 because it had not been widely implemented by the vendor community. Instead, the committee added a new virtual link layer, BACnet/SC with a completely different data communication model and architecture. However, this left deployments on the existing BACnet link layers with no security support. Section 6.8 discusses a few options for securing MS/TP and UDP/IPv4 BACnet link layer deployments.

Since the whole point of the BACnet standard is to foster multi-vendor interoperability, the standard includes material that provides a detailed interoperability functional architecture in the IAs and BIBBs. Section 6.9 discusses BACnet interoperability standards. BIBBs are used to define device profiles in several different device type categories. Vendors are permitted to claim that their devices implement one of these types if they meet the architectural criteria. However, an extension of the standard is also possible, and vendors can provide value-added enhancements as long as they are described in device's PICS, which is discussed in Section 6.7.8.

Concluding the chapter, Section 6.10 contains a few paragraphs on other building automation and control protocols commonly used in large commercial and public buildings. These protocols have legacy deployment in existing buildings and are likely to remain in place for the life of the building or until a deep remodel of the building with the replacement of the BAS. As such, interconnecting them with BACnet or through conversion to an IP protocol at the building gateway is a viable strategy for allowing the BEMS to access data and exercise control of equipment supporting the legacy protocols.

References

1 DOE. About the Commercial Buildings Integration Program [Internet]. [cited 2023 Jul 3]. Available from: https://www.energy.gov/eere/buildings/about-commercial-buildings-integration-program.

2 ASHRAE. Research Study Indicates BACnet Global Market Share over 60% [Internet]. [cited 2023 May 1]. Available from: https://bacnet.org/press-releases/research-study-indicates-bacnet-global-market-share-over-60/.

3 American Society of Heating. Refrigerating and Air-Conditioning Engineers. ANSI/ASHRAE Standard 135-2020. BACnet – A Data Communication Protocol for Building Automation and Control Networks. Atlanta: ASHRAE; 2020.

4 International Organization for Standardization. ISO 16484-5:2017. Building automation and control systems (BACS) – Part 5: Data communication protocol. Geneva: ISO; 2017.

5 American Society of Heating, Refrigerating, and Air-Conditioning Engineers. ANSI/ASHRAE Standard 135-2004. BACnet – A Data Communication Protocol for Building Automation and Control Networks. Atlanta: ASHRAE; 2004.

6 Newman H. M. BACnet: The Global Standard for Building Automation and Control Networks. New York: Momentum Press; 2013.

7 Electronic Industries Association (EIA)/Telecommunications Industry Association (TIA). Electrical Characteristics of Generators and Receivers for Use in Balanced Digital Multipoint Systems, Revision A, March, 1998.

8 Newman, BACnet: The Global Standard for Building Automation and Control Networks, p. 100.

9 Clause 9.5.4, 9.5.5, 9.5.6, and 9.5.7 ANSI/ASHRAE Standard 135-2020, pp. 106–121.

10 Steve Karg. Understanding BACnet MS/TP Encoding [Internet]. [cited 2022 Nov 18]. Available from: https://polarsoft.com/bex/papers/Understanding%20BACnet%20MSTP%20Encoding.pdf.

11 Newman, Section 8.1.2.1.1, BACnet: The Global Standard for Building Automation and Control Networks, pp. 124–126.

12 ISO/IEC 8824 (1990). Information technology – Open Systems Interconnection – Specification of Abstract Syntax Notation One (ASN.1) [Internet]. [cited 2023 Apr 19]. Available from: https://www.iso.org/standard/16288.html.

13 Clause 20.2, ANSI/ASHRAE Standard 135-2020, pp. 840–854.

14 Clause 21.6, ANSI/ASHRAE Standard 135-2020, pp. 919–922.

15 Clause 21.6, ANSI/ASHRAE Standard 135-2020, pp. 925–942.

16 Annex Q, ANSI/ASHRAE Standard 135-2020, pp. 1188-1199.

17 Annex X, ANSI/ASHRAE Standard 135-2020, pp. 1301–1310.

18 Clause 19.2, ANSI/ASHRAE Standard 135-2020, pp. 807–811.

19 Clause 20.2.1.4, ANSI/ASHRAE Standard 135-2020, pp. 841–842.

20 IEEE 754-2008 – IEEE Standard for Floating-Point Arithmetic [Internet]. [cited 2022 Apr 19]. Available from: https://ieeexplore.ieee.org/document/4610935.

21 ISO/IEC 8859-1. ISO/IEC 8859-1:1998 Information Technology – 8-Bit Single-Byte Coded Graphic Character Sets – Part 1: Latin Alphabet No. 1 [Internet]. [cited 2023 Apr 19]. Available from: https://www.iso.org/standard/28245.html.

22 ISO/IEC 10646. ISO/IEC 10646:2020 – Information Technology – Universal Coded Character Set (UCS) [Internet]. [cited 2024 Feb 19]. Available from: https://www.iso.org/standard/76835.html.

23 Clause 12.11, ANSI/ASHRAE Standard 135-2020, pp. 221–231.

24 Clause 12.56, ANSI/ASHRAE Standard 135-2020, pp. 542–565.

25 Clauses 12.2, 12.3, and 12.4, ANSI/ASHRAE Standard 135-2020, pp. 166–189.

26 Clauses 12.6, 12.7, and 12.8, ANSI/ASHRAE Standard 135-2020, pp. 191–212.

27 Clause 12, ANSI/ASHRAE Standard 135-2020, pp. 160–630.

28 Clause 12.24, ANSI/ASHRAE Standard 135-2020, pp. 308-314.

29 Clause 12.9, ANSI/ASHRAE Standard 135-2020, pp. 213–214.

30 Figure 12-5, ANSI/ASHRAE Standard 135-2030, p. 342.

31 Clause 12.28, ANSI/ASHRAE Standard 135-2020, pp. 341–350.

32 Clause 12.23, ANSI/ASHRAE Standard 135-2020, pp. 301–307.

33 Clause 21.6, ANSI/ASHRAE Standard 135-2020, pp. 892–903.

34 Clause 12.61, ANSI/ASHRAE Standard 135-2020, pp. 601–609.

35 US DOE. "Chapter 5: Increasing Efficiency of Building Systems and Technologies" in Quadrennial Technology Review: An Assessment of Building Systems and Technologies, September, 2015 [Internet]. [cited 2023 Apr 27]. Available from: https://www.energy.gov/sites/prod/files/2017/03/f34/qtr-2015-chapter5.pdf.

36 Clause 21.6, ANSI/ASHRAE Standard 135-2020, p. 913.

37 Clause 12.53, ANSI/ASHRAE Standard 135-2020, pp. 508–517.
38 Clause 12.54, ANSI/ASHRAE Standard 135-2020, pp. 518–531.
39 Clause 12.55, ANSI/ASHRAE Standard 135-2020, pp. 532–541.
40 Clause 13, ANSI/ASHRAE Standard 135-2020, pp. 631–724.
41 Clause 13.1, ANSI/ASHRAE Standard 135-2020, pp. 631–635.
42 Clauses 13.3 and 13.4, ANSI/ASHRAE Standard 135-2020, pp. 647–682.
43 Clause 12.12, ANSI/ASHRAE Standard 135-2020, pp. 232–239.
44 Clause 12.52, ANSI/ASHRAE Standard 135-2020, pp. 504–507.
45 Figure 13-1, ANSI/ASHRAE Standard 135-2020, p. 637.
46 Newman, Section 10.1.3.1.1.1, BACnet: The Global Standard for Building Automation and Control Networks, pp. 162–169.
47 ISO/TR 8509:1987. Information Processing Systems – Open Systems Interconnection – Service Conventions [Internet]. [cited 2023 Apr 29]. Available from: https://www.iso.org/standard/15732.html.
48 Clause 15, ANSI/ASHRAE Standard 135-2020, pp. 731–758.
49 Clause 16, ANSI/ASHRAE Standard 135-2020, pp. 759–776.
50 Clause 13, ANSI/ASHRAE Standard 135-2020, pp. 631–723.
51 Clause 14, ANSI/ASHRAE Standard 135-2020, pp. 725–730.
52 Clause 17, ANSI/ASHRAE Standard 135-2020, pp. 777–785.
53 Clause 21.4, ANSI/ASHRAE Standard 135-2020, pp. 871–880.
54 Clause 21.1, ANSI/ASHRAE Standard 135-2020, pp. 855–856.
55 Clause 20, ANSI/ASHRAE Standard 135-2020, pp. 828–854.
56 Clause 5, ANSI/ASHRAE Standard 135-2020, pp. 26–60.
57 Clause 21.6, ANSI/ASHRAE Standard 135-2020, p. 946.
58 Clause 21.1, ANSI/ASHRAE Standard 135-2020, pp. 952–953.
59 Newman. Section 3.9, BACnet: The Global Standard for Building Automation and Control Networks, pp. 42–45.
60 Annex AB, ANSI/ASHRAE Standard 135-2020, pp. 1377–1410.
61 Fette I., Melnikov A. The Websocket Protocol, RFC 6455 [Internet]. [cited 2023 Apr 18]. Available from: https://www.rfc-editor.org/rfc/rfc6455.html.
62 Rescorla E., Tschopfenig H., and Modadugu N. The Datagram Transport Layer Security (DTLS) Protocol Version 1.3, RFC 9147 [Internet]. [cited 2022 Apr 17]. Available from: https://www.rfc-editor.org/rfc/rfc9147.html.
63 Wikipedia. Subject Alternative Name [Internet]. [cited 2022 Apr 18]. Available from: https://en.wikipedia.org/wiki/Subject_Alternative_Name.
64 Cooper D., Santesson S., Farrell S., Boeyen S., Housley R. and Polk W. Internet X.509 Public Key Infrastructure Certificate and Certificate Revocation List (CRL) Profile [Internet]. [cited 2022 Apr 18]. Available from: https://datatracker.ietf.org/doc/html/rfc5280.
65 Volttron. Platform Security [Internet]. [cited 2023 Apr 18]. Available from: https://volttron.readthedocs.io/en/main/platform-features/security/volttron-security.html.

66 Clause 22.2, ANSI/ASHRAE Standard 135-2020, pp. 953–954.
67 Annex K, ANSI/ASHRAE Standard 135-2020, pp. 1069–1116.
68 Annex L, ANSI/ASHRAE Standard 135-2020, pp. 1118–1143.
69 Merz H, Hansemann T, Hübner C. Building Automation: Communication Systems with EIB/KNX, LON and BACnet. 2nd ed. Berlin: Springer; 2009.
70 KNX Association. KNX for Professionals [Internet]. [cited 2023 Jul 5]. Available from: https://www.knx.org/knx-en/for-professionals/index.php.
71 LonMark Interoperability Association. LonMark [Internet]. [cited 2023 Jul 6]. Available from: https://www.lonmark.org.
72 DALI Alliance. Digital Illumination Interface Alliance [Internet]. [cited 2023 Jul 6] Available from: https://www.dali-alliance.org.

7

Building Automation Stacks for Residential and Small Commercial Buildings

In contrast with the large commercial and public building market where BACnet is the dominant building automation protocol stack, the market for the residential and small commercial building market is fragmented across multiple wireless IoT sensor networks and application protocols. Building automation systems (BASs) have traditionally been expensive and therefore viewed as more appropriate for larger buildings where the initial cost could be justified in terms of savings on energy efficiency and maintenance, enhanced occupant comfort, need to control physical access to the building, etc. The exception is very high-end single-family residences or luxury condominiums where the cost of the building justifies the expense of adding a high-end, commercial grade BAS. For other residential and small commercial buildings, a common type of energy control system in the United States is a line voltage or 24-volt AC programmable thermostat for heating or air-conditioning that the occupant must program through a wall control. The thermostat communicates with the HVAC unit using a proprietary protocol, often analog or very low-level digital. Some homes in the United States may also have demand response systems in place for electric hot water heating and pool pumps, which allow the utility to curtail the load during times of peak demand, but mostly residential and small commercial buildings do not offer much in the way of a building energy management system (BEMS).

More recently, the advent of inexpensive smart IoT devices such as thermostats, plugs, light bulbs, and appliances that are typically managed through a "smart home" platform has provided additional tools for building energy management in residential and small commercial buildings. These devices use Wi-Fi, Zigbee, and Bluetooth to connect a supervisory control system in a cloud-based BAS platform with local controllers in the building. The cloud-based platforms are mostly run by a large technology companies like Google, Apple, Amazon, or Samsung.

Automating Building Energy Management for Accelerated Building Decarbonization: System Architecture and the Network Layer, First Edition. James Kempf.

Published 2025 by John Wiley & Sons, Inc.

While the primary focus of most smart home protocols, devices, and platforms is lifestyle management and not specifically building energy management, it is possible to build an energy management system for smaller buildings on top of such devices and protocols. For example, a smart home platform with scheduling functionality can control HVAC, plug loads, and even some appliances such as dishwashers, refrigerators, and ovens to avoid high time of use tariff periods when electricity from the grid may have a higher marginal carbon emission intensity. However, most smart home protocols do not support management of DERs such as on-site solar PV, storage, or EV changing, nor higher load appliances like hot water heaters and pool pumps.

The widespread deployment of smart home IoT solutions has been hampered by additional challenges. The existence of three separate wireless link layer protocols (Wi-Fi, Bluetooth Low Energy (BLE), and 802.15.4), and multiple network layer and IoT application layer stacks all designed to provide smart home IoT connectivity has been a source of confusion. Interoperability problems between the proprietary smart home platforms offered by the large technology companies and devices offered by smaller vendors have made buying and deploying smart home devices confusing and difficult for average consumers, professional developers, and building contractors, resulting in slow uptake of the technology. As a result, smart home technology has not yet scaled up to its true potential.

This chapter aims to clear up some of the confusion surrounding using smart home protocols for residential and small commercial BEMSs. BLE, as the WPAN protocol with the widest deployment, seems to be a natural choice for energy management application support. But energy management application protocols standardized by the Bluetooth SIG are mainly oriented toward monitoring the battery in personal devices such as cellphones, environmental sensing, controlling generic digital actuators, and monitoring analog sensors, and not specifically to BEMS applications such a load control. This may be partially due to the point-to-point nature of Bluetooth routing, which does not allow coverage of larger building areas without standardized bridging into a backbone network running the IP stack. Bluetooth has seen wider applicability as an adjunct protocol to the other smart home protocols for commissioning and device discovery, however. Section 7.1 discusses the BLE stack application support layer and a few applications that might be appropriate for a BEMS.

Zigbee is the smart home protocol with the broadest support for and deployment in energy management systems, and Section 7.2 covers Zigbee. Zigbee is a proprietary networking stack developed by the Zigbee Alliance (now the Connectivity Standards Alliance (CSA) [1]), from the network layer to the application layer, built directly on top of the 802.15.4 link layer. After an overview of the network architecture and the network layer, application layer, and cluster library, Zigbee clusters applicable to building a BEMS are discussed, and the *Demand Response and Load Control* (*DRLC*) cluster is explored in detail. Section 7.2 also briefly touches on the Zigbee Smart Energy profile which has been used by utilities to implement their advanced metering infrastructure (AMI). In addition, two interoperability specifications for Zigbee, the Zigbee Base Device Behavior (BDB) specification and Zigbee Direct, which uses BLE for device commissioning and discovery, are outlined.

The IP protocol provides a natural base for building an interoperable smart home protocol, and to address the fragmentation in the smart home IoT market, CSA has recently released a specification for a smart home IoT protocol defined on the IP stack. The standard, named Matter, uses the Thread mesh routing framework for providing mesh networking with 6LoWPAN as the network layer on the 802.15.4 link layer, and IPv6 on Wi-Fi and Ethernet, with BLE for device commissioning and discovery. Section 7.3 presents Matter and Thread. Most of the features Matter provides ensure that a device supporting Matter can be commissioned into and run on any large technology company's smart home platform that supports Matter, much as a laptop running IP can connect to and run over the Internet through any router connected to any Internet service provider's network. Matter is designed to make the user experience of deploying and managing a device supporting Matter the same on all platforms. Matter also provides standardized definitions for routing between 802.15.4 and the Wi-Fi and Ethernet building IP network, and gateways to Zigbee and Bluetooth. However, Matter has only recently been standardized and is still a work in progress. Most of the effort from the standardization group is being directed toward lifestyle applications such as controlling robotic vacuum cleaners and home media appliances such as TVs, but a few cluster libraries for controlling devices of interest for a BEMS such as HVAC are available in Version 1.2. The real potential for Matter to revolutionize BEMS in residential and small commercial buildings still lies untapped.

In Section 7.4, two other smart home protocols, Z-Wave and EnOcean, are briefly reviewed. Z-Wave is widely used in smart home applications and supports a thriving ecosystem, with over 3,300 certified interoperable products from various vendors. It operates on a mesh network topology and is generally considered strong in terms of security and reliability. EnOcean is known for its energy-harvesting capabilities and is often deployed in BASs to control lighting, HVAC, and other energy-consuming devices. EnOcean devices can operate without batteries, harvesting energy from their environment, such as light, temperature changes, or mechanical motion, to power their operation. Both protocols have found niches where they excel but are generally not interoperable with each other or other IoT protocols without the use of gateway devices. Finally, Section 7.5 summarizes the chapter.

7.1 Bluetooth Low Energy Application Layer

The BLE v5.3 specification [2] defines a collection of layers on top of the link layer for supporting applications, which were briefly mentioned in Section 3.6.1. These components are the Generic Attribute Profile (GATT), the Generic Access Profile (GAP), and the Attribute Protocol (ATT), shown in Figure 3.18. This section presents a short introduction to these stack layers. The GAP controls how devices become discoverable, are discovered, and establish connections, while the GATT describes the format and operations for data exchange between devices. The ATT is a low-level protocol that manages how data is structured, stored, accessed, and transmitted and acts like a combination of a database and the transport layer in the IP stack.

BLE defines standardized profiles for specific use cases on top of the GATT, GAP, and ATT application support layers. Profiles specify the structure of the information for use cases such as battery monitoring and are important for ensuring interoperability between devices and applications from different vendors. Most of the BLE profiles are for lifestyle or healthcare use cases, but although BLE has no profile specifically dedicated to energy control applications, there are a few profiles and services that could be used in a BEMS. These topics are also reviewed in this section. Despite the lack of standardized applications for energy management, BLE has been used widely for vendor-specific applications, and a vendor-specific application from Texas Instruments is discussed in this section.

While the base BLE protocol has only a few standardized applications defined that may be useful for BEMS applications, in 2018, the Bluetooth SIG introduced an entirely new networking stack called Bluetooth Mesh [3], defined on top of BLE, with an entirely new set of application support abstractions built on top of the GATT. The Bluetooth Mesh protocol uses mesh forwarding for transmitting messages rather than point-to-point forwarding like BLE, and security is not optional. In that respect, Bluetooth Mesh is like Zigbee and Matter, topics discussed in Sections 7.2 and 7.3, respectively. Recently, the Bluetooth Mesh Working Group has developed the Network Lighting Control (NLC) collection of profiles defining Mesh models for controlling lighting [4–8]. These profiles include the Smart Energy NLC profile [9] for energy monitoring which is the profile most relevant for energy management applications. Although space limitations prevent a thorough exploration of Bluetooth Mesh here, the series of video links below, produced by Ellisys a Bluetooth protocol analyzer vendor from the Bluetooth SIG website, provide an introduction to the important technical aspects of Bluetooth Mesh.

Bluetooth Mesh Introduction: Part 1

n the first video of the eries, Mohammad Afaneh overs the basics of luetooth Mesh terminol- gy, including nodes, mes- ages, the publish/subscribe ub/sub) interaction, mod- ls, and scenes. Pub/sub nteraction is covered in nore detail in Chapter 8, in ne context of the event- ased service architecture. Mohammad uses the exam- le of lighting fixture con- rols to illustrate how luetooth Mesh works.

Bluetooth Mesh Introduction: Part 2

In the second video of the series, Mohammad discusses basics of Bluetooth Mesh networking and the types of nodes in a Bluetooth Mesh network, including relay, low-power, friend, and proxy nodes. Mohammad also covers the Bluetooth Mesh stack architecture and network formation and packet forwarding in a Bluetooth Mesh network.

Bluetooth Mesh Introduction: Part 3

In the final video of the series, Mohammad presents how nodes are commissioned into the Bluetooth Mesh network, a process called "provisioning" in Bluetooth Mesh, and demos how a node is provisioned using the Ellisys protocol analyzer UI. He also describes Bluetooth Mesh security. Unlike BLE and classical Bluetooth, security is not optional in Bluetooth Mesh.

7.1.1 GAP and GATT Roles

The GAP defines the roles a node can take with respect to advertising and connection and were described briefly in Section 3.6.2.1. These roles are:

- Broadcaster – can send advertising packets but does not allow connections,
- Observer – scans for and receives advertising packets but does not initiate connections,
- Peripheral – advertises and accepts connections,
- Central – scans for advertisements and can initiate connections.

A Peripheral node cannot modify the properties of a connection, but it can request modifications from the Central node to which it is connected. Both the Peripheral and Central nodes can drop the connection for a variety of reasons.

The GATT and ATT also define roles for a node, but these roles are with respect to application profiles and are independent but compatible with GAP roles. The GATT/ATT roles are:

- Server – holds attributes relevant to the application profiles it supports and accepts ATT messages from the Client to read, write or update the values of the attributes. The Server is typically a sensor or lighting device with functions that implement a profile use case,
- Client – initiates ATT messages to the Server and receives responses from the Server. The Client is typically a control program running on a smartphone or other device that displays data from Servers and allows control actions on Server functions.

A single node may act as both a Server and a Client, depending on the nature of the applications it supports. For example, a smartphone might act as a Client when communicating with a smartwatch, but it might also act as a Server when communicating with a computer. Typically, the Server GATT/ATT role is held by a node with a Peripheral GAP role and a Client GATT/ATT role is held by a node with a Central GAP role, but these mappings are not fixed in the standard and a node can hold both at the same time. Since a Broadcaster and an Observer do not establish connections, they cannot take the Client and Server roles.

7.1.2 UUIDs for Identifying Attributes

The ATT and GATT use a 128-bit (16 byte) UUID to identify data structured as attributes. UUIDs are defined by the ISO/IEC 9834-8 standard [10]. Since a 16-byte UUID would take up more than half of the 27-byte link layer payload field, data objects that are part of a standardized profile or are specified in a Bluetooth specification use two additional formats, a 16-bit (2 byte) or 32-bit (4 byte) shortened form. The full UUID can be reconstructed by concatenating the shortened UUID in front of the official Bluetooth SIG suffix: 0000-1000-8000-00805f9b34fb. However, vendor defined and other nonstandardized UUIDs must use the long form.

7.1.3 ATT

The ATT structure data for application use cases as attributes. Attributes are the smallest data object defined by the ATT and are addressable through identifiers unique to the Server on which they are located. The GATT and ATT only communicate through attributes.

7.1.3.1 Attributes

Attributes are always held on the Server and are accessed and modified by the Client. The BLE specification defines the structure of attributes in an abstract form. The ATT can store attributes relevant to device configuration and other static information in nonvolatile memory while storing attributes that hold sensor data and data that can change or be modified by the Client in the device's main memory. Along with the actual data value, every attribute also includes metadata fields describing the attribute itself. Attributes consist of the following fields:

- Handle – a unique identifier for each attribute on the GATT/ATT Server, assigned by the Server when the attribute is added. The Handle is addressed by the Client to access the attribute. Handles are 16-bit values ranging from 0×0001 to 0×ffff and are guaranteed not to change. The Server typically assigns handles in numerically increasing order, but there can be gaps and Clients should not depend on numerical ordering when accessing handles,
- Type – a UUID that specifies the data type of the attribute Value field. The Server supports procedures to discover attributes based on the Type field. The designated type can correspond to an application type like Temperature, a higher level abstraction supported by the GATT like a service, or a type defined by a device vendor. For standardized attributes, the Bluetooth SIG specifies 16-bit or 32-bit UUID types, but vendor-defined types must use 128-bit UUIDs,
- Permissions – a collection of access rights and security settings on the attribute that define which Clients can access the attribute, what operations they can perform, and whether the attribute must be encrypted when transmitted over the air (OTA).

  Table 7.1 contains a list of permissions and security settings for ATT attributes,

Table 7.1 ATT Attribute Permissions and Security Settings

Permissions	Setting	Description
Access	None	Attribute cannot be read or written by the Client.
	Read	Attribute can be read by the Client.
	Write	Attribute can be written by the Client.
	Read and Write	Attribute can be read and written by the Client.
Authorization	None	No authorization is required by the Client to access.
	Required	Authorization is required by the Client to access.
Encryption	None	No encryption required (Security Mode 1, Level 1).
	Unauthenticated Keys	Connection must be encrypted but the key doesn't need to be authenticated. (Security Mode 1, Level 2).
	Authenticated Keys	Connection must be encrypted and the key need to be authenticated. (Security Mode 1, Level 3).

- Value – the actual data contained in the attribute. The size and format of the Value field depend on the attribute type in the Type field. The size of the Value field is limited to 512 bytes. The data value can be an application value like the setting on a thermostat, or it can be a structured data value used by the GATT or some other layer to define further abstractions on top of the ATT layer.

More details on the ATT attribute format and use can be found in the Bluetooth 5.3 specification in Volume 3, Part F, Section 3.2 [11].

7.1.3.2 ATT Protocol Exchanges

The ATT layer supports a collection of protocol messages for communicating attributes between the Client and Server. These messages are grouped into the following six exchange types:

- Commands – messages sent by the Client to the Server that do not require a Response,
- Requests – messages sent by the Client to the Server that require a Response,
- Responses – messages sent by a Server to a Client in response to a Request,
- Notifications – unsolicited messages sent by the Server to the Client that do not require a Confirmation,
- Indications – unsolicited messages sent by the Server to the Client that require a Confirmation,
- Confirmations – messages sent by the Client to the Server to confirm receipt of an Indication.

The operations implemented by the ATT messages include communicating information asynchronously and exercising control between the Client and Server, enabling the Client to discover attributes on the Server, and enabling the Client to read and write attributes on the Server. The ATT protocol messages are the basis on which GATT operations are built. Altogether, the ATT supports 31 different protocol messages in the exchange types above, but only 8 are relevant to the discussion of service discovery in Section 7.1.4.3. The four operations (Request/Response) are shown in Table 7.2. The GATT uses other ATT operations to read and write characteristics and to update Clients, but only the service discovery operations are discussed here. The ATT protocol is described in more detail in the Bluetooth 5.3 specification in Volume 3, Part F, Sections 3.3 and 3.4 [12].

7.1.4 GATT

The GATT defines a collection of higher level abstractions on top of the ATT which implement a three-level, hierarchical abstraction model. The GATT uses ATT attributes to implement the hierarchy and ATT protocol messages to implement

Table 7.2 ATT Operations Used in GATT Service Discovery

Operation	Request Parameters	Response Return Values	Function
Find-Attribute-By-Group-Type	• Attribute type UUID, • Starting handle, • Ending handle.	All attribute handles and values within the handle range whose type matches the UUID input.	Find handles and values of all attributes within a handle range that match an attribute type and the type is a grouping attribute as defined by a higher layer specification but the handle is not known.
Find-Attribute-By-Type-Value	• Attribute type UUID, • Attribute value, • Starting handle, • Ending handle.	A list of handle tuples for subranges where the attribute type and value match the inputs with the first element in the tuple being the subrange starting handle and the second element the ending handle.	Find the handle ranges of attributes that match an attribute type and an attribute value.
Find-Attribute-By-Type	• Attribute type UUID, • Starting handle, • Ending handle.	All attribute handles and values within the handle range whose type matches the UUID input.	Find handles and values of all attributes within a handle range that match an attribute type.
Find-Information	• Starting handle • Ending handle.	All attribute handles and values between the starting handle and ending handle.	Find attribute handles and values within a handle range regardless of attribute type.

operations called features. Profiles are the highest level, services are the next highest level, and characteristics are the lowest level. A profile is a collection of services that support a particular use case like lighting control or personal fitness. A service represents a collection of data objects that implement some function for a particular use case. A characteristic is a data value defined in a service along with metadata containing properties and configuration information about how the value is accessed and how the data value is represented.

Services act like object types in BACnet or classes in an object-oriented programming language, while characteristics are like properties in BACnet objects or instance variables in an object-oriented class. Each service and characteristic is identified by a UUID. A service definition can contain other services, mandatory characteristics, and optional characteristics. A characteristic represents a single data point, like an energy measurement from the energy monitor service, or a control point that allows a Client to modify the Server's operation with respect to the service, like a reset point in an energy monitor service that resets the service when written.

7.1.4.1 GATT Profiles

Profiles define how information is structured for a specific use case and how devices implementing the profile should communicate, helping to ensure interoperability between devices and applications from different vendors. A profile specification defines collections of services that implement a use case. In addition, it defines client and server roles, which roles devices must assume, and how they should behave in those roles. The profile specifies which GAP procedures are to be used for communication between devices. For example, it might dictate that devices use a specific advertising and scanning procedure for discovery and connection. Finally, a profile can specify some procedures around service discovery, such as which services and characteristics the client should look for, and what actions to take once those services and characteristics are discovered.

GATT profiles are defined in terms of GATT services and characteristics, and the service and characteristic constructs are defined as higher lever abstractions on top of ATT layer attributes. The Bluetooth SIG supports a collection of standardized profiles [13], but vendors are free to define their own profiles for new use cases. Device vendors may provide a programming language software development kit (SDK) customized to the profile to make developing applications more straightforward for developers, or developers may have to use the device SDK for the GATT directly to implement an application for the profile.

Figure 7.1 illustrates the GATT profile abstraction hierarchy, and Table 7.3 describes the ATT attribute types and the UUIDs that are used in constructing the GATT abstractions. Profiles consist of one or more services, each of which is declared separately. Services are either primary services at the top level of the

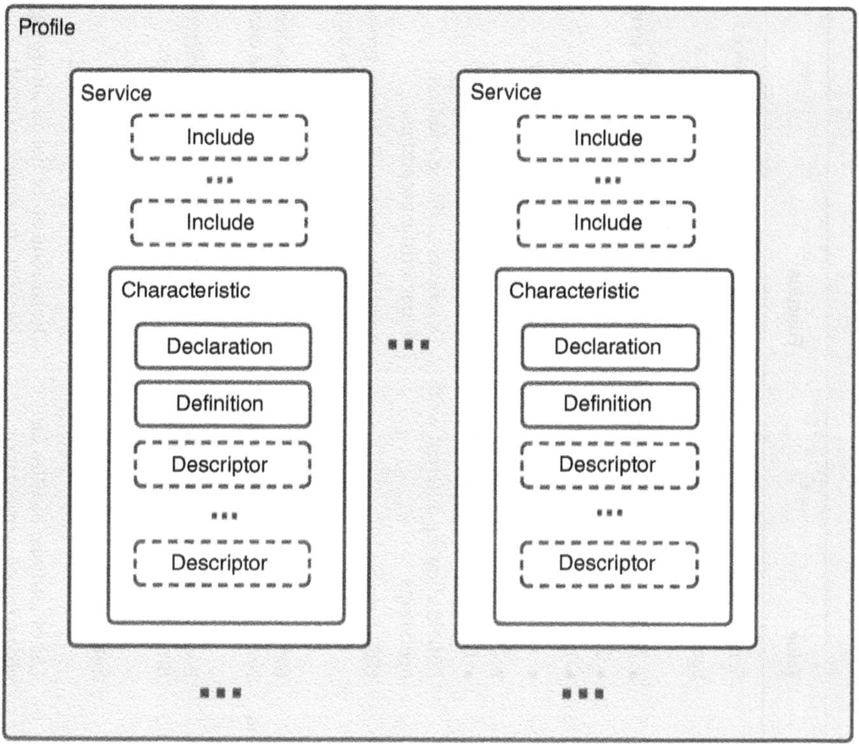

Figure 7.1 GATT Profile Hierarchy. *Source:* Adapted from [14]

hierarchy, declared in a required ≪*Primary Service*≫ type attribute, or secondary services, declared in a required ≪*Secondary Service*≫ type attribute. The value in both cases is the service type UUID. Secondary services are always nested within primary services.

The optional ≪*Include*≫ attribute can be used to reference other services including secondary services, and the Value field of the ≪*Include*≫ attribute contains the start attribute handle and end group attribute handle of the included service. The end group handle is the attribute handle of the last attribute in the included service definition. Note that a Server can support multiple services having the same type UUID, with each service declaration in a separate attribute having a separate attribute handle.

Characteristics consist of two required attributes: a declaration and the definition containing the actual characteristic data value itself. The characteristic attributes are positioned after the service declaration and any ≪*Include*≫ attributes. The declaration attribute has type ≪*Characteristic*≫ and is positioned before the definition attribute. It contains metadata describing the attribute. The declaration

Table 7.3 ATT Attributes Defining the GATT Abstractions

Attribute Type	UUID	Permissions	Value	Function
«Primary Service»	0×2800	Read	Service type UUID	Declaration for a primary service.
«Secondary Service»	0×2801	Read	Service type UUID	Declaration and definition for a secondary (included) service.
«Include»	0×2802	Read	• Included start service handle • End group handle • Included service type UUID	Definition linking a secondary service to a primary service.
«Characteristic»	0×2803	Read	• Access properties • Definition handle • Characteristic type UUID	Declaration and definition for a characteristic.
«Characteristic Extended Properties»	0×2900	Read	Bitfield allowing different write operations	Descriptor defining additional characteristic properties.
«Characteristic User Description»	0×2901	Implementation Specific	UTF-8 string	A variable sized UTF-8 string providing a textual description of a characteristic value (optional).
«Client Characteristic Configuration»	0×2902	Read, Write with authentication and authorization	Bitfield with characteristic Notification and Indication bits	Bitfield describing how a characteristic can be configured by a specific client (optional).
«Server Characteristic Configuration»	0×2903	Read, Write with authentication and authorization	Bitfield with characteristic Broadcast bit	Bitfield describing how a characteristic can be configured for the server (optional).
«Characteristic Presentation Format»	0×2904	Read	Single characteristic format	Declaration of format for the characteristic value (optional).
«Characteristic Aggregate Format»	0×2905	Read	List of attribute handles for characteristic presentation format declarations	Declaration of format for an aggregated characteristic value (optional).

attribute value consists of the characteristic's properties, the attribute handle of the definition attribute, and a UUID for the characteristic's type. Access properties is an 8-bit bitfield containing the operations and procedures that can be performed on the characteristic. The characteristic type UUID can be any standardized type from the Bluetooth Assigned Numbers specification [15] or a nonstandardized or vendor-specific type UUID. The definition attribute contains the attribute value in its Value field. The permissions on the definition attribute vary depend on the profile definition or some other higher layer specification. The attribute Type field contains the characteristic UUID from the declaration Value field.

The descriptor attributes in the characteristic are all optional and control various aspects of characteristic access and presentation. For example, the ≪*Characteristic User Description*≫ contains a UTF-8 documentation string describing the attribute. Descriptors are always stored after the characteristic declaration and definition attributes.

7.1.4.2 GATT Features

The GATT layer defines operations called features on top of the ATT protocol. These features fall into the following classes:

- Exchange MTU –a two-packet procedure allowing the ATT Client and Server to discover what the maximum packet size is that their buffers can handle,
- Service and Characteristic Discovery –allows the Client to discover services and characteristics on the Server, described in more detail in Section 7.1.4.3,
- Read Characteristics and Descriptors –several operations allowing the Client to read one or several characteristics and descriptors in two size classes (short and long) using either the characteristic's attribute handle to obtain a single attribute or the UUID to obtain all attributes of the corresponding type,
- Write Characteristics and Descriptors –several operations allowing the Client to write one or several characteristics and descriptors in two size classes (short and long) with or without Response,
- Server Initiated Updates –a Notification or Indication from the Server with the characteristic definition attribute handle and the current data value of the attribute. The Client sends no acknowledgement of receipt for a Notification but must acknowledge the Indication with a Confirmation.

7.1.4.3 GATT Service Discovery

When a Client and Server first initiate a connection, the Client has no knowledge of what services the Server supports. The client must perform a service discovery procedure to determine whether the Server supports the services and characteristics in which it is interested. The GATT builds its service and characteristic discovery features on top of the ATT operations described in Table 7.2.

The GATT has two features that support discovery of primary services:

- Discover all primary services – using *Find-Attribute-By-Group-Type*, discover all attributes that contain service declarations and definitions. This feature is used when the Client supports more than one service and needs to find out which of it's supported services are available on the Server,
- Discover primary services by service type UUID – using the *Find-Attribute-By-Type-Value* operation, find attribute subranges that correspond to a service declaration and definition.

Once the Client has found the services it is interested in, it can perform relationship discovery to determine any included services. This feature uses *Find-Attribute-By-Type* to search for all includes, by searching for attributes that match the ≪*Include*≫ attribute type. The attribute handles and values are returned.

Characteristics can also be discovered within a service using *Find-Attribute-By-Type*. The following two operations return information on characteristics:

- Discover All Characteristics of a Service – using *Find-Attribute-By-Type*, find all the characteristic declaration attribute handles and their values between the starting handle and ending handle of a service definition. The characteristic value includes the properties, characteristic definition handle, and characteristic type UUID,
- Discover Characteristics by Characteristic Type UUID – using *Find-Attribute-By-Type*, find all attribute handles and values for all characteristic declarations matching a specific type UUID. The *Find-Attribute-By-Type* operation returns all the attribute handles for characteristics within the service definition, and they are post-processed on the client so that only attributes matching the characteristic type UUID are returned.

The Client can further search for characteristic descriptors using *Find-Information*, by providing the attribute and handle range starting with the characteristic declaration handle plus 1 and ending with the end handle of the characteristic. All the handles in the range are returned, and the result is post-processed on the client to find the descriptors that match a particular descriptor type.

7.1.5 GATT Profiles, Services, and Characteristics for Energy Management Applications

The Bluetooth SIG has defined a few standardized GATT profiles and services that have functionality which could be useful for building an energy management application. However, most of these services and profiles are designed for monitoring and controlling the energy consumption of a connected personal device like a cellphone or earbuds powered by a battery and were not specifically designed for monitoring and controlling a building's AC appliances or DERs. They are missing some characteristics that would be useful in a BEMS such as frequency measurements and phase specification, which are important in AC power networks.

Standard profiles and services that could be useful in building a BEMS include the Time Profile 1.0 [16], the *Battery* service [17], the Environmental Sensing Profile 1.0 [18] which defines the *Environmental Sensing* service 1.0 [19], and the Automation IO Profile 1.0 [20] which defines the *Automation IO* service 1.0 [21]. In addition, the Bluetooth SIG GATT Specification Supplement document [22] and the Device Properties document [23] define a collection of standard characteristics independent from any service such as *Voltage*, *Electric Current*, and *Energy* that could be used to design services for interoperable energy management applications.

7.1.5.1 Time Profile

An energy management application like an electrical meter requires some date and time measurement embedded within it to provide accurate reporting of when energy measurements were made, or, for actuators, to schedule actions at the appropriate date and time. The GATT Time Profile 1.0 [16] requires the following three services for maintaining and implementing time reporting:

- *Current Time* –provides date and time information to other devices through three characteristics: *Current Time*, *Local Time Information*, and *Reference Time Information*,
- *Next daylight savings time (DST) Change* –contains the *Time with DST* characteristic with the *Date Time* field set to the date and time at which the next change to Daylight Savings Time occurs and the *DST Offset* field set to the offset from standard time for the next change,
- *Reference Time Update* –allows an external device to trigger an update of the system time from an external control point like a network time server by writing a command code to the *Time Update Control Point* characteristic and reading whether an update is pending and whether the last update succeeded or failed from the *Time Update State* characteristic.

There are also a few other characteristics for representing various sizes and precisions (number of bits) of time increments and other time values not associated with the Time Profile. Details on the Time Profile characteristics and other time-related characteristics can be found in the GATT Specification Supplement document [22].

7.1.5.2 *Battery* Service

BLE defines a *Battery* service 1.1 [17] primarily for management of batteries in Bluetooth-connected devices such as cellphones and earbuds. Nevertheless, the *Battery* service and characteristics could be useful in building a residential or

Table 7.4 Characteristics in the GATT *Battery* Service

Characteristic	UUID	Access Types	Function
Battery Level	0x2a19	Read, Notify*	An integer between 0 and 100, which indicates the percent charge of the battery.
*Battery Level Status**	0x2bed	Read, Notify, Broadcast*, Indicate*	A multi-field characteristic with a *Power State* field that represent the power state of a battery, an *Identifier* field to identify the service instance, the *Battery Level* field, and an *Additional Status* fields indicating if the battery requires service.
*Manufacturer Name String**	0x2a29	Read, Indicate	A UTF-8 string identifying the battery's manufacturer.
*Model Number String**	0x2a24	Read, Indicate	A UTF-8 string identifying the battery's model number.

commercial stationary battery energy management controller or reporting device. Mandatory and optional characteristics are shown in Table 7.4. In the Characteristic column and Access Types column, an asterisk after an entry indicates that the characteristic or access type is optional. The access types correspond to the ATT protocol operations discussed in Section 7.1.3.2.

The *Power State* field of the *Battery Level Status* characteristic is a 16-bit field of flags that indicate the state of the battery with respect to an external charging source, whether charging or discharging is occurring, a rough indication of charge level, and the reason for a charging fault if the last charging session ended in a fault. The *Additional Status* field of the *Battery Level Status* characteristic is an 8-bit field which indicates whether the battery had a fault and if it requires service. There are also a few other characteristics for representing various battery parameters not required by the Battery Service 1.1 specification. Details on the Battery Service characteristics and other battery related characteristics can be found in the GATT Specification Supplement document [22].

7.1.5.3 Environmental Sensing Profile

Although originally defined for sports and fitness use cases, the Environmental Sensing Profile 1.0 [19] defines a profile for an interoperable sensing device that can report on a wide variety of environmental parameters and could provide sensing for HVAC functions such as cooling and maintaining proper air quality. In addition to the *Environmental Sensing* service, which contains the characteristics for reporting on the environmental parameters, the Environmental Sensing profile can optionally support the *Battery* service, described in Section 7.1.5.2, and the

Device Information service [24]. The *Device Information* service contains several characteristics which provide a broad variety of information about the device, including the manufacturer's name, the device's model number, the device's serial number, and the firmware revision number, among others.

Characteristics that can be included in a device supporting the *Environmental Sensing* service are listed in Section 6.1.1 of the Bluetooth Assigned Numbers document [25]. A few that could be useful for HVAC sensing and management applications are:

- *Temperature* – the temperature in degrees Celsius,
- *CO2 Concentration* – the concentration of CO_2 in parts per million (ppm),
- *Humidity* – the relative humidity in percent,
- Particulate Matter – four characteristics that report the concentration of differently sized particulate matter, in kilograms per cubic meter:
  - *Particulate Matter – PM1 Concentration* – concentration of particulate matter less than 1 μm in diameter,
  - *Particulate Matter – PM2.5 Concentration* – concentration of particulate matter less than 2.5 μm in diameter,
  - *Particulate Matter – PM10 Concentration* – concentration of particulate matter less than 10 μm in diameter,
- *VOC Concentration* – the concentration of Volatile Organic Compounds (VOCs) in the air, expressed in parts per billion (ppb). VOCs are a variety of organic chemicals, which can be released as gases from paint, furniture, and other indoor items, that can be harmful to human health and cause discomfort.

7.1.5.4 Automation IO Profile

Finally, the Automation IO Profile 1.0 [20] and *Automation IO* service 1.0 [21] specifications define a standardized characteristic set for reading and writing to digital and analog inputs and outputs on a generic I/O device. This profile could be used to develop interoperable BEMS applications. The Automation IO profile contains only one service, the *Automation IO* service. The *Automation IO* service has three characteristics:

- *Digital* (UUID: 0×2a56) – the state of a digital signal as a collection of two bit registers that can be configured as an input or output and whether it supports Read, Write and Notify operations,
- *Analog* (UUID: 0×2a58) – the state of a single analog signal as a 16-bit integer that be configured as an input or output and whether it supports Read, Write, and Notify operations,
- *Aggregate* (UUID: 0×2a5a) – a collection of digital and analog signals from a device that provides a combined, single-point access to all the *Digital* and *Analog* characteristics included in the *Automation IO* service.

Unlike the other profiles and services reviewed so far, the *Automation IO* service also supports characteristic descriptors which are nonstandard, and some which are not optional. For example, the *Number of Digitals* descriptor provides the number of two-bit registers in the *Digital* characteristic with which it is associated. In addition, the characteristics for the *Automation IO* service are not listed in the GATT Specification Supplement, Assigned Numbers, or Device Properties documents. Instead, the Automation IO Service 1.0 document provides a detailed description of how the characteristics are accessed, what they represent, and their descriptors, but it does not provide their UUIDs. The UUIDs are specified in other documents. The Automation IO service UUID itself is listed in the Assigned Numbers document as 0×1815.

The *Automation IO* service can be used for implementing a BEMS application in a variety of ways. An *Analog* characteristic could be connected to sensors that measure energy use from different appliances like lighting, HVAC systems, and hot water heaters within the building. The results of these measurement could be input into an energy optimization and control algorithm in the BEMS, and *Digital* characteristics could then be used to control various devices to implement a flexible load program and reduce overall energy use. If the building incorporates DERs, the *Automation IO* service could monitor their output using *Analog* characteristics and control their operation using *Digital* characteristics.

7.1.5.5 Nonstandardized Services

In the absence of standardized profiles and services dedicated to energy management, vendors have stepped in with nonstandardized profiles. An example is the Texas Instruments service for energy metering [26] which dates from 2014. Figure 7.2 illustrates the Texas Instruments energy metering service definition and its characteristics. The energy metering service is defined as an extension of the CC2650 SensorTag board, and it runs as a GAP Peripheral and GATT Server.

The meter service is designed to provide measurements of voltage, current, power, power factor, and frequency from a multi-phase AC power source, like an electric utility connection, to a remotely monitoring Client. The UUID for the service is based on the Texas Instruments UUID used for the SensorTag service: f000e000-0451-4000-Bb000-000000000000. The characteristics are assigned UUIDs in increasing order, with 0×e001 through 0×e00C replacing the 0×e000 field in the service UUID. Characteristics that are read only include *Meter Properties*, *Meter Name*, and *Meter Version*. Characteristics that are read and write like the *Meter Reading Enable*, *Meter Reading Update Period*, *Meter Power On State*, and *Meter Phase Select* let the Client modify the meter's function in some fashion. For example, writing the *Meter Phase Select* characteristic changes the phase being monitored by the meter. Characteristics that are read and notify like the *Meter Phase Voltage*, *Meter Phase Current*, *Meter Phase Power*, *Meter Phase Power Factor*, and *Meter Phase Frequency*

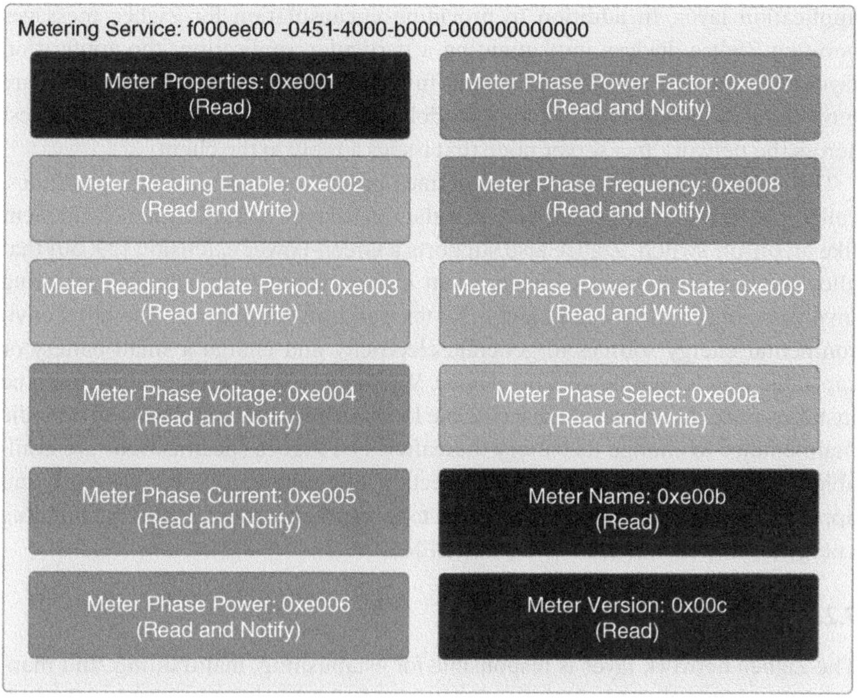

Figure 7.2 Texas Instruments Energy Metering GATT Service Characteristics. *Source:* Adapted from [26]

allow the Client to read basic measurements and for the Server to update the Client asynchronously about measurements when their values change. The Texas Instruments documentation provides a detailed description about how to customize the SensorTag firmware to include the metering service application firmware.

7.2 Zigbee

Zigbee [27] is a wireless communication protocol widely adopted for smart home and building automation use cases. It operates at the network and application layers, building upon the foundation provided by the IEEE 802.15.4 wireless physical and link layer standard. The Zigbee network layer adds multi-hop routing, network setup, and routing repair on top of the 802.15.4 link layer. These functions enable a low-power, low data rate network to deliver the level of adaptability and resilience required for BEMS application deployments. Unlike the IP stack, the Zigbee stack does not support a transport layer; however, it does support an application layer. In addition to providing encapsulation for Zigbee messages between Zigbee devices implementing a particular application, the application layer also supports some management functions for devices. Zigbee devices are built on a client/server interaction model, where a client node issues a request across the network to a server node that issues a reply to the client.

The Zigbee Cluster Library [28] defines groups of messages and attributes, called clusters, that implement a particular device type or application component like an on/off switch. Zigbee also supports a Green Power extension [29, 30] that allows a device to receive power from energy harvesting. Energy harvesting involves using vibrations, ambient light, temperature differentials, and other environmental energy sources to generate electricity and charge a small battery or capacitor which then powers the device. With energy harvesting, a device can be installed once in a relatively inaccessible location and does not need any periodic maintenance to change its battery thereafter. The Zigbee specifications are available from the CSA website [1]. This section discusses the Zigbee network and application layers, and clusters relevant to implementing and deploying building energy management applications in a BEMS.

7.2.1 Zigbee Network Architecture

The Zigbee network layer is responsible for establishing, maintaining, and managing the network. This includes mechanisms for joining and leaving the network, network address allocation, route discovery, routing, and security.

Zigbee networks consist of three different types of nodes:

- Zigbee Coordinator – responsible for admitting both end device and router nodes to the network and handling node departures. There is only one coordinator in a Zigbee network, and it plays the role of the 802.15.4 PAN coordinator by forming and maintaining the network. It selects the 802.15.4 channel for nodes to communicate and both the 16 and 64-bit PAN IDs, acts as a router, and must be an 802.15.4 full-function device. The coordinator can never go into sleep mode because it must always be ready to admit a new node to the network, so it needs to be connected into the building's grid powered network.
- Zigbee Router – after joining the network, a router admits other routers and end devices and can act as an 802.15.4 coordinator. It routes packets between end devices and other nodes and must also be an 802.15.4 full-function device. A regular router cannot go into sleep mode because it must always be ready to receive packets and route them to other nodes. A low-power router periodically switches off its radio like a low-power end device but must maintain synchronization with a regular router or the Zigbee coordinator, so it can be woken up when a packet needs routing. Regular routers need to be connected into the building's grid powered network; low-power routers can be battery powered.

- End Device – cannot admit other nodes to the network nor route and always communicates packets to and from its parent which must be either the coordinator or a router. An end device can be a reduced-function device and can go into sleep mode if it has no sensor data to report or is not expecting an actuation command. End devices can be powered by a battery.

Zigbee supports three different network architectures: star, peer-to-peer or mesh, and tree. The star and mesh architectures are shown in Figures 3.12 and 3.13, respectively, and are similar to the 802.15.4 architectures with the 802.15.4 PAN coordinator replaced by the Zigbee coordinator and the 802.15.4 coordinator replaced by the Zigbee router. The tree architecture is shown in Figure 7.3, with the Zigbee coordinator acting both as a gateway router to the building LAN and a Zigbee to IP gateway.

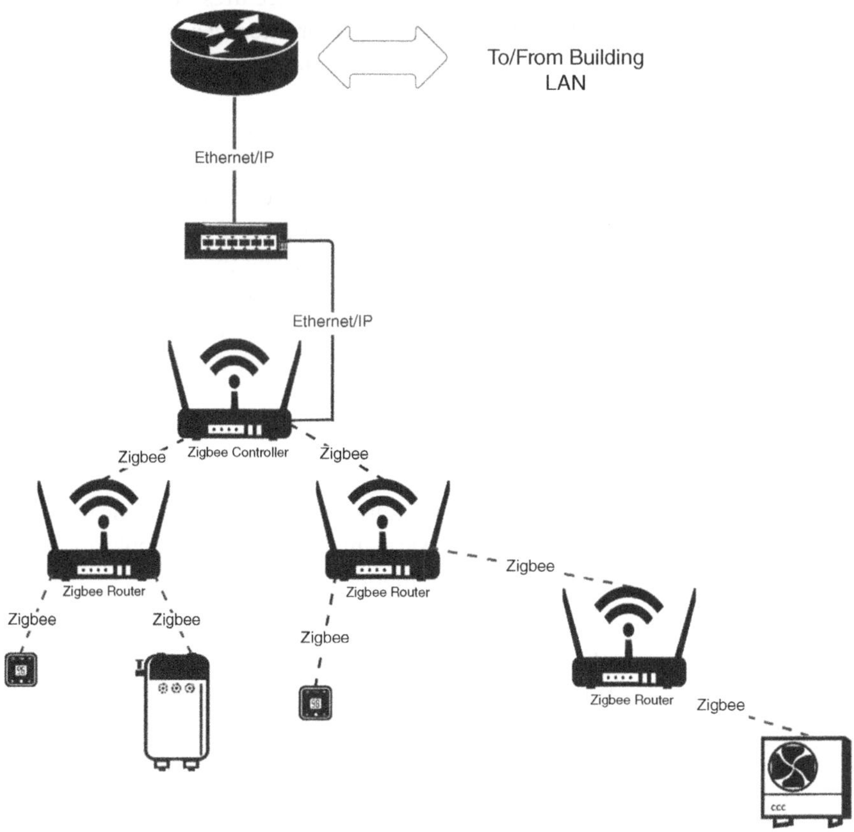

Figure 7.3 Zigbee Tree Network Architecture

The primary difference between the three architectures is how many parents a node can have and how packets are routed between routers. In the star architecture, there is only one router, the Zigbee coordinator, and all nodes route packets through the coordinator. No other router node is allowed to join the network. In the mesh architecture, multiple routers connect in a peer-to-peer fashion and route packets between each other, with the coordinator playing no special role in routing. Mesh routers can connect with multiple parent (upstream) nodes up to a limit. If a router becomes unavailable due to a drop in wireless connectivity or for another reason, other downstream routers can find a new route to ensure continuous communication using the Zigbee network layer routing protocol. In the tree architecture, all routers connect in a hierarchical fashion, with the coordinator at the top of the hierarchy. Each node, both router and end device, can only connect with one upstream node. If a router becomes unavailable, any child (downstream) routers and devices are disconnected from the network and must reassociate again to regain connectivity. In the star and tree architectures, the coordinator is the logical entity to connect into the building LAN for Internet connectivity. In the mesh architecture, a mesh router could also function as a gateway router into the building LAN.

A Zigbee network determines the routing topology to be one of the star, mesh, or tree architectures depending on the configuration of the coordinator and routers. The coordinator starts the process by admitting routers if a tree or mesh architecture is configured or only allowing end devices to enter if the star architecture is configured. In tree or mesh networks, routers entering the network exchange configuration messages with the coordinator or the router admitting them, determining whether they should use hierarchical or mesh routing. In star and tree networks, the coordinator and routers are permitted to use 802.15.4 beacons, but in mesh networks, 802.15.4 beacons are not allowed.

7.2.1.1 Zigbee Stack

Figure 7.4 illustrates the Zigbee stack. The Zigbee stack builds the application and network layers on top of the 802.15.4 physical and link layers. The application and network layers consist of the following components:

- Application Objects – end-user applications are implemented by device manufacturers through application objects. Application objects have 8-bit endpoint identifiers between 1 and 254, with 0 reserved for the Zigbee Device Object (ZDO) and 241 through 254 reserved for the CSA. Endpoint identifiers act like ports in the IP transport layer, and packets are delivered to an application object from the application support sublayer through the endpoint identifier,
- Application Framework – the environment in which Zigbee application objects are hosted, offering interfaces in support of the application objects,

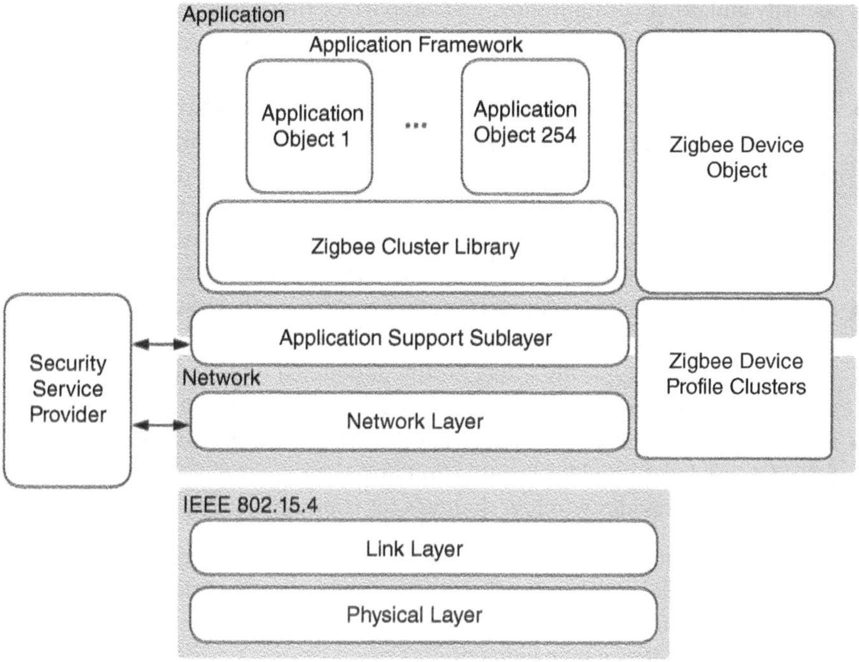

Figure 7.4 Zigbee Stack

- Zigbee Cluster Library – a library containing standardized groupings of attributes and messages called clusters, that support specific application use cases or node management functions. Clusters are identified by a 16-bit cluster identifier. The Zigbee Cluster Library is discussed in Section 7.2.7,
- Zigbee Device Object (ZDO) – a node management object that implements device initialization and network formation and performs other node management tasks such as security management and device discovery. The ZDO provides an interface between the application layer and the underlying Zigbee network. The ZDO is discussed in Section 7.2.5.
- Zigbee Device Profile Clusters – a collection of clusters that implement the lower level functions for the ZDO and interface with the application support sublayer and network layer. The clusters have cluster identifiers in a separate 16-bit identifier space from the Zigbee Cluster Library. The Device Profile Clusters is discussed in Section 7.2.6,
- Application Support Sublayer – the interface between the network layer and application objects and the ZDO. This includes interfaces for data transfer between the network layer and application layer, as well as messages supporting

security and other configuration, and support for transferring application layer messages between the client and server. The application layer is described in Section 7.2.4,

- Network Layer – supports the transport of packets between two or more Zigbee nodes on the same network. Toward that end, it adds a network layer header to an application layer message and routes the message to the destination node. The network layer is also responsible for maintaining routing tables if the node is a router and for performing other operations involved in forming and managing the network. The network layer is discussed in Section 7.2.2,
- Security Service Provider – supports authentication, authorization, and encryption services at the application and network layers, and is not part of every node's stack but is typically a specifically designated node in the network like the coordinator or a router. Security is discussed in Section 7.2.3,
- 802.15.4 Link Layer and Physical Layer – at the bottom of the stack, the 802.15.4 link layer provides framing and link layer security for Zigbee, and the physical layer provides wireless channels for communication.

The ZDO implements several functions that all receive messages over the same endpoint identifier, 0×00. Some of these messages are network layer messages, some are application sublayer messages, and some come from cluster libraries that support basic device management.

7.2.2 Zigbee Network Layer

The Zigbee network layer provides routing and delivery of Zigbee packets between nodes. The Zigbee stack extends the 802.15.4 addressing scheme across all wireless links within the PAN.

7.2.2.1 Zigbee Network Layer Header

Figure 7.5 illustrates the network layer header. The fields in the header have the following functions:

- Frame Control – flags that indicate the status of other fields in the header and control processing of the packet. Flags include the protocol version, packet type, whether route discovery should be used for this packet, whether network layer security is provided, and whether the 64-bit (8 byte) IEEE addresses and the source route subframe are present or not,

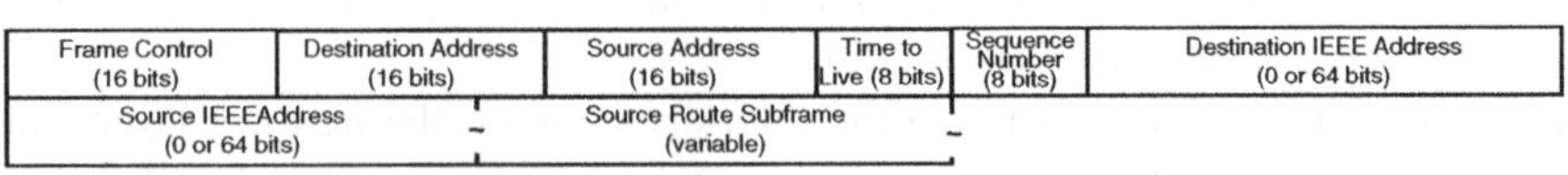

Figure 7.5 Zigbee Network Layer Header

- Destination and Source Address – the two-byte (16 bit) short address for the destination and source nodes,
- Time to Live – the number of hops before the packet should be discarded. This field is decremented on every hop through a Zigbee router,
- Sequence Number – incremented on each new packet sent from the source address. This field is used to prevent replay attacks,
- Destination and Source IEEE Address – the 8 byte (64 bit) IEEE addresses of the source and destination nodes. These fields are only present if their respective flags are set in the header,
- Source Route Subframe – a list of node addresses in the path from the source to the destination constituting a source route through which the packet should be routed. This field is only present if the respective flag is set in the header.

The network layer header is followed by the message payload, which can be of variable size. The total packet size is required to be less than 128 bytes including network and application headers and the payload.

7.2.2.2 Zigbee Network Layer Packet Types

The Frame Control packet type field for network packets can indicate one of three message types: data messages, inter-PAN messages, or command messages. Data messages are used for carrying application layer message payloads between nodes. Inter-PAN messages allow Zigbee devices to exchange messages with devices not in the same PAN without requiring them to form or join the same Zigbee network. This functionality is especially useful for low-power devices that implement the Green Power (energy harvesting) specification since it avoids the energy expenditure of joining and maintaining connectivity to the PAN, and is described in Annex G of the Zigbee specification [31].

The command type packets carry traffic for network management and maintenance. Network command packet types include the following:

- *Route Request* – broadcast by a node to discover a route to another node within the network, initiating the route discovery process,
- *Route Reply* – sent in response to a Route Request to indicate the route to the destination node and unicast back to the originator of the Route Request,
- *Route Record* – contains the complete route from the source to destination for a source route, and is usually sent preceding the actual data packets when source routing is employed,
- *Leave* – sent by a node to leave the network or instruct another device to leave, triggering the network leave procedure. Sending this message ensures that a node departing the network has a graceful exit and does not leave connections hanging,

- *Rejoin Request* – sent by a node to rejoin the network after it has left or been disconnected, initiating the rejoin procedure,
- *Rejoin Response* – allows or disallows the node to rejoin the network based on various conditions such as network security settings,
- *Link Status* – exchanged between nodes to report on link quality,
- *Network Status* – reports errors or other conditions on the network layer, for example, if a link fails while a node is forwarding a unicast packet or an address conflict is detected,
- *Network Report* – sent by a device to the network manager node (coordinator or parent router) to report on issues such as PAN ID conflicts or wireless channel conditions,
- *Network Update* – sent by the network manager node (coordinator or parent router) to initiate a change in configuration of all devices in the network,
- *End Device Timeout Request* – sent by a node that periodically sleeps to its router parent informing the parent of when the timeout ends.
- *End Device Timeout Response* – sent by a router parent that previously received an *End Device Timeout Request* informing the node whether it has accepted the timeout value and what its capabilities are,
- *Link Power Delta* – allows a node to communicate to another node or the parent the difference in wireless signal strength between the optimal signal strength and the signal strength of the last received packet, enabling the receiving node to optimize its transmissions and thus its power use,
- *Network Commissioning Request* – sent by a node that wants to join or rejoin a network. This message can also be used to request key provisioning for security,
- *Network Commissioning Response* – sent by the recipient of a *Network Commissioning Request* message to inform the node of its network commissioning request status.

Details of the Zigbee network layer command message types can be found in Section 3.4 of the Zigbee specification [32].

7.2.2.3 Zigbee Addressing

Zigbee uses both the permanent IEEE 64-bit address and an ephemeral 16-bit link layer address from the 802.15.4 link layer specification as the network layer addresses. In a Zigbee network, the Zigbee coordinator is responsible for network formation and assigning 16-bit network layer addresses to nodes when they join the network. During network formation, the coordinator selects a radio channel, defines the PAN ID, and assigns itself a 16-bit network address (usually 0×0000). When a new node wants to join the network, it communicates its 64-bit address using a *Network Commissioning Request* message to the coordinator if the network uses a star or tree architecture, or a router, if the network uses the mesh

Table 7.5 Zigbee Broadcast Addresses

16 bit address	Destination
0xffff	All nodes in the PAN.
0xfffe	Reserved.
0xfffd	Nodes that can receive packets in idle mode, typically connected to the building power network.
0xfffc	All routers and the Zigbee coordinator.
0xfffb	Low-power routers only.
0xfff8 – 0xfffa	Reserved.

architecture. The coordinator or router assigns the node a unique 16-bit address within the network. These addresses are unique only within a particular network defined by the PAN ID.

Zigbee also supports broadcast and multicast transmission. Table 7.5 lists the Zigbee broadcast addresses. Broadcast addresses are used for node address discovery and network management. Broadcast and multicast forwarding uses flooding rather than routing, with routers sending a packet having a broadcast or multicast address to all nodes in their routing tables.

Multicast addresses, which Zigbee calls group addresses, are formed when an application on the node requests a 16-bit group address from the application support sublayer. The application support sublayer sets up the address and begins listening for packets on the address. Nodes maintain a group address table with each entry containing a group address and the list of destination endpoint identifiers of application objects on the node to which messages arriving for the group are addressed. The application support sublayer filters the packets based on whether an application object has subscribed to the group address. Remote configuration of group addresses is accomplished through messages implemented by the *Group* cluster in the Zigbee Cluster Library. Group addresses can be particularly useful for BEMS applications. For example, in an intelligent lighting system, lights in an area can be grouped together so that they can be controlled simultaneously based on occupancy data or ambient light conditions.

7.2.2.4 Zigbee Routing

Zigbee routing depends on whether the network architecture is organized according to the star, tree, or mesh topology. In a star topology network, all communication goes through the coordinator. There is no complex routing involved as end devices only communicate directly with the coordinator. In a tree topology network, nodes only communicate with their parent or child nodes, following the hierarchical structure of the tree. Messages can be forwarded through several devices from child to parent upward toward the coordinator or from parent to child downward toward an end device. Routing in a tree topology network is also very structured and does not allow for direct communication between arbitrary pairs of nodes.

Routing in a mesh topology network is more complex since multiple paths between nodes are possible. Zigbee mesh topology supports two types of routing:

- On Demand Routing – On demand routing is used by Zigbee to discover the best path between devices. When a router receives a packet whose destination address is not in its routing table, it starts a route discovery process, broadcasting a *Route Request* message to the all-nodes broadcast address (0×ffff). Neighboring routers forward the request until the destination is reached or until the packet reaches a router with the destination route in its routing table. The end recipient then sends a *Route Reply* message back to the originating node, forming a route. Routers along the return path record the route in their routing tables for future reference.
- Source Routing – Zigbee also uses source routing, where the entire path from source to destination is defined at the source node and included in the header of the packet. This method can be more efficient in terms of network overhead but requires the source node to know the complete route.

Zigbee nodes maintain neighbor tables containing information about neighboring devices within their radio range, and periodically prune the entries in the neighbor table if, for example, a sleeping device does not report back within its reported timeout period. Routers and coordinators also maintain routing tables that contain information about the next hop on the path to a particular destination. Existing routes may be maintained through periodic updates or rediscovered as needed. If a route breaks, error messages are used to notify affected devices. Zigbee also considers link quality in its routing decisions. Metrics such as signal strength can be used to choose the best path among multiple available routes.

7.2.2.5 Low-Power Routers

Low-power routers are defined as routers operating on batteries for multiple years by regularly powering off their radios. Low-power routers in Zigbee offer a way to extend the network coverage while conserving energy and are designed to operate with minimal energy consumption. In contrast, full-function routers always have their receivers on and need to be grid powered.

A low-power router only communicates with its parent and children. It does not participate in multi-hop routing, further reducing energy consumption. When a low-power router wants to join the network, it must find the coordinator or a neighboring full-function router that will accept low-power child nodes.

The low-power router then communicates its low-power status during the join process, allowing the parent device to accommodate it. Low-power routers implement a sleep/wake cycle to conserve energy. They communicate their wake periods to their parent, allowing the parent to buffer messages intended for the low-power router while it is asleep. The parent then communicates the buffered messages when the low-power router wakes up.

While low-power routers can have child nodes, they do not generally participate in relaying messages for other nodes in the network. This is a significant distinction from regular Zigbee routers, and it means that low-power routers are less suitable for use in mesh topologies where extensive routing is required. Low-power routers are most suitable for network topologies where the routing demands are relatively simple, and where energy conservation is a priority. They can be an effective way to extend network coverage into areas where power from the building power network is unavailable, but physically changing a battery frequently may be difficult. However, the energy-saving features of low-power routers may introduce latency due to their sleep/wake cycles.

7.2.3 Zigbee Security

Zigbee supports two security architectures depending on the network architecture: a centralized architecture with a Trust Center providing configuration and mapping to the coordinator for the star and tree topologies, or a distributed architecture for a mesh topology network. The Trust Center is the central point for security management in a centralized architecture. It handles authentication and key management for all nodes joining the network. In the distributed architecture, authentication and key management for nodes joining the network are handled through the node's parent router. Zigbee uses the AES-128 standard for encryption and AES-CCM* [27, 33] for encryption and authentication on both the network and application layers.

The main security components in a Zigbee network include the following:

- Network Key – a shared key used by all nodes on the network for network-level encryption and authentication,
- Link Key – a key used between two nodes for application-level encryption and authentication,
- Packet Counter – a counter that helps in protecting against replay attacks,
- Security Control Field – a field in the packet that defines the security level and security attributes applied to the packet.

Unlike the IP stack in which end to end security is provided at the network and transport levels, the Zigbee stack provides hop by hop security at the network level through a shared network key and end to end security at the application level through a key shared between the two end nodes.

At the network level, hop-by-hop encryption and authentication are performed on messages by encrypting the message as it passes through a router and attaching a message integrity code using the network key. The receiving router or the coordinator decrypts the message and re-encrypts it using the same network key and checks and recalculates the message integrity code before the packet is forwarded. However, because the source and destination address are part of the 802.15.4 link layer header, they remain unencrypted. The link key is used to encrypt and authenticate messages end to end between the source and destination nodes. This prevents intermediate nodes from reading the content. Link keys are only shared between the communicating nodes.

Key provisioning of the network and link keys on a node joining a Zigbee network generally follows a two-step process. First, the Trust Center in the centralized architecture or parent router in the distributed architecture and the joining node are provisioned with a link key for encrypting application communication end to end; then the shared network key is provisioned over the secure channel for securing hop-by-hop network communication.

Depending on which version of the Zigbee standard the joining devices support, one of the four different link key provisioning protocols is used:

- Static key joining – Prior to Revision 21 of the Zigbee standard, this was the only supported method of key provisioning. The key for securing communication with the Trust Center was either pre-provisioned and published, or derived from a secret install code pre-provisioned on the joining node and Trust Center.
- Static key replacement in centralized security networks – In Revision 21, the joining node was required to replace its pre-provisioned key with a new key using the Trust Center link key update mechanism.
- Joining in a distributed security network – In Revision 21, distributed security networks were added and no key update was required.
- Dynamic key derivation – In Revision 23, the latest version of the standard, dynamic key exchange using elliptic curve Diffie–Hellman or Simple Password Exponential Key Exchange (SPEKE) [34] key derivation was introduced. This allows the Trust Center or parent router and the joining node to dynamically generate a key using public key cryptography based on a pre-provisioned pass phrase.

Note that if one of the published static keys is used in any of the first three methods rather than a secret install code, the network has a window of vulnerability during the initial key provisioning since an attacker can easily gain access to the published shared key and decrypt communications.

For encrypting application level traffic end-to-end between two nodes, the nodes negotiate a link key. The Trust Center can be utilized in the centralized architecture to establish a symmetric pass phrase between the two nodes first to allow authentication as described in Section 4.98 of the Revision 23 specification [35].

In the distributed architecture, Revision 23 states that the two sides use a higher level protocol for establishing a link key without parent router involvement as a Trust Center, but does not provide any details of the actual protocol, leaving it up to the application.

Two nodes that have taken specific security measures to secure communication between them form a Trusted Link. The security measures include the following:

- using one of the key management protocols described above to provision both nodes with network and link keys,
- performing mutual authentication between the devices,
- establishing encryption and authentication on packets exchanged between the two nodes,
- updating link keys depending on the network policy about key lifetimes.

Trusted Links are typically used in scenarios where enhanced security is required, such as communication between a smart meter and a utility company's data concentrator within a Zigbee Smart Energy network.

7.2.4 Zigbee Application Layer and Application Support Sublayer

The Zigbee application layer consists of the application support sublayer, the application framework, the objects implementing the applications, and the ZDO. Zigbee defines the basic interaction between two nodes using a client-server model, where the application object on the client node makes requests to the server node and the corresponding application object on the server node replies. The application framework is the environment in which application objects, including the ZDO, are hosted on the application layer. The application layer and application support sublayer provide an interface with the network layer for communication of the Zigbee Cluster Library command messages. The ZDO provides an interface to the node and network management functions supported by the ZDO.

Prior to the release of Zigbee 3.0 in 2016, Zigbee devices implementing a certain class of applications, like lighting or smart energy management, were defined within an application profile. Application profiles were assigned a 16-bit identifier and contained standardized descriptions of messages, message formats, and processing actions including basic device behavior. However, when devices defined according to different profiles tried to interoperate on the same network, they often ran into problems. With Zigbee 3.0 (now called Zigbee Pro), the Basic Device Behavior specification, discussed in Section 7.2.11.1, standardized the basic behaviors and procedures for a device to be a "good citizen" on a Zigbee network, and application profiles were deprecated. The specifications for application specific messages and attributes were moved into clusters in the cluster library.

With Revision 23, Zigbee has only three profiles: Zigbee 3.0 (0×0104), Smart Energy (0×0109, discussed in Section 7.2.7.4), and Green Power for devices powered by energy harvesting (0×7f02). The Zigbee 3.0 profile includes messages and attributes from the previous application specific profiles such as home automation and lighting in clusters as part of the Zigbee Cluster Library.

The application support sublayer implements four types of messages:

- data type messages are used for transporting payloads implemented by a cluster library running in an application object,
- acknowledgement type messages are sent to acknowledge the receipt of an application support sublayer data message,
- command type messages are sent to perform certain security functions and are delivered to the ZDO for processing,
- Inter-PAN type messages are sent between different PANs and can also be used for device commissioning before a device is part of a PAN. They are described in Annex G of the Zigbee specification [31].

The application support sublayer also contains a collection of configuration parameters for configuring various aspects of messaging with other nodes, and a collection of state variables that hold the current state of the application layer. Notable among these parameters are the group table, which contains entries recording group addresses to which the node has subscribed, and the binding table, which contains entries with the application layer bindings to other nodes.

7.2.4.1 Application Support Sublayer Message Header

Figure 7.6 shows the generalized application support sublayer message header. The header is followed by the variable-sized application support sublayer payload. Which fields are present depends on the message type and addressing.

The fields have the following functions:

- Frame Control for the Application Support Sublayer – flags that indicate the status of other fields in the header and control application support sublayer processing of the packet. More details on the Frame Control field are given below,
- Destination Endpoint – the 8-bit endpoint identifier of the application object on the destination node that is to receive the application packet. The destination endpoint 0×00 addresses the packet to the ZDO, a destination endpoint 0×01-0×fe

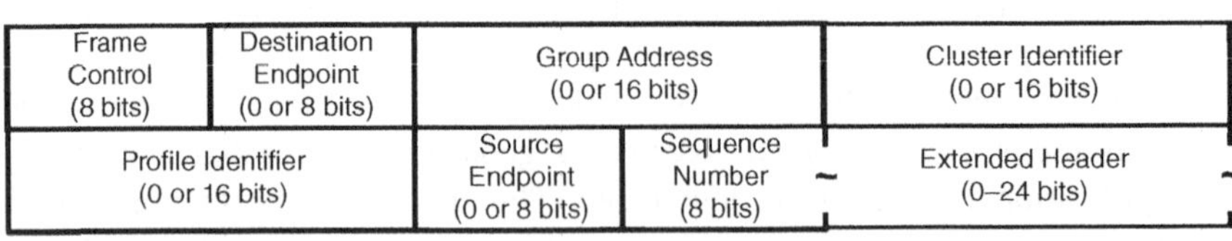

Figure 7.6 Zigbee Application Support Sublayer Header

addresses the packet to an application object, and a destination endpoint of 0×ff (broadcast) addresses the packet to all active endpoints except endpoint 0×00. If the Group Address field is present, the Destination Endpoint field is omitted since the node's group table contains the destination endpoints to which the message should be delivered. This field is also omitted in command type packets,

- Group Address – if the Frame Control field indicates that the delivery mode is for group addressing, the 16-bit group address appears in this field; otherwise, the field is absent. If the group address is present, the packet is delivered to all endpoints in the group address's group table entry,
- Cluster Identifier – the 16-bit cluster identifier implementing the message. This field is only present in data and acknowledgement type messages,
- Profile Identifier – the 16-bit profile identifier indicating the application for which the message is intended. Application and device profiles, with the exception of three profiles described in the introduction to this section, have been deprecated in Revision 23. This field can only be present in data and acknowledgement type messages,
- Source Endpoint – the 8-bit source endpoint identifier indicating the 8-bit source endpoint of the application object on the source node that sent the message. This field is only present in data and acknowledgement type messages,
- Sequence Number – an 8-bit sequence number set to 1 for the first message sent and incremented for each subsequent message sent. This field is used to prevent processing of duplicated messages,
- Extended Header Subframe – a variable-sized field up to 24 bits used to control message fragmentation. This field is only present in data and acknowledgement type messages.

The header is followed by a variable-sized application support sublayer payload. If the message is a data message, the payload content originates from a cluster library. If the message is a command message, the payload content is from the ZDO. Acknowledgement messages have no payload.

The actual format of the message header is controlled by the Frame Control field. The Frame Control field is shown in Figure 7.7. The fields are the following:

- Frame Type – indicates which of the four types the message contains. The codes for the message types are:
  - 00 – data,
  - 01 – command,

Frame Type (2 bits)	Delivery Mode (2 bits)	ACK Format (1 bit)	Security (1 bit)	ACK Required (1 bit)	Ext. Header present (1 bit)

Figure 7.7 Application Support Sublayer Frame Control Field

  - 10 – acknowledgement,
  - 11 – Inter-PAN.
- Delivery Mode – indicates what type of delivery the message requires. The codes for the delivery types are:
  - 00 – unicast delivery,
  - 01 – reserved,
  - 10 – broadcast delivery,
  - 11 – group addressing.
- ACK Format – set to 1 in data acknowledgement messages, indicating that the destination endpoint, cluster identifier, profile identifier and source endpoint fields are present in the header, and 0 in command acknowledgment messages indicating that these fields are absent,
- Security – 1 if the frame requires security processing, and 0 if not,
- ACK Required – 1 if an acknowledgement message is required, and 0 if not,
- Ext. Header present – 1 if the packet header includes an Extended Header field, and 0 if not.

7.2.4.2 Application Support Sublayer Command Message Types

Command message types supported by the application support sublayer fall into two classes: security-related messages and messages for tunneling or relaying packets to other nodes. All other application support sublayer messages are data messages that carry either Zigbee Cluster Library commands for both applications and node and network management or Zigbee Device Profile Clusters messages. Table 7.6 contains a complete list of the application support sublayer command message types. Messages that can be sent from or to the Trust Center or a router are sent to a Trust Center in the centralized security architecture and a router in the distributed security (mesh) architecture.

7.2.4.3 Binding

Binding simplifies application delivery by making routing and application message processing more efficient. The binding information is stored in a binding table, which contains entries recording the source node's short network address, the source endpoint, the cluster identifier, and one or more destination node short network addresses or group addresses for the binding. The entry can in addition optionally include other destination endpoint identifiers if the destination address field is a node address and not a group address. Binding allows the nodes to execute specific functions, such as controlling or querying a device's status. For example, a temperature sensor node could be bound to a thermostat node, allowing the thermostat to receive temperature readings directly and adjust heating or cooling. In addition, if wireless reception allows it, two bound nodes can communicate directly without having to go through other nodes.

Table 7.6 Application Support Sublayer Command Message Types

Message	Code	Function
APS_CMD_TRANSPORT_KEY	0×05	Securely transport network and link keys to nodes.
APS_CMD_UPDATE_DEVICE	0×06	Inform the Trust Center about a change in the status of a device, such as when it joins or leaves the network. Usually sent by the device's parent node.
APS_CMD_REMOVE_DEVICE	0×07	Sent by an authorized device such as the coordinator or a router to instruct the Trust Center or router to remove a device from the network security associations.
APS_CMD_REQUEST_KEY	0×08	Request cryptographic keys from the Trust Center or router.
APS_CMD_SWITCH_KEY	0×09	Sent by the Trust Center or router to indicate that a device or all devices should switch to a new active network key.
APS_CMD_TUNNEL	0×0E	Tunnel non-Zigbee packets between Zigbee nodes in an application support sublayer tunnel.
APS_CMD_VERIFY_KEY	0×0F	Sent to the Trust Center or routing by a node joining the network to verify its updated link key.
APS_CMD_CONFIRM_KEY	0×10	Sent by the Trust Center or router to a node to confirm the updated link key.
APS_CMD_RELAY_MESSAGE_DOWNSTREAM	0×11	Sent by the Trust Center to relay a message through a parent router to a joining node.
APS_CMD_RELAY_MESSAGE_UPSTREAM	0×12	Sent by an unauthorized joining node to relay a message through a parent router to the Trust Center.

7.2.5 Zigbee Device Object

Unlike the IP stack, the Zigbee stack performs certain network management functions related to the node, like address discovery, at the application layer through the ZDO [36]. These functions are implemented with clusters from the Zigbee Device Profile Clusters [37]. Messages are delivered from the application support sublayer to the ZDO through the destination endpoint 0×00.

Section 2.5 [36] in the Zigbee specification defines how other application objects interface with and utilize the services of the ZDO. The functions of the ZDO related to node, service, and network management include the following:

- Node and Service Discovery – provides mechanisms to discover the short address of another node given its IEEE 16-bit address and vice versa, as well as other information about the node including the application services that the node offers,

- Binding Management – manages binding and unbinding requests. However, the application support sublayer manages the binding table,
- Network Management – handles node role assignment by determining the role a node plays in the network, i.e. whether it operates as an end node, router, or Zigbee coordinator. If the node is a Zigbee coordinator, network management is responsible for initializing and setting up the network by selecting the radio channel, assigning addresses, and managing routing,
- Security Management – establishes secure links with other nodes, and, if the node is a Zigbee coordinator in a star or tree topology network or a router in a mesh topology network, manages the secure joining of new nodes to the network, functioning as the security service provider,
- Group Management – manages inclusion and removal of application objects for group addressing within the node under application control. However, the application support sublayer manages the group table directly.

While the Zigbee Cluster Library mostly contains clusters concerned with application use cases, the ZDO implements some of its functions utilizing clusters in the Zigbee Cluster Library. These clusters provide additional messages and attributes for managing the node and its relationship to other nodes in the network. Device Profile Clusters are discussed in the next section and Section 7.2.7.3.

7.2.6 Zigbee Device Profile Clusters

The bulk of the functionality provided by the ZDO is implemented by the Zigbee Device Profile Clusters. These clusters also have interfaces into the application support sublayer and the network layer. Unlike clusters in the Zigbee Cluster Library, clusters in the Zigbee Device Profile Clusters only define messaging commands, and they do not define any attributes for state maintenance. All Zigbee devices are required to implement the Zigbee Device Profile Clusters, which are described in Section 2.4 of the Zigbee Specification [37].

As with any cluster, the Device Profile Clusters support two classes of services: client services and server services. The client services implement the transport of node and service discovery requests, binding and unbinding requests, and network management requests from the client to server and the receipt of responses from the server. The client services also support a restricted mode, in which the node only accepts a request depending on the source address, and encryption of the incoming message. The server services implement the processing of client node and service discovery requests, binding and unbinding requests, and network management requests, and the transmission of responses with the results of the processing back to the client.

The Zigbee Device Profile Clusters described in the Revision 23 specification support collections of clusters having the following functions:

- Node and Service Discovery – implements messages to discover information about other nodes including their addresses and what services they offer,
- Bind, Unbind and Binding Management – implements messages for binding and unbinding of a source and destination node, and clearing the binding table,
- Network Management – implements messages requesting various network management functions from another node such as obtaining the routing table, obtaining the binding table, etc.,
- Security Management – implements messages involved in security such as key negotiation, etc.

Note that these cluster collections correspond to the functionality provided by the ZDO to the application layer discussed in Section 7.2.5, except for the lack of a cluster implementing group management functionality in the Zigbee Device Profile Clusters. Group management is implemented in the ZDO by the Zigbee Cluster Library *Groups* cluster.

7.2.7 Zigbee Cluster Library

The Zigbee Cluster Library contains clusters that support particular use cases, both application use cases, such as energy management, and use cases related to managing the node and network interaction on the device. A cluster is a group of messages and attributes that implement functions specific to a use case. The messages are commands sent from the client to the server and from the server to the client. In a type 1 cluster, the client initiates messages to the server, whereas in a type 2 cluster, the server initiates messages to the client. The attributes are data values which represent a physical quantity or state. A cluster is assigned a standardized 16-bit cluster id within the Zigbee Cluster Library. Message type codes and attributes for clusters are specified in the Zigbee Cluster Library standards' document [28].

7.2.7.1 Cluster Library Message Header

Cluster library messages are included in the payload of application support sublayer messages and have their own header, as shown in Figure 7.8. The fields include the following:

- Frame Control – the Frame Control field is an 8-bit bit field with bits set to one indicating how to process the packet, see below for a brief explanation of the contents,

Frame Control (8 bits)	Manufacturer Code (0 or 16 bits)	Transaction Seq. Num. (8 bits)	Command Code (8 bits)

Figure 7.8 Zigbee Cluster Library Message Header

Message Scope (2 bits)	Manufacturer Specific (1 bit)	ACK Format (1 bit)	Direction (1 bit)	Disable Default Resp. (1 bit)	Reserved (2 bits)

Figure 7.9 Zigbee Cluster Library Frame Control Field

- Manufacturer Code – a manufacturer code for proprietary extensions. This field is absent if the appropriate bit in the Frame Control field is not set,
- Transaction Seq. Num. (Sequence Number) – a sequence number set to the same value for all packets in a transaction. The same value is used for a query and all response packets generated in response to the query. Incremented at the beginning of a new transaction,
- Command Code – either a common command code delivered to all clusters or a cluster specific command code, depending on the value of the appropriate bit in the Frame Control field.

Figure 7.9 shows the fields in the Frame Control Field. Only the first six bits are used, and the rest are reserved and must be set to zero. The fields have the following functions:

- Message Scope – indicates the message scope. If the bits are set to 0b00, then the scope is global including manufacturer proprietary clusters. If the bits are set to 0b01, then the message is local to a specific cluster,
- Manufacturer Specific – indicates that the message is a proprietary, manufacturer specific message if set and that the manufacturer code is in the header,
- ACK Format – the message has been sent from the server to the client if set; otherwise, the message has been sent from the client to the server,
- Disable Default Resp. (Response) – disables the default response if set unless there is an error. If bit not set, the default response is returned.

7.2.7.2 Cluster Library Global Message Types

The cluster library supports 22 global message types that are delivered to all clusters on a node. As mentioned in Section 7.2.7.1, a global cluster library message is indicated in the cluster library message header Frame Control Field by setting the first two bits of the field to 0b00. These messages support general classes of operations of interest to all clusters. Most of the commands are associated with attributes: reading and writing attributes, discovering the data type and access control of attributes, configuring attribute reporting, and reporting attribute values at periodic intervals. Other messages include messages to query and report back the commands received and sent by a node. The *Default Response* is used to return a status code indicating the success or failure of messages received with the same transaction identifier. The status codes are listed in Table 2–12 of the Zigbee Cluster Library specification [38], and cluster library global messages are described in Section 2.5 [39] of the specification.

7.2.7.3 Clusters for Node and Network Management

In addition to clusters supporting application use cases, the cluster library contains a collection of clusters that provide node support for all applications. These clusters include:

- *Basic* – contains attributes such as Zigbee Cluster Library version number, manufacturer id, etc., describing the device, and one command to perform a factory reset of the device when it is received, changing all the attributes back to their original values when the device left the factory. This cluster also includes an attribute allowing the device to be disabled when set,
- *Power Configuration* – attributes containing information about the device's power source (battery, grid, etc.) and messages for configuring under- and over-voltage alarms,
- *Device Temperature Configuration* – attributes for determining the device's internal temperature and for configuring alarms when the temperature exceeds the device's operating range,
- *Identify* – attributes and commands to switch the device into identification mode, causing a light to flash or some other physical signal, indicating to an installer which device has been addressed. The command returns a reply with identifying information to the sender. This cluster cannot be disabled,
- *Group* – contains a collection of messages for remote management and commissioning of groups. Messages are included for discovering the group membership of a node, adding a group with an optional group name, requesting information on a group by name, removing a group and removing all groups. Note that configuration of group addresses from application objects running in the node's application framework uses the application support sublayer binding mechanism, not this cluster, to define a new group address. The group table is also managed by the application support sublayer, not this cluster.

The Zigbee Cluster Library includes a few other clusters that address advanced node and network management functions like OTA upgrade, commissioning, and tracking the received signal strength of a device's wireless signal.

7.2.7.4 Clusters for Energy Management Applications

The Zigbee Cluster Library contains a wealth of clusters that are applicable to BEMS applications. The Zigbee Cluster Library specification documents the BEMS applicable clusters in chapters reflecting higher level application categories such as measurement and sensing, lighting, HVAC, closures (doors, windows, and blinds), smart energy, and appliances. In addition, the cluster library specification has chapters defining clusters supporting interoperability with BACnet, allowing the BEMS device layer in larger commercial and public buildings to incorporate both Zigbee and BACnet devices. Such networks can be particularly useful in a building where an existing BACnet BAS is in place, but additional devices are required to implement additional sensor monitoring or control. Installing a few Zigbee devices could be a lower cost option than upgrading the entire BACnet network.

Some clusters provide basic functionality for energy control and monitoring devices, such as modeling a switch. These clusters are:

- *Scenes* – a scene is a predefined set of attribute values for one or more nodes. The *Scenes* cluster allows these settings to be saved, recalled, and removed, simplifying complex multi-device configurations,
- *On/Off* – allows devices to be changed between an activated and a deactivated state,
- *On/Off Switch Configuration* – manages the behavior of on/off switching devices, useful in controlling lighting and other appliances for energy efficiency,
- *Level* – controls nodes that have variable levels, such as dimmable lights or motorized blinds,
- *Level Control for Lighting* – extends the *Level* cluster with attributes and commands specifically for controlling a lighting device,
- *Time* – enables time synchronization across devices, useful for scheduling energy management operations,
- *Alarms* – sends alerts or alarms related to energy management, such as consumption thresholds being reached.

In addition, the Zigbee Cluster Library specification includes a chapter with descriptions of clusters for measuring various physical parameters, some of which could be useful in energy management applications.

More complex clusters provide functions useful for deploying in a BEMS. These include:

- *Demand Response and Load Control* – manages energy demand by sending messages to devices to reduce energy consumption during peak times. Section 7.2.8 examines this cluster in more detail,
- *Price* – conveys real-time pricing information to energy consuming devices, allowing them to make decisions about energy use based on pricing,

- *Metering* – monitors energy consumption through smart meters, reporting variables like current power, total consumption, and demand,
- *Electrical Measurement* – enables measurement of electrical parameters such as voltage, current, frequency, and power factor,
- *Power Profile* – provides information about the expected power consumption profile of a device, allowing better scheduling and energy management,
- *Prepayment* – helps manage credit and energy consumption rates when energy prepayment services are involved,
- *Thermostat* – controls HVAC systems to manage energy use efficiently,
- *Pump Configuration and Control* – manages the configuration and control of pumps integral to pools, HVAC and water heating systems,
- *Shade Configuration* – allows control over motorized shades for regulating heat and light,
- *Appliance Control* – provides a collection of messages for controlling appliances. Additional clusters involving appliance control include *Appliance Identification*, *Appliance Events and Alerts*, and *Appliance Statistics*.

7.2.8 *Demand Response and Load Control (DLRC)* Cluster

An example of a cluster applicable to BEMS applications is the *DRLC* cluster, specified in Section 10.3 of the Zigbee Cluster Library document [40]. Chapter 10 of the Cluster Library specification additionally defines other clusters for energy management. This cluster is useful for implementing control of flexible load devices in residential and small commercial buildings and is similar to the BACnet *Load Control* object for large commercial and public buildings discussed in Section 6.7.5.2 of Chapter 6. The *DRLC* cluster is a type 2 cluster, with the server originating messages to the client, and has cluster identifier 0×0701.

Figure 7.10 shows the system architecture of a residential building energy control network, in which the devices and residential energy management gateway use the *DRLC* cluster to manage load, controlled through a cloud-based BEMS run by an aggregator. An alternative is for the utility's demand response service to talk directly to the residential energy management gateway. The *DRLC* cluster is a type 2 cluster due to the gateway acting as a proxy for the upstream BEMS or utility demand response service, and having to store state or do translations from the utility demand response signaling into the message format of the *DRLC* cluster. The *DRLC* server receives event notifications from the upstream demand response event originator, records them in a table, and conveys the event notifications to the appropriate *DRLC* client devices downstream.

The *DRLC* cluster has provisions for grouping devices into classes to simplify demand response event management. For example, hot water heating, HVAC equipment, and pool pumps are three types of high energy consuming devices

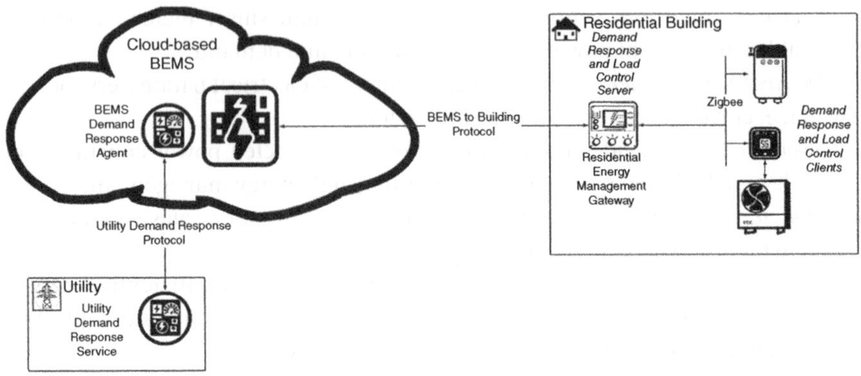

Figure 7.10 Architecture of the *DRLC* Cluster

that can be grouped into separate classes. These devices have separate time of use demands, which do not fully overlap. Hot water heating typically needs to be scheduled to run during the early morning and late afternoon hours outside of the peak grid load, so water is available for morning showers and evening dishwashing. Pool pumps have no constraints on when they can be run, so they can be scheduled for times when power is generally cheaper and has lower marginal carbon emissions. HVAC equipment needs to be scheduled to run outside of grid peak periods, usually in the late afternoon and early evening, when power can be very expensive and marginal carbon emissions are high. But the building can be precooled or preheated somewhat above or below the normal setpoint during the afternoon when grid marginal carbon emissions are lower due to high solar content so that the temperature stays in the comfortable zone during the evening hours. Separating these devices into three separate classes allows the BEMS to accommodate the preferences of the building occupants.

Because some of the commands in the cluster contain timestamps, the cluster assumes that the gateway and devices have access to real-time clocks which they can synchronize. If a device does not have a real-time clock, it should ignore all timestamps with the exception of a "Start Now" message. However, the device is still required to approximate duration in order to estimate the duration of a requested demand response event. In addition, all temperature fields in messages are in units of Celsius, so measurements in Fahrenheit need to be converted before being sent.

7.2.8.1 *DRLC* Client and Server Attributes

The *DRLC* server has no attributes. The *DRLC* client supports the attributes shown in Table 7.7. All client attributers are mandatory and have read–write access by default. Section 10.3.3.2 in the Zigbee Cluster Library specification [41] describes

Table 7.7 DRLC Client Attributes

Attribute ID	Attribute Name	Attribute Type	Range	Default Value
0×0000	*UtilityEnrollmentGroup*	Unsigned 8-bit integer	0×00-0×ff	0×00
0×0001	*StartRandomizationMinutes*	Unsigned 8-bit integer	0×00–0×3c (0–60)	0×1e(30)
0×0002	*DurationRandomizationMinutes*	Unsigned 8-bit integer	0×00–0×3c (0–60)	0×00
0×0003	*DeviceClassValue*	Unsigned 16-bit integer	0×0000–0×ffff	Not Applicable

the client attributes. These attributes can be configured on the *DRLC* client to control how it responds to demand response events.

The attributes have the following functions:

- *UtilityEnrollmentGroup* – an identifier for grouping devices into arbitrary groups. For example, a utility may run separate programs for residential and commercial customers and use *UtilityEnrollmentGroup* to differentiate between the two. When the *DRLC* server sends out demand response events, it can target these to specific *UtilityEnrollmentGroups* by setting the Utility Enrollment Group field in the *Demand Response Event* message announcing the event to the group at which the event is targeted, allowing for more granular control over the demand response actions of the connected devices. The value 0×00 (the default value) indicates that the device is a member of all groups,
- *StartRandomizationMinutes* – used to introduce a degree of randomness in the actual start time of a demand response event at the client side. The attribute specifies a time window in minutes, within which the client randomly selects a start time for the event. This randomization helps to prevent simultaneous activation or deactivation of multiple devices, thereby avoiding sudden spikes or drops in electrical load. For example, if a demand response event is scheduled to start at 12:00 PM and *StartRandomizationMinutes* is set to 10 minutes, the client device will randomly choose a start time between 12:00 PM and 12:10 PM to begin the demand response event,
- *DurationRandomizationMinutes* – introduces a level of randomness in the duration of the demand response event. The attribute specifies a time window in minutes, within which the client will randomly select the actual duration for the event, helping to mitigate sudden spikes or drops in electrical load when multiple devices end a demand response event at the same time. For example,

if a demand response event has a planned duration of 30 minutes and *DurationRandomizationMinutes* is set to 5 minutes, the client will randomly extend or shorten the event duration by up to 5 minutes. This means the event could last anywhere between 25 and 35 minutes,

- *DeviceClassValue* – a bitmap whose fields specifying what class of device the client represents. Like *UtilityEnrollmentGroup*, this identifier allows the event to be selectively targeted at specific devices. Although the attribute is marked as having read–write access, the device can refuse to change the value by setting the status field of the global cluster library message write attribute record to NOT_AUTHORIZED. Table 7.9 in Section 7.2.8.3 indicates the valid bitmap settings. A device like an in-home display that monitors events for all device types should set this attribute to 0×ffff. The *DeviceClassValue* and *UtilityEnrollmentGroup* attributes are used together by the client to decide which *Load Control Event* and *Cancel Load Control Event* messages to accept and which to ignore.

7.2.8.2 *DRLC* Client and Server Message Types

The *DRLC* client and server message types are listed in Table 7.8. The Command Code values for these messages appear in the Command Code field of the cluster library message header shown in Figure 7.8. All messages are mandatory to implement.

Clients are expected to maintain an event table with a minimum of three events. If additional events are received that do not match existing events and there is no space in the event table, the client should ignore the new events and retrieve them later using the *Get Scheduled Events* message when there is space in the event table.

7.2.8.3 *DRLC* Server *Load Control Event* Message

The *DRLC* cluster defines message formats for messages between the client and server. Many of the fields in the messages use an encoding to reduce the number of bits in the value, and some of the fields are optional. In the descriptions of the fields below, if a field is not designated as optional, it is mandatory.

Table 7.8 DRLC Client and Server Message Types

Message	Code	Sender	Response Message
Load Control Event	0×00	Server	*Report Event Status*
Cancel Load Control Event	0×01	Server	*Report Event Status*
Cancel All Load Control Events	0×02	Server	*Report Event Status*
Report Event Status	0×00	Client	Not Applicable
Get Scheduled Events	0×01	Client	*Load Control Event*

Issuer Event ID (32 bits)	Device Class (16 bits)	Utility Enrollment Group (8 bits)	Start Time (32 bits)	Duration (16 bits)	Criticality Level (8 bits)	Cooling Temp. Offset (8 bits)
Heating Temp. Offset (8 bits)	Cooling Temp. Setpoint (16 bits)	Heating Temp. Setpoint (16 bits)	Ave. Load Adjustment % (8 bits)	Duty Cycle (8 bits)	Event Control (8 bits)	

Figure 7.11 *DRLC* Server *Load Control Event* Message Body Format

The *DRLC* server *Load Control Event* message is issued to clients when the server receives notification from an upstream demand response event originator that an event has been scheduled. As indicated in Table 7.8, the clients respond with a *Report Event Status* message. Figure 7.11 shows the message format for the *DRLC* server *Load Control Event* message. The fields define the following values:

- Issuer Event ID – an event identifier assigned by the upstream demand response event originator uniquely identifying the event. The contents can be a random identifier or a UTC timestamp. If two events overlap, the newer event should have a larger identifier than the older. If the Device Class and/or the Utility Enrollment Group overlaps between the two events, then the newer event supersedes the older,
- Device Class – a bit-encoded field representing the class of the device to which the event applies. Multiple bits can be set, and each set bit indicates that the corresponding device class must participate in the event. The impacted device class can be different from the controlling device which is the target of the *DRLC* message. For example, a thermostat is the controlling device for a heat pump, but the heat pump would be the impacted device. The DRLC message is directed to the thermostat, which acts on behalf of the heat pump. The device classes only represent impacted devices. Bit flag locations from the left side of the field and corresponding device types are listed in Table 7.9,
- Utility Enrollment Group – used together with the Device Class field to group devices into utility programs or geographic areas. For example, a utility might have one demand response program for residential users and one for commercial and industrial users, with different Utility Enrollment Group values. A value of 0×0 means the message applies to all classes,
- Start Time – the UTC time for the start of the event. A value of 0×00000000 indicates "Start Now". If the event is retransmitted, it should continue to use "Start Now" for the start time but adjust the duration. The *DRLC* server should record the actual start time when storing the event in its event table,
- Duration – the event duration in minutes. The maximum value is 1440 which corresponds to one day,
- Criticality Level – defines the level of criticality for the event. Table 7.10 lists valid values. The action taken by a *DRLC* client can depend entirely on this field or can take other fields into consideration. The criticality level of 0×00 and

Table 7.9 Device Class Bit Flags for *DRLC* Server *Load Control Event* Message

Bit Location	Device Class
0	Air Conditioner, Heat Pump or Furnace
1	Resistance Heaters
2	Hot Water Heater
3	Pool Pump or Spa
4	Smart Appliance
5	Irrigation Pump
6	Managed Commercial and Industrial Loads
7	Residential Plug (On/Off) Loads
8	Outside Lighting
9	Inside Lighting
10	Electric Vehicle Charging/Discharging
11	Generation Systems

Table 7.10 Criticality Level for DRLC Server Load Control Event Message

Criticality Level	Description	Participation
0×01	Green	Voluntary
0×02	1	Voluntary
0×03	2	Voluntary
0×04	3	Voluntary
0×05	4	Voluntary
0×06	5	Voluntary
0×07	Emergency	Mandatory
0×08	Planned Outage	Mandatory
0×09	Service Disconnect	Mandatory

0×ff are reserved for future CSA use, and 0×0a-0×0f are reserved for definition by the upstream demand response event originator. The Green event, level 0×1, is especially useful for building decarbonization since it indicates that the grid electricity comes from nonrenewable sources. Levels 0×02-0×06 indicate that the upstream provider requires increasingly higher levels of load reduction. Level 0×07 indicates that the grid is experiencing an emergency and all nonessential loads should be shed. What constitutes an essential load is defined by the utility. Levels 0×08 and 0×09 indicate that electricity from the grid will shut down and could be used by the residential energy management gateway to switch to a residential battery or other local backup,

- Cooling Temp. Offset (Optional) – the offset to apply to the normal cooling setpoint, in units of +0.1 °C, when the event starts,
- Heating Temp. Offset (Optional)– the offset to apply to the normal heating setpoint, in units of −0.1 °C, when the event starts,
- Cooling Temp. Setpoint (Optional)– the cooling setpoint in units of 0.1 °C,
- Heating Temp. Setpoint (Optional)– the heating setpoint in units of 0.1 °C,
- Ave. Load Adjustment % (Optional) – the maximum energy use defined as a percentage of the *DRLC* client's average energy use. The field ranges from −100% – +100%, with −100% indicating complete load shed, 0% indicating average energy use, and +100% indicating double the average energy use. A value of 0×80 indicates that the field is not used,
- Duty Cycle (Optional) – the maximum percentage of time the *DRLC* client should remain switched on for the duration of the event. For example, a value of 0×32 (50) indicates that the client should remain on half the time. The value ranges from 0 to 100 (0×00-0×64), and a value of 0×ff indicates that the field is not used. A Duty Cycle of three minutes is recommended for most devices and an off period must precede an on period. Some devices (e.g., heat pumps) may be restricted in how quickly they can switch between on and off states,
- Event Control – a bitfield with control options for randomizing the response to the event. Bit 1 indicates that the start time should be randomized, while bit 2 indicates that the duration should be randomized. The client uses randomization attributes together with this field to determine if the start and stop time, and duration should be randomized. By default, *DRLC* clients will randomize start times but not durations, with the start and end times randomized by the same amount for all demand response participants, so that the events are equal length for all participants. For shedding load quickly (Criticality Level 0×07), start randomization is not applied. Duration randomization is applied to slowly add load back to the grid, to avoid issues with instantaneous overload.

The heating and cooling temperature offsets and setpoints are encoded as unsigned 8-bit integers. To calculate the offset in °C, divide the offset field value by 10. The offset is limited to 0–24.5 °C. For example, a field value of 0×0a (10) is equivalent to 1 degree Celsius. The sign is determined by whether the field is a cooling offset (sign is +) or a heating offset (sign is −) The setpoint field can have values between 0×00 and 0×fe (0–25.4 °C), with 0×ff indicating that the field is not used.

The setpoints are typed as signed 16-bit integers. To calculate the setpoint, divide the setpoint field value by 100 for the setpoint temperature. For example, if the heating setpoint is 0×0708 (1800 decimal), dividing by 100 gives a setpoint temperature of 18 °C. The range of the field is 0×954d–0×7fff (−273.15 – 327.67 °C). A value of 0×8000 indicates that the field is not used.

Even though they are listed as optional, the offset and setpoint fields are required in every message. If a field is not used, it should be set to the respective value indicating that the field is not used (0×ff for offsets and 0×8000 for setpoints).

7.2.8.4 *DRLC* Server *Cancel Load Control Event* Message

The *DRLC* server *Cancel Load Control Event* message is issued by the server when the upstream demand response event originator has cancelled an event. The clients respond with a *Report Event Status* command indicating that the event has been cancelled. If a client receives a *Cancel Load Control Event* message after the event has ended, it must respond with a *Report Event Status* message with the Event Status field set to "Rejected – Invalid Cancel Command (Undefined Event)". All fields are mandatory.

Figure 7.12 illustrates the message body format. The Issuer Event ID, Device Class, and Utility Enrollment Group are defined as in the *Load Control Event* message described in Section 7.2.8.3. The Device Class and Utility Enrollment Group fields must have the same value as in the *Load Control Event* message with the Issuer Event ID matching the ID that established the event. If the Device Class and/or Utility Enrollment Group fields do not match for a particular device, the command should be ignored.

The remaining fields have the following values:

- Cancel Control – a bitmap in which only the first bit is currently used. If the bit is set to 0 and the event is in progress, the randomization applied to the event creation should be overridden and the event should be terminated immediately. If the bit is set to 1, the original randomization should be used to terminate the event.
- Effective Time – deprecated, must be set to 0×00000000 indicating that the cancellation is effective immediately.

7.2.8.5 *DRLC* Server *Cancel All Events* Message

When the server wants to cancel all load control events, it issues the *Cancel All Events* message. The format of the message is shown in Figure 7.13. The message consists of a single field, the Cancel Control bitfield. The Cancel Control field has

Issuer Event ID (32 bits)	Device Class (16 bits)	Utility Enrollment Group (8 bits)	Cancel Control (8 bits)	Effective Time (32 bits)

Figure 7.12 DRLC Server Cancel Load Control Event Message Body Format

the same value as in the *Cancel Event* message described in Section 7.2.8.4. The clients respond with a *Report Event Status* message indicating that all events have been cancelled.

Cancel Control (8 bits)

Figure 7.13 *DRLC* Server *Cancel All Events* Message Body Format

7.2.8.6 *DRLC* Client *Report Event Status* Message

Figure 7.14 shows the format for the *DRLC* client *Report Event Status* message body. The *DRLC* client sends this message to the server whenever there is a change in state for an active event. The Issuer Event ID, Criticality Level Applied, Cooling Temp. Setpoint Applied, Heating Temp. Setpoint Applied, Ave. Load Adjustment % Applied, Duty Cycle Applied, and Event Control fields are the same as those in the server side *Load Control Event*, described in Section 7.2.8.3, except these fields contain the actual values applied by the client. The client can also set the fields to one of the values indicating that it was not used, for example, 0×80 for the Ave. Load Adjustment % Applied.

The fields that are not the same as in the *Load Control Event* message are:

- Event Status – a code indicating the status of the event. Possible codes are listed in Table 7.11. If a client issues an "Opt-Out" or "Opt-In" status (codes 0×04 and 0×05) for an event that is eventually cancelled, the event is recorded as cancelled at the scheduled time. If a client issues an "Opt-Out" or "Opt-In" status for an event that is not cancelled, the event is listed as partially completed based on the last *Report Event Status* message sent with codes 0×08 and 0×09. If a client sends a *Report Event Status* message with a status of "Rejected – Invalid Cancel Command (Undefined Event)" (code 0×fd), the optional fields should report the values indicating that the field should be ignored, for example, 0×80000 for Heating Temp. Setpoint Applied,
- Event Status Time – the UTC time at which the event status being signaled occurred,
- Signature Type – the type of the signature in the Digital Signature field. A value of 0×00 indicates that there is no signature and that the field is absent. A value of 0×01 indicates that the signature is calculated using an elliptic curve digital signature algorithm,
- Digital Signature – an optional field containing a digital signature for verifying the message.

Issuer Event ID (32 bits)		Event Status (8 bits)	Event Status Time (32 bits)	Criticality Level Applied (8 bits)	Cooling Temp. Setpoint Applied (16 bits)
Heating Temp. Setpoint Applied (16 bits)	Ave. Load Adjustment % Applied (8 bits)	Duty Cycle Applied (8 bits)	Event Control (8 bits)	Signature Type (8 bits)	Digital Signature (336 bits)

Figure 7.14 *DRLC* Client *Report Event Status* Message Body Format

Table 7.11 Event Status Codes for *DRLC* Client *Report Event Status* Message

Event Status Code	Description
0x01	Event Received.
0x02	Event Started.
0x03	Event Completed.
0x04	Opt-Out.
0x05	Opt-In.
0x06	Event Cancelled.
0x07	Event Superseded.
0x08	Event Partially Completed with Client Opt-Out.
0x09	Event Partially Completed with Client Opt-In.
0x0a	Event Completed – Client Didn't Participate due to a Previous Opt-Out.
0xf8	Rejected – Invalid Cancel Command (Default).
0xf9	Rejected – Invalid Cancel Command (Invalid Effective Time).
0xfb	Rejected – Event Received after It Expired (i.e., current time > Start Time + Duration).
0xfd	Rejected – Invalid Cancel Command (Undefined Event).
0xfe	Rejected.

The Zigbee Cluster Library specification does not provide many details on how to calculate the digital signature. It does say that the Matyas–Meyer–Oseas hash function specified in Annex B.4 of the base Zigbee Specification [42] should be used in place of SHA-1 and specifies how to calculate the signature, but it does not say how the key pair should be provisioned nor which elliptic curve algorithm should be used. The *Key Establishment* cluster in Chapter 10 of the Zigbee Cluster Library specification [43] and Annex C in the Zigbee Smart Energy Standard [44] profile specification discusses certificate provisioning for certificate-based shared key establishment and could be used for guidance in the absence of any further information.

Response messages to the *DRLC* server should be delayed for a random amount of time between 0 and 5 seconds to avoid network congestion due to all clients reporting at the same time. On the server side, duplicate *Report Event Status* messages should be ignored. Duplicates can be identified by the Sequence Number field in the Zigbee network layer header discussed in Section 7.2.4.1.

Minimum Start Time (32 bits)	Number of Events (8 bits)	Issuer Event ID (32 bits)

Figure 7.15 *DRLC* Client *Get Scheduled Events* Message Body Format

7.2.8.7 *DRLC* Client *Get Scheduled Events* Message

The *DRLC* client *Get Scheduled Events* message provides a way for clients to retrieve messages from the server. The format for the *Get Scheduled Events* message body is shown in Figure 7.15. The Minimum Start Time field is the UTC starting time for events on the server from when the client would like unrecorded events. The Number of Events field indicates the maximum number of events the server should return. A value of 0×00000000 in the Minimum Start Time field has no special meaning, but a value of 0×00 in the Number of Events field indicates that all events from the Minimum Start Time to the present time should be returned.

The Issuer Event ID field is optional, and a value of 0×ffffffff indicates that the server should not use this field in deciding which events to return. The Issuer Event ID is used to identify unique demand response events for the *DRLC* server to return and acts as a starting point. The server responds with events that have an Issuer Event ID greater than or equal to the specified ID, effectively filtering out older or irrelevant events. When both the Issuer Event ID and Minimum Start Time fields are used, the *DRLC* server filters events to meet both criteria.

The client issues this message to retrieve messages that may have been dropped due to not having any space in the event table or if it wants to verify the events in its event table. It can also be used when the client joins the network or reboots due to loss of power or is reset to obtain any pending events from the DRLC server. If no matching events are available, the server returns the cluster library global message *Default Response* with a status code set to NOT_FOUND.

7.2.9 Example *DRLC* Cluster Event Message Exchange Scenarios

This subsection presents two examples of message exchange scenarios between the BEMS, residential energy management gateway *DRLC* server, and an HVAC thermostat client to illustrate how the *DRLC* cluster handles load control. The first example shows a successful demand response event, the second shows a demand response event that is overridden by a building occupant. In the diagrams, the residential building energy management gateway uses Zigbee to communicate load control events received from the BEMS to the HVAC units under its control. The protocol used to communicate between the BEMS and the residential building energy management gateways under its supervisory control is left unspecified but is discussed in more detail in Chapters 8 and 9.

7.2.9.1 Successful Demand Response Event Executed

The first scenario is illustrated in Figure 7.16. The exchange starts with the BEMS receiving a new demand response event from the utility having a unique Issuer Event ID. The Utility Enrollment Group for the event indicates that the event applies to residential customers and that it requires HVAC units to turn down their power during the period from 4 PM to 7 PM that day which is during summer. The BEMS notifies all the energy gateways in the residential buildings under its control through the Internet.

The residential energy management gateway receives the event and records it in its event table. Normally, at that time during the summer, the HVAC unit setpoint is 72F and cooling. The residential energy management gateway sends a DLRC cluster library server-side Load Control Event to the HVAC thermostat in the building with the Issuer Event ID sent by the BEMS. The Start Time field is set to 4 PM, and the Duration field is set to 180 minutes (3 hours), so the event will conclude at 7 PM. The Cooling Temp. Offset field is set to +4F, indicating that the thermostat should increase the cooling setpoint to 76F for the duration of the event. The Event Control field has both bits 1 and 2 set to 1, indicating that the thermostat should randomize both the start time and duration of the event. All other optional fields are set to their ignore values.

The thermostat receives the Load Control Event message, then chooses a random start time and duration within its start time and duration windows that are the values for its *StartRandomizationMinutes* and *DurationRandomization Minues* attributes, respectively, and inserts the event into its event table. The thermostat returns a *Report Event Status* message with the time the message was received, the cooling temperature setpoint for the duration of the event, and a status code of 0×00, "Event Received."

At the chosen randomized start time, it moves the setpoint up to 76F, reporting back to the building energy management gateway about the start of the event including the randomized start time with a *Report Event Status* message having code 0×02, "Event Started". After the randomized duration has expired, the thermostat decreases the setpoint to 72F and reports back to the building energy management gateway about the event terminated with a *Report Event Status* message having code 0×03, "Event Completed". The building energy management gateway then reports the results back to the BEMS.

7.2.9.2 Demand Response Event Terminated Due to Occupant Override

The second scenario showing a demand response event terminated due to occupant override is illustrated in Figure 7.17. The first part of the message exchange is exactly as in Figure 7.16, until the event starts. Shortly, after the HVAC cooling setpoint is turned up to 76F, the occupant notices that the air-conditioning has gone off, so they turn the setpoint back down to 72F. The thermostat reports the

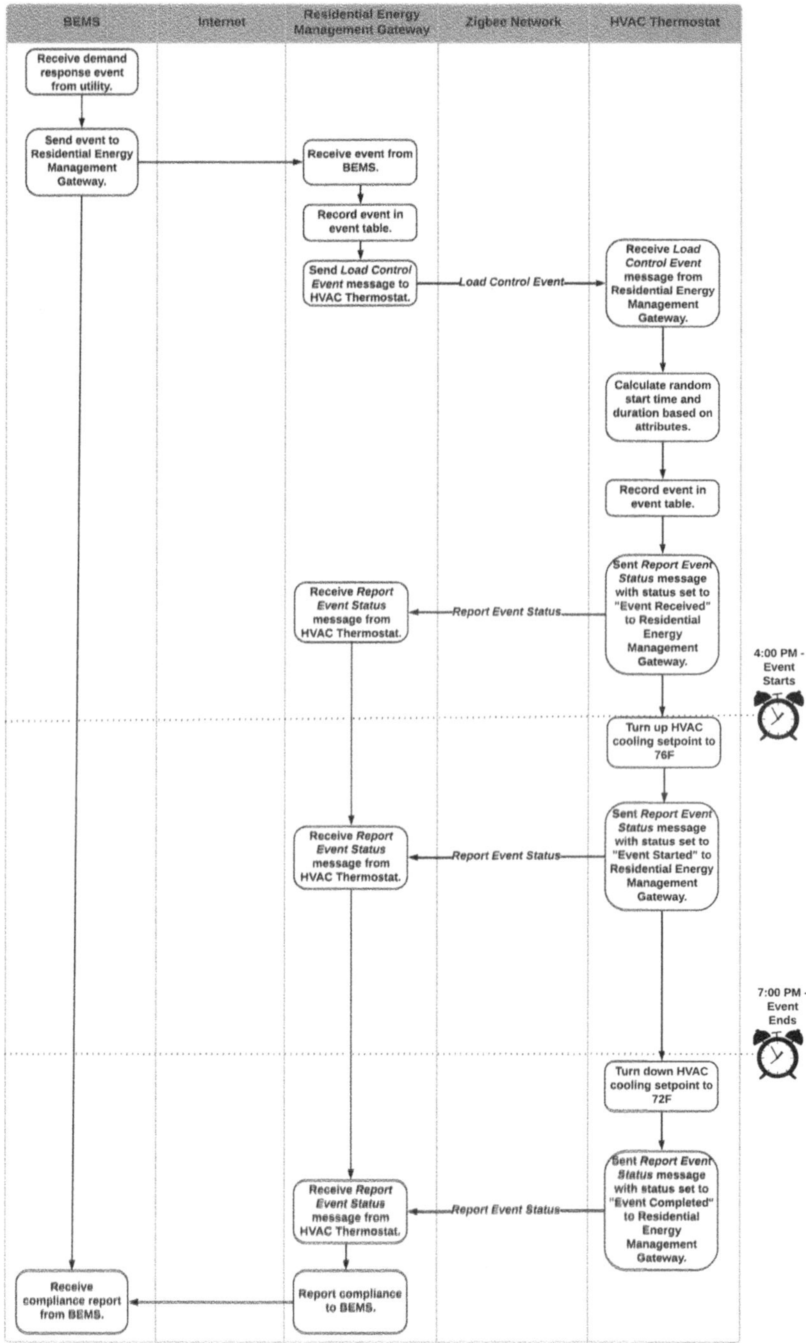

Figure 7.16 Successful DLRC Load Control Event Message Exchange

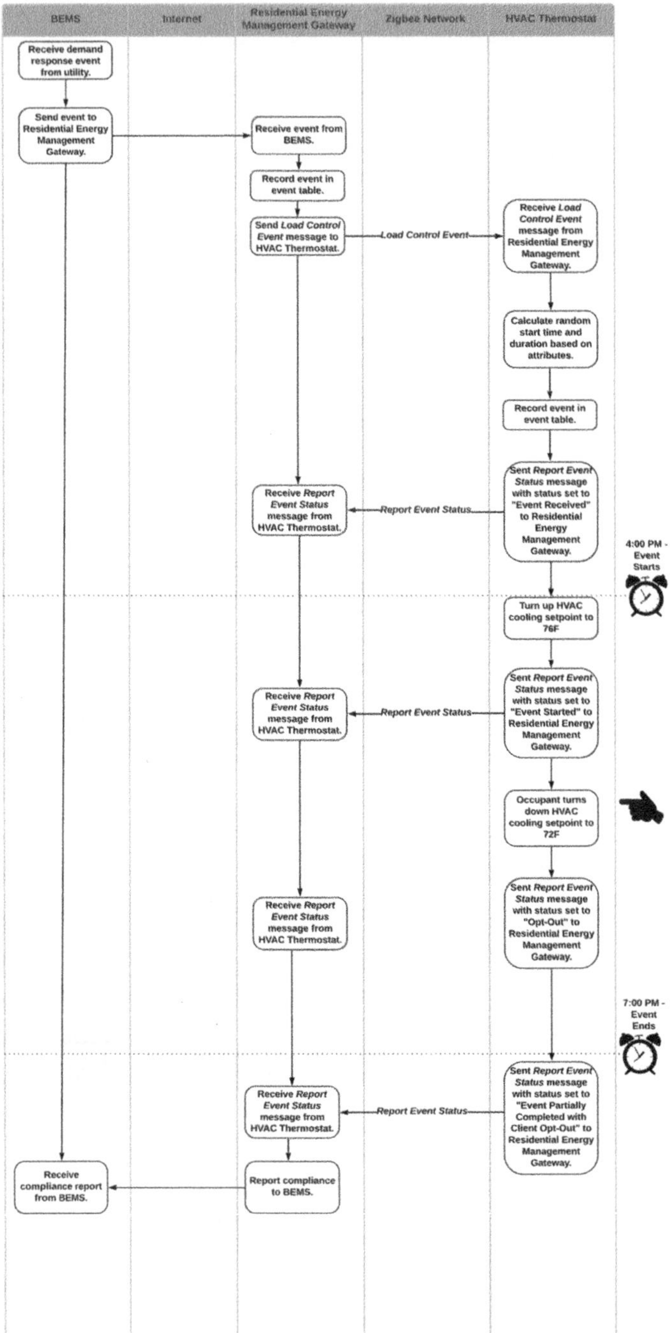

Figure 7.17 Terminated DLRC Load Control Event Message Exchange

override back to the building energy management controller with a *Report Event Status* message, having the Event Status field set to 0×04, "Opt-Out", the Cooling Temp. Setpoint Field set to 72F, and the time set to the time at which the occupant chose to turn down the thermostat setpoint.

When the event concludes at 7 PM, the thermostat sends a *Report Event Status* message to the building energy management controller with the Cooling Temp. Setpoint Field set to 72F, the time set to 7 PM, and the Event Status set to "Event Partially Completed with Client Opt-Out". The building energy management controller sends a report back to the BEMS as in Figure 7.16.

7.2.10 Smart Energy Profile

The CSA still maintains one application profile independent from the Zigbee Cluster Library and the Basic Device Behavior specifications, the Zigbee Smart Energy Standard [44] (profile identifier 0×0109). Although the Smart Energy profile is separately defined, it uses clusters from the cluster library. The Smart Energy profile has been deployed by a number of electric utilities in their AMI, including Pacific Gas and Electric in California and BC Hydro in Canada, to provide commercial and retail customers real-time information on the electricity use. While smart energy clusters from the cluster library can be used independently from the Smart Energy profile, the Smart Energy profile may be useful for BEMS applications if the local utility has deployed it. The Smart Energy profile specifies clusters required for devices and standard networking practices for smart energy applications in residential or commercial buildings. Device interfaces in the Zigbee Smart Energy Standard are based on the cluster specifications in the Zigbee Cluster Library. The *DLRC* cluster is an example.

The application domains of the Smart Energy Standard include metering, pricing, scheduling, messaging, demand response, and load control. Procedures and protocols are specified for commissioning and managing such networks, sometimes called Home Area Networks (HANs), and specifications are provided for operating in both credit and prepayment modes. The specification also includes information on inter-PAN networking, which may be required to connect a HAN into an energy service provider network. The intent of the specification is to standardize clusters and network operating procedures allowing interoperability among Zigbee devices produced by different manufacturers and among services provided by different smart energy service providers.

7.2.11 Interoperability Specifications

The CSA has defined two specifications to improve interoperability between Zigbee devices deployed on the same network but implementing different applications and to enable device initialization and network formation between Zigbee and Bluetooth. The Zigbee PRO Base Device Behavior (BDB) specification [45] is the latest version of the device interoperability specification originally introduced with Zigbee 3.0. Zigbee Direct [46] contains a one-way (Zigbee to Bluetooth) interoperability specification with BLE, so Zigbee network layer packets can be transmitted over the Bluetooth wireless and link layers rather than the 802.15.4 wireless and link layers.

7.2.11.1 Zigbee Base Device Behavior (BDB) Specification

Prior to Zigbee 3.0, the basic Zigbee device behavior differed depending on the application profile. This hindered the ability of devices implementing different application profiles to interoperate on the same network. With Zigbee 3.0, the base behavior expected of a device was standardized in the Zigbee BDB specification, the latest publication of which is the Zigbee PRO BDB specification [45]. The BDB specifies the environment and a set of standardized procedures for initialization, commissioning, reset, and key provisioning for security to ensure interoperability between devices implementing different applications on the same network.

The BDB specifies the process by which a device initializes itself within the network and the expected behaviors of the three roles – Coordinator, Router, or End Device – that devices can play which were discussed in Section 7.2.1. The default values of attributes for the different device roles and any necessary setup for the device's application are also specified. The BDB defines a standard procedure for joining an existing Zigbee network, including scanning for available networks, selecting a network based on criteria such as signal strength or PAN ID, and authenticating with the network if required. If the device is acting in the coordinator role, the BDB describes how to form a new Zigbee network by selecting an appropriate channel, setting a network PAN ID, and initializing security keys. The BDB also specifies provisions for network security, including processing and distributing encryption and authentication keys.

One important aspect of the BDB is the specification of a standardized procedure for commissioning, the process by which a device is enrolled and integrated into an existing network. Commissioning may be performed when a node is not on a network, on a network but not bound to another node, or on a network and bound to another node. It includes network joining, configuration for security, and establishing connections with other devices within the network implementing the same application. Commissioning ensures that the device is not only physically installed but is also logically integrated and able to communicate with other devices. The BDB defines four different commissioning modes for devices:

- Network Formation – This commissioning mode is specific to Zigbee coordinators and routers. It defines the process by which a coordinator and routers form a new Zigbee network. The coordinator or router node selects the radio channel, establishes the PAN ID, initializes security settings, and sets up other necessary parameters to create the network infrastructure. Once the network is formed, other devices can join using the network steering mode.

- Network Steering – This mode is focused on joining a device to a network and requires no user interaction other than to set the criteria for network discovery. Network steering must be supported by all Zigbee devices. The device first scans the available networks and attempts to join them based on predefined criteria such the PAN ID. Network steering helps in scaling commissioning by allowing devices to find and join an appropriate network with minimal manual intervention.
- Touchlink Commissioning – This mode allows Zigbee devices to be commissioned based on proximity and is used for devices that are in close physical range. Touchlink commissioning is triggered by a user action such as pressing a button. Once a responding device is found, the initiating device proceeds to authenticate the responder using pre-defined shared security keys. After the devices have authenticated to each other, they establish a link and finalize the connection.
- Finding and Binding – Finding allows devices within the network to find the short 16-bit network address based on the IEEE 64-bit address, while binding establishes two unidirectional logical connections as discussed above in Section 7.1.7.2. The BDB standardizes the finding and binding commissioning mode across application profiles.

Finally, to ensure that devices from different manufacturers abide by the BDB, a base device test specification was defined, which any device that is advertised as Zigbee PRO compliant must pass. By adhering to the BDB specification, devices implementing different applications can interoperate on the same network, making network construction, configuration, and commissioning much simpler for building network engineers and building managers.

7.2.11.2 Zigbee Direct

Every laptop computer and smart phone today has a wireless interface supporting BLE and Wi-Fi. In contrast, almost no devices outside of IoT devices have an IEEE 802.15.4 interface. This introduces a barrier to setting up a Zigbee network, particularly a mesh network, and interacting locally with the devices on the network, particularly for commissioning. In recognition of this fact, the CSA developed Zigbee Direct [46], a specification for enabling a device with a BLE interface to perform Zigbee device commissioning and to send and receive Zigbee packets over the Bluetooth wireless link to a Zigbee device with dual wireless interfaces on Bluetooth and IEEE 802.15.4. The Zigbee Direct specification calls the Bluetooth-only device a Zigbee Virtual Device (ZVD) and the dual stack device a Zigbee Direct Device (ZDD) . However, a Zigbee Direct deviced is not a full featured Zigbee to Bluetooth gateway since Bluetooth application packets are not routed between the Zigbee and Bluetooth networks.

Figure 7.18 illustrates the stack architecture for the ZVD and ZDD. As shown in the figure, the ZVD and the ZDD communicate packets over BLE using a point-to-point Bluetooth tunnel. Zigbee Direct specifies three Bluetooth GATT services:

- Zigbee Direct GATT Tunnel Service (Optional) – sets up and maintains a Bluetooth tunnel between the ZVD and the ZDD. Only nodes with the coordinator or router role are permitted to create a Tunnel Service,
- Zigbee Direct GATT Security Service – sets up a secure BLE session using the BLE security protocols described in the Zigbee Direct specification. The service does not use any BLE-specified security,
- Zigbee Direct GATT Configuration Service – allows the ZVD to control a ZDD on the Zigbee network after a secure BLE session having the correct authorization level has been established.

The BLE GATT layer is described in more detail in Section 7.1.4.

In addition to the three BLE GATT services, Zigbee Direct also specifies the *Zigbee Direct Configuration* cluster. The *Zigbee Direct Configuration* cluster provides an interface for configuring Zigbee Direct from a Zigbee application.

The main use for Zigbee Direct is for provisioning a ZDD over the ZVD's BLE link and commissioning it onto a Zigbee network. If the ZDD has its BLE interface turned on but is not yet a member of a Zigbee network, the ZDD sends out BLE advertisements which are received by ZVD. The ZVD then provisions a secure BLE link to the ZDD. When the link is established, the ZVD uses the Zigbee Direct Commissioning service to commission the ZDD, enabling it to join an existing network. If the ZDD is already commissioned, the ZVD joins the same Zigbee network as a router, and then sets up a tunnel through the Zigbee Direct Tunnel service. Once the tunnel is up, Zigbee packets can be exchanged between the ZVD and devices on the Zigbee network. More information on Zigbee Direct is available in the Zigbee Direct Specification [46] and in an article on about Zigbee Direct on the Silicon Labs website [47].

7.3 Matter

As discussed in the introduction to this chapter, smart home IoT technology has had difficulty scaling up to the mainstream market due to several issues. The existence of three different wireless link layer technologies (Bluetooth, Wi-Fi, and 802.15.4) has made setting up an IoT network challenging. Most smart home IoT stacks are targeted at a single wireless link layer, with interoperability features for the others, if that. Bluetooth, for example, has no standardized interoperability with Wi-Fi or 802.15.4/Zigbee except to go through a dual stack device with a vendor-specific application that routes from one network to the other.

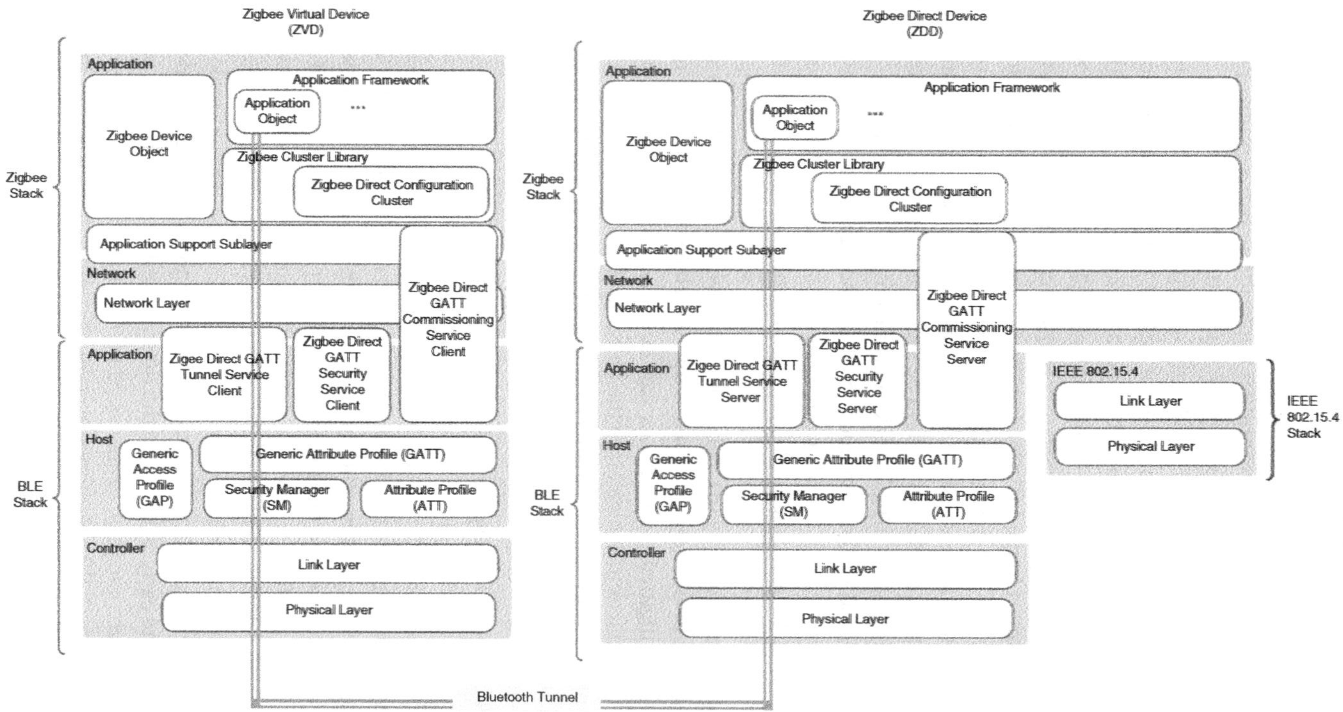

Figure 7.18 Zigbee Direct Stack Architecture for ZVD and ZDD

Zigbee has standardized Zigbee Direct for interoperability with Bluetooth but only for forwarding Zigbee packets, and it does not specify a standardized router between Zigbee and an Internet protocol network running Wi-Fi or Ethernet. Connecting a smart home IoT device into a home IoT network is not "plug and play" like connecting a laptop running an IP stack into a home network.

In addition, the smart home IoT stacks are all proprietary, from the network layer to the application layer, so they do not benefit from the mass of developer experience and open source software that has accumulated around the IP stack. As a result, applications typically come prepackaged with the device and are developed by the hardware vendor. Security is also challenging for users to configure. While all the smart home IoT protocols have security features in their design specifications, whether and how to enable them when setting up an IoT network is not specified in the standards. Since the standards specify security as optional in many cases and setting up the network to be secure is often difficult, many users do not bother leaving home IoT networks a prime target for attackers.

These challenges led to the introduction of technology platforms from major technology companies such as Samsung SmartThings [48], Apple Home (or HomeKit as the APIs are called) [49], Google Home [50], and Amazon Alexa Smart Home [51] to simplify the user experience of setting up and configuring a home/small business building IoT network. These technology platforms simplify the consumer's job by providing a hub through which a smart home IoT network can be set up, configured, and devices can be commissioned. The hub allows smart home IoT devices to connect into the building IP network through a proprietary router provided by the platform company. From there, the device can connect into an Internet cloud service run by the technology platform vendor. The hub also provides a convenient commissioning graphical user interface for devices through a Web application, especially for security.

But because the smart home IoT protocol design specifications do not cleanly separate out the stack layers, devices with software developed for one vendor's platform cannot connect into a network formed with a hub from another vendor, even if both devices are running the same wireless link layer and the same smart home IoT protocol. Smaller companies offering smart home devices must choose for which vendor's platform to develop, or offer the same device with software for multiple platform vendors at additional expense. This has resulted in market fragmentation, reducing the ability of smart home IoT technology to scale to the mainstream market like PCs, laptops, cellphones, and network gear did during the 1990s, 2000s, and early 2010s when the Internet and wireless Internet took off.

In an effort to overcome these obstacles, Matter, initially known as Project CHIP (Connected Home over IP), was announced in December 2019 as a joint initiative under the Zigbee Alliance. The project was a collaborative effort involving the major tech companies with smart home platforms and the Zigbee Alliance to create a new, open standard for smart home devices defined on top of the IP stack and to increase interoperability among them. In late 2020, the draft specifications for Matter were circulated among consortium members. In May 2021, the Zigbee Alliance rebranded itself as the CSA to reflect its broader mission and Project CHIP was also renamed as Matter during this period to signify its applicability beyond just home environments to potentially any "matter" that can be connected.

Around the same time in 2021, the CSA announced the first certification programs for Matter, which included comprehensive testing of devices to ensure that they met the new connectivity standard. By 2021, many leading technology companies and device manufacturers had announced their support for Matter, indicating that future products would be designed with Matter compatibility in mind. Companies beyond the original founding members of the CSA, including those in lighting, HVAC, and home security, also expressed support for the new standard. Version 1.0 of the Matter specification was released by the CSA in September 2021, and Version 1.2 was released in October 2023. The latest version, Version 1.3, was released in May 2024. The rest of this section discusses Matter 1.2.

Unlike Zigbee, Matter maintains a clean separation between the core platform, libraries supporting basic device operations including application specific devices, and libraries supporting specific smart home use case applications. The Matter specifications reflect this separation in three documents:

- Matter Core Specification – defines technical specifications including security and commissioning to enable an interoperable application layer solution for smart home devices over IP [52],
- Matter Device Library Specification – defines technical specifications for Matter device types. This includes both system level operations that all devices need like commissioning and operations for application specific devices like lights [53],
- Matter Application Library Specification – defines technical specifications for specific smart home use case applications like HVAC, lighting controllers, etc. [54].

The specifications are available from the CSA website [1]. In addition, an open source implementation of the Matter SDK is available on Github [55] together with documentation on how to use it [56].

7.3.1 Thread

The designers of Matter decided to use the Thread mesh routing framework with IPv6/6LoWPAN for interfacing with the 802.15.4 link layer. 6LoWPAN is the adaptation of IPv6 to the 802.15.4 low-power, lossy, wireless PAN that is discussed in Chapter 4. The advantage of Thread is that it provides fast and reliable routing over 6LoWPAN and 802.15.4 and specifies standardized border routers to IPv6 networks on Wi-Fi and Ethernet. Thread is described in the Thread specification [57] which is available from the Thread Group website [58].

As with Matter, Thread was developed by a group of technology companies concerned about the performance and scalability of existing mesh routing protocols on 802.15.4. The Thread Group, the official organization responsible for developing and maintaining the Thread specification, was founded in July 2014 as a collaborative effort among industry leaders with smart home IoT products including Google's Nest Labs, Samsung, and ARM. The group's main objective was to develop a reliable, cost-effective, and low-power wireless mesh networking protocol specifically designed for home automation. In July 2015, the Thread Group released the first version of the Thread specification. In November 2015, the Thread Group launched the Thread product certification program to ensure that any products built on the Thread specifications would be interoperable and secure. By late 2015 to early 2016, the first Thread-certified products began to appear on the market.

In May 2016, Google's Nest Labs took a significant step by releasing OpenThread [59], an open-source implementation of the Thread networking protocol. OpenThread has made it easier for developers and manufacturers to adopt the Thread standard by providing a fully functional, platform-independent code base that can be run on a variety of hardware platforms. OpenThread is available from Github [60]. Prior to the adoption of Thread by Matter in 2020, however, Thread's momentum was still primarily within specific niches of home automation, such as lighting and thermostats. Applications and devices running other protocols, particularly Zigbee, Bluetooth, and Z-Wave, covered a much broader area of building automation and were much more widely available. With the incorporation of Thread into the Matter specification, interest has picked up in Thread as an key component of the Matter network.

7.3.1.1 Thread Node Roles

The Thread specification classifies nodes based on their ability to route traffic. The specification defines the following two node types:

- Minimal Thread Device – can act only as an end node, only maintains a connection with its parent which is a full Thread device, and cannot forward messages for other devices or become a router. Minimal Thread devices typically have less memory and consume less power than full Thread devices,
- Full Thread Device – can act as an end node or a router. Full Thread devices communicate with other nodes, can participate in routing, and have child devices. Full Thread devices are generally more complex, require more memory, and often run on grid power since they cannot ever sleep.

Within these two types, the Thread specification defines different roles for nodes depending on their responsibilities for maintaining the network. The following node roles, important for Matter, are defined in the Thread specification:

- Router – a node that can forward packets on behalf of other nodes in the network. They form the backbone of the mesh network and can also act as parent nodes to other nodes,
- Border Router – a gateway router that connects the Thread network to other IP-based networks running over Wi-Fi or Ethernet. The border router performs IP routing functions between the Thread network and external networks, enabling end-to-end IP connectivity. A Thread network may contain one or more border routers, and multiple border routers in different Thread networks can join together to cooperate within a Thread domain,
- Leader – a router that is responsible for managing network-wide resources and functionality, such as assigning router id s and handling address assignments. A Thread network always has only one leader, and the role is transferred to another router if the current leader goes offline,
- Router-Eligible End Device – a node that has the potential to upgrade to a router but currently operates as an end device due to network topology or wireless connectivity conditions. They can become routers if the network conditions change,
- Full End Device – an end device dedicated to a particular application use case that communicates only through its parent which is a router. An end device will never become a router,
- Sleepy End Device – a low-power end device that is usually in a sleep mode and only wakes up periodically to communicate with their parent.

There are two additional node roles, synchronized sleepy end devices and Bluetooth end devices, that are not important to Matter. Synchronized sleepy end devices are treated like sleepy end devices, and Matter uses Bluetooth in a way that does not require the Bluetooth end device role.

7.3.2 Matter Network Architecture

Matter supports two basic network topologies, a mesh or peer-to-peer architecture for 802.15.4 and a star or infrastructure architecture for Wi-Fi. The mesh for 802.15.4 is shown in Figure 3.13, while the star architecture for Wi-Fi is shown in Figure 3.4. The Matter specification calls the former the 802.15.4 topology and the latter the Wi-Fi topology. The Matter specification uses the term star topology to designate a collection of 802.15.4 mesh networks tied into the building LAN infrastructure backbone typically through a Wi-Fi access point. The Matter star architecture is shown in Figure 7.19. The Matter specification calls the 802.15.4 Thread networks peripheral networks and the Wi-Fi network a hub network.

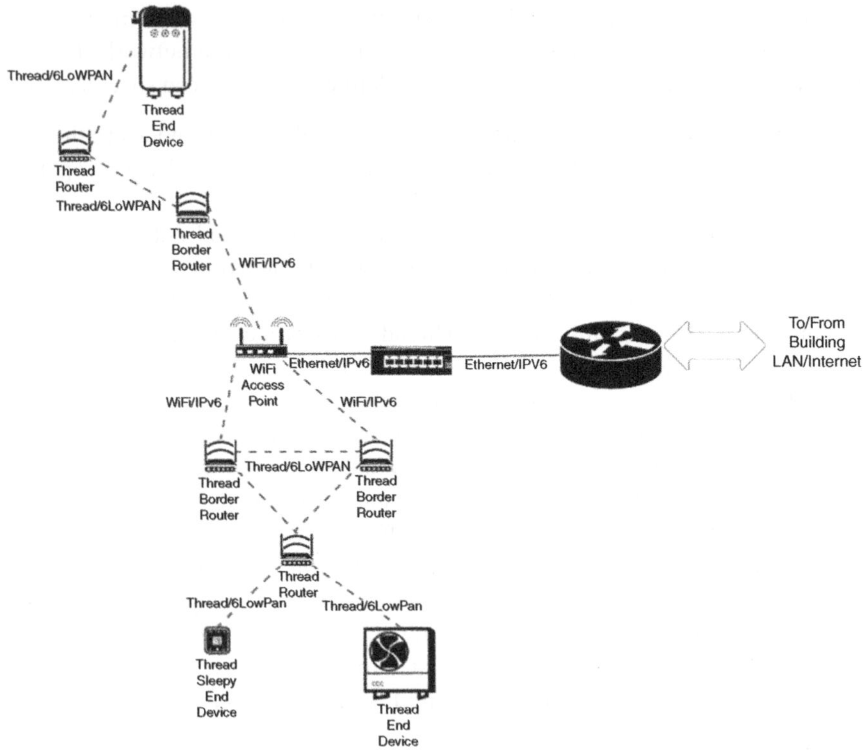

Figure 7.19 Matter Star Network Architecture

Matter networks can operate in the absence of an Internet connection, and Matter supports communication across multiple IPv6 subnets. If there is only one link available, all devices are connected into a single Matter network. The link can be any one of the three link technologies supported for operational communication: Thread/802.15.4, Wi-Fi, or Ethernet. For Wi-Fi and Ethernet, the Matter link can span multiple Wi-Fi or Ethernet links provided the links are bridged. Any number of peripheral networks can be joined with a hub network through Thread border routers.

Matter also specifies a bridge architecture that supports devices running Zigbee and other IoT network protocol standards on non-Matter networks to expose their interfaces to devices on the Matter network. Since building control systems change very slowly, this allows existing Zigbee and other building automation protocols to continue in use until they need to be replaced. The Matter bridging architecture is described in Section 9.12 of the Matter 1.2 specification [61].

7.3.3 Matter Stack

Figure 7.20 illustrates the Matter stack. The components in the stack marked with the Matter logo in the upper left corner (three inward facing arrows) are part of Matter, while those marked with the Bluetooth logo (Norse rune "B") are part of Bluetooth and the component marked with the Thread logo (script "T") is Thread. The figure shows four different link layers on which Matter runs: Wi-Fi, Ethernet, 802.15.4, and Bluetooth. Matter uses IPv6 for operational communications between devices over Wi-Fi, Ethernet, and 802.15.4. For commissioning, Matter uses IPv6 over Wi-Fi and Ethernet and defines the Bluetooth Transport Protocol (BTP) for commissioning over the Bluetooth GATT. Bluetooth is not used for operational communications, and 802.15.4 is not used for commissioning.

Because Matter runs over IPv6, it can in principle run over any physical and link layer on which IPv6 is available, although the current Matter standard only specifies Ethernet, Wi-Fi, and 802.15.4. Matter uses standard IP routing protocols for routing over Ethernet and Wi-Fi. For Bluetooth, routing is simple since the Peripheral can only send to the Central and vice versa. On 802.15.4, Matter uses the Thread mesh routing framework. However, Matter does not use all of Thread's

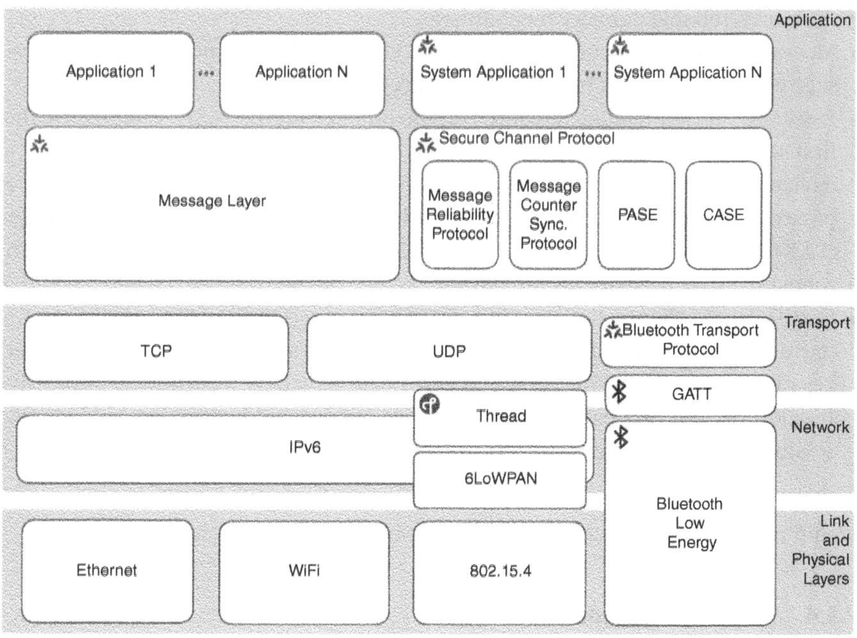

Figure 7.20 Matter Stack

features. It specifies its own security, service discovery, and identifier scheme and its own procedures for device commissioning and discovery.

At the transport layer, Matter uses the UDP and TCP transport protocols on IPv6. On Bluetooth, Matter defines BTP [62] on top of the Bluetooth GATT. BTP provides a TCP-like reliable, connection-oriented stream abstraction to the Matter application layer for commissioning. Matter uses the same approach to achieving reliable delivery over UDP by defining the Message Reliablity Protocol (MRP) on top of UDP as part of the Secure Channel Protocol sublayer.

At the application layer, Matter defines two general sublayers, the Message Layer and the Secure Channel Protocol. The Message Layer defines the basic packet format and fields for messaging between devices on a Matter network and is responsible for formatting traffic to/from applications from/to the network, and taking traffic from and submitting traffic to the transport layer. The Secure Channel Protocol acts as a control plane for secure communication between devices. The Secure Channel Layer contains the following subprotocols:

- MRP – as mentioned previously, the MRP [63] provides delivery confirmation for messages delivered over UDP, providing a reliabile delivery channel abstraction. Reliable delivery can be enabled on an individual message basis as needed by the application. MRP is not used over TCP or BTP since these protocols have their own reliable delivery mechanisms,
- Message Counter Synchronization (Sync.) Protocol [64] –synchronizes the message counters between nodes to prevent replay attacks,
- Passcode Authenticated Session Establishment (PASE) [65] –establishes the first session between a new device being commissioned into the network and a device that performs commissioning for the network. The protocol utilizes a passcode shared out-of-band together with passcode authenticated key exchange (PAKE). PAKE is discussed in Chapter 5, Section 5.5.1 about the Wi-Fi WPA robust security network architecture,
- Certificate Authenticated Session Establishment (CASE) [66] –establishes a mutually authenticated session between two nodes and generates cryptographic keys for the session using an X.509 certificate issued by a certificate authority. In addition, the two participants exchange operational parameters such as the session identifier and parameters for the MRP.

Most developers will access the Matter stack through an SDK interface and would not have to deal with formatting messages to the stack layers directly.

7.3.4 Matter Data Model

While Zigbee does not formally define abstractions for its data elements, the Matter data model defines a collection of abstractions for data elements that are independent of the underlying stack layers, in addition to the usual fundamental types like unsigned eight-bit integers. However, the specification does not specify how the data is stored, encoded or what operations are allowed on the data, so they are not, strictly speaking, data types. The following collection of abstractions is defined to standardize how Matter devices interact:

- Fabric – a collection of Matter devices sharing a trusted root. The trusted root is the root CA that provides X.509 node operational certificates (possibly through intermediate CAs) used to commission devices into the network. The full fabric reference consists of the public key of the root CA and a 64-bit fabric id unique within the root CA, but operational discovery uses a 64-bit compressed fabric id whose calculation is described in Section 4.3.2.2 of the Matter 1.2 specification [67],
- Node – a unique resource that is addressable on the network and has a unified set of functions and capabilities providing a distinct service or application. Usually, a node corresponds to a physical device or a logical instance simulating a physical device, but it is possible for a physical device to have more than one node, for example, a light fixture with multiple lights. A node is the highest element in the data model that is uniquely addressable. Nodes are identified by a 64-bit operational node id and an X.509 node operational certificate public key,
- Endpoint – an instance of an application or utility function provided on a node to implement the service or application offered by the node. Endpoints are identified with 16-bit endpoint number,
- Device Type – defines a set of requirements for an endpoint in terms of the clusters it should support to enable interoperability with devices from multiple vendors. Although Matter calls these identifiers device types, they are more endpoint types. Device types are identified by a 32-bit device type id and standardized device types are specified in the Matter Device Library Specification [53],
- Cluster – the lowest independent functional element in the data model. Clusters contain data type definitions, attributes, commands, and events implementing utility or application functions and are defined by the cluster specification. Clusters can also implement a form of inheritance like object-oriented languages, with the parent being a base cluster and the child being a derived cluster. Clusters are identified by a unique 32-bit cluster id and standardized application clusters are specified in the Matter Application Cluster Specification [54] while utility clusters are specified directly in the base Matter Specification [52],
- Attribute – stores state information or configuration settings related to a cluster's functionality. Attributes can also be used for configuration settings, to query a device for configuration or state, or, if they are writable, can be written to control a device's function. Attributes are defined by a 32-bit id unique within the cluster,
- Command – a message that can be exchanged between the client side device and server side device to change configuration settings, query state, and perform other functions. Commands are defined by a 32-bit id unique within the cluster,

- Event – a record of some change in state or communication that occurred in the past. An event allows state changes to be tracked, and are collected into a log of records. Each record contains the event number, timestamp, priority, and any data fields. Events are defined by a 32-bit id unique within the cluster.

As with Zigbee, standardized cluster definitions in Matter facilitate interoperable communication and interaction between applications and devices from different vendors on the Matter network. By defining a standardized, common set of clusters, Matter aims to ensure interoperability between IoT devices from different manufacturers. The Matter data model is specified in Chapter 7 of the Matter 1.2 specification [68].

7.3.5 Matter Fabrics and Nodes

The Matter data model defines a fabric as both an administrative domain and a security domain for a collection of devices. As an administrative domain, the Matter fabric provides a way to manage and control the devices on the fabric. A Matter fabric can be used for such administrative tasks as assigning permissions to devices, configuring security policies, and deploying software updates. As a security domain, the Matter fabric provides a way to protect the devices on the fabric from unauthorized access and attacks. Devices use encryption and authentication to protect communications on the fabric.

Each Matter device is issued a Node Operational Credentials (NOCs) and an operational node id tied to the NOC when it is commissioned into the network by a commissioner. The NOC allows a node to cryptographically identify itself within the fabric by binding the operational node id through the public-private key pair associated with the NOC. The certificate is attestable by the root of trust, the root CA, that issued the NOC or that is the root for a chain of intermediate CAs terminating in the CA that issued the NOC. The operational node id is removed on factory reset or the dissolution of the fabric.

Before a device is initially commissioned into a fabric, the device contains no operational certificate, and no operational identities in the form of fabric ids and node ids. Since some operations required for commissioning expect a fabric id or a node id prior to the device being fully commissioned, a special fabric id, 0×fffffff, is reserved for bootstrapping. This primordial fabric id associates a set of initial access control privileges with any communications that are part of the initial commissioning process.

More information on how Matter uses scoped names to designate fabrics and nodes, and the types of identifiers can be found in Sections 2.4 through 2.6 of the Matter 1.2 specification [69].

7.3.6 Device Type Classification

The Matter data model contains a high-level device type classification along a few dimensions. A device type can be either a utility device type, supplying system services for implementing node and network management, or an application device type, supplying services implementing a particular application such as fan control or lighting. A utility device type can only specify utility clusters such as *Basic Information*, and may represent the physical device. There can also be more than one utility device type on a node. The most common utility device type is Root, which contains the critical utility clusters for commissioning, credential management, and other functions on every Matter node.

The application device type implements a particular application and is the most common type of device type in Matter. An application device type is scoped to a particular endpoint, and supports clusters that implement the application functionality. Examples of application device types with application to BEMS include Fan, Thermostat, and On/Off Light. Application device types are further classified as simple and dynamic. A simple application device type implements local control that is persistent, independent, and unsupervised and are the most common device types. The Fan device type is an example of a simple device type. The dynamic application device type implements intelligent and supervisory services, such as commissioning, monitoring, trend analysis, scheduling, and central management. Dynamic device type endpoints are usually found on a central controller where a smart control application manages simple device types in control devices. Dynamic device type endpoints can support complex relationships between other endpoints on the same and other devices.

There are a few other device type classifications used in the Matter Device Library specification that are not mentioned in Chapter 7 of the Matter specification where the Matter device model is documented. If an endpoint only includes client or server clusters, it can be classified as a client or server device type. A composed device type is an endpoint that consists of two or more other device types. A multiple device type endpoint is a composed device type in which the other devices all have the same device types. The Matter Device Library also specifies some specialized device types for Zigbee interoperability and bridging. Composed and multiple device types are described in Chapter 9 of the Matter specification [70].

7.3.7 Matter Interaction Model

The Matter interaction model specifies how a node accomplishes actions by communicating with another node through an encrypted Matter network. An interaction is composed of one or more transactions. Within a transaction, actions constitute the actual messages between the nodes. The node that initiates the transaction is called the initiator; the responding node is the target. The initiator is typically the node running the client cluster, while the target is running the server cluster, but there are some exceptions. All messages in a transaction contain the same transaction id, so the receiving node can process them properly.

Matter supports the following types of transactions:

- Read – a request for attribute or event data and returns a report of the attribute or event data or a status code indicating the error,
- Subscribe – creates a subscription to reports on changes in cluster attributes or on events and returns a subscription id and the minimum and maximum interval between reports,
- Write – modifies cluster attribute values on the receiver and returns a list of paths to the written attributes and status codes,
- Invoke – invokes cluster specific commands and returns a list of command-specific responses or status codes.

The read, subscribe, and write transactions are similar to the Zigbee Cluster Library global message types discussed in Section 7.2.7.2. More information on the Matter interaction model is available in Chapter 8 of the Matter 1.2 specification [71].

7.3.8 Groups and Paths

As with Zigbee, Matter supports multicast messaging through the group construct. A group is a collection of nodes that are targets for the same action. The nodes all share a 16-bit group id. Unlike Zigbee, however, Matter leverages Thread's IPv6 multicast routing for routing group messages, making groups much more efficient. All group members share a multicast address for the group.

When accessing resources in a message, Matter uses a construct called a path to designate their location. The path establishes the location of the resource on a node, like a link in a web browser. Paths may also use groups or wildcard operators to address resources on multiple nodes or multiple clusters with a single message. Wildcards are designated by simply omitting the respective field from the path, whereas including a group id is used to indicate a specific group. Groups and wildcards increase the efficiency of network communications by not requiring multiple messages to perform the same action on different nodes, and improve the human perception of the result. For example, if a lighting controller sends separate commands to all the fixtures in a room, there will be a noticeable delay between the fixtures switching off.

Paths can have one of the following forms:

```
path = node endpoint cluster ( attribute | command | event )
path = group cluster ( attribute | command | event )
```

The fields in the path are the corresponding identifiers; for example, a cluster id for `cluster`. The `endpoint` and `cluster` fields can also contain wildcards for selecting more than one node.

7.3.9 Security

A variety of cryptographic algorithms are used to provide security services in Matter. Shared key cryptography is used for confidentiality protection, integrity protection, and data origin authentication. Matter 1.2 uses AES-CTR [72] for data in transit on the network if confidentiality protection only is required. If data authenticity is additionally required, Matter uses AES-CCM [73] for confidentiality protection, integrity protection, and data origin authentication. PAKE [74] is used for shared key provisioning during commissioning prior to the operational certificate being installed on the node, and Elliptic Curve Diffie–Hellman (ECDH) after a certificate is installed. The SHA-256 hashing algorithm is specified for computing cryptographic hashes. Matter 1.2 also includes a number of security features that are designed to protect devices and networks from denial-of-service attacks, man-in-the-middle attacks, and replay attacks. Matter does not specify the message format for encryption and authentication but rather defines an SDK interface for securing messages [75].

7.3.10 Device On-boarding

On-boarding of a device into a Matter network consists of three steps: commissionable discovery, commissioning, and operational discovery. Commissionable discovery occurs before the devices are commissioned and designed to function in an environment where the device has yet to be provisioned with security credentials for the Matter fabric it is trying to enter. Commissioning provisions the device with security credentials for the fabric. Operational discovery occurs when other devices need to find the new device to connect with it. The sections below provide more detail on these steps.

7.3.10.1 Commissionable Discovery

Commissionable discovery is initiated between a newly arrived device, called the Commissionee, and the Matter fabric node responsible for commissioning devices into the network, called the Commissioner. Matter offers three methods for commissionable discovery: BLE, Wi-Fi "soft" access point, and DNS service discovery. All commissionable discovery methods are required to provide a 12-bit statistically unique discriminator to identify the device, and can optionally include a vendor id, product id, and additional data. Section 5.4.2.4 in the Matter 1.2 standard has more details on the format of commissionable device advertisement information [76].

In the BLE method, the Commissionee advertises its presence using BLE advertisement frames. The Commissionee functions as a BLE Peripheral, while the Commissioner serves in the Central role. Initially, the device advertises at intervals between 20 and 60 msec for the first 30 seconds after activation. After that, the advertising frequency decreases to between 150 and 1500 msec. The Commissioner does not perform active scanning but passively scans three BLE advertising channels with minimal overlap with the Wi-Fi spectrum to avoid interference. The Commissionee includes the discriminator and other information in the advertisement frames. BLE commissioning is described in Section 5.4.2.5 of the Matter 1.2 specification [77].

The soft Wi-Fi access point (soft-AP) method requires the Commissionee to act as a Wi-Fi access point without uplink connectivity. The Commissionee uses a specific SSID format that includes the discriminator, vendor id, and product id , which allows the Commissioner to identify and associate with the device over IPv6. Once the connection is established, both sides configure a unique IPv6 link-local address and use DNS service discovery for further communication. The Commissioner scans Wi-Fi channels 1, 6, and 11 but is not restricted to them. Soft-AP commissioning is described in Section 5.4.2.6 of the Matter 1.2 specification [78].

Matter leverages DNS service discovery [79] for both commissionable and operational discovery. To utilize DNS service discovery in commissionable discovery, the Commissionee must have access to Ethernet or already be part of another Wi-Fi or Thread network. In general, DNS service discovery employs IPv6 multicast, but because multicast is limited on Thread networks, Thread uses a modified form of DNS service discovery involving a DNS proxy. The Commissionee advertises its host name and service instance name, both uniquely constructed, in the DNS SRV records, which are then used by the Commissioner for discovery. The Commissionee includes its discriminator and other information in DNS TXT records. For more on commissionable discovery with DNS service discovery including a few examples, see Section 4.3.1 in the Matter 1.2 specification [80].

7.3.10.2 Commissioning

The commissioning process starts after successful completion of commissionable discovery. The Commissioner connects to the Commissionee over an unsecured connection and reads various identification parameters from the Commissionee's advertisement source. The Commissioner can also obtain the onboarding payload through near-field connectivity, a scanned QR code, user input, or through some other out-of-band means. Note that since the connection is insecure at this point, care needs to be exercised to not use a method of configuring the onboarding payload that would expose it to snooping since it also contains a passcode used for PASE.

Secure shared keys are then established between the two devices using the PASE protocol, described briefly in Section 7.3.3, and PAKE. These keys are then used for encrypting and authenticating communication between the Commissioner and Commissionee. The Commissioner configures various parameters like the wireless regulatory information, local timezone, and others in the Commissionee's clusters. Device attestation is performed using a device attestation certificate and a certificate chain embedded in the device to ensure that the Commissionee is a certified Matter device. Section 6.2.2 in the Matter 1.2 specification describes how the device attestation certificate is used in attestation [81].

The Commissioner and the Commissionee perform a series of steps to establish the NOCs for the Commissionee. At the request of the Commissioner, the Commissionee generates a certificate signing request and sends it to the Commissioner. The Commissioner sends the certificate signing request to either the fabric root CA or intermediate CA to obtain the NOC. The NOC contains the node's operational id. The Commissioner then installs these credentials in the Commissionee's *Operational Credentials* cluster. Subsequently, the Commissionee's access control lists (ACLs) are configured by the Commissioner and, if the Commissionee is not already on the operational network, it is provisioned to switch to the operational network using the *Network Commissioning* cluster.

7.3.10.3 Operational Discovery

Once the Commissionee node is on the operational network, administrative control shifts to the designated Administrator, which could be the Commissioner node itself. The Administrator node role is responsible for configuring and managing the network, including adding or removing devices and setting up security policies. The Commissionee node registers itself for operational discovery using DNS service discovery with the operational network's DNS server. For operational discovery, the Commissionee constructs the host name from its link layer address, and its service instance name by concatenating the fabric id and node id. The Administrator performs operational discovery on the Commissionee and the Commissionee and Administrator use the CASE protocol, discussed briefly in Section 7.3.3, together with their NOC private keys to generate shared keys using ECDH for securing future administrative traffic. The process is finalized when the Administrator sends a command indicating successful completion of the commissioning process to the Commissionee. The new node's device and service record is available in the network's DNS service for other nodes to discover.

Chapter 5 of the Matter 1.2 specification contains a complete specification of all commissioning cases [82]. In addition, operational discovery is described in Section 4.3.2 of the Matter 1.2 specification [83].

7.3.11 Matter Utility Clusters

As with Zigbee, Matter defines utility clusters to provide state and operations supporting the basic operation and management of nodes and services. Utility clusters are a standardized set of clusters that handle tasks such as device configuration, diagnostics, and network management. Some of the clusters that play a role in commissioning are mentioned above in Section 7.3.10. Unlike application clusters, which are designed for specific use cases, utility clusters provide the basic building blocks required for device operation, security, and integration of a device into the Matter fabric.

Table 7.12 contains a list of the utility clusters, their cluster ids, and a brief description of their function. In some cases, multiple clusters implement different aspects of the same operation or function, such as commissioning or diagnostics. These are shown in the table grouped with the operation or function of the cluster group. Note that the *Label* cluster does not have a cluster id because it only serves as a base cluster for the *Fixed Label* and *User Label* clusters and is never implemented in a device on its own. Of particular interest are the clusters implementing bridging information, proxies, and over-the-air update. These are all functions specific to a particular kind of device, with enough memory, processing capacity, and grid power to serve these more resource consumption functions on the Matter fabric. More on utility clusters can be found in Chapters 9 and 11 of the Matter 1.2 specification [70, 84]. The Matter Application Cluster Specification 1.2 contains a few additional utility clusters, such as the *Groups* cluster [54].

7.3.12 Matter Energy Management Clusters

There are a few clusters defined in Matter 1.2 that would work well in a BEMS, many of which have been ported from Zigbee:

- *Scenes* – as with Zigbee, Matter supports scenes in a dedicated cluster. A scene is a set of stored values of attributes for one or more clusters on the same end point as the *Scenes* cluster,
- *On/Off* – attributes and commands for turning an appliance on and off,
- *Level Control* – a collection of clusters for controlling the level of an appliance, like the heat output of a heater or light output of a lighting fixture,
- *Mode Select* – a base cluster for controlling an appliance that has different modes, like heating and cooling on a heat pump,
- *Switch* – a cluster modeling a switch device with a data type and attribute indicating different types (latching switch, etc.),

Table 7.12 Matter Utility Clusters

ID	Name	Operation	Description
0×0003	*Identify*		Provides an endpoint to active a physical identifier like a flashing light.
0×001d	*Descriptor*		Provides information about the device's clusters, attributes, events, and actions. Replaces the Zigbee Device Object.
0×0028	*Basic Information*		Provides basic information about the origin and identity of the node, such as its name, manufacturer, and model number.
0×003e	*Operational Credentials*		Supports data types, attributes, and actions for attestation, constructing the certificate signing request, and storing the NOC.
0×001e	*Binding*		Create and manage bindings between Matter nodes, including managing the binding table.
0×001f	*AccessControl*		Exposes a view of the node's ACL which embodies the policy enforcing access control on a node's endpoints and clusters.
0×0038	*Time Sync*		Attribute for reading a node's current time, and a command allowing an Administrator to set the time, timezone, and daylight savings time.
0×0004 0×003f	*Groups* *GroupKeyManagement*	Group Management	Provide actions to create, manage, and delete groups, manage the keys for groups, and manage the group table.
0×0030 0×0031 0×003c	*General Commissioning* *Network Commissioning* *Administrator Commissioning*	Commissioning	Provide common commissioning operations for commissioning a device into a fabric, commissioning a device onto a specific type of network, and triggering a node to be commissioned by a new Administrator.

(*Continued*)

Table 7.12 (Continued)

ID	Name	Operation	Description
0x002e	*Power Source Configuration*	Power Monitoring	Attribute with a list of power sources on the device, and attributes for each endpoint describing the power source.
0x002f	*Power Source*		
0x002b	*Localization Configuration*	Localization	Attribute for configuring locale and data types and attributes allowing configuration of units and time representation to match locale.
0x002c	*Time Format Localization*		
0x002d	*Unit Localization*		
n/a	*Label (base cluster)*	Labelling	Base cluster for endpoint labelling and clusters for adding a read-only label or a read–write label to an endpoint.
0x0040	*Fixed Label*		
0x0041	*User Label*		
0x0039	*Bridged Device Basic Information*	Bridging Information	Collection of attributes describing a non-Matter device and the actions that can be performed on it.
0x0025	*Actions*		
0x0043	*ProxyDiscovery*	Proxy Discovery	Support for subscription proxies which allow nodes with minimal amount of memory to handle an arbitrary number of subscriptions.
0x0042	*ProxyConfiguration*		
0x0044	*ValidProxies*		
0x0032	*Diagnostics Logs*	Diagnostics	Support for unstructured logs, and subscriptions to event-based logging for various device subsystems.
0x0033	*General Diagnostics*		
0x0034	*Software Diagnostics*		
0x0035	*Thread Network Diagnostics*		
0x0036	*Wi-Fi Network Diagnostics*		
0x0037	*Ethernet Network Diagnostics*		
0x0029	*OTA Software Provider*	Over-the-air Software Update	Provide information on available software updates and advertise a software update.
0x002a	*OTA Software Requestor*		

- Measurement and Sensing Clusters – a collection of clusters for measuring temperature, occupancy, illuminance, etc., which could be useful in providing input data to a BEMS,
- Lighting Control – a collection of clusters allowing fine grained control over lighting,

- HVAC Control – a collection of clusters controlling various aspects of HVAC systems,
- Closure Control – a collection of clusters controlling the open-closed state of windows and doors. These could be useful for controlling natural lighting and ventilation.

Complete documentation for Matter clusters is available in the Matter Application Cluster Specification [54].

7.3.12.1 *Fan Control* Cluster

The *Fan Control* cluster is a simple example that could be of use in a BEMS. In particular, the Fan device type and *Fan Control* cluster illustrate how the Matter device model represents various device functions.

Figure 7.21 illustrates a schematic layout for part of the Matter cluster architecture for a hypothetical fan controller device. Such a controller could be connected to a whole house fan, the fan on a forced air heating system, or a variable air volume box on a rooftop space conditioning unit. At the top of the data model hierarchy is the whole device and the node, addressable through its IPv6 address. Endpoint 0 supports the root device type and contains the utility clusters for implementing basic node and network management. Two such clusters are shown: the *Basic Information* cluster and the *Access Control* cluster. The root device type is one of the six utility device types described in Chapter 2 of the Matter Device Library specification: root node, power source, OTA requester, OTA provider, aggregator, and bridged node.

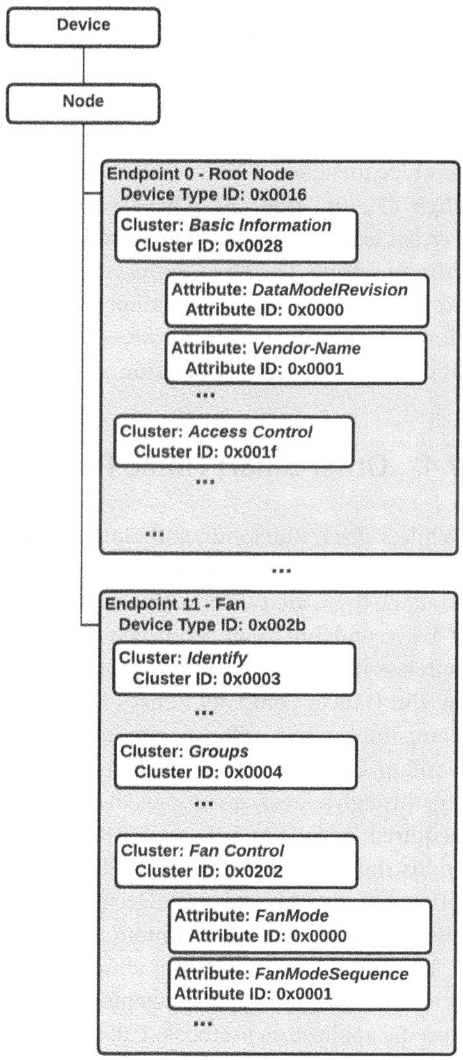

Figure 7.21 Example of a Matter Fan Controller Device

Details about the clusters supported by these endpoints are given in the base Matter Specification.

After the utility device types, the endpoints include different device types specific to the application use case. The Fan device type only supports three clusters: *Identity*, *Groups*, and *Fan Control*. The *Identity* and *Groups* clusters support device identification through some physical means and the device's participation in groups. Groups could be used to include fan control into control of a larger system, like the rooftop space conditioning unit. The *Fan Control* cluster supports no commands but only data types and attributes which can be read and written to monitor or change the fan's state. These attributes provide indications of whether the fan controller supports particular kinds of actions, for example, rocking, together with an attribute indicating the current setting.

The BEMS can turn the fan off or on by setting the *FanMode* attribute, which has type unsigned 8-bit integer, to one of its settings: *Off* (0), *Low* (1), *Medium* (2), *High* (3), or *Auto* (5). These settings control other attributes, such as the *PercentSetting* attribute, which indicates the percent of maximum speed at the current setting. The *Fan Control* cluster is documented in Section 4.4 of the Matter Application Cluster Specification [85]. The fan controller can also support additional application specific clusters such as *On/Off* and *Thermostat* for other types of control over the fan operation.

7.4 Other Smart Home Protocols

While Zigbee, Bluetooth, and Matter are the smart home protocols with the most deployment momentum (or potential deployment momentum in the case of Matter), there are two other smart home protocols that deserve mention, namely Z-Wave and EnOcean. Both consisting of a completely custom stack from the wireless physical layer to the application layer. Z-Wave was originally developed by the Danish company Zensys in 1999. Around 2005, several large American companies began offering products and formed the Z-Wave Alliance [86] to develop and promote the stack and maintain the specifications. After transitioning through a few acquisitions, the Z-Wave business and intellectual property was acquired by Silicon Labs in 2018 [87]. EnOcean was spun out of the German industrial automation company Siemens in 2001, and in 2008, the companies that offered products based on EnOcean formed the EnOcean Alliance [88] to promote the EnOcean stack and maintain the specifications [89].

Like Zigbee, with which it competes, the Z-Wave application layer supports a wide variety of energy management applications through very granular use case-specific application protocols, called command classes. Examples of energy-related command classes include the *Climate Control* command class for controlling the

HVAC system, the *Demand Control* command classes for controlling loads in response to a utility demand response signal, and the *Energy Production* command class for monitoring and controlling DERs. Z-Wave builds a network and application layer stack on top of the ITU-T G.9959 low data rate wireless networking standard [90] for the physical and link layers. The Z-Wave network layer supports both the peer-to-peer/mesh and star network architectures from Chapter 3. Z-Wave has not been deployed as widely as Zigbee by utilities, but mostly by private real estate developers, hotels, and other accommodations, and individuals interested in monitoring their home energy use. The link in the box to on this page points to a page from the Silicon Labs website with their Z-Wave 100 Technology Training Series, which contains links to videos on Z-Wave technical features.

Introduction to Z-Wave

Unlike the other QR code links in the book, this link is to a collection of short (between 3 and 15 minute) videos on the Silicon Labs website about technical features of Z-Wave. The videos cover the physical and link layers, mesh network architecture and routing, Z-Wave applications, creating and managing a Z-Wave network, mesh network performance, and security, among other topics.

EnOcean's distinguishing feature is that the wireless physical and link layer protocols use very little power so that devices can be powered by energy harvesting, like the Zigbee Green Power profile. EnOcean is used for wireless switches to controlling lighting in some commercial and industrial buildings, but mostly for wireless sensors measuring temperature, humidity, CO_2 and other environmental parameters. Since sensors only need to wake up periodically and transmit their values to a central point, they do not need to remain constantly active to receive commands making them a good match with EnOcean's low-power wireless protocol. In 2012, EnOcean's stack was ratified by ISO as standard ISO/IEC 14543-3-10 [91].

7.5 Summary

In this chapter, we discussed three smart home IoT protocols suitable for building a BEMS in a residential or small commercial building: Bluetooth, Zigbee, and Matter. Bluetooth is the most widely deployed protocol, with availability on millions of smart phones, laptops, and other consumer electronic devices. Bluetooth supports a few standardized services that could be used to implement a BEMS, and more proprietary services are available from specific manufacturers, but Bluetooth's role in BEMS for residential and small commercial buildings is primarily as a protocol to discover and commission devices into the network for other protocols such as Zigbee and Matter. With completion of development on Bluetooth Mesh, this situation may change as Bluetooth Mesh has many characteristics similar to Zigbee and Matter that make it a more attractive protocol for operational device monitoring and control in a BEMS than the basic BLE protocol.

Zigbee is the protocol with the most available devices and applications appropriate for incorporating into a BEMS. Zigbee runs on the 802.15.4 link layer, and Zigbee's Smart Energy profile has even been used by utilities to implement their AMS. This chapter examined the *DRLC* Zigbee cluster, a member of Zigbee's application cluster library that is an appropriate base for implementing a demand response and flexible load program in a BEMS. Matter has only recently been developed to bring the interoperability and ease of configuration and use of the IP stack to smart home IoT devices. Matter is the successor to Zigbee, substituting the Thread mesh networking protocol and the IPv6 stack for Zigbee's proprietary networking stack, and expanding link layer operational connectivity to include Wi-Fi and Ethernet, as well as 802.15.4, and BLE for discovery and commissioning.

Matter is designed to overcome the incompatibility between different large technology vendors' smart home platforms, simplifying the experience of consumers, installers, and others that want to utilize the technology. As an example of a BEMS application that could be built on Matter 1.2, this chapter examined how the Matter *Fan Control* cluster could be used in a fan controller. Matter is still a work in progress as far as energy management applications are concerned, so it is still too early to say whether Matter will succeed in its goals. The recent introduction of an 802.15.4 wireless interface on Apple's iPhone 15 Pro indicates that the large technology companies are starting to take Thread and Matter more seriously [92].

Two other protocols with strong support from device manufacturers were also briefly discussed, Z-Wave and Enocean. If Matter succeeds in unifying the smart home IoT community, these protocols and many others will likely become legacy protocols. But they as well as Zigbee are likely to see a long tail on their phaseout because unlike laptops and smartphones, buildings and their automation systems typically have a long lifetime.

References

1 Connectivity Standards Alliance. Building the Foundation and Future of the IoT [Internet]. [cited 2022 Jul 10]. Available from: https://csa-iot.org/.

2 Bluetooth-SIG. Bluetooth Core Specification, Bluetooth Special Interest Group, Core Specification Working Group, Version 5.4, January, 2023.

3 Bluetooth-SIG. Bluetooth Mesh Protocol, Bluetooth Special Interest Group, Draft Specification, Version 1.1, May, 2023.
4 Bluetooth-SIG. Occupancy Sensing NLC Profile [Internet]. [cited 2023 Jul 14]. Available from: https://www.bluetooth.com/specifications/specs/occupancy-sensor-nlc-profile/.
5 Bluetooth-SIG. Ambient Light Sensor NLC [Internet]. [cited 2023 Jul 14]. Available from: https://www.bluetooth.com/specifications/specs/167782/.
6 Bluetooth-SIG. Basic Lightness Controller NLC [Internet]. [cited 2023 Jul 14]. Available from: https://www.bluetooth.com/specifications/specs/basic-lightness-controller-nlc-profile/.
7 Bluetooth-SIG. Basic Scene Selector NLC [Internet]. [cited 2023 Jul 14]. Available from: https://www.bluetooth.com/specifications/specs/basic-scene-selector-nlc-profile/.
8 Bluetooth-SIG. Dimming Control NLC [Internet]. [cited 2023 Jul 14]. Available from: https://www.bluetooth.com/specifications/specs/dimming-control-nlc-profile/.
9 Bluetooth-SIG. Energy Monitor NLC Profile [Internet]. [cited 2023 Jul 14]. Available from: https://www.bluetooth.org/DocMan/handlers/DownloadDoc.ashx?doc_id=566173.
10 ISO/IEC. ISO/IEC 9834-8, Information Technology – Procedures for the Operation of Object Identifier Registration Authorities – Part 8: Generation of Universally Unique Identifiers (UUIDs) and Their Use in Object Identifiers [Internet]. [cited 2023 Jul 17]. Available from: https://www.iso.org/standard/62795.html.
11 Volume 3, Part F, Section 3.2, Bluetooth Core Specification, Version 5.4, pp. 1404–1409.
12 Volume 3, Part F, Sections 3.3 and 3.4, Bluetooth Core Specification, Version 5.4, pp. 1409–1450.
13 Bluetooth-SIG. Specifications [Internet]. [cited 2023 Jul 13]. Available from: https://www.bluetooth.com/specifications/specs/.
14 Volume 3, Part G, Figure 2.8, Bluetooth Core Specification, Version 5.4, p. 1471.
15 Bluetooth-SIG. Assigned Numbers, Bluetooth Special Interest Group, July, 2023.
16 Bluetooth-SIG. Time Profile 1.0 [Internet]. [cited 2023 Jul 25]. Available from: https://www.bluetooth.com/specifications/specs/time-profile-1-0/.
17 Bluetooth-SIG. Battery Service [Internet]. [cited 2023 Jul 25]. Available from: https://www.bluetooth.com/specifications/specs/battery-service/.
18 Bluetooth-SIG. Environmental Sensing Profile 1.0 [Internet]. [cited 2023 Jul 27]. Available from: https://www.bluetooth.com/specifications/specs/environmental-sensing-profile-1-0/.
19 Bluetooth-SIG. Environmental Sensing Service 1.0 [Internet]. [cited 2023 Jul 27]. Available from: https://www.bluetooth.com/specifications/specs/environmental-sensing-service-1-0/.
20 Bluetooth-SIG. Automation IO Profile 1.0 [Internet]. [cited 2023 Jul 27]. Available from: https://www.bluetooth.com/specifications/specs/automation-io-profile-1-0/.

21 Bluetooth-SIG. Automation IO Service 1.0 [Internet]. [cited 2023 Jul 27]. Available from: https://www.bluetooth.com/specifications/specs/automation-io-service-1-0/.

22 Bluetooth-SIG. GATT Specification Supplement, Bluetooth Special Interest Group, Supplemental Specification, Version 10, June, 2023.

23 Bluetooth-SIG. Device Properties 3 [Internet]. [cited 2023 Jul 27]. Available from: https://www.bluetooth.com/specifications/specs/mesh-device-properties/.

24 Bluetooth-SIG. Device Information Service 1.2 [Internet]. [cited 2023 Jul 27]. Available from: https://www.bluetooth.com/specifications/specs/device-information-service/.

25 Section 6.1.1, "Assigned Numbers", p. 194.

26 Texas Instruments. Bluetooth Low Energy for Energy Monitoring [Internet]. [cited 2023 Jul 23]. Available from: https://www.ti.com/lit/pdf/tidua18.

27 Connectivity Standards Alliance. Zigbee Specification, Revision 23, Zigbee Document 05-3474-23, March 15, 2023.

28 Connectivity Standards Alliance. Zigbee Cluster Library, Connectivity Standards Alliance Document 07-5123, Revision 8, December, 2019.

29 Connectivity Standards Alliance. Zigbee PRO Green Power Feature Specification: Basic Functionality Set, Version 1.1.1, Zigbee Document 14-0563-18, January, 2019.

30 Connectivity Standards Alliance. ZigBee PRO Green Power feature Specification, Version 1.0a, Revision 26, ZigBee Document 095499, May, 2014.

31 Annex G, Zigbee Specification, Revision 23, pp. 610–620.

32 Section 3.4, Zigbee Specification, Revision 23, pp. 289–319.

33 Federal Information Processing Standards. FIPS 197: Advanced Encryption Standard (AES), Information Technology Laboratory, National Institute of Standards and Technology, Gaithersburg, MD, May, 2023. https://doi.org/10.6028/NIST.FIPS.197-upd1.

34 Wikipedia. SPEKE [Internet]. [cited 2023 Aug 21]. Available from: https://en.wikipedia.org/wiki/SPEKE.

35 Section 4.98, Zigbee Specification, Revision 23, pp. 525–526.

36 Section 2.5, Zigbee Specification, Revision 23, pp. 213–233.

37 Section 2.4, Zigbee Specification, Revision 23, pp. 121–213.

38 Table 2-12, Zigbee Cluster Library, Revision 8, pp. 2-55–2-57.

39 Section 2.5, Zigbee Cluster Library, Revision 8, pp. 2-9–2-42.

40 Section 10.3, Zigbee Cluster Library, Revision 8, pp. 10-68–10-94.

41 Section 10.3.3.2, Zigbee Cluster Library, Revision 8, pp. 10-78–10-79.

42 Annex B.4, Zigbee Specification, Revision 23, p. 534.

43 Section 10.7, Zigbee Cluster Library, Revision 8, pp. 10-227–10-273.

44 Connectivity Standards Alliance. Zigbee Smart Energy Standard, Zigbee Document: 07-5356-21, Version 1.4, Revision 21, June, 2017.

45 Connectivity Standards Alliance. PRO Base Device Behavior Specification, v3.0.1, Zigbee Document 16-02828-012, September, 2021.

46 Connectivity Standards Alliance. Zigbee Direct Specification, Connectivity Standards Alliance Document 0-27688-037, November, 2022.
47 Silicon Labs. Introducing Zigbee Direct [Internet]. [cited 2023 Aug 25]. Available from: https://docs.silabs.com/zigbee/latest/zigbee-direct/#zigbee-direct-provisioning.
48 Samsung. Smarter Life with SmartThings [Internet]. [cited 2023 Sep 17]. Available from: https://www.samsung.com/us/smartthings/.
49 Apple. The Foundation for a Smart Home [Internet]. [cited 2023 Sep 17]. Available from: https://www.apple.com/home-app/.
50 Google. Smart Home Automation from Google [Internet]. [cited 2023 Sep 17]. Available from: https://home.google.com/welcome/.
51 Amazon. Alexa Smart Home [Internet]. [cited 2023 Sep 17]. Available from: https://www.amazon.com/alexa-smart-home.
52 Connectivity Standards Alliance. Matter Specification, Version 1.2, May, 2023.
53 Connectivity Standards Alliance. Matter Device Library, Version 1.2, May, 2023.
54 Connectivity Standards Alliance. Matter Application Cluster Specification, Version 1.2, May, 2023.
55 Nordic Semiconductor. Matter [Internet]. [cited 2024 March 2]. Available from: https://github.com/nrfconnect/sdk-connectedhomeip.
56 Connectivity Standards Alliance. Welcome to Matter's Documentation! [Internet]. [cited 2024 March 2]. Available from: https://project-chip.github.io/connectedhomeip-doc/index.html.
57 Thread Group. Thread Specification, Version 1.3, July, 2022.
58 Thread Group. Thread [Internet]. [cited 2023 Sep 15]. Available from: https://www.threadgroup.org/.
59 Google. OpenThread [Internet]. [cited 2023 Sep 15]. Available from: https://openthread.io/.
60 Google. OpenThread [Internet]. [cited 2024 Mar 2]. Available from: https://github.com/openthread.
61 Section 9.12, Matter Specification, Version 1.2, pp. 500–510.
62 7, Matter Specification, Version 1.2, pp. 203–220.
63 Section 4.11, Matter Specification, Version 1.2, pp. 141–152.
64 Section 4.16, Matter Specification, Version 1.2, pp. 195–203.
65 Section 4.13.1, Matter Specification, Version 1.2, pp. 156–162.
66 Section 4.13.2, Matter Specification, Version 1.2, pp. 162–184.
67 Section 4.3.2.2, Matter Specification, Version 1.2, pp. 103–104.
68 Chapter 7, Matter Specification, Version 1.2, pp. 355–424.
69 Sections 2.4–2.6, Matter Specification, Version 1.2, pp. 51–57.
70 Chapter 9, Matter Specification, Version 1.2, pp. 469–562.
71 Chapter 8, Matter Specification, Version 1.2, pp. 425–468.
72 NIST. NIST SP 800-38A Recommendation for Block Cipher Modes of Operation: Methods and Techniques [Internet]. [cited 2023 Sep 23]. Available from: https://nvlpubs.nist.gov/nistpubs/Legacy/SP/nistspecialpublication800-38a.pdf.

73 NIST. NIST SP 800-38C Recommendations for Block Cipher Mode of Operation: The CCM Mode for Authentication and Confidentiality [Internet]. [cited 2023 Sep 23]. Available from: https://nvlpubs.nist.gov/nistpubs/Legacy/SP/nistspecialpublication800-38c.pdf.

74 Taubert T., Wood C.A. SPAKE2+, an Augmented PAKE [Internet]. [cited 2023 Sep 23]. Available from: https://datatracker.ietf.org/doc/pdf/draft-bar-cfrg-spake2plus-02.pdf.

75 Chapter 3, Matter Specification, Version 1.2, pp. 63–82.

76 Section 5.4.2.4, Matter Specification, Version 1.2, pp. 246–249.

77 Section 5.4.2.5, Matter Specification, Version 1.2, pp. 249–251.

78 Section 5.4.2.6, Matter Specification, Version 1.2, pp. 251–255.

79 Cheshire S., Krochmal M. DNS-Based Service Discovery, RFC 6763 [Internet]. [cited 2023 Sep 19]. Available from: https://www.rfc-editor.org/rfc/rfc6763.

80 Section 4.3.1, Matter Specification, Version 1.2, pp. 87–103.

81 Section 6.2.2, Matter Specification, Version 1.2, pp. 282–302.

82 Chapter 5, Matter Specification, Version 1.2, pp. 221–278.

83 Section 4.3.2, Matter Specification, Version 1.2, pp. 103–107.

84 Chapter 11, Matter Specification, Version 1.2, pp. 601–868.

85 Chapter 4, Matter Application Cluster Specification, Version 1.1, pp. 211–268.

86 Z-Wave Alliance. The Internet of Things is Powered by Z-Wave [Internet]. [cited 2023 Jul 14]. Available from: https://z-wavealliance.org/.

87 Wikipedia. Z-Wave [Internet]. [cited 2023 Jul 28]. Available from: https://en.wikipedia.org/wiki/Z-Wave.

88 EnOcean Alliance. EnOcean Alliance – Building Smarter Connectivity – EnOcean Radio [Internet]. [cited 2023 Jul 28]. Available from: https://www.enocean-alliance.org/.

89 Wikipedia. EnOcean [Internet]. [cited 2023 Jul 28]. Available from: https://en.wikipedia.org/wiki/EnOcean.

90 ITU-T. G.9959: Short range narrow-band digital radiocommunication transceivers – PHY, MAC, SAR and LLC layer specifications [Internet]. [cited 2023 Jul 25]. Available from: https://www.itu.int/rec/T-REC-G.9959-201501-I/en.

91 ISO/IEC. ISO/IEC 14543-3-10:2020: Information Technology – Home Electronic Systems (HES) Architecture – Part 3-10: Amplitude Modulated Wireless Short-Packet (AMWSP) Protocol Optimized for Energy Harvesting – Architecture and Lower Layer Protocols [Internet]. [cited 2023 Jul 28]. Available from: https://www.iso.org/standard/80934.html.

92 Tuohy J. P. Here's Why Apple Put a Thread Radio in the iPhone 15 Pro [Internet]. [cited 2023 Sep 25]. Available from: https://www.theverge.com/2023/9/14/23871781/apple-iphone-15-pro-thread-radio-bluetooth.

8

Application Protocols, APIs, and Architectures

Designing abstractions that hide irrelevant details and expose only those details that are important for the functional role of a software component is the key to making a complex software system easier to maintain and upgrade. BEMS developers working in high-level programming languages should not have to be concerned with the details of how to construct messages in an application layer protocol and how to send and receive messages across the network. And they should not have to worry about how to reach into devices to access registers or otherwise deal with low-level software details.

The best way to hide these details is to provide developers with a well-designed, application-oriented API. An API oriented toward the application domain implemented by the software component behind the API abstracts away the protocol and the details from the calling side. It provides a high level interface suitable for a developer to call. A well-designed API exposes just those details that application developers need to be concerned with to write their applications. An additional benefit of APIs is that if they are written correctly, they can hide the difference between different manufacturer's devices and even different software services used to support the BEMS, enabling interoperability and application portability.

APIs act as the glue between distributed system components, allowing them to be composed exactly like procedures in a large monolithic program. A BEMS can then be built from modular functional components consisting of devices and distributed services, each of which has a dedicated function expressed through its API. These functional elements can be deployed and maintained in a scalable fashion. In this chapter, the standards for writing application-specific APIs, the protocols that underlie them, and the two main application architectures – client–server and event-based – that APIs enable in a distributed system like a BEMS are the topics for discussion.

Automating Building Energy Management for Accelerated Building Decarbonization: System Architecture and the Network Layer, First Edition. James Kempf.

Published 2025 by John Wiley & Sons, Inc.

Section 8.1 examines the basic question: what is an API? Using a few simple examples, this section explains the key functions of APIs in simplifying the construction of a BEMS out of a network of modular distributed services and devices and discusses how modern APIs evolved. Section 8.2 explores the topic of application architectures, with detailed discussions of the two most common distributed system architectures, the client–server architecture and the event-based architecture. Protocols are up next, and Section 8.3 describes HTTP [1] and MQTT [2], the two protocols typically associated with APIs for the client–server and event-based architectures, respectively, as well as the two data formats, JavaScript Object Notation (JSON) [3] and YAML [4], used to represent both an API specification and data formatted for the API. A few other alternative protocols for event-based architectures are also briefly discussed.

Since the function of APIs is to abstract out details unnecessary from the application standpoint, standards for API definition languages help structure API definitions in a way that makes their function clear to application developers and additionally can be automatically translated into programming language code for the sender and receiver. Standard languages for designing and specifying APIs are presented in Section 8.4. For the client–server architecture, the REST API [5] design pattern is the basis for most modern client–server protocol APIs. Several REST API standards were developed on top of HTTP in the 2010s. The most widely deployed REST API standard is OpenAPI 3.0/3.1 [6], discussed in Section 8.4. API standards are still in development for event-based architectures, but the AsyncAPI standard [7], which is based on OpenAPI 3.0, is promising and is also presented in Section 8.4. Section 8.5 presents a few alternative API frameworks that have achieved some amount of deployment momentum. Finally, since security is critical for APIs, Section 8.6 discusses the components supporting the implementation of a zero-trust architecture for services built on APIs. These components include lower level system components such as VPNs and TLS, protocol features in HTTP and MQTT for security, IAM frameworks that have grown up in parallel with the two API standards, and how to specify security for API operations in the OpenAPI and AsyncAPI specification languages.

8.1 What Is an API?

An API provides a programmable interface as an abstraction of some complex functionality. A simple analogy is an electrical plug and socket. The plug and socket act as an abstraction of the AC electrical system. A user simply plugs in a light in order to use power and does not have to be concerned with the details of how the AC power system works. In the same way, an HVAC control service running in a BEMS does not need to be concerned with the details of how to obtain the temperature from a thermostat if it accesses the temperature through an API. The thermostat may need to read hardware registers or perform some other hardware-specific operations or computations to obtain the temperature. The API provides a *ReadTemperature* abstraction behind which the details of the actual access operation are hidden and can vary from one thermostat model to another without requiring the code in the BEMS HVAC control service to change. A well-designed API should only provide the caller with exactly what is necessary to fulfill its specified design, and not add any unnecessary additional details.

The application layer network protocol provides a common language for communicating a request between the sender and the receiver. However, translating requests from programming languages to network protocols requires a lot of boilerplate code. APIs also provide an additional layer of abstraction away from the details of translating programming language data types into a protocol-specific format, sending them over the wire to the receiver, and reconstructing the data types on the receiver side. The operations of translating programming language representations of data into protocol representations and vice versa are called serializing and deserializing, respectively. These operations are usually coded into a reusable program library since there is much commonality between the basic serializing and deserializing operations for different high-level API operations.

Figure 8.1 illustrates the API protocol and application code stack. The networking is based on IP stack, with IP as the network layer and UDP and TCP/TLS as the transport layers. On top of the transport layer, an application layer protocol implements the API requests, responses, and notifications, using DNS to look up the host and service names/ports for the service. As mentioned in the introduction to this chapter, HTTP [1] is the main application protocol used for services designed according to the client–server architecture. The default port for HTTP running over TCP is port 80, while for HTTP running over TLS (HTTPs), it is port 443. MQTT [2] is widely used for services in the IoT event-based architecture. The default port for MQTT running over TCP is port 1883, while for MQTTs running over TLS, the default port is 8883.

The application programming language provides a library for serializing and sending, and deserializing and receiving API communications. Note that this library requires no knowledge of the process by which the recipient actually fulfills the request, it just needs to be concerned with translating the data types in the request from the programming language form to the application protocol and back. Finally, the service-specific application code sits on top of the API library and implements the particular functionality required for the application.

8.1.1 REST and RPC

The term REST, stands for "representational state transfer," and was proposed by Roy Fielding in Chapter 5 of his PhD dissertation entitled "Architectural Styles and the Design of Network-Based Software Architectures." [8]. Fielding introduced the concept of REST as a set of constraints aimed at creating scalable, stateless, cacheable, and uniform interfaces for complex distributed software systems. Fielding

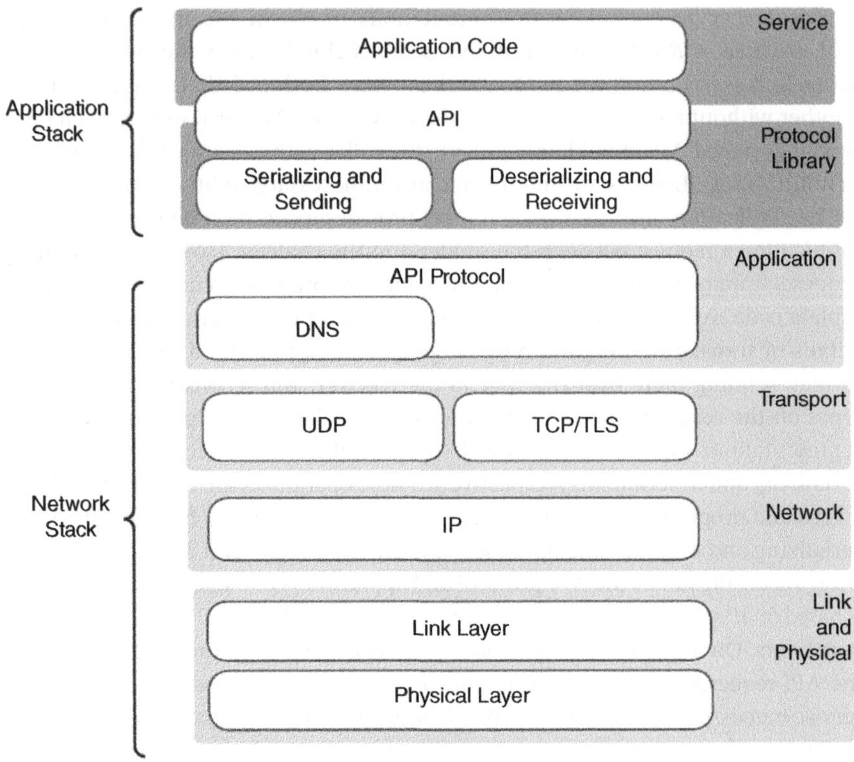

Figure 8.1 API Protocol and Application Code Stack

was motivated to develop REST as an alternative to the prevailing architectural style at the time for enterprise distributed systems, namely, remote procedure call (RPC). Fielding's original work was based on the following five principles:

1) Each request must contain all information needed to understand and process the request. There is no session state maintained on the receiver.
2) REST encourages a separation of concerns, where the receiver holds the resources and the business logic while the sender handles the user interface (UI) and presentation.
3) REST mandates that the API should have a standard, uniform interface for interacting with resources, typically through create, read, update, and delete (CRUD) operations.
4) Responses must explicitly state whether they can be cached or not to increase performance.
5) Intermediate components such as load balancers and caches can be introduced without affecting the system's basic behavior.

These principles were designed to encourage the development of "thin" clients that only hold code for following links and displaying the results. The results were returned exclusively as a web page formatted in HTML [9] containing a collection of data values and hypertext links. The client could then track down further data values and pages by following the links as the user clicked on them. An additional advantage of REST, according to Fielding, was that no documentation was required for the client–server interaction, as the returned results were self-documenting.

The REST architectural style contrasted with the RPC style that dominated the design of distributed enterprise systems prior to Fielding's work. In this model, the designers create an interface specification for the sender and receiver that defines the nature of the interaction between the two sides. The application code then uses the API in a manner similar to procedure calls in local code, except an RPC protocol maps the function calls of normal code onto network calls. RPC was primarily designed for "thick" clients, where the clients need to perform more processing than simply rendering a web page and conveying user clicks on links.

In the 2000s, Simple Object Access Protocol (SOAP) [10] RPC was developed to support RPC between a sender and receiver communicating over the Internet. SOAP is a web-based RPC protocol that uses the eXtensible Markup Language (XML) [11] to define structured data. XML provides a standardized way to describe data, enabling the easy sharing of structured data across the network between a node requesting data and a node replying. While HTML focuses on the presentation and layout of text and other media documents, XML focuses on the content and structure of the data.

During the late 2000s and early 2010s, JavaScript [12] began to gain in popularity for programming applications in web browsers. JavaScript applications are known as single-page applications because, in contrast to a linked sequence of web pages, the application runs on a single page on the browser and downloads resources as required to render a new display. JavaScript uses JSON [3] for representing data rather than XML. JSON features a simple key/value format with grouping using braces such as C++ structures, allows nesting, and includes data types such as strings, Booleans, numbers, and arrays. Linking is accomplished by using a uniform resource locator (URL) or uniform resource identifier (URI). JSON gained enormous popularity as a data format due to its simplicity, ease of use, and compatibility with web technologies. In contrast, XML specifications use a tag-based structure that has no predefined data types, requires more code to parse, and is more difficult for developers to read and understand.

Today, web application developers use the terms "REST" or "RESTful," indicating that the API is REST-like but does not fully obey Fielding's five principles, to describe an API that abides by most of Fielding's principles but uses JSON instead of hypertext for representing data. For many applications, such as building energy management, the second principle – that the server holds the business logic while the client just performs the presentation – no longer holds. These "thick" clients do their own computation, such as an energy control service that fetches data from device controllers using a REST API and uses it for load control calculations. The downside is that REST/JSON APIs are no longer self-documenting. They require documentation describing the JSON data representation. XML remains an option for data representation in the API standards but is used mostly for older APIs.

Introduction to XML

While most new REST APIs use JSON for data serialization, older APIs that specify XML are still in widespread use. The series of seven videos accessible through the link above introduce XML and provide more detail on how to use it, including creating XML schema. The first 10-minute video provides an overview of XML. XML remains an option in contemporary REST APIs including in the OpenAPI and AsyncAPI standards.

8.2 Application Architectures

The application architecture provides an overall design framework within which individual services for the BEMS discussed at a high level in Chapter 2 are designed and implemented. There are two application architectures in widespread use today: the client–server architecture and the event-based architecture. This section provides an overview of the client–server and event-based architectures. Note that a distinction should be made between the application architecture and the protocol supporting the API. Protocols such as HTTP are classified as client–server protocols because they use request-response semantics, but it is possible to implement an event-based architecture using an application API that compiles into HTTP. This topic is discussed in more detail in Section 8.4.3.

8.2.1 Client–Server Architecture

The client–server architecture is one of the most widely-used application architectures in computing, and REST APIs were originally defined for the client–server architecture. In this architecture, there are two roles, the client and the server:

- The client initiates requests for data or actions to the server and synchronously waits for a response.
- The server hosts resources, such as databases, files, and physical resources, such as sensors and actuators. The server listens for incoming client requests over a

network, processes them, and returns the appropriate response, possibly controlling an actuating device as a side effect if requested by the client.

Clients and servers can exist as separate hardware devices or as software instances, depending on the system design and use case.

In a client–server architecture, resource-intensive tasks are generally performed on the server side to offload the computational burden from the client, and communication follows a synchronous, request-response pattern. This is particularly important in energy control systems, where servers can handle complex calculations related to load balancing, energy optimization, running machine learning models, and real-time analytics. The interaction between the client and server is governed by the application layer protocol. The application layer protocol provides a common language for both parties to express the requests and responses. The architecture allows for scalability, as multiple clients can interact with a single server, and additional servers can be added to distribute the load, since the server can become a bottleneck if a single server cannot handle the incoming requests efficiently. Most cloud-based BEMS use autoscaling to automatically increase the number of running server instances when a deterioration in response time is seen by the monitoring software.

8.2.2 Event-Based Architecture

The event-based architecture is an application interaction architecture that focuses on the production, detection, consumption, and reaction to events. Unlike request-response models such as the client–server architecture, event-based systems are driven by changes in state or events monitored from outside or within the BEMS, and communication between the producer of an event and the consumer is asynchronous. This makes the event-based architecture particularly suitable for applications that use periodic event reporting, such as a BEMS.

There are two roles in an event-based architecture: the producer and the consumer:

- The producer emits events that represent changes in state.
- The consumer reacts to events posted by the producer.

Producers and consumers are loosely coupled, allowing for a high degree of flexibility in system design. Most event-based protocols use message brokers as intermediaries between producers and consumers. These brokers manage queues where events are stored until they are consumed, allowing the delivery of events to be asynchronous. The asynchronous nature of event-based systems enables high-throughput, low-latency processing.

Event-based systems can easily scale due to their asynchronous nature and loose coupling between components. Additional consumers and producers can be added to manage increased workloads, usually with no change to the existing producers and consumers. Most event-based systems include features such as event buffering and message queuing, to protect against system failures in the message broker. If a consumer is temporarily unable to process events, the message broker can store them until the consumer is back online.

Many event-based systems are based on a publish–subscribe (pub-sub) model. The pub-sub model introduces additional roles for the producers, as publishers, and consumers, as subscribers. Subscribers express interest in certain events which the publishers emit. This model also employs a message broker to manage the distribution of events. One advantage of pub-sub systems is the ability to filter events at the broker level based on topics, attributes, or content that subscribers have expressed interest in. Subscribers receive only the events that match their expressed interests, reducing the amount of network traffic processing the subscribers must do.

One challenge in event-based systems is maintaining the correct sequence of events, particularly in distributed environments. Mechanisms such as timestamps or sequence numbers may be used to ensure that events are processed in the correct order. Event-based architectures are also a good basis for real-time analytics and monitoring, as events can be analyzed and visualized as they occur. Stream processing engines, together with real-time analytics engines coupled to an IoT message bus, are also available from the major cloud providers as a service. Real-time analytics can be particularly useful in energy control systems, where real-time data can be critical for decision-making.

8.3 API Data Representation and Protocols

Three language support components are required to support an API standard:

- A formal way to name and locate resources on the receiver that the sender wants to retrieve, modify, or otherwise operate on.
- A protocol into which the sender's commands and data and the receiver's responses can be serialized.
- A language for specifying the API that can easily be compiled into code for high-level programming languages that implement the low-level details of the application protocol, and that can also be used to specify complex data structures.

For REST and event-based APIs, resources are named and located using URIs and URLs, protocols for on-the-wire representation include HTTP and MQTT, and API specification languages include JSON and YAML. These three components are the topics of this section.

8.3.1 URIs, URNs, and URLs

Client REST APIs use resources to identify the target of their requests on the server. Resources are named using universal resource identifiers (URIs). RFC 3986 [13] contains a specification for the URI format. A URI is a character string that uniquely identifies a physical or virtual resource. The characters in the string are from the ASCII character set consisting of digits, letters, and symbols. Symbols are reserved for specific uses according to the following scheme:

```
reserved = generic-uri-delimiters | sub-uri-delimiters
generic-uri-delimiters = : | / | ? | # | [ | ] | @
sub-uri-delimiters = ! | $ | & | ' | ( | ) | * | + | , | ; | =
```

Generic URI delimiters are used in URI definition schemes to delimit hierarchy levels or different components, for example, between directories and subdirectories in a file pathname, while sub-URI delimiters are used to define subsections of a URI. If one of the reserved characters or a non-printing character is used as part of a URI name, the character must be inserted with a percent-encoding mechanism. The percent character (`%`) is followed by the two hexadecimal digits of the character's ASCII code; for example, `%20` indicates the ASCII space character.

A URI begins with a case-insensitive scheme name that starts with a letter and is composed of digits, letters, and the characters plus (+), minus (-), or period (.). The scheme defines the format of the scheme-specific identifier in a scheme specification document. The identifier is separated from the scheme name by a colon character (:). Standardized URI schemes are registered with IANA, and a list of standardized schemes is available on the IANA web page [14], along with links to their specification documents.

Most URIs are one of two subtypes:

- Uniform Resource Names (URNs) – an identifier that names a resource but provides no information about its location. URNs typically begin with the scheme designator `urn`,
- Uniform Resource Locators (URLs) – an identifier that includes a means of locating the resource. These typically begin with the name of the protocol used to access the resource and include a location, such as a server name or IP address. URLs are the ubiquitous links entered into browser address bars to access a website.

An example of a URI that is neither a URN nor a URL is the tag URI scheme [15]. The tag scheme starts with the scheme identifier `tag` and has the following format (items in square brackets (*[ ]*)are optional):

```
tag: taggingEntity , date : specific [ # fragment ]
```

where

taggingEntity – the domain name or email address of the entity that made the tag.
date – the date the tag was made, in *4digitYear [– 2digitMonth [– 2digitDay]]* format,
specific – the namespace-specific part of the tag.
fragment – an optional fragment identifier referring to some subpart of the specific.

An example of a tag is as follows:

```
tag:example.com,2023-10-10:geb-book:chapter-8#
section-8.3.1
```

which refers to the current section of this book.

An example of a URI that is a URN is the following:

```
urn:uuid:6e8bc430-9c3a-11d9-9669-0800200c9a66
```

The naming scheme follows the Version 1 Universally Unique Identifier (UUID) specification [16].

URLs are the most common type of URI and are the identifiers used to designate resources in web browsers. They are displayed in the browser bar above the web page, indicating from where the page was fetched, and appear at the bottom of a web page when the mouse cursor hovers over a link. URLs contain the primary access mechanism and a location in addition to naming the resource.

URL syntax and semantics are specified in RFC 1738 [17]. The general syntax for a URL is[1]:

```
scheme :// host : port / url-path
```

Here, the *scheme* component is the URL's scheme name, such as `http`. The *host* component is the name of the host on which the resource is located, which can be the host's DNS fully qualified domain name, a host name on the local network, or the same machine on which the request was issued, such as `localhost`, or even an IPv4 or IPv6 address. The *port* is optional and specifies the port number. If the

1 RFC 1738 also specifies the user name and password as part of the location, but this was deprecated by RFC 3986 as it poses a security risk if the message is unencrypted and intercepted by an unauthorized party. In REST APIs, access is controlled by the techniques discussed in Section 8.5.

port number is omitted, then the default port for the protocol is assumed. The *url-path* component consists of a hierarchical pathname with the subcomponents separated by forward slashes (/) similar to a Linux file pathname. In REST APIs, the *url-path* component can also contain parameters, but these are discussed in Section 8.4.2.

8.3.2 HTTP and REST APIs

From 1988 through 1991, the HTTP was developed primarily as an application layer network protocol for document transfer. HTTP was originally intended to convey a collection of documents formatted primarily in HTML and linked through hyperlink references [18] from a server machine to a client running a program called a user agent, which later became known as the browser. The documents were located using URLs and the hyperlinks were also formatted as URLs. HTTP went through several iterations in the 1990s until the development of HTTP 1.1 in 1999, which is specified in RFC 2616 [19]. In the 2010s, HTTP 2 was developed and standardized to provide some efficiency improvements over HTTP 1.1, including asynchronous connections, binary as opposed to text-based formatting, and header compression. HTTP 2 is standardized in RFC 9113 [20], and is used in the non-REST, RPC-based API framework gRPC [21] and for implementing asynchronous, event-based messaging systems.

Introduction to gRPC

This short video introduces the gRPC protocol framework. Unlike previous RPC frameworks, gRPC operations can be either synchronous or asynchronous, and gRPC uses Protobuf as a language-neutral, platform-neutral extensible mechanism for serializing structured data. The video compares gRPC with REST for writing service APIs.

In the 2010s, HTTP 1.1 developed into a general-purpose application protocol, the protocol of choice for implementing REST APIs with JSON for data representation as described in Section 8.1.1. REST APIs are mostly concerned with providing a programmatic interface to web services and not with moving and referencing documents. APIs allow applications running on clients to access data and exercise control through web services in a standardized way. With REST APIs, the web model has been extended from passive document fetching to a means for applications to communicate in distributed systems.

Figure 8.2 contains a schematic overview of the interactions in a client–server, HTTP REST API. On the client side, web clients using desktops, laptops, and mobile phones can access server resources through

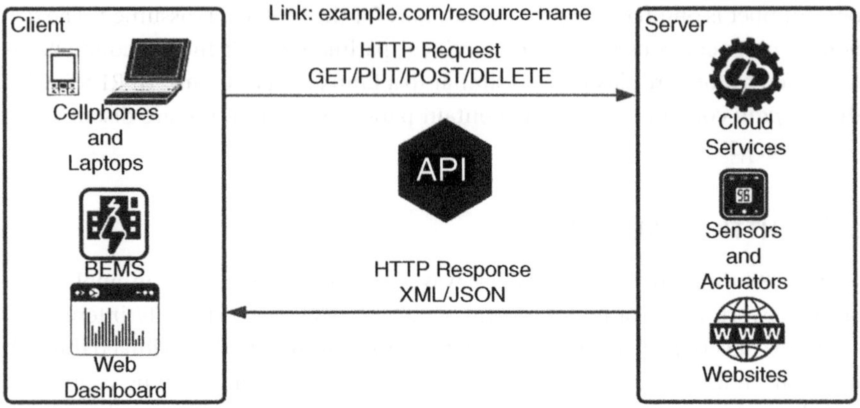

Figure 8.2 HTTP REST API Architecture

single-page JavaScript applications running in a browser. Additionally, services written in any programming language running directly in an operating system process can use REST APIs to communicate among themselves. The applications utilize REST APIs to access server resources using HTTP GET, PUT, POST, and DELETE requests. If the request returns a reply, the reply is formatted as XML or JSON. Cloud services, sensor and actuator devices, and websites can all act as servers.

8.3.2.1 HTTP Message Format

As shown in Figure 8.2, HTTP messages are of two types: a request sent from the client to the server asking for some data or action and a response sent by the server to the client returning the data or reporting on the status of executing the action. HTTP request and response messages are encoded in ASCII text and share a general overall structure:

- A single start line at the top of the message. In requests, the start line describes the request made by the client. In responses, the start line describes the status of the request,
- An optional set of headers with metadata for the message, specifying characteristics of the message or describing the body,
- A single blank line, indicating that the metadata is complete,
- An optional body including the data associated with the request or returned as the response. The size of the body is specified by a header in the header section if the body length is not self-describing.

HTTP has many features supporting media transfer and other web-centric applications, but only those important in the REST API standards are reviewed in this section.

8.3.2.2 HTTP Headers in REST APIs

The HTTP headers in the header section provide metadata describing the request or response and details on how the sender should interpret the body. They also contain other information about the request or response, such as the host name of the sending host.

Some common headers in REST requests are the following:

- `Host` – the host name of the client host,
- `Accept` – the data format types that the client can accept in the response. Examples for REST APIs are `application/xml` for XML and `application/json` for JSON,
- `Authorization` – the credentials for authenticating the client with the server. The header name is followed by the credential type then the credential itself. The most common credential type is `Bearer`, and the credential is an IAM token, called a bearer token in the HTML documentation. IAM is discussed in Section 8.6,
- `Content-Length` – the size of the request body in bytes,
- `Content-Type` – the data format type of the body data,
- `Cookie` – used for transferring authorization information in some REST APIs.

REST response headers commonly seen are the following:

- `Content-Length` – indicates the size of the response body in bytes,
- `Content-Type` – specifies the data format type of the response body data,
- `Location` – used in redirection or when a new resource has been created and is discussed in more detail in the following.

Other headers include those controlling access to other locations, specifying caching, allowing the two sides to negotiate the version of the protocol in use, and controlling access to web intermediaries called proxies. A full list of headers, along with short descriptions, is available at the following reference [22].

The `Location` header is used in HTTP responses to indicate the URL of a newly created resource or to signal that the client should perform a redirection to another URI. In a `201 Created` response following a `POST` request, the `Location` header specifies the URI where the newly created resource can be found. For example, after successfully creating a new user record in a database, the server might respond with `201 Created` status and a `Location` header pointing to the new user's URI, such as the following:

```
Location: https://api.example.com/users/123
```

The value of the `Location` header should be an absolute URI that provides the full path to the resource, including the scheme (`http`/`https`), domain name, any necessary path, and URL parameters if applicable.

For redirection, the `Location` header is included in responses with status codes that indicate the client should look for the resource at a different URI. For example, with a `301 Moved Permanently` status code, the `Location` header would contain the new URI where the requested resource is located. The client is expected to follow the URI for subsequent requests.

8.3.2.3 HTTP Request Methods

HTTP client requests must include a method name to indicate what action the server should take on the resource specified by the URL. These methods have the following functions:

- `GET` – fetches a representation of the resource. In a REST API, GET must be an idempotent operation, meaning that making a request on the same URL multiple times produces the same result if the resource has not changed. The request retrieves data but does not modify it or have any other side effect,
- `PUT` – updates a resource at a known URL with a new set of data or creates a new resource if it does not exist. In a REST API, this method must also be idempotent,
- `POST` – creates a new resource or triggers operations that do not fit into the other methods,
- `PATCH` – partially modifies the resource identified by the URL, only affecting the fields in the request body,
- `DELETE` – removes the resource identified by the URI. In a REST API, it is an idempotent operation. If the resource is already deleted or not found, the operation should return an appropriate response code.

HTTP supports a few other methods, but they are uncommon in REST APIs.

8.3.2.4 HTTP Status Codes

The server returns a status code as part of the response message. HTTP divides status codes into different ranges depending on the overall nature of the status, with the response body providing more details:

- 100s – an informational response,
- 200s – the request was successful,
- 300s – the request should be redirected,
- 400s – the request encountered a client error,
- 500s – the request encountered a server error.

Common responses seen in REST APIs are the following:

- Successful responses – `200 OK`, `201 Created`, `202 Accepted`,
- Redirection – `301 Moved Permanently` with URL indicating where the client should send future requests,

- Client errors - `400 Bad Request`, `401 Unauthorized`, `403 Forbidden`, `404 Not Found`,
- Server errors - `500 Internal Server Error`.

There are many other status codes, especially in the 400 range, to provide clients with fine-grained information about what may have caused the error, and some HTTP server implementations have added additional, nonstandard status codes. A complete list of the standard status codes can be found in this reference [23].

8.3.2.5 Simple Example HTTP REST Request/Response

An example of a simple request is an HVAC controller service that requests the temperature from a thermostat and a response from the thermostat with the temperature. The URI specifies the building, room, and thermostat number (point) for which the response is desired. The format of the request and response does not follow any REST API standard such as OpenAPI 3.0, but rather provides an overview of the major features of an HTTP REST API request and response.

Here is the request:

```
GET /building20/room442/thermostat505/
temperature HTTP/1.1
Host: example.com
Authorization: Bearer <access-token>
Accept: application/json
```

The first line includes the method type, `GET`, and the URI for the thermostat with location information, Building 20, Room 442, thermostat #505, and the desired resource, temperature. The headers specify the host from which the request originated, provide a bearer token for IAM, and specify the data format acceptable in the response, namely JSON.

The server replies with the following:

```
HTTP/1.1 200 OK
Content-Type: application/json
Content-Length: 27

{
  "temperature": 22.5,
  "unit": "Celsius"
}
```

The start line indicates that the response was successful and follows the HTTP 1.1 standard. The headers include the type of the response (JSON) and length (27 bytes). The response itself contains the temperature, 22.5, and the unit type, Celsius.

8.3.2.6 Tools for HTTP

The `curl` (Command URL) command [24] can be used to test a server running HTTP without requiring a web browser. The URL and parameters are included on the command line. Distributions are available for many different operating systems and architectures [25], and `curl` supports many different protocols besides HTTP/HTTPS, including MQTT. The `httping` (HTTP Ping) command [26] can be used to measure the latency and throughput of a web server and is commonly used to measure the performance of servers in large software as a service (SaaS) systems such as a BEMS, for ensuring that service performance matches the agreed upon service metrics.

8.3.3 MQTT

The MQTT protocol is a lightweight messaging protocol designed for low-bandwidth, high-latency, or unreliable networks. Developed initially for constrained devices and low-bandwidth, high-latency communication, it has gained widespread adoption in various applications beyond its original scope, including the IoT and building energy control systems. Most of the major cloud providers support an MQTT service for IoT connectivity from edge sites to the central cloud deployment of a service. The protocol works on a publish-subscribe model that facilitates one-to-many communication mediated through a central MQTT broker. This broker is responsible for managing sessions, authenticating clients, and ensuring that messages reach their intended recipients. MQTT has gone through versions 3.1, 3.1.1, and 5.0, and many public cloud services support all versions. The latest version of MQTT is Version 5 [27], and all versions of the MQTT specification are published and maintained by OASIS [2] .

8.3.3.1 MQTT Architecture

Although MQTT is widely used to implement communication in the asynchronous event-based system architecture, the protocol itself actually follows the synchronous client–server model. A broker acts as an intermediary for nodes that are publishing events related to certain topics and nodes that have expressed interest in those topics. The publishing nodes are clients to the MQTT broker acting as a server, while the subscribing nodes are servers for the MQTT broker acting as a client. A node can additionally act as both a publisher and a subscriber; for example, a thermostat that publishes the room temperature at regular intervals and in addition subscribes to set point updates from a load control service. Topics serve as a kind of "address" for messages and allow for hierarchical structuring. The broker also handles quality of service (QoS) levels, ensuring that messages are delivered according to their configured reliability levels. At the system level, communication between the publishing nodes and the subscribing nodes is often called a message bus in analogy with the system bus connecting the main processor with various output devices in computer hardware since communication flows in one direction: from the publishing nodes to the subscribing nodes.

Figure 8.3 schematically illustrates the basic publish–subscribe MQTT architecture. In the figure, a collection of smart energy devices at a building site to the left publishes periodic updates on their operation to an MQTT broker in the middle under a list of topics related to their functions. Services responsible for specific functions in the BEMS on the right subscribe to these topics, and the MQTT broker pushes the updates to the services when they become available. For example, the BESS periodically publishes its state of charge under the topic "Battery State of Charge." The MQTT broker then pushes the BESS updates to the Generation Control service, where they are input to the decision making algorithms to control battery charging and discharging. In addition, the Historian and Dashboard Backend services have subscribed to all topics since the Historian is responsible for maintaining accurate time series records of all smart device data in its database, and the Dashboard Backend needs to display an up-to-date graph of smart energy device data on the Web Application Dashboard to the building administrator or homeowner.

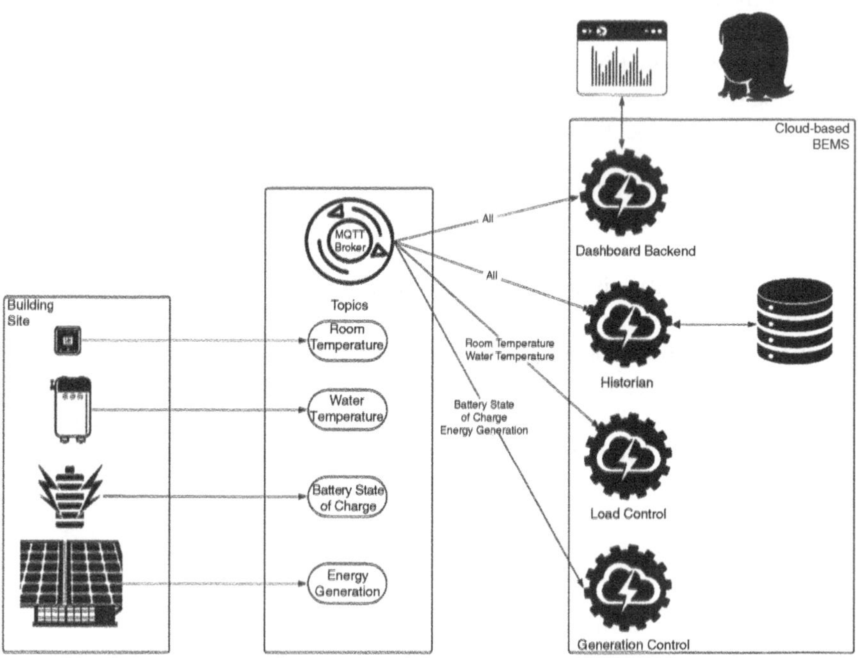

Figure 8.3 MQTT Publish–Subscribe Message Bus Architecture

8.3.3.2 MQTT Message Format

MQTT is a binary protocol that uses TCP/TLS for transport. Figure 8.4 shows the overall format of the MQTT message. The fields in the message are listed as follows:

- Control Field – basic 8-bit packet header containing two subfields: the 4-bit Packet Type field and the 4-bit Flags field,
- Remaining Length – an 8–32-bit field specifying the length, in bytes, of the rest of the packet after the Remaining Length field,
- Variable Header – a variable length field whose contents depend on the packet type. Some packet types do not have a variable length header,
- Payload – a variable length field with the message payload. The MQTT payload is typically text based, and MQTT uses UTF-8 strings to represent text, prefixed by a 2-byte string length.

Table 8.1 contains a few message type codes, the direction in which they are sent, and a brief description. MQTT Version 5 has a total of 15 message types, see the Version 5 specification for a complete list [27]. Each of the *CONNECT*, *SUBSCRIBE*, and *UNSUBSCRIBE* client-to-broker messages has a corresponding acknowledgement message from the broker to client, that can contain a status code. For the *PUBLISH* message, the acknowledgement sequence depends on the QoS level, see Section 8.3.3.3.

The 4-bit Flags field is reserved and must be set to either 0b0000 or 0b0010, depending on the message type [28]. The exception is the *PUBLISH* message. In the *PUBLISH* message, the bits are set by the sender as follows:

- Retain (bit 0) – the client sets this bit to 1 in the client-to-broker message if the client wants the broker to delete any existing messages and replace them with new ones,
- QoS (bits 1 and 2) – indicates the QoS level desired, see Section 8.3.3.3 for a discussion of QoS,
- Dup (bit 3) – valid on both the client-to-broker and broker-to-client messages. If not set, indicates that this is the first time the message has been sent. If set, indicates that the message is a redelivery. The flag is set independently by the broker on the outgoing message from the flag value on the incoming message.

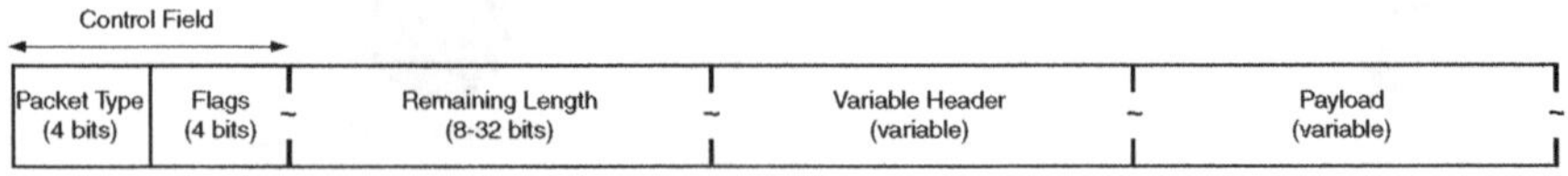

Figure 8.4 MQTT Message Format

Table 8.1 MQTT Message Type Codes

Message Name	Code	Direction	Description
CONNECT	1	Client to Broker	Connection request
CONNACK	2	Broker to Client	Connect acknowledgment
PUBLISH	3	Client to Broker or Broker to Client	Publish message
PUBACK	4	Client to Broker or Broker to Client	Publish acknowledgment (QoS = 1)
PUBREC	5	Client to Broker or Broker to Client	Publish received (QoS = 2 delivery part 1).
PUBREL	6	Client to Server or Server to Client	Publish release (QoS = 2 delivery part 2).
PUBCOMP	7	Client to Broker or Broker to Client	Publish complete (QoS = 2 delivery part 3).
SUBSCRIBE	8	Client to Broker	Subscribe request.
SUBACK	9	Broker to Client	Subscribe acknowledgment.
UNSUBSCRIBE	10	Client to Broker	Unsubscribe request.
UNSUBACK	11	Broker to Client	Unsubscribe acknowledgment.
DISCONNECT	14	Client to Broker or Broker to Client	Disconnect notification.
AUTH	15	Client to Server or Server to Client	Authentication exchange.

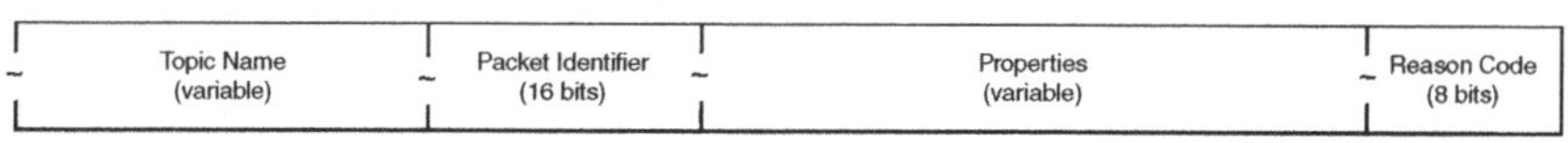

Figure 8.5 MQTT Packet Variable Header Field Format

The format of the Variable Header field is shown in Figure 8.5. Of the message types shown in Table 8.1, only packets with types *CONNECT*, *CONNACK*, *PUBLISH*, *PUBACK*, *PUBREC*, *PUBREL*, *PUBCOMP*, *SUBSCRIBE*, *SUBACK*, *UNSUBSCRIBE*, *UNSUBACK*, *DISCONNECT*, and *AUTH* have a Variable Header. Common subfields in the variable header include the following:

- Topic Name – a variable-sized subfield that contains the topic name formatted as discussed in Section 8.3.3.4. This field is only present in the *PUBLISH* message,
- Packet Identifier – a 16-bit subfield that holds a unique integer only included in messages having types *PUBACK*, *PUBREC*, *PUBREL*, *PUBCOMP*, *SUBSCRIBE*, *SUBACK*, *UNSUBSCRIBE*, *UNSUBACK*, and *PUBLISH* if the QoS value is not 0. This field is discussed in more detail below,

- Properties – a variable-sized subfield that contains a list of properties for the message, discussed in more detail below,
- Reason Code – an 8-bit subfield sent only in the *CONNACK*, *PUBACK*, *PUBREC*, *PUBREL*, *PUBCOMP*, *DISCONNECT*, and *AUTH* messages. The Reason Code indicates the status of the message request to the sender. There are 162 possible Reason Codes, with 0 indicating that the request was successful. In addition, the *SUBACK* and *UNSUBACK* messages contain a list of one or more Reason Codes in the Payload field.

Some message types, such as *CONNECT*, have additional fields in the Variable Header.

The Packet Identifier is used for matching sent messages with responses. A recipient includes the packet identifier in its acknowledgment response or responses, after which the identifier can be reused. Packet identifiers for *PUBLISH*, *SUBSCRIBE*, and *UNSUBSCRIBE* are drawn from a unified pool, and the same packet identifier cannot be used by two outstanding (unacknowledged) messages at the same time.

The Properties subfield begins with the list length encoded as a variable byte integer; the algorithm for the encoding is described in Section 1.5.5 of the standard [29]. In Section 2.2.2.2, the standard specifies 42 different property types. Each property entry in the list begins with a property type code, again encoded as a variable byte integer, followed by the property value itself, which can be any of the various types described in Section 2.2.2.2 [30].

Finally, the last field in the packet is the Payload field. Of the message types in Table 8.1, only the *CONNECT*, *SUBSCRIBE*, *SUBACK*, *UNSUBSCRIBE*, and *UNSUBACK* types are required to have a payload. The payload is optional for the *PUBLISH* message type, but if it is included, it contains a description of the event with data. In the *SUBSCRIBE* message, the payload contains the topic to which the sender wants to subscribe.

8.3.3.3 Quality of Service

An important feature of MQTT is its QoS levels, which allow for different message delivery guarantees. The choice of QoS level depends on the application requirements. For example, a real-time monitoring system may require a higher QoS level to ensure reliable data delivery, while a less critical application, such as a weather station, may only require lower reliability delivery. The higher the QoS level, the more bandwidth and processing power is required for message delivery.

There are three QoS levels in MQTT corresponding to different QoS flag field values:

- QoS 0 (QoS Flags = 0b00) – At Most Once: in this level, the message is delivered based on the best-effort principle, meaning no acknowledgment is required. The message is sent a single time, and there is no guarantee that it will be received by the recipient. This level is the fastest but least reliable method of delivery.

- QoS 1 (QoS Flags = 0b01) – At Least Once: the message is assured to reach the recipient at least once. When the broker receives the message, it sends an acknowledgment back to the sender. If the acknowledgment is not received within a specified timeframe, the sender can resend the message. While this level ensures that the message is delivered, it could lead to duplicate messages at the receiver end.
- QoS 2 (QoS Flags = 0b10) – Exactly Once: this level is the most reliable but also the most resource-intensive level. It ensures that the message is delivered exactly once by using a four-step handshake between the sender, broker, and receiver. This involves multiple acknowledgment steps to guarantee that a single copy of the message is delivered and processed and has the highest processing requirement of all the QoS levels.

Note that the QoS Flags value of 0b11 (3) is reserved and must not be used. For QoS levels 1 and 2, the *PUBLISH* message must include a packet identifier in the Variable Header field of the packet. This identifier is included in all the acknowledgment messages between the sender and recipients. For levels 1 and 2, the sender maintains a copy of the message until the acknowledgment handshake is complete and deletes it thereafter. The acknowledgment message sequence depends on the QoS level. For QoS level 0, the message is not acknowledged, while for QoS level 1, a single *PUBACK* acknowledgment message is returned to the sender. For QoS level 2, Figure 8.6 illustrates the message exchange between the publisher and broker and the broker and subscriber. The four-step process between the sender of a *PUBLISH* message and the receiver is as follows:

- The sender sends a *PUBLISH* message with a packet identifier to the broker.
- The broker stores the message and responds with a *PUBREC*, acknowledging the receipt of the *PUBLISH* message.
- The sender then sends a *PUBREL* message to the broker, indicating the message is ready for further processing, for example, sending to subscribers.
- The broker sends the message to the subscriber with a *PUBLISH* message, and sends a *PUBCOMP* message to the sender indicating that the message has been delivered. The sender deletes the message.
- The subscriber and broker exchange a *PUBREC/PUBCOMP* message sequence exactly the same as the sequence between the sender and broker.
- The subscriber then processes the message or distributes it further and sends a *PUBCOMP* message to the broker when it is finished.
- Both the broker and the subscriber delete the message, completing the four-step handshake.

8.3.3.4 Topic Format

A publisher creates a topic by sending a message to a previously unused topic name. The topic is a structured string that serves as an address for a category of messages. The publisher sends updated messages to the topic name when events related to the

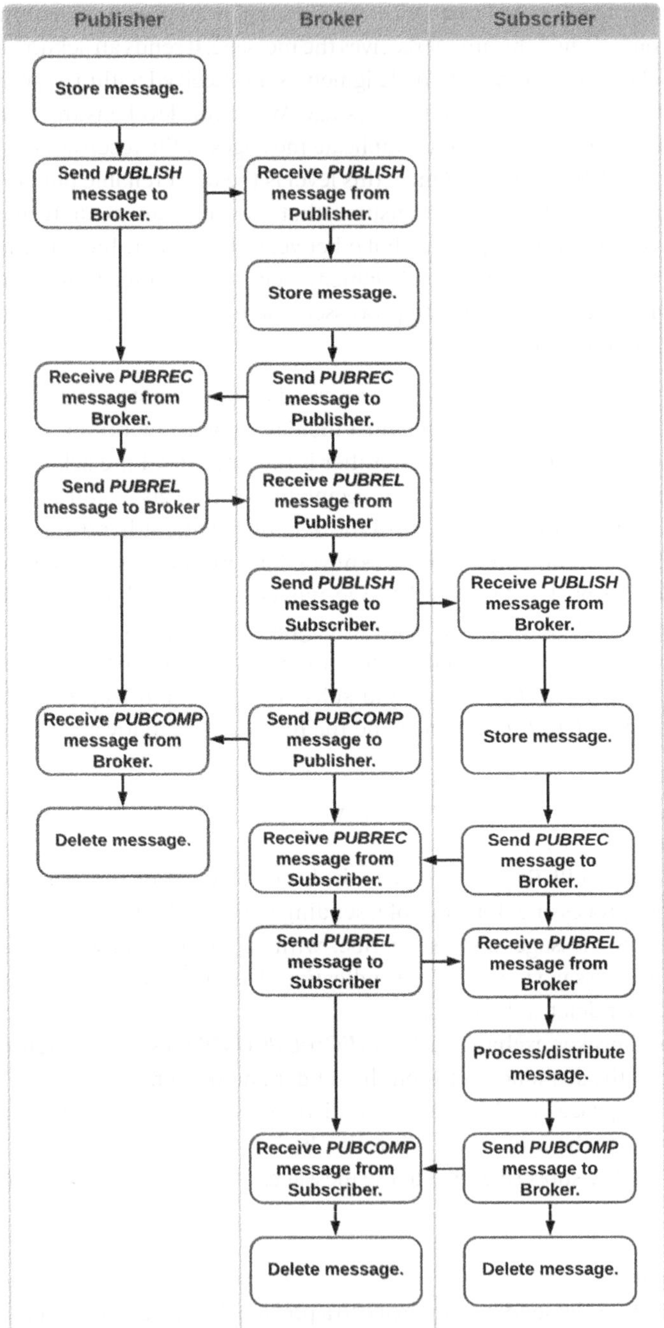

Figure 8.6 MQTT Acknowledgement Sequence for QoS = 2

topic occur and the broker then forwards updates to all clients that have subscribed to the topic. Other publishers can send messages to the topic name as well, and multiple subscribers can also subscribe to it. Each subscriber will receive a separate message every time a publisher publishes a new message on the topic. Topics are included in a Topic subfield of the Variable Header field in the *PUBLISH* message and in the payload for the *SUBSCRIBE* message. The MQTT does not support any message for obtaining a list of topics from the broker, but topic lists can be implemented using HTTP or some other out-of-band means if required.

The topic namespace is hierarchical and segmented using a forward slash (/) as a delimiter, like in a URI. For example, a topic could be `building1/floor2/thermostat505/temperature`, indicating that messages to the topic come from thermostat number 505 in Building 1 on the second floor and contain the current temperature. The hierarchical structure allows for more refined subscriptions, as clients can subscribe to wildcard topics. For instance, subscribing to `building1/floor2/+/temperature` would allow a client to receive messages on the current temperature from all the thermostats on the second floor of Building 1. The hash character (#) can be used to wildcard multiple components of the topic name; for example, `building1/#/temperature` subscribes to all the `temperature` updates from all thermostats in Building 1. Once a message is published to a topic, the broker forwards the message to all subscribed clients, in accordance with their specified QoS levels.

8.3.3.5 MQTT Cloud Services and Open Source Software

Even though MQTT is not as widely used as HTTP, there are many choices available for implementation. One possible implementation strategy is to use a managed MQTT service from a cloud provider. If the BEMS is being deployed in a cloud, then this choice provides advantages if the cloud provider offers such a service. The primary advantage is the ease of scalability. Cloud services are designed to handle the scale-up of workloads with minimal effort, eliminating the need for businesses to manage complex MQTT infrastructure in-house. Security features, such as TLS and various authentication methods, are often built-in and maintained by the cloud provider. Integration with other cloud services, such as data storage, analytics, and machine learning, is typically straightforward, allowing the BEMS to utilize these services. Additionally, managed services usually come with robust monitoring and maintenance features, eliminating the need to manage broker uptime, patches, and updates.

The primary drawback is cost, especially for large-scale deployments. Data privacy is another concern since the BEMS data is being entrusted to a third-party provider. In some cases, this could be a regulatory issue. Vendor lock-in is also a potential disadvantage since migrating to a different service later can be complicated and costly. Finally, cloud services might offer less flexibility in terms of customization and control over the MQTT broker compared to a self-hosted solution, as the service parameters and features are predefined by the cloud provider.

Two of the major cloud providers offer MQTT services coupled with a collection of broader IoT-managed services. Amazon Web Services (AWS) provides a managed service for IoT applications called AWS IoT Core, which includes support for the MQTT protocol [31] . This service allows connected devices to interact with cloud applications and other devices. AWS IoT Core supports MQTT versions 3.1.1, 3.1, and 5, as well as WebSockets for MQTT over WebSockets. Microsoft Azure offers a suite of IoT services, with Azure IoT Hub being the central service that allows bi-directional communication between IoT applications and the managed devices. Azure IoT Hub provides native support for the MQTT protocol version 3.1.1, with version 5.0 in trial [32].

Open-source software is also available for MQTT. The Mosquitto broker is a widely used open-source MQTT broker developed by the Eclipse Foundation [33]. Written in the C programming language, it has a lightweight footprint and high performance. Mosquitto supports all MQTT protocol versions, including 3.1, 3.1.1, and 5.0. One of the strengths of Mosquitto is its cross-platform support. It can be run on various operating systems, including Windows, macOS, and Linux, and it also offers Docker images for easy containerization and deployment. Mosquitto provides client libraries that facilitate the development of MQTT clients. These libraries are also written in C and are called `libmosquitto`. In addition to the native C libraries, Mosquitto client libraries are available for several programming languages through the Paho Eclipse open-source project [34]. The client libraries can also be used on the cloud providers' managed MQTT services, though additional configuration is required.

WebSockets

The video at the link above contains a short video explaining the basics of WebSockets. WebSockets is an alternative for on-site BEMS applications and devices that need to stream data to the BEMS supervisory control applications in the cloud. The video contrasts WebSockets with HTTP and HTTP polling. WebSockets are one of the alternative protocols available through the AsyncAPI 3.0 specification.

8.3.4 Other Message Bus Protocols for Event-Based Systems

While MQTT is the most widely used protocol for IoT systems, there are several other protocols that can be used for building event-based message bus systems. These include the following:

- Advanced Message Queuing Protocol (AMQP) – a protocol designed for interoperability between different messaging middleware systems, often used in enterprise-level applications [35].

- Kafka – a custom binary TCP-based protocol designed for the Apache Kafka distributed streaming database, suitable for real-time analytics and monitoring [36].
- RabbitMQ – a widely used open-source message broker that supports AMQP, MQTT, and other message bus protocols. It supports MQTT through an MQTT plugin [37].
- WebSockets – a protocol providing full-duplex communication between two nodes over a single, long-lived connection, commonly used for real-time web applications [38].
- HTTP/2 – while not traditionally a message bus protocol, HTTP/2's server push capabilities can be adapted to create event-driven architectures [39].

While these protocols are not typically used in IoT systems, they may have some advantages if a BEMS is combined with enterprise software, such as asset tracking or real-time customer analytics.

8.3.5 Data Representation Languages

API standards use data representation languages to represent data schema generated and consumed by the API and to compile the API into code for serializing and deserializing data into a programming language. Many APIs use JSON, which is specified in RFC 8259 [3], as their base representation format and most recently developed API standards also support representation in YAML [4]. The YAML format is closer to that of a programming language and, therefore, easier for programmers to use when writing an API. The YAML specification defines a mapping between YAML and JSON. This subsection examines these two representation formats.

8.3.5.1 JSON

JSON is a lightweight data-interchange format that is easy for people to read and write and easy for machines to parse and generate. While it originated as a serialization format for JavaScript object data structures (hence the name), JSON is now a

JSON Schema

One way to ensure that incoming JSON data is properly structured is to use JSON schema, just as the schema for an SQL database defines the database structure. OpenAPI and AsyncAPI both utilize JSON schema in their `schema` sections, but JSON schema can be used in other places, for example to validate that a file is properly structured before reading it in. The video link above provides a short introduction to JSON schema.

language-independent data serialization format with program language support in all major languages. It is primarily used to transmit data between a server and a client in REST APIs as an alternative to XML since JSON is less verbose than XML. Syntactic efficiency often results in better performance and less bandwidth consumption.

JSON objects are composed of key-value pairs, where the keys are strings, and the values can be strings, numbers, Booleans, objects, or arrays. An object begins with a left brace ({) and ends with a right brace (}). Strings are designated by a sequence of characters in double quotes for example: `"this is a string."` Numbers have the usual format, with floating point numbers having a decimal point and integers having none. Booleans consist of the lowercase words `true` or `false`. An array is an ordered list of values, beginning with a left bracket (`[`) and ending with a right bracket (`]`), for example: `[33, 55, 66]`. Key-value pairs within objects and values within arrays are separated by commas, with the comma omitted after the last pair or value in the collection.

Objects can be nested within other objects, for example:

```
{
    "person": {
        "name": "John Doe",
        "contact": {
            "email": "john.doe@example.com",
            "phone": "123-456-7890"
        }
    }
}
```

Values in an array or object are not restricted to one type, as in the following:

```
{
  "name": "Eve",
  "age": 25,
  "isEmployed": true,
  "children": ["John", 2, "Sara", 1]
}
```

JSON does not offer native support for date/time types, binary data, or comments, and these must be represented using standard data types such as strings or numbers.

8.3.5.2 YAML

The YAML [4], which stands for "YAML Ain't a Markup Language" (a recursive acronym), is a human-readable data serialization standard that all major programming languages support. In contrast with JSON, YAML files are formatted more like a programming language and less like data, making it easier to write and understand API specifications. YAML documents are often used for configuration files, but they can also be used for data exchange between languages with different data structures. Data structures supported by YAML include scalars (e.g. strings, integers, and floats), collections (lists and arrays), and mappings (hash tables or dictionaries. YAML also supports complex nesting of these types. The format uses indentation whitespace to indicate nesting, providing a visual representation of a data hierarchy.

In contrast with JSON, YAML supports syntax allowing comments, which is crucial for configuration files since it allows developers to annotate the purpose or function of a particular piece of code. YAML also does not require keys to be quoted as strings. Only data values that are strings need to be quoted. Additionally, YAML supports advanced features such as anchors for designating a chunk of code that can be reused elsewhere and aliases for referencing anchors to reuse the code. These features allow for a more modular representation of complex data structures by reducing repetition. While YAML's human-readable format is an advantage, the designation of whitespace as part of the language semantics can lead to issues. For example, because YAML relies on indentation whitespace to represent nesting, a misplaced space can alter a data structure definition or trip up the parser. YAML also does not accept tabs, requiring an editor that converts tabs to whitespace. Many modern configuration management tools use YAML, and it is the standard for writing REST API specifications.

The following is an example of how YAML represents key-value pairs:

```
name: "John Doe"
age: 30
email: "john.doe@example.com"
```

Nested fields are indented using two or more spaces. The following definition shows how to represent the JSON `person` object example from Section 8.3.5.1 in YAML:

```
person:
  name: "John Doe"
  contact:
    email: "john.doe@example.com"
    phone: "123-456-7890"
```

YAML represents collections in either a list format or an array format. The array format uses the same syntax as in JSON, a comma-separated list of values surrounded by square brackets. Alternatively, the list format puts each item on a separate line beginning with a hyphen:

```
languages:
  - "Python"
  - "JavaScript"
  - "Go"
```

Objects can also be included in lists and arrays.

YAML has some convenient formatting options to make reading and writing YAML code files simpler. It uses the hash character (#) at the beginning of the comment string, and comments can appear anywhere in the line but are terminated at the end of the line. Multi-line strings can be handled in different ways, such as using forward angle bracket character (>) to strip newlines or pipe character (|) to preserve them:

```
stripExample: >
  This string will span
  multiple lines but will
  be converted to a single
  line.

preserveExample: |
  This string will span
  multiple lines and
  preserve the newlines.
```

In both cases, the string is terminated by a single blank line, or by the indentation of the next key-value definition.

Repetition of the same definition in multiple places can be reduced by using anchors, represented by the ampersand character (&) together with an anchor tag to establish the anchor, and aliases, represented by the star character (*):

```
flowers:
  - fuchsia
  - &prefer lily
  - dandelion
  - *prefer
  - camelia
```

The YAML parser converts this into a list with the anchor value substituted for the alias:

```
flowers:
  - fuchsia
  - lily
  - dandelion
  - lily
  - camelia
```

If the anchored text is a collection of key-value pairs rather than a scalar data value, the reference must include two forward angle brackets (<<) together with the anchor tag, to deference the anchor for the key-value pairs to be merged into the mapping:

```
defaultAddress: &defaultAddress
  street: "123 Main St"
  city: "Anytown"

alice:
  <<: *defaultAddress
  zipcode: "12345"

bob:
  <<: *defaultAddress
  zipcode: "67890"
```

In the above example, `defaultAddress` is defined as an anchor and is then dereferenced in the definitions for `alice` and `bob` using aliases. The two backward angle brackets (<<) indicate that the anchor should be merged into the mapping.

8.4 API Standards

API standards provide a structured way for application developers to build and maintain APIs and communicate their functions to other developers who might want to use the API. While there are many RESTful API standards for client-server systems, OpenAPI 3.0/3.1 [6] stands out as the standard with the strongest deployment momentum. No single API standard has yet achieved broad deployment momentum for event-based systems, however, primarily because event-based architectures are a more recent development. The AsyncAPI specification [7] shares many of the basic syntactic elements of OpenAPI 3.0/3.1, has an open-source sponsor in the Linux Foundation and seems a promising candidate for an event-based API standard.

8.4.1 An Example Application

In the following sections, an example of an API for a BEMS application is progressively developed from a simple client–server architecture to a more complex, real-time event-based architecture to illustrate how APIs simplify the design of a BEMS. The application is an HVAC load controller with two operations:

- Get Temperature – reads the temperature from a thermostat in the client–server architecture or subscribes to real-time updates of temperature changes published by thermostats in the event-based architecture,
- Set Temperature Setpoint – sets the setpoint on a thermostat for a specific time to reduce electricity use.

Figure 8.7 illustrates the architecture of the client–server HVAC load control service. The HVAC Load Controller retrieves temperature data from and controls the setpoints on multiple thermostats in different rooms of a building on the campus of a school or commercial enterprise. Each thermostat has a number, as does the room and the building. The HVAC Load Controller acts as the client and periodically polls the thermostats, which act as the servers, for their current temperature. The HVAC Load Controller also modifies the setpoint of the thermostats

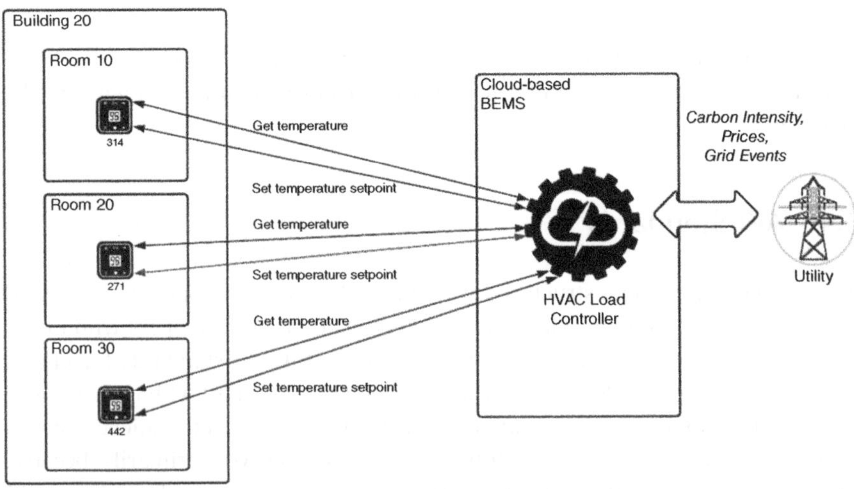

Figure 8.7 Example Architecture for a Client–Server HVAC Load Control Service

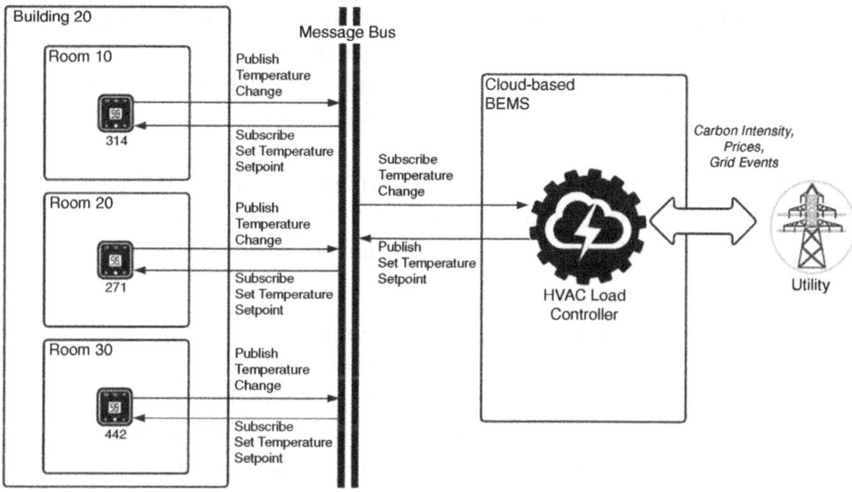

Figure 8.8 Example Architecture for an Event-based HVAC Load Control Service

to reduce electricity use in response to a notification from the utility of a need to reduce load or an indication that the marginal carbon emissions intensity has increased above a limit specified by the building operations manager.

In Figure 8.8, the event-based HVAC load control service architecture is shown. In this system, the thermostats publish temperature changes to the message bus using an event-based protocol such as MQTT and subscribe to updates of their setpoints. The HVAC Load Controller subscribes to temperature change messages from the thermostats and publishes temperature setpoints to the thermostats. Both architectures have their place in a well-designed BEMS, though often one or the other may be better suited to a particular application. In this case, the event-based architecture is a better match because it eliminates the need for the HVAC Load Controller to poll for temperature changes.

8.4.2 OpenAPI

One of the most widely used client–server REST API specification standards is OpenAPI 3.0. OpenAPI evolved out of the Swagger [40] REST API specification and Swagger still provide a collection of web-accessible and open-source tools for creating and maintaining REST APIs and their documentation. OpenAPI has gone through several minor versions, but the most widely deployed version is OpenAPI 3.0 [6]. OpenAPI 3.1 [41] is the latest version and contains some additional features which are discussed in Sections 8.4.2.3 and 8.4.2.4 but the overall design of the language remains unchanged. An OpenAPI specification provides a detailed outline of the API, describing its endpoints and the types of requests and responses it can handle, including parameter data and other related data. An OpenAPI specification allows for detailed data validation and type checking, including checking of strings against regular expression patterns, providing APIs with type safety.

An OpenAPI document contains the following required and optional sections formatted as keys at the top level in JSON or YAML, with each section defined as an object in the specification. The following are a subset of commonly used top-level sections:

- `openapi` – a required single line section contains the key `openapi` together with the version of OpenAPI used to author the document,
- `info` – a required section that provides metadata about the API, such as its title and version. The `info` section serves as a header for the API document,
- `servers` – an optional section that specifies the servers and any additional resource path components where the API can be accessed, for example, http://example.com/ v2.0.0. It can include multiple URLs if the API is available across different environments, such as staging and production or with and without TLS,
- `paths` – a required section that describes the available API endpoints (minus the `server` part) and HTTP methods `get`, `post`, `put`, and `delete`. Each path item defines the parameters of the API endpoint, the request body, and the responses.
- `components` – an optional section that defines various elements that can be reused within the body of the document to avoid repetition. Reusable data types, schemas for request bodies and response bodies, and parameters, security schemas, and schemas for many other parts of the API can be defined in the `components` section, and referenced in multiple places with the API definition using `$ref`,
- `security` – an optional section that defines security schemes for the API by referencing schemes previously defined in the `components/securitySchemes` section of the document. The `security` key can also be present in other sections of the document, for example, under `paths/get`. When defined at the root level, it applies globally across all API endpoints unless overridden by an individual operation-level security field. IAM schemes for OpenAPI 3.0 are discussed in Section 8.6.

OpenAPI identifiers, which are part of the API definition, can be in mixed upper- and lowercase for an identifier that combines multiple words, and camelCase (first word lowercase; second word uppercase) or snake_case (two words separated by an underscore) are commonly used.

One of the advantages of using OpenAPI 3.0 is its support for API versioning. By convention, API versioning is handled by including the version number in the URL. It can be included in either the `servers` part or in the `paths` part if multiple versions are included in the same API definition file. An example of the former is as follows:

```
servers: https://example.com/v2.0.0
```

Developers are encouraged to use semantic versioning on their APIs, where the version number consists of three numbers separated by periods (.) [42]. The first number is the major version, which is incremented when there are incompatible changes that require application clients of the API to change their code, such as deleting an operation from the API. The second number is the minor version, incremented for new features that are backward-compatible, such as adding an operation. The third number is the patch version, incremented for backward-compatible bug fixes. The OpenAPI 3.0 specification contains no mechanism to automatically include the version number from the document `version` field into a path so the version number needs to be kept in sync by hand or using automated tooling.

8.4.2.1 Example OpenAPI HVAC Controller API

Figure 8.9 contains a YAML definition for the HVAC controller API discussed in Section 8.4.1. The OpenAPI definition is very simple and only contains two operations in the `paths` section:

- `temperature` – an HTTP `GET` method that returns the current temperature of the thermostat connected to the HVAC system, in degrees Celsius,
- `setTemperatureSetpoint` – an HTTP `POST` method that sets a time at which the HVAC setpoint should change to a different temperature, in degrees Celsius.

Unstructured comments are formatted by starting the comment with a hash (#) character and are not included in the definition.

OpenAPI also supports a few keys to provide structured documentation that is included in the API definition. The optional `tags` key allows request definitions to be grouped when displayed by development tools such as syntax-directed editors. In Figure 8.9, the `temperature` request is tagged with `monitoring`, and the `setTemperatureSetpoint` is tagged with `control`. These requests could be displayed with other requests whose functions are similar at a high level. While tags have no semantic impact on the generated code, they do allow developers to understand which API requests perform similar functions more quickly. OpenAPI also includes two keys for documenting a specific API item: `description` and

```
openapi: 3.1.0
info:
  title: HVAC Load Control API
  version: 1.0.0
  description: API to control and monitor an HVAC system.
paths:
  /temperature:
    get:
      tags:
        - monitoring
      description: Retrieve the current temperature from the HVAC system.
      responses:
        '200':
          description: Current temperature retrieved successfully.
          content:
            application/json:
              schema:
                $ref: '#/components/schemas/temperatureResponse'
  /setTemperatureSetpoint:
    post:
      tags:
        - control
      description: Set temperature setpoint along with the time for load control.
      requestBody:
        content:
          application/json:
            schema:
              $ref: '#/components/schemas/setTemperatureRequest'
      responses:
        '201':
          description: Setpoint created successfully.
        '400':
          description: Invalid input, setpoint or time format is incorrect.
components:
  schemas:
    temperatureResponse:
      type: object
      properties:
        temperature:
          type: number
          format: float
          description: Current temperature in degrees Celsius.
        timestamp:
          type: string
          format: date-time
          description: Date and time at change in RFC 3338 format.
    setTemperatureRequest:
      type: object
      properties:
        setpoint:
          type: number
          format: float
          description: Setpoint temperature in degrees Celsius.
        time:
          type: string
          format: time
          description: Starting time for the setpoint in RFC 3338 format.
```

Figure 8.9 OpenAPI 3.0 Definition for the Simple HVAC Load Controller

summary. The description key is optional and can contain a verbose description of the operation, path, or other key under which it is located. The text can be formatted in Markdown [43], a text-based formatting language tools that can be displayed in a browser and is used by GitHub and other code development. The description text is often displayed in syntax-directed editors. The summary key, not shown in Figure 8.9, is also optional and is intended for a short descriptive string without formatting.

In the paths section, the individual URIs for the API requests are listed, followed by their by the definitions of their methods. A URI can be followed by multiple definitions for any of the four HTTP methods, with the method type indicated by the keyword. In the case of Figure 8.9, the only method types defined for the HVAC load controller are get for the /temperature request and post for the /setTemperatureSetpoint request. The requestBody and responses keywords indicate the start of the definitions for the request and response bodies, respectively. The content key appears after the declaration of the requestBody or responses and signals the start of the encodings and schemas of the request or response section. The sections for the different encoding types follow, and multiple encoding type sections are possible. An encoding type section is introduced with the MIME type as the key. For example, in Figure 8.9 after the definition of the application/json encoding type, an additional encoding of application/xml could be declared. The encoding type specifier must be a valid MIME type encoding, and a list is found in the IANA MIME type registry [44] . The encoding type is followed by the schema key and a definition of the request body or responses in the encoding type of the section.

In the figure, the temperature response body schema and the setTemperatureSetpoint request body schema are defined in the components section. These definitions use the $ref key to refer to a definition located elsewhere in the file, in another file, or at a remote link on a website through a URI. In this case, a relative URI, indicated by the hash (#) character at the beginning, directs the reference to a subsection of the components/schemas section in the same file. The schemas defined in the components/schemas section can be referenced elsewhere in the OpenAPI document where data schemas are required, such as in request bodies, response bodies, or even in parameters and headers. The definitions for the temperatureResponse and the setTemperatureRequest bodies for the two API requests translate into JSON object definitions. The JSON object corresponding to OpenAPI's schema and schemas sections is an extended subset of the IETF JSON Schema Specification Wright draft (now expired) [45]. Some common fields in the schema object are:

- type – specifies the data type, such as integer, string, object, array, etc,
- format – further describes the data format, such as int64, int32, etc. to provide additional parameter verification,

- `properties` – for type object, this defines the properties that the object can have, along with their data types,
- `required` – specifies which properties are mandatory,
- `items` – for type array, this defines the schema for the items within the array,
- `enum` – enumerates the acceptable values for a string or integer type,
- `default` – specifies a default value for an optional key,
- `minimum` and `maximum` – defines the minimum and maximum acceptable values for integer or number types,
- `description` – provides a description of the schema.

The `components/schemas` definitions for both bodies illustrate how additional constraints can be put on `string` parameters to ensure they are properly typed. The `format` key specifies a date-time format in the `temperature` response body and a time format in the `setTemperatureSetpoint` request body. The format follows the RFC 3339 specification [46] for the UTC date-time and full-time encodings: *yyyy:mm:ddThh:mm:ssZ* fore date-time or *yyyy:mm:dd:Thh:mm:ss±hh:mm* for full-time The first string specifies the date and time format in the UTC time zone, the second in a time zone offset by plus or minus *hh* hours and *mm* minutes. For example, the string "2023:11:15T08:00:00-08:00" specifies 8 AM Pacific Standard Time (8 hours behind UTC) on November 15, 2023. The time format drops the date part before the letter T. OpenAPI supports constraints on other types as well, see the specification for a complete list [41].

8.4.2.2 Parameterizing Endpoint Definitions

Parameters define values in the API's interface that can change on each call and they can appear in various parts of the API definition. Parameters can be categorized into four types depending on where they appear in the API call: path, query, header, and cookie. In an OpenAPI definition, parameters are specified in the `parameters` section of the endpoint definition. The `parameters` key is followed by a YAML list of all the parameters in the call. Each list item consists of the parameter name (key `name`) followed by specifiers defining the parameter. The specifiers include the key `in` followed by where the parameter is located in the request (one of `path`, `query`, `header`, or `cookie`), the key `required` followed by whether the parameter is required or optional (value Boolean), the key `schema` followed by the parameter's schema, and optional documentation following the key `description`, along with a few others.

Path parameters are part of the endpoint URI and are defined within curly braces ({ }) in the API endpoint's path. They are used to identify a specific resource or a collection of resources. For example, the HVAC controller can specify that a thermostat point identifier be provided, as shown in Figure 8.10. The `thermostats` path item specifies that a specific thermostat point identifier needs to be provided,

```
paths:
  /thermostats/{pointName}/temperature:
    get:
      tags:
        - monitoring
      description: Retrieve the current temperature from a specific thermostat
      parameters:
        - name: pointName
          in: path
          required: true
          description: Alphanumeric point identifier for the thermostat.
          schema:
            type: string
            pattern: '^B[0-9]{1,3}R[0-9]{1,3}-[0-9]{1,3}'
      responses:
  ...
```

Figure 8.10 Example of an OpenAPI 3.0 Path Parameter

with `{pointName}` being the point identifier. The `in` key specifies that the parameter must be in the path. The parameters section `required` key specifies that this parameter is not optional; it must be provided on every request. The `pointName` parameter's data type is then defined in the `schema` subsection as a string with the `pattern` key specifying a regular expression for matching the point identifier.

OpenAPI specifies the use of JavaScript-style regular expressions for defining string formats to obey particular patterns, allowing the strings to be type-checked when code is generated or executed for the API. In this case, the regular expression indicates that the string consists of three fields:

- An initial field beginning with the letter `B` followed by between one and three digits.
- A field beginning with the letter `R` followed by between one and three digits.
- A field beginning with the hyphen character (`-`) followed by between one and three digits.

This string names the thermostat point using the building number, room number, and thermostat identification number as the point name. An example of a URL illustrating how this path parameter might be used is the following:

```
https://hvac.example.com/thermostats/B12R240-567/temperature
```

which stands for Building 12, Room 240, Thermostat 567.

The `pattern` key and regular expressions can be used anywhere a `type: string` parameter is specified to achieve stronger type checking if a specific format for the string, such as `date-time`, is not defined. For a more detailed explanation of regular expression matching, see the Mozilla Foundation web page on the topic [47].

```
paths:
  /temperature:
    get:
      tags:
        - monitoring
      description: Retrieve the current temperature, choosing units.
      parameters:
        - name: units
          in: query
          required: true
          description: >
              The units for temperature measurement.
              Choose between Fahrenheit and Celsius.
          schema:
            type: string
            enum: ['Fahrenheit', 'Celsius']
      responses:
...
```

Figure 8.11 Example of an OpenAPI 3.0 Query Parameter

Query parameters appear after a question mark (`?`) in the URL at the path component to which they apply and are used to filter, sort, or provide additional context to the API request. Figure 8.11 illustrates an OpenAPI 3.0 definition for the HVAC load controller with a query parameter. Here, the `temperature` call requires specification of the units using one of the strings `Farhenheit` or `Celsius`. The `enum` key specifies that the query parameter can only come from a limited set of strings, with the set specified as a JSON array containing the strings. An example of a URL illustrating how this query parameter might be used is the following:

`https://hvac.example.com/temperature?units=Celsius`

Multiple query parameters are separated by an ampersand (`&`) character. For example, if the example in Figure 8.11 also included the date and time as a query parameter, the URL would look like the following: .

`https://hvac.example.com/temperature?time=2023:11:15T08:00:00-08:00&units=Celsius`

Header parameters are included in the HTTP header of the request and can be used for tasks such as authentication or setting request metadata. The HVAC API could use a header parameter to specify that the `Authorization` header is required with a token. Cookie parameters are also part of the HTTP header but are specifically used to pass a cookie to the API. The HVAC API could use the cookie parameter to enable logging and then retrieve the log later. However, header parameters and cookie parameters are not used as frequently in OpenAPI definitions as path and query parameters.

8.4.2.3 Building Asynchronous Calls with Callbacks and Webhooks

While OpenAPI and HTTP primarily support the synchronous client–server system architecture, OpenAPI has two features allowing the construction of an asynchronous, event-based architecture on top of the basic client–server communication model of HTTP: callbacks and webhooks. These features can be used to build asynchronous APIs on top of the HTTP request-response model. A callback is an endpoint definition that is included in the definition of another API endpoint in the `paths` section. The endpoint definition is passed as an argument to the endpoint in which it is defined, and the callback is intended to be executed on the server side. A webhook is a more general mechanism whereby a server-side application can notify a client-side application when a new event of interest to the client has occurred on the server, but the URL, which is hosted by the client side, needs to be passed to the server by some out of band mechanism.

A callback allows the API to specify that the server should initiate an HTTP request to the callback endpoint hosted on the client, creating a return channel between the server and client independent from the client-to-server call. Callbacks are operation-specific and defined within the `paths` endpoint to which they are related. They document possible asynchronous outgoing calls from the server side to the client side. Each callback has a name and describes a request that the server may make to the client. Parameters in the callback URL can be defined using runtime expressions, which refer to values that will be available during the operation's execution. A callback is included in an endpoint definition by defining a JSON object with a parameter for the callback endpoint identifier component. The actual callback definition is in the `callbacks` definition section which is at the same level as the `requestBody` and `responses` keys, followed by named definitions for the callbacks. The definitions follow the same syntactic pattern as for the endpoint definitions in the `paths` section, except the URL is typically parameterized with the values substituted for the parameters when the server makes the call and must include the callback endpoint identifier component.

An example of how callbacks can be used in the HVAC load controller API is shown in Figure 8.12. The `setTemperatureSetpoint` operation from Figure 8.9 has been modified to include the `temperature` request as a callback from the thermostat in the `callbacks` section rather than as a separate client-initiated call as in Figure 8.9. The thermostat reports the temperature back to the HVAC load controller only when it changes, rather than requiring the HVAC control controller to periodically poll. The `requestBody schema` section of the `setTemperatureSetpoint post` operation includes an `allOf` keyword indicating that all the items in the following list should be included in the request body. The first item on the list is the `$ref` for the setpoint temperature and setpoint time. The request body for the temperature and time data, `setTemperatureRequest`, are referred to with `$ref` and have been omitted

```
paths:
  /setTemperatureSetpoint:
    post:
      tags:
        - control
      description: Set temperature setpoint.
      requestBody:
        description: |
          Target setpoint (content defined in components section) with callback
          for asynchronous temperature change reporting.
        required: true
        content:
          application/json:
            schema:
             allOf:
               - $ref: '#/components/schemas/setTemperatureRequest'
               - type: object
                 properties:
                    callbackUrl:
                      type: string
                      format: uri
                 required:
                    - callbackUrl
      responses:
        '201':
          description: Setpoint and callback created.
      callbacks:
        temperature:
          '{$request.body#/callbackUrl}/temperature':
            post:
              summary: Real time update (content defined in components section).
              requestBody:
                content:
                  application/json:
                    schema:
                      $ref: '#/components/schemas/temperatureResponse'
              responses:
                '200':
                  description: Successfully received temperature update

    ...
```

Figure 8.12 Example Thermostat Control API with OpenAPI 3.0 Callback

since they are the same as in Figure 8.9. The second item on the list is a JSON object definition for the callback URL. The schema definition of the `callbackUrl` key's value follows. The `format` keyword with `uri` value constrains the string to URI format, while the `required` key is followed by an array of the required parameters, in this case, just the `callbackUrl` parameter.

The `allOf` ensures that when the `post` request is sent by the HVAC controller, the request body will consist of all key/value pairs in a single JSON object like the following:

```
{
    "setpoint": 24.0,
    "time": "08:00:00-8:00",
    "callbackUrl": "http://hvac-controller.example.com"
}
```

The `responses` section indicates that the server should return the `201 Created` response when the setpoint and callback are created.

The `callbacks` section defines all callbacks declared in the endpoint request bodies in the `paths` section. Within the `callbacks` definition section, `temperature` is the name of the callback definition and is a convenience locator. The URL path indicates the resource hosted on the client side for the real-time temperature update with a runtime expression:

```
'{$request.body#/callbackUrl}/temperature'
```

This runtime expression indicates that the callback URL is based on the parameters of the `setTemperatureSetpoint` operation and is evaluated at runtime when the client sends the request. The expression `$request.body` refers to the body content of the outgoing request, while the # indicates a pointer to the request body JSON object. The field `callbackUrl` specifies the key in the JSON object to retrieve, and the `temperature` indicates the resource hosted on the client for receiving the request. For the request body JSON object given above, the server retrieves the following:

```
"http://hvac-controller.example.com/temperature"
```

The callback HTTP method, `post` in this case, and any schema for the request and response bodies are defined after the callback URL. In Figure 8.12, the request body, sent by the server, is formatted according to the `temperature` schema in the components section, which is the same as in Figure 8.9. The response body, returned by the client, consists of a simple `200 OK`. Runtime expressions can be used to refer to other parts of the client request, for example query parameters. More information on runtime expressions and callbacks can be found on the Swagger website [48].

.Webhooks were added in the OpenAPI 3.1 specification [49] and are not directly tied in the API definition to a URL hosted on the client as callbacks are. Webhooks are defined independently of the URL on which the client will receive them. They are defined with the `webhooks` key at the top level of the API specification, the same level as `paths` and `components`. The URL must be communicated from the client to the server by some out of band mechanism, for example, an HTTP `post` that the client uses to configure the callback URL on the server. In addition, while the API specification can define what a webhook should look like, it does not dictate how the server should implement the triggering logic for that webhook, so the callback cannot be generated from the API specification. That logic is part of the server's custom code base.

Figure 8.13 illustrates how to define asynchronous temperature reporting from Figure 8.12 using `webhooks` instead of `callbacks`. In this case, the callback URL is passed as part of the `setTemperatureSetpoint` request body, just as in Figure 8.12, but the server requires custom code to extract it and set up the callback to report the temperature change to the client, and the client must

```
paths:
  /setTemperatureSetpoint:
    post:
      # Same as callback-based API.
webhooks:
   temperature:
     post:
       summary: Real time update (content defined in components section).
       requestBody:
         content:
           application/json:
             schema:
               $ref: '#/components/schemas/temperatureResponse'
       responses:
         '200':
           description: Successfully received temperature update
        ...
```

Figure 8.13 Example Thermostat Control API with OpenAPI 3.1 Webhook

explicitly set the URL parameter from the request generation code. However, the API could choose to pass the URL in another operation if necessary.

8.4.2.4 Building Streaming APIs with Server-Sent Events

In addition to callbacks and webhooks, OpenAPI 3.0 and 3.1 support another way of building asynchronous APIs on top of the synchronous HTTP request model. The HTTP return type `text/event-stream` defines server-sent events, a standard allowing the server to push data to the client over an open HTTP connection [50]. Server-sent events are an alternative to using callbacks and webhooks or an asynchronous protocol such as MQTT or Websockets. With server-sent events, the client initiates a connection by making a regular HTTP request to the server. However, the request header includes the HTTP header `Accept: text/event-stream`.

If OpenAPI is used to specify the request, the response body definition includes a content encoding type of `text/event-stream` which is translated into the `Accept` header:

```
...
  responses:
    '200':
       content:
         text/event-stream:
            <schema describing response data for one event>
```

indicating to the server that the client is expecting a stream of events.

The server keeps the connection to the client open and sends event data as it is generated or received. The event stream consists of simple text data, formatted according to the schema in the response body definition. Event records can have the following lines:

- A line beginning with `data:` containing the event data,
- Optionally, a line beginning with `id:` to set the event identifier,
- Optionally, a line beginning with `event:` to specify the event type, with the default being `message`,
- Two newline characters to mark the end of the event record.

The server can also periodically include a line beginning with `retry:` followed by the number of milliseconds to wait before retrying in case the connection breaks.

OpenAPI does not include any support for specifying that the server code generated from the API specification should include the `id:` and `event:` lines in an event response, so the code to generate them needs to be added by hand. However, lines beginning with `id:` and `event:` can be added after the `example` or `examples` key, to document that server code modifications are required; for example:

```
responses:
  '200':
    content:
      text/event-stream:
        schema:
          type: string
          example: |
            id: 1
            event: temperatureUpdate
            data: {"temperature": 15.0, "time":
            "2023:11:30T05:17:03-08:00"}

            id: 2
            event: temperatureUpdate
            data: {"temperature": 16.0, "time":
            "2023:11:30T06:20:42-08:00"}
```

On the client side, the client must also keep the connection open and a thread running to receive the event and dispatch it. This requires the client code to be written in an asynchronous fashion, even though the protocol used to implement the API is not asynchronous. If the connection is closed, the client automatically attempts to reconnect to the server after a default time-out. The server can also suggest a reconnection time using the `retry:` line in the event data. If an event identifier is included in the message stream definition, the client sends the identifier of the last received event with the reconnection request, allowing the server to continue sending events from where it left off. The server indicates the end of the event stream by simply closing the connection.

8.4.2.5 Tools for OpenAPI

Tools are also available to generate code for client libraries and server stubs in different programming languages. The generated code is mostly boilerplate for serializing and deserializing data from and to the programming language data types, formatted into HTTP requests and responses. Some example tools include the Swagger editor [51], a syntax-directed editor available either as a web application running in a browser or as a standalone download that enables writing or importing OpenAPI specifications. It offers real-time error checking, a preview of the API documentation, the ability to run a test version of the API right in the editor, and the generation of client and server stubs in different programming languages. Postman [52] is another widely used tool for API definition and testing. Postman supports defining APIs and uploading OpenAPI specifications to generate a collection of API endpoints for easy testing.

8.4.3 AsyncAPI

While it is possible to build an event-based system using callbacks, webhooks, and server-side events with OpenAPI, the underlying protocol is still HTTP. An alternative for defining asynchronous APIs is AsyncAPI. AsyncAPI provides a YAML- and JSON-based language for describing asynchronous, event-based APIs similar to OpenAPI for synchronous REST APIs. In addition, AsyncAPI can be translated into a number of underlying asynchronous protocols. Since there is currently no single, universally agreed upon application layer protocol for systems designed around an asynchronous event-based architecture such as HTTP for client–server based systems, AsyncAPI supports multiple application layer protocols, including MQTT, AMQP, Kafka, HTTP/2, and WebSockets. AsyncAPI also provides more syntactic and semantic support for asynchronous API definition than OpenAPI but uses the same basic YAML and JSON definition language structure and data definition schema. AsyncAPI has gone through two versions, 2.0 and 3.0, with substantial semantic differences between the two. This section focuses on AsyncAPI 3.0 [53] since the semantics of operations are clearer and it is more recent.

An AsyncAPI specification defines various elements such as channels, messages, and operations to describe the behavior and data formats of an event-based system. Many of the sections in an AsyncAPI specification have a format very similar to OpenAPI. An AsyncAPI specification is written in JSON or YAML and contains the following required and optional sections at the top level as keys whose subsections become objects in the specification:

- `asyncapi` – string that specifies the AsyncAPI specification version that the document is written in,
- `info` –provides metadata about the API. Subkeys include `title`, `version`, and `description`, among others, which provide context for the API,
- `servers` (optional)– contains a list of server name subkeys (for example, `production`, `testing`) with specifications of the servers using subobjects such as `url`, `protocol`, `description`, and `port`,
- `channels` (optional) – defines the communication channels available in the API for sending and receiving messages. Each channel corresponds to a specific topic to which clients publish and subscribe on the message bus. Subkeys are the topic names, and under each topic name, subkeys include `messages`. The values for this subkey contains a list of the message formats for events. The `parameters` subkey includes specifications for the format of any parameterized parts of the topic name,
- `operations` (optional) – defines specific named operations in subsections. Subkeys for the operations include `channel` which ties the operations back to the channel on which they occur, `action` which can be either `send` or `receive` depending on whether the API client publishes or subscribes, `messages` which contains a list of message formats for the operation messages, and `traits` which is used to include additional message data specific to the underlying event-based protocol,
- `components` (optional) – data schema definitions for parameters and message bodies, security schemes, message traits, and other reusable components, formatted as in OpenAPI 3.0/3.1,
- `tags` (optional) – a list of tags containing additional metadata, formatted as in OpenAPI 3.0/3.1,

CloudEvents

An alternative to using APIs to describe how event-based services produce and consume data is to specify the schema of the event data and the context within which the data is generated and consumed. CloudEvents is a specification designed to describe event data in a common way, to provide interoperability across event-based protocols and services, and is a project within the Cloud Native Computing Foundation (CNCF). The QR code above links to a video describing CloudEvents. CloudEvents seeks to facilitate the development of event-driven architectures by standardizing the way events are described, making it easier for developers to produce and consume events across various services without needing to know the details of the services.

- `externalDocs` (optional) – contains a link to external documentation and a description of the documentation.

The AsyncAPI is still under development, but having a well-defined way to specify and communicate APIs and compile them into stub code for asynchronous, event-based systems should simplify designing and building event-based systems, just as OpenAPI 3.0 has for client–server systems. However, the designers of AsyncAPI have a more difficult task because they must design a specification language that targets multiple backend protocols, and not just HTTP. The code generation tools must also accommodate the multi-protocol nature of the event-based protocol ecosystem.

8.4.3.1 Example AsyncAPI HVAC Controller API

Figure 8.14 illustrates an asynchronous, event-based API written in AsyncAPI 3.0 for the HVAC load control service. This API specification is a modification of the synchronous HVAC load control API from Figure 8.9, incorporating the asynchronous callback/webhook functionality from Figure 8.12 and Figure 8.13 directly into the API. The `components` section is shown in Figure 8.15.

The `asyncapi` and `info` sections are similar to OpenAPI. For the `servers` section, AsyncAPI uses the `servers` key to specify the connectivity details. The `servers` key in AsyncAPI contains subsections named representatively with descriptions of the broker or messaging servers where the asynchronous protocol is hosted. Under each subsection name, the `host` subkey specifies the server's host name, and, optionally, a port parameter name using the parameter format, `{port}`, if the protocol can utilize multiple ports. The definition of the port parameter, `port`, in this case, follows the schema describing the port parameter. In this case, the port type is specified as an `enum` limited to the two MQTT port numbers. The protocol type is specified using the `protocol` key, in this case, `mqtt`. The server's name in Figure 8.14 is the public Eclipse Foundation MQTT test broker, test.mosquitto.org, running the open-source Mosquitto MQTT broker [54].

The `paths` section in OpenAPI and the `channels` section in AsyncAPI serve similar roles, namely, they describe the available endpoints or channels to which clients can send messages. Both APIs allow the paths or channels to be parameterized, although AsyncAPI only allows path parameters, and the schema of the parameters is much more restricted. A major difference is that OpenAPI defines the HTTP methods GET/PUT/POST/DELETE as subkeys of the endpoint URI while AsyncAPI defines the operations under a separate top-level section, `operations`, since most event-based protocols do not have predefined operation names such as HTTP's operation names. AsyncAPI requires that the topic name be specified in the `address` section under the relevant channel name since the channel name need not be the same as the topic name. The string following the `address` subkey is the actual topic name to which messages are published or subscribed.

```yaml
asyncapi: 3.0.0
info:
  title: HVAC Load Control and Temperature Monitoring Asynchronous API.
  version: 1.0.0
  description: API to control and monitor an HVAC system.

servers:
  test:
    host: test.mosquitto.org:{port}
    protocol: mqtt
    description: Test broker.
    variables:
      port:
        description: Secure connection (TLS) is available through port 8883.
        default: '1883'
        enum:
          - '1883'
          - '8883'

channels:
  hvacTemperatureMonitoring:
    description: Channel for monitoring real-time temperature readings.
    address: temperatureMonitoring/thermostat/{pointName}
    parameters:
      pointName:
        $ref: '#/components/parameters/pointName'
    messages:
      temperatureMeasured:
        $ref: '#/components/messages/temperatureMessage'
  hvacLoadControl:
    description: Channel for receiving temperature setpoint instructions.
    address: loadControl/setSetpoint/thermostat/{pointName}
    parameters:
      pointName:
        $ref: '#/components/parameters/pointName'
    messages:
      setTemperatureSetpoint:
        $ref: '#/components/messages/setTemperatureSetpointMessage'

operations:
  getTemperature:
    action: receive
    summary: Receive the temperature from thermostat.
    channel:
      $ref: '#/channels/hvacTemperatureMonitoring'
    traits:
      - $ref: '#/components/operationTraits/mqtt'
  setSetpoint:
    action: send
    summary: Set the thermostat setpoint.
    channel:
      $ref: '#/channels/hvacLoadControl'
    traits:
      - $ref: '#/components/operationTraits/mqtt'

components:
    ...
```

Figure 8.14 Example AsyncAPI 3.0 YAML Definition for a Simple HVAC Controller API

```
components:
  parameters:
    pointName:
      description: A thermostat point name.
      enum: ['B01-R01-THM01', 'B01-R02-THM01', …]
  messages:
    temperatureMessage:
      name: temperatureMessage
      summary: Report of temperature and time upon change.
      payload:
        $ref: '#/components/schemas/temperaturePayload'
    setTemperatureSetpointMessage:
      name: setTemperatureSetpointMessage
      summary: Set the temperature setpoint starting at date at the given time.
      payload:
        $ref: '#/components/schemas/setTemperatureSetpointPayload'

  schemas:
    temperaturePayload:
      type: object
      properties:
        temperature:
          type: number
          format: float
          description: Current temperature in degrees Celsius.
        timestamp:
          type: string
          format: date-time
          description: Date and time at change in RFC 3338 format.
    setTemperatureSetpointPayload:
      type: object
      properties:
        setpoint:
          type: number
          format: float
          description: Setpoint temperature in degrees Celsius.
        time:
          type: string
          format: time
          description: Starting time for the setpoint in RFC 3338 format.
  operationTraits:
    mqtt:
      bindings:
        mqtt:
          qos: 1
```

Figure 8.15 AsyncAPI Components, Schemas, and Traits for a Simple HVAC Controller API

In Figure 8.14, the `channels` key defines two channels:

- `hvacTemperatureMonitoring` – a collection of channels with address:

  `temperatureMonitoring/thermostat/{pointName}`

  to which thermostats publish real-time updates when the temperature changes, and the BEMS HVAC load controller subscribes to obtain the updates,

- `hvacLoadControl` – a collection of channels with address:

  `loadControl/setSetpoint/thermostat/{pointName}`

to which the BEMS HVAC load controller publishes changes in the thermostat setpoint, and to which thermostats subscribe to receive setpoint change commands.

The HVAC load controller sends set point control commands to and receives temperature changes from each thermostat on a separate channel, indicated by the channel parameter `pointName`.

The channel name is defined after the `address` key, and the format of the thermostat point name parameter `pointName` in the channel address is specified in the `components/parameters` section. The `messages` section declares message names that can be sent or received on this channel. Under each message name, the message is defined. In Figure 8.14, the `hvacTemperatureMonitoring` channel declares the `temperatureMeasured` message, and the `hvacLoadControl` channel declares the `setTemperatureSetpoint`. Both messages are defined in the `components/messages` section under the `components` top-level key and referenced in the message definition using `$ref`.

The `operations` key declares a top-level section in which each operation has a name subkey, followed by the operation definition. The two operations declared in Figure 8.14 are `getTemperature` for receiving temperature updates and `setSetpoint` for changing the setpoint to accomplish load control. The `action` key specifies which direction the client takes for this action and can be either `send` if the client sends (publishes) on this channel or `receive` if the client receives (subscribes). For the HVAC load control API in Figure 8.14, the API is defined for the cloud-based HVAC Load Controller service, so the `getTemperature` operation is `receive` and the `setSetpoint` operation is `send`.[2] If the API were defined for the thermostat, the opposite actions would hold. The `channel` key specifies the channel on which this operation should take place, using the `$ref` syntax to refer to a named channel definition in the `channels` section. The `traits` key under `operations` provides a way to define protocol-specific features for the operation. The `traits` sections in Figure 8.14 refer to a definition for MQTT in the `components/operationTraits` section `mqtt` subsection.

The `components` top-level section contains subkeys that define reusable components for referencing in other parts of the document, exactly as for OpenAPI. The `components` section in Figure 8.15 illustrates the components for the event-based HVAC Load Control API. The figure contains the following definitions:

- `parameters` –the format for the thermostat channel parameter names,
- `messages` –the format for messages referenceable from the `messages` subkey of the `channels` section,
- `schema` – defines data schema definitions referenceable throughout the document,

2 The AsyncAPI 3.0 specification encourages defining a separate API for the publisher side and subscriber side rather than using the same API to specify both the publisher side and subscriber side.

- `operationTraits` – defines protocol-specific parameter values for the underlying event-based messaging protocol.

The `parameters` section in Figure 8.15 contains the definition of the `pointName` parameter that is a component of the channel address. Unfortunately, AsyncAPI does not provide the same level of rich semantic support for defining parameters that OpenAPI does. The possible definitions include `enum` and `examples`, among a few others. The `enum` key, which can be used to define a limited set of values that the parameter can take, was used in this case to define all the thermostat point names under the control of the HVAC Load Controller, of which only two are shown here. The `examples` key allows representative examples of the parameter to be defined.

The `messages` section in Figure 8.15 contains definitions for the `temperatureMessage` and `setTemperatureSetpointMessage` messages. Each definition consists of a name after the `name` key, documentation after the `summary` key, and definitions for the payloads in the `payload` section. Both definitions refer to the `schema` section for the data schema definitions.

In the `schemas` section of Figure 8.15, two different subkeys are defined for the `messages/payload` sections of the message definitions:

- `temperaturePayload` –the data schema for the payload of the `temperatureMessage` message body definition,
- `setTemperatureSetpointPayload` –the data schema for the payload of the `setTemperatureSetpointMessage` message body definition.

The schemas for the two messages match the schemas for the OpenAPI simple HVAC load control API specification in Figure 8.9.

Finally, in the `components/operationTraits` section, parameters for different underlying event-based application protocols are defined. The subsections are the names of the event-based protocols, and bindings for multiple protocol names can be included. The parameter binding values appear under the `bindings` subkey. In Figure 8.15, the `mqtt` binding is specified, with `qos` of 1 for MQTT messages, corresponding to at least one delivery semantics. AsyncAPI also supports a `messageTraits` section where common sections for messages, like message headers can be defined, see the AsyncAPI specification for details [53]. This is required for some event-based protocols like Kafka.

8.4.3.2 Tools for AsyncAPI

The open-source tooling ecosystem is not as well developed for AsyncAPI as for OpenAPI, but there are options available. The Linux Foundation AsyncAPI project supports a syntax-directed editor called Async API Studio that supports both AsyncAPI 2.6 and 3.0 [55], and Swagger's next-generation editor also supports

AsyncAPI 2.6 in addition to OpenAPI 3.1 [56], but neither have built-in code generation for AsyncAPI like the Swagger editor does for OpenAPI. Both editors are web applications that run in a web browser. The AsyncAPI project also supports a command line interface (CLI) called `asyncapi` with distributions that run on Linux, Windows, and MacOS and can be installed locally. The `asyncapi` tool provides a variety of functions, including code generation and generation of models [57]. Backend code generators are open-source projects and exist for JavaScript, Python, Java, C#, and Go, as well as templates for documentation generation in Markdown, HTML, and other documentation languages. Models are typed data schema for a particular language, and an even broader selection of languages is supported for models. The `asyncapi` tool uses a backend code or document generator which takes a template format that can be customized for particular needs if the current set of templates does not meet them.

8.5 Other API Definition Frameworks

A few other API definition frameworks based on different design premises than OpenAPI and AsyncAPI have gained some amount of deployment. One of the most widely used is GraphQL [58]. GraphQL is a query language for APIs and a runtime for executing those queries by using a type system for the data specified in the API. As the name implies, GraphQL's design was motivated by database query languages such as SQL. Unlike the REST approach, which typically involves accessing multiple endpoints to retrieve different pieces of data, GraphQL allows clients to request what they need and nothing more directly from a single endpoint. The efficiency of data retrieval is achieved through a query syntax that enables specifying fields on objects, nesting queries, and even fetching related data with minimal additional code, making GraphQL an attractive choice for APIs that are primarily concerned with accessing data. GraphQL also offers real-time updates through subscriptions, and its type system helps validate queries against defined data schema. Designed and open-sourced by Facebook in 2015, GraphQL has since been adopted by a wide range of companies and projects seeking more flexibility and efficiency in cloud services networking. An active developer community has grown up around GraphQL, with an annual conference in the fall.

JSON API [59] is a specification for building APIs using JSON and is designed to minimize both the number of requests and the amount of data transmitted between clients and servers. By establishing detailed conventions for actions such as fetching and modifying resources, JSON API streamlines API development and fosters standardized communication for client–server APIs. The specification covers various aspects of API interaction, including requests, responses, data structure, and the management of relationships between resources. Its format includes provisions for handling metadata, errors, and pagination, promoting API consistency and interoperability. While JSON API is not as widely used as OpenAPI, some open-source energy management platforms use it for internal communication between services.

Rest.li [60] is an API framework designed and implemented by LinkedIn for the development of asynchronous, event-based RESTful web services. While it adheres to REST principles for standard operations on resources through resource-oriented URLs and HTTP methods, Rest.li additionally supports asynchronous, event-driven interactions. By incorporating features such as event subscriptions and callbacks, Rest.li enables clients to handle asynchronous operations. It also supports complex data models, strong typing, and selective data fetching to simplify development.

8.6 Zero Trust Security for APIs and Application Protocols

Building a BEMs based on a zero trust architecture requires a layered approach. As discussed in Chapter 2, APIs and application layer protocols depend on firewalls to keep out unwanted traffic and ACLs and node security groups to regulate traffic between the BEMS services on the internal subnets. In addition, the BEMs services depend on the network layer and transport layer mechanisms discussed in Chapter 5 for security services such as confidentiality protection, integrity protection, and data origin authentication. These lower layer services are typically provided by VPNs at the network layer or mutual TLS (mTLS) at the transport layer, which were also discussed in Chapter 5. HTTP and MQTT provide application protocol layer security mechanisms for IAM on fine-grained URL resources and topics. The OpenAPI and AsyncAPI standards contain syntactic support for incorporating these application protocol-level security mechanisms into API definitions. This section discusses API and application protocol security.

8.6.1 VPNs

VPNs are used to connect a building gateway at a remote site to a cloud gateway over the Internet. A VPN ensures that the communication is encrypted and authenticated to prevent eavesdropping and man in the middle attacks for traffic on the public Internet, but it does not provide end-to-end authentication and encryption for nodes that are parties to the communication. Since packets are simply tunneled between the two gateways, the gateways are authenticated to each other, but the nodes behind the gateways are not. Between nodes on the site local network and a cloud BEMS deployment or among the nodes at the site and among the cloud BEMS nodes themselves, identity management and encryption are needed to authenticate the communicating parties and to ensure confidentiality. This prevents an attacker from breaking into the local network through the hard outer shell of the firewall and then moving laterally within the internal network, snooping on unencrypted traffic, and accessing services without authorization.

8.6.2 Mutual TLS

Mutual TLS (mTLS) is commonly used between nodes in cloud services systems for identity management, data origin authentication, and encryption because it requires both parties in a communication transaction to authenticate. With mTLS, both nodes must present X.509 certificates to verify their identity, in contrast with how TLS is typically used for e-commerce and other web applications where only the server needs to present a certificate. Mutual TLS was briefly discussed in Section 5.11. With mTLS, the communicating nodes are provided with confidentiality protection, integrity protection, data origin authentication, and forward secrecy at the transport layer. While mTLS does not provide authorization down to the level of specific HTTP resources or MQTT topics, it does provide authentication of the node and will ensure that an attacker who manages to gain access to a node cannot propagate their attack further without access to the node's private key.

Using mutual TLS between all nodes does come with a deployment cost, however. Nodes must be provisioned with public key certificates, which typically requires deploying the infrastructure for a certificate authority locally, in contrast with using the public key infrastructure for generating server certificates, as is typical for e-commerce and web applications. Procedures need to be in place to rotate certificates and public keys periodically to reduce the likelihood of compromise. To prevent a compromised node from using a leaked certificate and private key, a procedure must be in place to identify leaked certificates and invalidate them.

As discussed in briefly Chapter 5, many web applications use a CRL or the OCSP for this purpose. A CRL is a list of certificates that have been revoked, available from a domain's issuing CA, that is circulated periodically to all nodes in the domain. RFC 5280 contains a specification of the CRL profile format [61] Prior to checking the signature of a certificate on an incoming TLS connection, the node checks whether the certificate is on the CRL and rejects the connection if so. Since CRLS can get quite lengthy for domains that handle traffic from outside sources, OSCP was developed to provide a more dynamic method of handling revocation requests. A node can query the domain local certificate authority using OCSP about the revocation status of a certificate it receives in a TLS connection request [62].

The infrastructure deployment effort required for mutual TLS can be reduced by using the Spiffe specifications [63] for identity management and the Spire open-source implementation of Spiffe [64]. Spiffe, which stands for Secure

Production Identity Framework for Everyone, specifies a URI format and an identity document format for identifying workloads. The URI format consists of the `spiffe` scheme name, followed by a trust domain name and a path specifying the workload, for example `spiffe://`ca.example.com/bems/hvac-controller. Two identity document formats are supported. The X.509 certificate identity document format can be used for mTLS identity management. The certificate contains information identifying the workload, including the URI and other information. The JSON Web Token (JWT) format supports application protocol IAM, for example, in HTTP API requests. JWTs are discussed in detail in Section 8.6.4.3. The specifications are designed so that cross-domain authentication is possible, and several major public cloud providers offer a Spiffe service in conjunction with their base IAM systems.

8.6.3 Application Protocol Identity and Access Management

While VPNs and TLS support basic security services, they do not provide fine-grained IAM services. TLS authenticates one node to another, but typically, a TLS certificate does not provide any authorization information within the certificate defining the permissions a client service has to access resources on a server. For some services, the access permissions can be inferred from the node's host name, but particularly if the API call is coming from a UI with a human user logged in, the host name is not sufficiently granular to determine the access permissions. Both HTTP and MQTT support features for more fine-grained authorization.

8.6.3.1 HTTP Identity and Access Management

REST HTTP APIs utilize a few mechanisms based on the underlying HTTP protocol to prove the right to access a particular resource. One way is to embed the identity and access control information into the URL. For example, user name and password credentials can be included as part of the node name:

```
https://<username>@<password>.example.com/resource
```

or as query parameters:

```
https://example.com/resource?username=<username>&password=<password>
```

This method should never be used with unencrypted HTTP, and even with HTTPs, a risk exists if the building site or cloud deployment requires an HTTP proxy, as the credentials could be stolen if an attacker compromises the proxy since the proxy may need to decrypt the TLS traffic to forward traffic further.

While there are many authorization schemes for web applications in general [65], REST APIs predominantly use the HTTP Authorization header.

The HTTP `Authorization` header provides a way to specify and check fine-grained IAM to server resources [66]. The client embeds credentials in the request using an `Authorization` header and the HTTP server checks the credentials according to a particular authorization method. If the credentials cannot be verified, access to the resource is rejected with a `401 Unauthorized` response. The `Authorization` header has the following format:

```
Authorization <authorization-scheme> <authorization-credentials>
```

The basic authentication scheme uses the `Basic` keyword for the *<authorization-scheme>* value along with a *<username>:<password>* for the *<authorization-credentials>*. The user name and password are separated by a colon character (:). In addition, both the user name and password must be base 64 encoded [67]. Basic authentication is an alternative to passing the authentication information in the URL, and has the same risk of compromise through a HTTP proxy. Bearer authentication uses the `Bearer` keyword for the authorization scheme. The authorization credentials are encapsulated in a token which the client includes in the header. The compromise risk is limited by requiring the token to be periodically refreshed, unlike a user name and password.

Other alternatives are to use a custom header or a custom authorization scheme with the `Authorization` header or to repurpose one of the other HTTP headers for authorization. Section 8.6.4.1 discusses custom schemes in the context of API keys, where they are frequently used.

8.6.3.2 MQTT Identity and Access Management

In MQTT, topics are like URLs in HTTP since they provide access to information on events and issue commands posted by publishing clients to subscribing clients, so any fine-grained IAM method must focus on topic IAM. There are three areas of permission required on topics:

- The allowed topics or wildcarded topics to which a particular client can publish or subscribe,
- The allowed operation: publish or subscribe,
- The allowed QoS level: 1, 2, or 3.

As with any identity and access control scheme, MQTT IAM requires the client to present an attestation of their identity and a policy with respect to their access permissions for these three areas. The proof of identity typically must be provided by the client directly, but the permissions can be stored on an IAM server and accessed by the broker when the client first attempts to establish a session, similar to how Radius is used in Wi-Fi networks, as discussed in Section 5.3.3. Note that while an obvious candidate for authenticating the client's identity might seem to be the TLS certificate, using the TLS certificate from the transport level at the application level is a layer violation and, therefore, not architecturally sound. In addition, multiple clients using the same TLS certificate may be running on the same node. A separate attestation of identity at the application layer is required.

8.6.3.3 MQTT *CONNECT/CONNACK* IAM

MQTT sessions start when the client and broker exchange *CONNECT*/*CONNACK* messages. In MQTT 5.0, the *AUTH* message was added to provide a way for the client and broker to negotiate a longer authentication exchange if required. Two mechanisms are specified for a client to authenticate and prove access authorization for specific topics and operations in the *CONNECT/CONNACK* exchange:

- For simple authentication, include the user name and password in the Payload of the *CONNECT* message and indicate the presence of those values in the message by bits in a Connect Control bitfield located in the Variable Header field.
- For extended authentication, include the authentication method and authentication data properties in the Variable Header Properties subfield of both the *CONNECT* and *CONNACK* messages, and exchange *AUTH* messages if more authentication data is required.

Figure 8.16 shows the 8-bit Connect Control subfield in the *CONNECT* Variable Header that indicates what items are included in the payload. The bit numbers in the field are shown above the bits. From the right the fields are as follows:

- Reserved (bit 0) – reserved, always set to 0,
- Clean Session (bit 1) – indicates whether the connection starts a new session (bit set to 1) or continues with a previous session (bit set to 0),
- Will (bit 2) – when set, indicates that a "will" is present in the Payload field. A will is a message included in the *CONNECT* Payload field that is published by the MQTT broker if it detects that the client has disconnected unexpectedly [68]. The will message serves as a notification mechanism, alerting subscribed clients about the unexpected disconnection of a peer,
- Will QoS (bits 3 and 4) – indicates the QoS of the Will packet. 0x0 corresponds to a QoS of 0, 0x1 to a QoS of 1, and 0x10 to a QoS of 2,

7	6	5	4 3	2	1	0
User Name (1 bit)	Password (1 bit)	Will Retain (1 bit)	Will QoS (2 bits)	Will (1 bit)	Clean Session (1 bit)	Reserved (1 bit)

Figure 8.16 MQTT *CONNECT* Message Connect Control Bitfield

- Will Retain (bit 5) – if the Will bit is 0, this bit must also be 0. If this bit is set to 1, the Will message will be retained when sent; otherwise, it will be deleted,
- Password (bit 6) – if set, indicates the presence of a password subfield in the Payload field. The password is encoded as binary data,
- User Name (bit 7) – if set, indicates the presence of a user name in the Payload field. The user name must be encoded as a UTF-8 string.

Simple authentication requires only user name and password values in the Payload of the *CONNECT* message, indicated by setting the User Name and Password bits in the bitfield. Since the password can be included without a user name if only the password flag is set, the password field can hold a token for token-based authentication, an API key, or any other IAM credentials. If the authentication credentials are acceptable to the broker, it returns a *CONNACK* message with a Reason code of 0x00; otherwise, it returns a code of 0x86, indicating a bad user name and/or password.

MQTT 5.0 introduced a mechanism for performing extended authentication during the *CONNECT/CONNACK* exchange, which is illustrated in Figure 8.17. The authentication method (property type code 0x15) and the authentication data (property type code 0x16) properties are included in the Variable Header Properties subfields of the *CONNECT*, *CONNACK*, and *AUTH* messages to facilitate a challenge-response authentication exchange between the client and the broker. The client initiates the process by including the authentication method property in the *CONNECT* message. The authentication method property specifies the name of the authentication method as a UTF-8 string. The client can optionally include an authentication data property with initial binary data if it is required by the requested authentication method. If the broker does not support the authentication method, it returns a *CONNACK* message with one of the Reason error codes.

If more data is required, the client and broker exchange *AUTH* messages with the authentication name and data properties in the *AUTH* message header with Reason code 0x18, indicating the authentication exchange should continue in the *AUTH* messages. Once the authentication completes, the broker sends a *CONNACK* message to the client. The *CONNACK* message header must include the authentication method property with the same method name as the client sent in the *CONNECT* message and optionally may contain an authentication data property containing any access credentials required for authorizing further traffic. The *CONNACK* message includes one of the following Reason codes depending on the outcome of the authentication:

- 0x00 – authentication completed successfully,
- 0x8c – bad authentication method,
- 0x87 – not authorized,

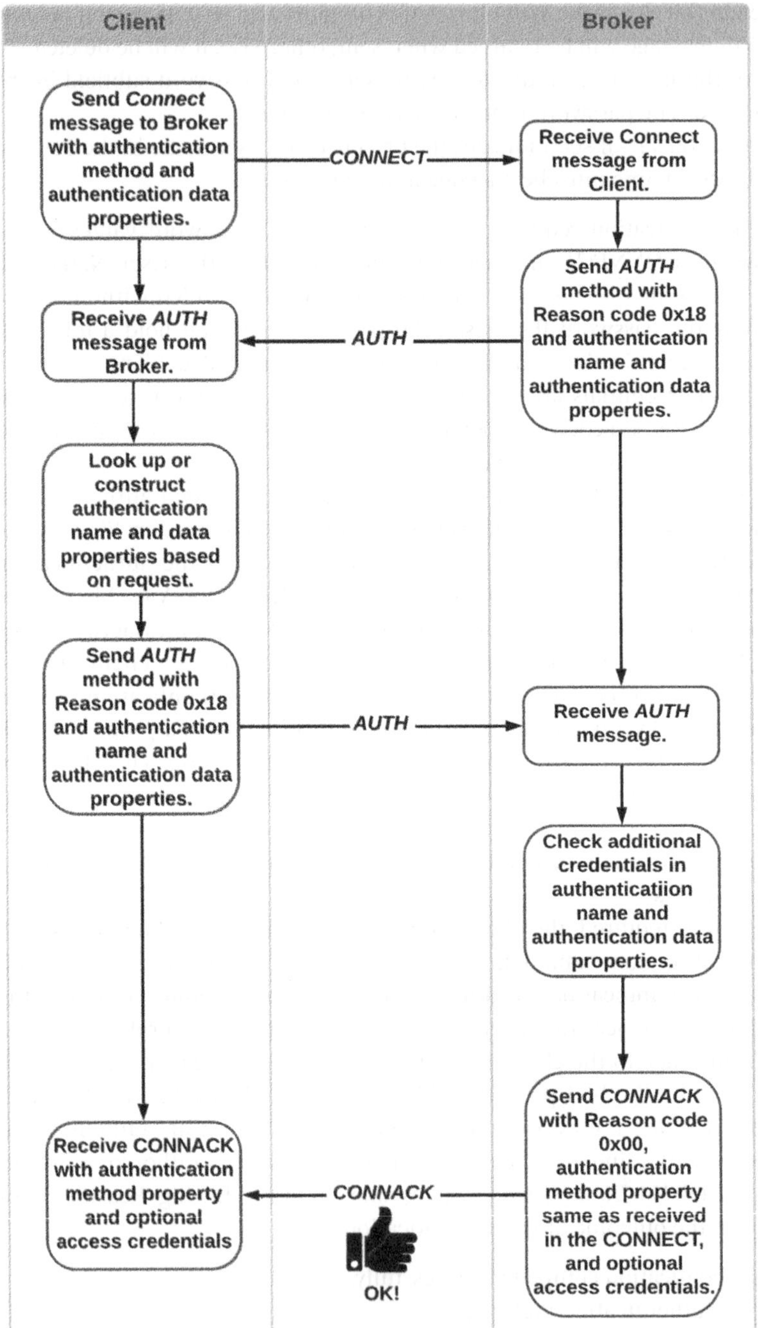

Figure 8.17 MQTT Advanced Authentication Exchange

- 0x80 and above – codes involving protocol errors that can be used by the broker to prematurely terminate an authentication session if some error is detected.

Codes 0x8c or 0x87 are returned if the client's proposed authentication method is not supported by the server. MQTT 5.0 also provides additional mechanisms for reestablishing IAM credentials at any time, see Section 2.12.1 of the MQTT 5.0 specification [69].

8.6.3.4 MQTT Topic Access Filters

Topic access filters for permitted topics can be communicated either in data sent to the client by the broker in the authentication data field, indicating what topics the broker's policy has authorized for the client, or can be sent from the client to the broker in the password Payload value or authentication data field if the client is presenting some form of authentication credential, such as a token issued by a third party IAM server. The filters can include wildcard characters (* and #) to indicate that the client is allowed to access multiple subtopics under a single topic header. The filters can additionally include the operations (publish or subscribe) the client is authorized for and which QoS levels the client is allowed to use for the permitted topics. The MQTT 5.0 specification document contains no format for specifying policy, so any policy implementation is vendor specific.

8.6.3.5 Allow/Deny Responses on MQTT *Publish* and *Subscribe* Messages

Once the client is authenticated, it can start publishing and subscribing to topics allowed by the access policy and contained in the access credentials returned during the authentication process. If a client publishes on topics to which it is allowed access, the returned *PUBACK* or *PUBREC* message includes a Reason code of 0x00 in its Variable Header field. For subscriptions, the *SUBACK* message will similarly include a Reason code of 0x00 in its Payload field if the topic is allowed. If, however, the client attempts to publish on a topic for which it is not authorized, the broker will return a Reason code of 0x87, not authorized, in the *PUBACK*, *PUBREC*, and *SUBACK* messages. The broker also has the option of disconnecting the client if it tries to publish on a topic where it does not have permission.

A few Reason codes can be used to inform subscribing clients about their requests for a specific QoS level. If the client is authorized to subscribe to the topic, the *SUBACK* message can return one of the following Reason codes indicating that the subscription was accepted for a QoS level:

- 0x00 – the subscription was accepted with a QoS level of 0,
- 0x01 – the subscription was accepted with a QoS level of 1,
- 0x02 – the subscription was accepted with a QoS level of 2.

The broker can downgrade the QoS level on an accepted subscription if the client requests but is not authorized for a higher level, then inform the client in the *SUBACK* of the QoS level at which the message was published.

However, for publishing clients, there is no similar set of informative Reason codes indicating how their requests for a particular QoS level are handled. Assuming the client is authorized to publish to a particular topic, the broker can downgrade the QoS level and accept the publication at the downgraded level. Alternatively, the broker can issue a Reason code of 0x87 indicating that the client is not authorized to publish at that QoS level, but it is up to the client to determine that the problem was not lack of permission on the topic but lack of permission for the QoS level.

8.6.4 IAM Frameworks

Application-level protocols depend on an IAM framework to provide an attestation of a client's identity and an access control policy about what resources or topics a client is allowed to access. Unlike web applications, where a human user can provide input for identity attestation, most BEMS services need to operate in an automated fashion without the benefit of human input. The major cloud providers have custom IAM systems for IoT services that provide a fully managed solution offering regulatory compliance to customers, but these systems lock the service into the vendor's cloud platform. Other solutions that can be built on top of protocols and infrastructure frameworks developed for web applications are discussed in this section.

8.6.4.1 API Keys

API keys bootstrap API security by requiring a human user or a service with an established, attestable identity to log into a website and obtain an opaque token, which can then be used to access the API. The server that runs the website issuing the token maintains the link between human user's or services's identity, the token, and the set of access permissions attached to it and enforces these permissions on API clients. The API client must pass the API key in every API call.

With HTTP, there are a variety of ways to pass the key. Some older APIs pass the API key as a request body parameter. Another way is to use a URL query parameter, as with the user name and password discussed in Section 8.6.3.1. For example:

```
https://example.com/resource?api-key=<api-key>
```

Most modern APIs pass the key as a header parameter. The most widely deployed way is to use a custom header. Custom headers must be prefixed with `X-` to indicate that the header is nonstandard. The most frequently used custom header is the `X-API-KEY` custom header with an API key included as the parameter, for example:

```
X-API-KEY <api-key>
```

Another common choice is to use the `Authorization` header The API key can be passed using the `Basic` scheme by base 64 encoding the API key and providing it as the *<authorization-credential>* instead of a user name and password. Although the `Bearer` authorization scheme was specifically developed for tokens, some APIs use it with API keys. Others use a custom *<authorization-scheme>* name in the `Authorization` header, such as `Api-key` or `X-API-KEY`. A less common way is to pass the API key as a `Cookie` header parameter.

Some MQTT IoT services use API keys, typically with basic authentication. A user name is included in the user name Payload field, and the API key is included in the password field. Alternatively, the API key may be included in the user name field and a token used in the password field. The advantage of the API key is that it decouples the authentication credential for the IoT service from the user's more long-lived password and can be given a more selective, fine-grained set of access permissions. The API key could also be passed in the authentication data property with extended authentication.

There are a few best practices associated with API keys that can help maintain good security [70]. The API key embodies the application or user's identity and access rights, so it needs to be unique, random, and not guessable, similar to a UUID. API keys can use digits, alphabetic characters, and special characters such as period (`.`) and question mark (`?`). Since an API key is similar to a password, it should never be stored in clear text in a database on the API server but rather hashed. An incoming API key on an API REST request can then be hashed and compared with the value in the database. However, this presents a usability problem since the server's website will not have a clear text copy of the API key after it has been generated. For this reason, most server web pages that issue API keys provide a warning that the key cannot be regenerated. Another technique is to prefix the API key with a unique sequence of letters. The server can then display the prefix on a web page if the user wants to modify the permissions associated with the key. API keys should be scoped to logical groups of API calls and not issued with permissions for every API call offered by the service. API calls should also be rate-limited, so if a key is compromised, an attacker cannot launch a denial-of-service attack on the service.

One drawback with API keys is that most services issue them with unlimited lifetimes unless they are revoked, which can increase the risk of compromise. As with any IAM credential sent over the public Internet, the lifetime of an API key should be limited, and the API client should be required to refresh the key periodically in case the key is compromised, using a credential that is not exposed to the public Internet. Infrastructure support for automating API key refresh using a private credential is not strong, and refreshing the key typically requires a human user to log into a website and obtain a new key. While service-to-service IAM can, in principle, be anchored in an API key, since the regeneration of the keys on a periodic basis requires human input, using API keys for a BEMS with many services is impractical.

8.6.4.2 OAuth 2.0

OAuth 2.0 is a token-based IAM framework that was developed for web application IAM, where a human user running a browser client is trying to access a resource owned by another party on a resource server. OAuth 2.0 is specified in RFC 6749 [71] and maintained through the OAuth 2.0 website [72]. An authorization server controls access to all resources. The authorization server issues a token to the client which allows the client to access the resource and can revoke issued tokens if there is evidence of compromise, so the authorization server and the resource server must trust each other. The access token, which is exposed to the public Internet in API calls, has a bounded lifetime and can be granted to the client together with a refresh token, which the client keeps private. The client uses the refresh token to periodically obtain a new access token when the lifetime expires.

OAuth 2.0 defines a grant type as a standardized method involving a series of actions on the part of the client user and resource owner and protocol exchanges between the client, the authorization server, and the resource server, in which the client acquires an access token and presents it to the resource server to gain access. OAuth 2.0 supports four different types of grants: authorization code, Proof Key for Code Exchange (PKCE) [73], device code, and client credentials. In addition, there are two grant types that have been deprecated because they are less secure: implicit grant and resource owner credentials. Since OAuth 2.0 was initially designed primarily to support human-in-the-loop use cases, all but the client credentials grant type require human interaction on the part of the client user or resource owner. However, the device code grant type, which was specifically developed for commissioning smart home devices, and the client credentials grant type can be deployed for authorizing devices and services in an automated fashion.

Figure 8.18 illustrates the device code grant-type flow. The device code grant type is specified in RFC 8628 [74] and was added in 2019 to allow smart home devices to obtain an access token and an optional refresh token. The device code flow works similarly to the Bluetooth Passkey Entry authentication, described in Section 5.6.5. The process starts when the human user initiates commissioning on the device. The device requests a device code from the authorization server's device authorization URI. The authorization server returns the device code, along with a user code, its verification URI, the lifetime that the verification remains outstanding, and an optional time interval between polling requests.

The device displays the user code, verification URI, and verification lifetime to the human user on a display and prompts the user to visit the verification URI and

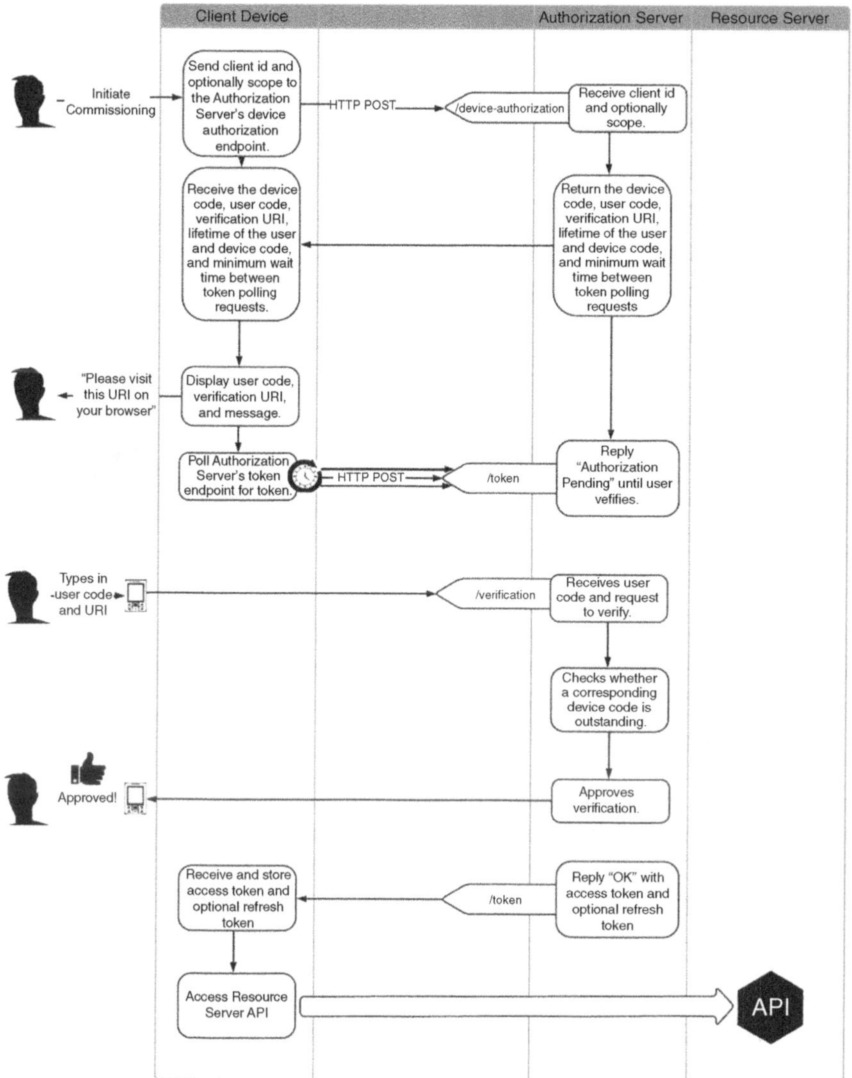

Figure 8.18 Device Code Grant Type for OAuth 2.0

approve the request. The human user has until the verification lifetime expires to approve the request. The device then begins polling the authorization server's token endpoint for the token, and the authorization server returns an authorization pending indication. Meanwhile, the human user visits the verification URI on their mobile phone or laptop and types in the user code when prompted.

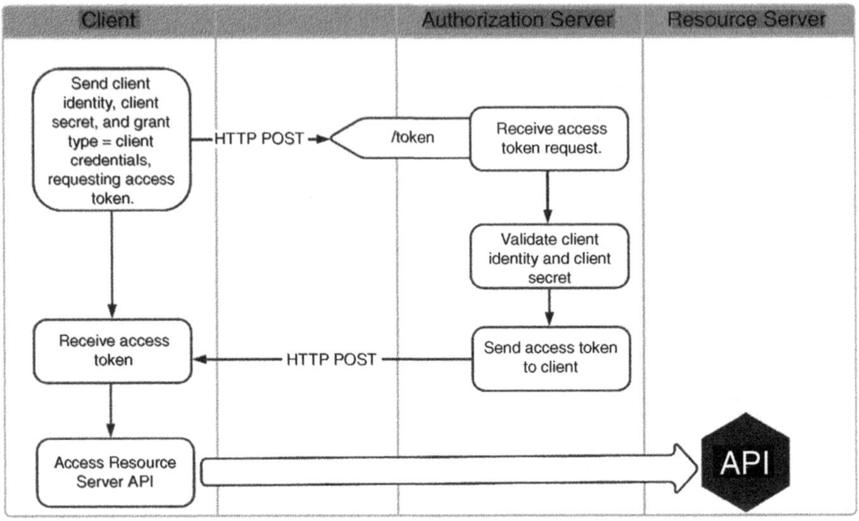

Figure 8.19 OAuth 2.0 Client Credentials Protocol Flow

The authorization server approves the request, and on the next poll, it returns an access token and optional refresh token to the device. The device then presents the access token when it accesses the API.

While the device code flow works well for devices that have a display to prompt the human user, for service-to-service or device-to-service authorization where there is no human user in the loop, the client credentials grant-type flow is the only flow that works. Figure 8.19 illustrates the protocol exchanges for the client credentials flow. The client requests a token by sending an HTTP `POST` to the authorization server with its client identity, client secret, and an indication that the flow is the client identity flow. The authorization server validates the client's credentials then returns the access token by HTTP `POST`. The client can access the resource on the resource server using the token until the token expires. One disadvantage of the client credentials flow is that it requires a prearranged shared secret, namely the client password, between the authorization server and the client application. The other OAuth 2.0 grant-type protocol exchanges are designed so that no prearranged secret is required since a human user provides the approval for the token request.

8.6.4.3 JSON Web Tokens (JWT)

The OAuth 2.0 tokens can be opaque bearer tokens similar to API keys but they are usually formatted as a JWTs, pronounced "jot". JWTs are specified in RFC 7519 [75], and the specification is maintained through the jwt.io website [76]. JWTs contain three sections that are all base 64 URL encoded: the header, payload, and signature. The header and payload sections are both JSON objects, while the signature provides authentication on the header and payload. In OAuth 2.0 authentication exchanges, the token is typically not encrypted if the exchange runs over TLS, but for exchanging information privately, encryption can be applied as well.

The header contains the control information for the token. For OAuth 2.0 JWTs, the following properties are commonly included in the header:

- `alg` – the signing algorithm used for the token. Common values are `hs256`, `es256`, and `rs256`, indicating the shared key HMAC, public key elliptic curve, and RSA signing algorithms, respectively. The signing algorithm is used to sign a SHA-256 digest of the token's header and parameter sections,
- `typ` – the token type, usually set to `JWT`,
- `kid` – if a public key signature algorithm is used for the signature, `kid` provides the key identifier for retrieving the public key from the authorization server. The key identifier is frequently included when the authorization server supports multiple signing keys, especially in systems that rotate keys.

The header can also include other, less commonly used properties, but the number of properties and size of their values is typically limited to keep the JWT size compact.

If a shared key algorithm such as HMAC is used to sign the JWT, the key must be pre-shared between the client and server. If one of the public key algorithms is used, OAuth 2.0 supports a lightweight certificate in a format called JSON Web Key (JWK), which is a way for a resource server to obtain the public key of the client without incurring the administrative overhead of having to utilize a certificate and deploy a certificate authority. The JWK and JSON Web Key Set (JWKS) specifications are defined in RFC 7517 [77]. The authorization server exposes a JWK endpoint from which the resource server can look up the client's key identifier provided in the `kid` property if it is included in the header. If the client has more than one key, the authorization server will return a JWKS set, and the resource server can look up the key identifier in the set.

The payload contains the access claims asserted by the token holder. There are three types of claims defined in the payload:

- Registered claims – a set of predefined claim properties that are not mandatory but recommended for interoperability. The registered claims are discussed in more detail below,
- Public claims – claims that are either defined in the IANA JWT Registry [78] or are defined as a URI whose scheme implements a collision-resistant namespace,
- Private claims – custom claims created to share information between parties that agree on using them and are neither registered nor public.

Registered claims are defined in RFC 7519 and consist of the following properties and their values:

- `iss` (Issuer) – who issued the JWT,
- `sub` (Subject) – the subject of the JWT, which is usually an abbreviated but unique name identifying the service, device, or user that is presenting it,
- `aud` (Audience) – the recipients that the JWT is intended for, which could be the service or device that the client is attempting to access,
- `exp.` (Expiration Time) – the expiration time on or after which the JWT must not be accepted for processing in NumericDate format [46],
- `nbf` (Not Before) – the time before which the JWT must not be accepted for processing in NumericDate format,
- `iat` (Issued At) – the time at which the JWT was issued, in NumericDate format,
- `jti` (JWT ID) – a unique identifier for the JWT.

Among the public claims, three that can be used in a BEMS are `name`, `admin`, and `scope`. The `name` property is a string with a longer, human-readable name, such as `"Building 12, Room 200, Thermostat 322"`. The `admin` property is a Boolean indicating whether the holder has administrative permissions on the resource server. The `scope` property is a space-separated string indicating the resources and their access permissions granted to the holder. For example:

```
"publish:temperature subscribe:setpoint"
```

might be used to define the permissions for a thermostat in the HVAC load control service defined in Section 8.4.3. The permissions on URL resources for an OpenAPI/HTTP client or topics for an AsyncAPI/MQTT client could also be defined in the private claims section.

The signature is calculated by forming a base 64 URL encoding the header and payload separately and concatenating the encoded values with a period. Base 64 URL encoding requires first performing standard base 64 encoding in non-padding mode, then substituting an underscore (_) for URL-reserved characters such as plus (+), minus (-), and forward slash (/). The concatenated string is then hashed with the hashing algorithm to create a digest, signed using the signature algorithm, and base 64 URL encoded to form the signature section. The final JWT is constructed by concatenating all three base 64 URL encoded sections separated by a period:

```
<base 64 URL encoded header> . <base 64 URL encoded
payload> . <base 64 URL encoded signature>
```

The JWT can then be used in API calls.

For HTTP APIs, the client includes the JWT as the `<authorization-credentials>` in the `Authorization` header with the `Bearer` `<authorization-scheme>`:

```
Authorization Bearer <JWT>
```

Upon receipt, the server checks the signature and enforces the permissions in the `scope` string or private claims.

While RFC 6749 states that OAuth is explicitly designed for HTTP, it is possible to use it with MQTT. In basic authentication, the JWT can be included as the Payload field password value with the user name value set to the same value as the `sub` property in the JWT header. In extended authentication, the JWT can be included in the authentication data property, and the authentication method property can be set to `OAuth 2.0`. No further *AUTH* message exchanges should be required as the JWT should be sufficient to authenticate the client and provide access permissions unless the permissions are too large to fit into a single JWT. This may require the server to request any additional claims in an *AUTH* message follow-up.

8.6.4.4 Open Source Tools Supporting OAuth 2.0

OAuth 2.0 support is available through a wide variety of open-source tools. One of the most comprehensive is Keycloak [79]. Keycloak is a comprehensive open-source IAM system that offers many features, such as user federation, identity brokering, and social login. Social login is described in more detail in the next section. Keycloak simplifies deploying secure applications and services by requiring little to no additional code. OAuth2-Proxy is a reverse proxy that provides authentication with Google, GitHub, or other authentication providers, which can validate and control access to various resources [80]. Gluu is an IAM platform that supports OAuth 2.0 using low-code programming techniques [81]. Both Keycloak and Gluu support other IAM standards, such as OIDC, UMA, SAML, and FIDO in addition to OAuth 2.0.

8.6.4.5 Other Token-Based Frameworks

A few other token-based IAM frameworks are in common use, especially in enterprise environments. OpenID Connect (OIDC) [82] is an authentication layer on top of OAuth 2.0 that provides user authentication and single sign-on (SSO) functionality. OIDC uses JWTs to secure the identity tokens passed back from an identity provider to a client application, and includes standardized scope identifiers for access control permissions. SSO allows a user to log in once and reuse their token on multiple websites. OIDC is sometimes called "social login" because many websites display a list of OIDC identity providers, such as Google and Facebook, on their login page in addition to a standard user name and password form, allowing users to log in with one of the OIDC providers and not have to set up a user name and password at the site.

Security Assertion Markup Language (SAML) [83] is an XML-based framework for exchanging authentication and authorization data between an identity provider and a service provider. SAML is widely used for SSO between intra-net web applications in enterprise environments. Finally, Web Services Federation (WS-Federation) [84] is a specification that allows different security realms to broker information on identities, identity attributes, and authentication. It uses the concept of federation to enable interoperable SSO and is often used in enterprise deployments involving Microsoft products. Each framework serves slightly different deployment use cases. While OIDC and OAuth 2.0 are commonly used in modern web and native applications, SAML and WS-Federation are often found in enterprise environments with legacy systems.

8.6.5 Specifying IAM in OpenAPI and AsyncAPI Definitions

The OpenAPI and AsyncAPI standards support syntax allowing API designers to specify what sort of security to use in an API. During code generation, the specifications are then translated into code in the implementation program language that constructs the protocol messages in the application protocol using the specified IAM.

8.6.5.1 OpenAPI Security

In OpenAPI, security schemes are defined by the API developer in the `components` section using the `securitySchemes` key. This section contains a list of named security schemes for reference elsewhere in the document. Within a named scheme, the `type` key indicates which type of IAM to use. Valid type names are `apiKey`, `http`, `oauth2`, `openIdConnect`, and `mutualTLS` (for OpenAPI 3.1). For each type, keys specify scheme-specific parameters. In addition, each scheme can have a `description` key followed by a string documenting the scheme.

The `apiKey` type requires the `in` key whose value specifies the location of the API key (header parameter, query parameter, or cookie), and the `name` key specifies the name of the parameter if it is a header or query parameter. For example:

```
hvac_iam:
    type: apiKey
    name: X-API-KEY
    in: header
```

defines a scheme called `hvac_iam` of type `apiKey`, which is passed through the custom header `X-API-KEY`.

The `http` type requires the additional `scheme` key, which defines the authorization scheme and must be a standard scheme name from the IANA HTTP

Authentication Scheme registry [85]. An optional `bearerFormat` key can be included, indicating the format of the token for the `Bearer` scheme. Using HTTP authorization for the previous example gives the following:

```
hvac_iam:
    type: http
    scheme: bearer
    bearerFormat: JWT
```

The `oauth2` type has the most complicated set of additional specifications. The `flows` key indicates which of the grant-type flows are required. OpenAPI 3.1 supports implicit, authorization code, resource owner credentials, and client credentials, and has not yet added the device code type or deprecated the implicit and resource owner types. Under the `flows` section, the allowed flows are listed, identifying the flow, and additional required information is provided. An authorization URL is included as the value of the `authorizationUrl` key, and the URL value can be relative to the servers declared in the `servers` section at the beginning of the API definition. The `tokenUrl` key provides a similar URL from which the API can fetch the access token, and the optional `refreshUrl` key provides a URL for fetching a refresh token. Finally, the `scopes` key provides a list of named scopes with a description of the scope.

Returning to the previous example, here is the HVAC load controller `securitySchemes` definition using the OAuth 2.0 client credentials flow:

```
hvac_iam:
    type: oauth2
    flows:
       clientCredentials:
          authorizationUrl: /authorization
          tokenUrl: /token
          refreshUrl: /refresh
          scopes:
             read:temperature: Read
             temperature from thermostats
             write:setpoints:  Write
             setpoint to thermostats
```

Once the security schemes are defined, they can be used in other parts of the specification such as in the `paths` section.

The `security` section is used in the `paths` section to indicate that one or more security schemes should be applied. The schemes are included as a YAML list of key-value pairs after the scheme name. The values of the scheme keys are a list of the scope names to be applied to the scheme. Here is an example of adding security to the HVAC load control API `temperature` operation from Figure 8.9 using the OAuth 2.0 `hvac_iam` scheme defined above:

```
paths:
  /temperature:
    get:
      summary: Get current temperature
      description: Retrieve the current temperature from the HVAC system.
      security:
          - hvac_iam:  [read:temperature]
      responses:
          '200':
          ...
```

In this case, only the `read:temperature` scope is needed.

8.6.5.2 AsyncAPI Security

AsyncAPI uses a similar approach to defining security schemes as OpenAPI. Security schemes are defined in the `securitySchemes` subsection of the `components` section. The format of the scheme definitions in the user-defined scheme sections is identical to OpenAPI, with the same key names. The `description` key can also be included, with a short description of the security scheme. The primary difference from OpenAPI is in scheme types, which are values of the `type` key. AsyncAPI supports more types than OpenAPI, and the `http` and `apiKey` types are slightly different:

- `userPassword` –user name and password transferred in a protocol-specific manner,
- `X509` – mutual TLS using X.509 certificates,
- `symmetricEncryption` – end-to-end encryption using a shared key cryptosystem,
- `asymmetricEncryption` – end-to-end encryption using a public key cryptosystem,
- `apiKey` – API key for non-HTTP protocols, with the location specified by the `in` key. Valid values for the `in` key are `user` or `password`, indicating that the API key is passed in the user name or the password field, respectively,
- `http` – HTTP `Authentication` header with an IANA standard scheme such as `basic` designated as the value of the `scheme` key. If the `scheme` key value is `bearer`, the `bearerFormat` key can also be included with the value being the format of the bearer token, as in OpenAPI,
- `httpApiKey` – HTTP authentication with the API key in a query, header, or cookie parameter specified with the `in` key for the location of the API key and the `name` key for the name of the query or header parameter as in OpenAPI. Valid values for the `in` the key are `query`, `header`, or `cookie`,

- `plain` – authentication based on the Simple Authentication and Security Layer (SASL). SASL is a framework for providing security with pluggable authentication and encryption described in RFC 4422 [86]. The plain type uses SASL plain text authentication.
- `scramSha256` and `scramSha512` – these two schemes use the Salted Challenge Response Authentication Mechanism (SCRAM) mechanism [87] [88] for salted SHA256 and SHA512 hashed passwords with SASL to obscure the actual password value,
- `gssapi` – authentication based on the Generic Security Service API (GSS-API), another framework for providing security services in a portable manner [89].

The `plain`, `scramSha256`, `scramSha512`, and `gssapi` schemes are primarily to support the enterprise security needs of the Kafka publish–subscribe system. Lane [90] has a guide to the different types of Kafka security and how they map them into the AsyncAPI security types.

In AsyncAPI, the security schemes are applied at the `operations` level using the same `security` key as for OpenAPI. The MQTT HVAC load control API discussed in Section 8.4.3 can define an API key security scheme in the `securitySchemes` section as:

```
securitySchemes
    api_key:
    type: apiKey
    in: password
```

This uses MQTT basic authentication to pass the API key through the *CONNECT* password Payload field.

The security on the `getTemperature` operation in the HVAC load control API in Figure 8.14 is specified as:

```
operations:
  getTemperature:
    action: receive
    security:
      api_key: []
    summary: Receive the temperature from
    thermostat.
    channel:
      $ref: '#/channels/hvacTemperature
      Monitoring'
    traits:
      - $ref: '#/components/operation
      Traits/mqtt'
    ...
```

8.7 Summary

The original concept of the Internet application layer envisioned different application layer protocols for every application, often using a binary encoding. But after development of the World Wide Web and HTML, application designers began to use HTTP as a common protocol for multiple client-server applications and MQTT and other event-based protocols for event-based/publish-subscribe applications. Built on top of a common application protocol, APIs provide a layer of abstraction more tailored to the application design than the underlying application protocol. An API also hides unnecessary details from the application developer exposing only what the developer needs to accomplish the API's job.

This chapter presents a comprehensive look at application layer protocols, the APIs defined on top of them, and the application architectures that can be designed with functional components built on top of the protocols and APIs. Modern API technology evolved out of RPCs and Section 8.1.1 discusses the similarities and differences between the two. Updated RPC technology is still used in today's architectures, most notably in gRPC [21]. Section 8.2 contains a discussion of the most commonly used architectural design patterns in distributed systems, namely the client–server and event-based patterns.

Section 8.3 discusses the structured data representation languages and protocols used to define APIs for the client–server and event-based architectures. URIs, URNs, and URLs are one representation for named objects and these are discussed in Section 8.3.1. These naming schemes are primarily used with HTTP. The HTTP protocol and its use in REST APIs is the topic of Section 8.3.2. MQTT is a common choice for event-based APIs in IoT systems, primarily due to its lightweight, high-performance nature. MQTT is discussed in Section 8.3.3, but other event-based protocols are in common use so a short discussion of other message bus protocols is included in Section 8.3.4. Standardized languages for representing APIs are based on the JSON and YAML data representation languages. These data representation languages are discussed in Section 8.3.5.

The API standards themselves are then discussed, using an API for an HVAC load control service running in a cloud-based BEMS with thermostats at a remote building site introduced in Section 8.4.1, as an example. Section 8.4.2 uses OpenAPI 3.0/3.1 standard to develop an API design on top of HTTP from a synchronous client–server API through an asynchronous event-based API using callbacks, webhooks, and server-sent events. Section 8.4.3 uses the AsyncAPI standard to define an asynchronous event-based API. Section 8.5 briefly examines a few other API standards that have achieved some amount of deployment, namely, GraphQL, JSON API, and Rest.li.

Security is critical to any IoT system, and zero trust security requires attention to threat mitigation through proper design and deployment of technology features at every layer of the BEMS stack. How to secure lower layers with VPNs and mutual TLS is presented in Sections 8.6.1 and 8.6.2, followed by a detailed exploration of IAM for application layer security across the HTTP and MQTT protocols in Section 8.6.3. The API keys and OAuth 2.0 IAM frameworks for providing IAM security services at the API layer are discussed in detail in Section 8.6.4, and open-source tools for OAuth 2.0 and other token-based IAM frameworks that may be of interest in building a BEMS are mentioned in the section. Finally, the chapter concludes with guidance and a few examples of integrating IAM into OpenAPI and AsyncAPI definitions in Section 8.6.5.

References

1 Fielding R., Ed., Nottingham M., Ed., and Reschke J., Ed. HTTP Semantics, RFC 9110 [Internet]. [cited 2023 Sep 27]. Available from: https://www.rfc-editor.org/rfc/rfc9110.html.

2 OASIS. MQTT Specifications [Internet]. [cited 2023 Sep 27]. Available from: https://mqtt.org/mqtt-specification.

3 Bray T. The JavaScript Object Notation (JSON) Data Interchange Format, RFC 8259 [Internet]. [cited 2023 Oct 2]. Available from: https://datatracker.ietf.org/doc/html/rfc8259.

4 Yaml.org. YAML™ Specification Index [Internet]. [cited 2023 Oct 5]. Available from: https://yaml.org/spec.

5 Wikipedia. REST [Internet]. [cited 2023 Sep 29]. Available from: https://en.wikipedia.org/wiki/REST.

6 Open API Initiative. OpenAPI Specification v3.0.3 [Internet]. [cited 2023 Oct 8]. Available from: https://spec.openapis.org/oas/v3.0.3.

7 Linux Foundation AsyncAPI Project. Welcome to the AsyncAPI Docs! [Internet]. [cited 2023 Sep 27]. Available from: https://www.asyncapi.com/docs.

8 Fielding R.T. Architectural Styles and the Design of Network-Based Software Architectures. University of California, Irvine; 2000.

9 W3C Foundation. HTML Standard [Internet]. [cited 2023 Oct 2]. Available from: https://html.spec.whatwg.org.

10 W3C Foundation. SOAP Specifications [Internet]. [cited 2023 Oct 2]. Available from: https://www.w3.org/TR/soap.

11 W3C Foundation. Extensible Markup Language (XML) 1.0 (Fifth Edition) [Internet]. [cited 2023 Oct 2]. Available from: https://www.w3.org/TR/xml.

12 Flanagan D. JavaScript: The Definitive Guide. 7th ed.Sebastopol: O'Reilly Media; 2020.

13 Berners-Lee T., Fielding R., and Masinter L. Uniform Resource Identifier (URI): Generic Syntax, RFC 3986 [Internet]. [cited 2023 Oct 3]. Available from: https://www.rfc-editor.org/rfc/rfc3986.html.

14 IANA. Uniform Resource Identifier (URI) Schemes [Internet]. [cited 2023 Oct 3]. Available from: https://www.iana.org/assignments/uri-schemes/uri-schemes.xhtml.

15 Kindberg T. and Hawke S. The 'tag' URI Scheme, RFC 4151 [Internet]. [cited 2023 Oct 13]. Available from: https://datatracker.ietf.org/doc/html/rfc4151.

16 Wikipedia. Uniform Resource Name [Internet]. [cited 2023 Oct 14]. Available from: https://en.wikipedia.org/wiki/Uniform_Resource_Name.

17 Berners-Lee T., Masinter L., and McCahill M. Uniform Resource Locators (URL), RFC 1738 [Internet]. [cited 2023 Oct 3]. Available from: https://www.rfc-editor.org/rfc/rfc1738.

18 Mozilla Foundation. Evolution of HTTP [Internet]. [cited 2023 Sep 27]. Available from: https://developer.mozilla.org/en-US/docs/Web/HTTP/Basics_of_HTTP/Evolution_of_HTTP.

19 Fielding R., Gettys J., Mogul J. et al. Hypertext Transfer Protocol -- HTTP/1.1, RFC 2616 [Internet]. [cited 2023 Nov 10]. Available from: https://datatracker.ietf.org/doc/html/rfc2616.

20 Thomson M., Ed. and Benfield C., Ed. HTTP/2, RFC 9113 [Internet]. [cited 2023 Nov 10]. Available from: https://www.rfc-editor.org/rfc/rfc9113.

21 grpc.io. gRPC [Internet]. [cited 2023 Nov 9]. Available from: https://grpc.io.

22 Wikipedia. List of HTTP header fields [Internet]. [cited 2023 Oct 4]. Available from: https://en.wikipedia.org/wiki/List_of_HTTP_header_fields.

23 Wikipedia. List of HTTP status codes [Internet]. [cited 2023 Oct 4]. Available from: https://en.wikipedia.org/wiki/List_of_HTTP_status_codes.

24 curl.se. curl.1 the man page [Internet]. [cited 2023 Nov 30]. Available from: https://curl.se/docs/manpage.html.

25 curl.se. Releases and Downloads [Internet]. [cited 2023 Nov 30]. Available from: https://curl.se/download.html.

26 Ubuntu. httping - measure the latency and throughput of a webserver [Internet]. [cited 2023 Nov 30]. Available from: https://manpages.ubuntu.com/manpages/trusty/man1/httping.1.html.

27 OASIS. MQTT Version 5.0 [Internet]. [cited 2023 Oct 9]. Available from: https://docs.oasis-open.org/mqtt/mqtt/v5.0/os/mqtt-v5.0-os.html.

28 Section 2.1.3 Flags, MQTT Version 5.0 [Internet]. [cited 2024 March 17]. Available from https://docs.oasis-open.org/mqtt/mqtt/v5.0/os/mqtt-v5.0-os.html#_Toc3901023.

29 Section 1.5.5 Variable Byte Integer, OASIS MQTT Version 5.0 [Internet]. [cited 2024 March 17]. Available from https://docs.oasis-open.org/mqtt/mqtt/v5.0/os/mqtt-v5.0-os.html#_Toc3901011.

30 Section 2.2.2 Properties, OASIS MQTT Version 5.0 [Internet]. [cited 2024 March 17]. Available from https://docs.oasis-open.org/mqtt/mqtt/v5.0/os/mqtt-v5.0-os.html#_Toc3901027.

31 Amazon Web Services. MQTT [Internet]. [cited 2023 Oct 11]. Available from: https://docs.aws.amazon.com/iot/latest/developerguide/mqtt.html.

32 Microsoft. Communicate With an IoT Hub Using the MQTT Protocol [Internet]. [cited 2023 Oct 11]. Available from: https://learn.microsoft.com/en-us/azure/iot/iot-mqtt-connect-to-iot-hub.

33 Eclipse Foundation. Eclipse Mosquitto™: An open source MQTT broker [Internet]. [cited 2023 Oct 11]. Available from: https://mosquitto.org.

34 Eclipse Foundation. Eclipse Paho [Internet]. [cited 2023 Oct 11]. Available from: https://projects.eclipse.org/projects/iot.paho.

35 AMQP.org. AMQP: Advanced Message Queuing Protocol [Internet]. [cited 2023 Oct 10]. Available from: https://www.amqp.org.

36 Apache.org. Kafka Protocol Guide [Internet]. [cited 2023 Oct 10]. Available from: https://kafka.apache.org/protocol.html.

37 Rabbitmq.com. RabbitMQ: Easy to Use, Flexible Messaging and Streaming [Internet]. [cited 2023 Oct 10]. Available from: https://www.rabbitmq.com.

38 Fette I. and Melnikov A. The WebSocket Protocol, RFC 6455 [Internet]. [cited 2023 Oct 10]. Available from: https://datatracker.ietf.org/doc/html/rfc6455.

39 Thomson M. and Benfield C. HTTP/2, RFC 9113 [Internet]. [cited 2023 Oct 10]. Available from: https://datatracker.ietf.org/doc/html/rfc9113.

40 Swagger.io. API Development for Everyone [Internet]. [cited 2023 Oct 5]. Available from: https://swagger.io.

41 OpenAPI Initiative. OpenAPI Specification v3.1.0 [Internet]. [cited 2023 Nov 10]. Available from: https://spec.openapis.org/oas/v3.1.0.

42 Semver.org. Semantic Versioning 2.0.0 [Internet]. [cited 2023 Oct 11]. Available from: https://semver.org.

43 McFarlane J. CommonMark Spec [Internet]. [cited 2023 Dec 13]. Available from: https://spec.commonmark.org.

44 IANA. Media Types [Internet]. [cited 2024 March 19]. Available from https://www.iana.org/assignments/media-types/media-types.xhtml.

45 Wright A., Ed., Andrews H., Ed., and Hutton B., Ed. JSON Schema: A Media Type for Describing JSON Documents [Internet]. [cited 2023 Oct 5]. Available from: https://datatracker.ietf.org/doc/html/draft-bhutton-json-schema-00.

46 Klyne G. Date and Time on the Internet: Timestamps, RFC 3339 [Internet]. [cited 2023 Oct 7]. Available from: https://datatracker.ietf.org/doc/html/rfc3339.

47 Mozilla Foundation. Regular Expressions [Internet]. [cited 2023 Oct 19]. Available from: https://developer.mozilla.org/en-US/docs/Web/JavaScript/Guide/Regular_expressions.

48 Swagger, Callbacks [Internet]. [cited 2023 March 19] Available from https://swagger.io/docs/specification/callbacks.

49 Open API Initiative. OpenAPI Specification v3.1.0, Webhooks [Internet]. [cited 2023 Oct 8]. Available from: https://spec.openapis.org/oas/v3.1.0#oasWebhooks.

50 World Wide Web Consortium. Server-Sent Events [Internet]. [cited 2023 Nov 30]. Available from: https://html.spec.whatwg.org/multipage/server-sent-events.html.

51 Swagger.io. Swagger Editor [Internet]. [cited 2023 Oct 5]. Available from: https://swagger.io/tools/swagger-editor.

52 Postman.com. Postman API Platform [Internet]. [cited 2023 Oct 5]. Available from: https://www.postman.com.

53 Linux Foundation. AsyncAPI 3.0.0 [Internet]. [cited 2023 Nov 7]. Available from: https://www.asyncapi.com/docs/reference/specification/v3.0.0-next-major-spec.15.

54 Eclipse Foundation. Mosquitto [Internet]. [cited 2023 Oct 20]. Available from: https://test.mosquitto.org.

55 Linux Foundation. AsyncAPI Studio [Internet]. [cited 2023 Oct 21]. Available from: https://studio.asyncapi.com.

56 Swagger. Swagger Editor [Internet]. [cited 2023 Oct 21]. Available from: https://editor-next.swagger.io.

57 Linux Foundation. AsyncAPI CLI [Internet]. [cited 2023 Oct 21]. Available from: https://www.asyncapi.com/docs/tools/cli.

58 GraphQL Foundation. GraphQL [Internet]. [cited 2024 March 20]. Available from https://spec.graphql.org

59 JSONAPI.org. Latest Specification (v1.1) [Internet]. [cited 2024 March 20]. Available from https://jsonapi.org/format.

60 LinkedIn. Rest.li Protocol [Internet]. [cited 2024 March 20] Available from https://linkedin.github.io/rest.li/spec/protocol.

61 Cooper D., Santesson S., Farrell S., et al. Internet X.509 Public Key Infrastructure Certificate and Certificate Revocation List (CRL) Profile [Internet]. [cited 2023 Nov 7]. Available from: https://www.ietf.org/rfc/rfc5280.txt.

62 Santesson S., Myers M., Ankney R., et al. X.509 Internet Public Key Infrastructure Online Certificate Status Protocol - OCSP, RFC 6960 [Internet]. [cited 2023 Nov 7]. Available from: https://www.rfc-editor.org/rfc/rfc6960.html.

63 Cloud Native Computing Foundation. Secure Production Identity Framework for Everyone (SPIFFE) [Internet]. [cited 2023 Nov 7]. Available from: https://github.com/spiffe/spiffe/blob/main/standards/SPIFFE.md.

64 Cloud Native Computing Foundation. Spiffe Runtime Environment (SPIRE) [Internet]. [cited 2023 Nov 7]. Available from: https://github.com/spiffe/spire.

65 Mozilla Foundation. Authorization Schemes [Internet]. [cited 2023 Oct 29]. Available from: https://developer.mozilla.org/en-US/docs/Web/HTTP/Authentication#authentication_schemes.

66 Mozilla Foundation. Authorization [Internet]. [cited 2023 Oct 29]. Available from: https://developer.mozilla.org/en-US/docs/Web/HTTP/Headers/Authorization.

67 Kumar A. Base64 Encoding: What Is It? How Does It Work? [Internet]. [cited 2023 Oct 30]. Available from: https://builtin.com/software-engineering-perspectives/base64-encoding.

68 Sections 3.1.3.2-3.1.3.3, MQTT Version 5.0, OASIS [Internet]. [cited 2023 Nov 2]. Available from: https://docs.oasis-open.org/mqtt/mqtt/v5.0/os/mqtt-v5.0-os.html#_Toc3901069.

69 Section 2.12.2, MQTT Version 5.0, OASIS [Internet]. [cited 2023 Nov 2]. Available from: https://docs.oasis-open.org/mqtt/mqtt/v5.0/os/mqtt-v5.0-os.html#_Re-authentication.

70 Lingappa R. Best Practices for Building Secure API Keys [Internet]. [cited 2023 Nov 4]. Available from: https://www.freecodecamp.org/news/best-practices-for-building-api-keys-97c26eabfea9.

71 Hardt D. The OAuth 2.0 Authorization Framework, RFC 6749 [Internet]. [cited 2023 Nov 3]. Available from: https://datatracker.ietf.org/doc/html/rfc6749.

72 OAuth.net. OAuth 2.0 [Internet]. [cited 2023 Sep 27]. Available from: https://oauth.net/2.

73 Sakimura N., Bradley J., and Agarwal N. Proof Key for Code Exchange by OAuth Public Clients, RFC 7636 [Internet]. [cited 2023 Nov 5]. Available from: https://datatracker.ietf.org/doc/html/rfc7636.

74 Denniss W., Bradley J., Jones M., and Tschofenig H. OAuth 2.0 Device Authorization Grant, RFC 8628 [Internet]. [cited 2023 Nov 6]. Available from: https://datatracker.ietf.org/doc/html/rfc8628.

75 Jones M., Bradley J., and Sakimura N. JSON Web Token (JWT), RFC 7519 [Internet]. [cited 2023 Nov 6]. Available from: https://datatracker.ietf.org/doc/html/rfc7519.

76 Okta. Introduction to JSON Web Tokens [Internet]. [cited 2023 Nov 6]. Available from: https://jwt.io/introduction.

77 Jones M. JSON Web Key, RFC 7517 [Internet]. [cited 2023 Nov 6]. Available from: https://datatracker.ietf.org/doc/html/rfc7517.

78 IANA. JSON Web Token (JWT) [Internet]. [cited 2023 Nov 6]. Available from: https://www.iana.org/assignments/jwt/jwt.xhtml.

79 Cloud Native Computing Foundation. Keycloak [Internet]. [cited 2023 Nov 7]. Available from: https://www.keycloak.org.

80 OAuth 2.0 Proxy Project. oauth2-proxy/oauth2-proxy [Internet]. [cited 2023 Nov 7]. Available from: https://github.com/oauth2-proxy/oauth2-proxy.

81 Gluu, Inc. Web, Mobile Modern Authentication Your Way [Internet]. [cited 2023 Nov 7]. Available from: https://gluu.org.

82 OpenID Foundation. What are OpenID Specifications? [Internet]. [cited 2023 Nov 7]. Available from: https://openid.net/developers/specs.

83 OASIS. Security Assertion Markup Language (SAML) V2.0 Technical Overview [Internet]. [cited 2023 Nov 7]. Available from: http://docs.oasis-open.org/security/saml/Post2.0/sstc-saml-tech-overview-2.0.html.

84 OASIS. Web Services Federation Language (WS-Federation) Version 1.2 [Internet]. [cited 2023 Nov 7]. Available from: http://docs.oasis-open.org/wsfed/federation/v1.2/os/ws-federation-1.2-spec-os.html.

85 IANA. Hypertext Transfer Protocol (HTTP) Authentication Scheme Registry [Internet]. [cited 2023 Nov 7]. Available from: https://www.iana.org/assignments/http-authschemes/http-authschemes.xhtml.

86 Melnikov A., Ed. and Zeilenga K., Ed. Simple Authentication and Security Layer (SASL), RFC 4422 [Internet]. [cited 2023 Nov 8]. Available from: https://datatracker.ietf.org/doc/html/rfc4422.

87 Newman C., Menon-Sen A., Melnikov A., and Williams N. Salted Challenge Response Authentication Mechanism (SCRAM) SASL and GSS-API Mechanisms, RFC 5802 [Internet]. [cited 2023 Nov 8]. Available from: https://datatracker.ietf.org/doc/html/rfc5802.

88 Hansen T. SCRAM-SHA-256 and SCRAM-SHA-256-PLUS Simple Authentication and Security Layer (SASL) Mechanisms, RFC 7677 [Internet]. [cited 2023 Nov 8]. Available from: https://datatracker.ietf.org/doc/html/rfc7677.

89 Linn J. Generic Security Service Application Program Interface, Version 2, RFC 2078 [Internet]. [cited 2023 Nov 8]. Available from: https://datatracker.ietf.org/doc/html/rfc2078.

90 Lane D. Describing Kafka security in AsyncAPI [Internet]. [cited 2023 Nov 8]. Available from: https://dalelane.co.uk/blog/?p=4409.

9

Managing DERs and the Utility Connection

So far, the control protocols discussed in this book are primarily focused on loads. The BACnet *Load Control* object in Chapter 6 and the Zigbee *Demand Response and Load Control* cluster in Chapter 7 provide a means to modify load in response to a utility signal. This offers the building manager and the BEMS an opportunity to reduce load when the grid is under stress and carbon emissions from gas-fired peaker plants are at a maximum. The BEMS can also increase the load at other times when grid solar and wind generation is at a maximum and carbon emissions are lower to implement flexible load policies such as precooling and preheating below or above the usual setpoints, allowing reduction of power use from the grid during critical times. However, another possibility if solar PV panels, a battery energy storage system (BESS), a local wind generator, or an electric vehicle (EV) with bidirectional charging – collectively called DERs – are installed in the building is to schedule the loads to draw from local renewable generation when the grid is under stress, or otherwise to schedule the use of locally generated renewable power to reduce building operational carbon emissions or energy costs.

For example, if the utility has a time-of-use (TOU) tariff, the solar PV-generated power stored in the battery can be scheduled to serve building loads during the peak TOU period to reduce energy costs or to discharge to the grid if the utility provides compensation for energy discharged to the grid. If the building is on a utility plan where additional payment is possible for energy discharged to the grid during a peak grid event, the battery can be scheduled to discharge if the utility signals a peak event, earning the owner compensation. Finally, in the case of a grid outage, a locally installed battery connected to solar PV provides backup for essential loads in the building, and communication with the battery allows the state of charge to be monitored during the outage if network service is still available, providing an added measure of resiliency. If the building can remain

Automating Building Energy Management for Accelerated Building Decarbonization: System Architecture and the Network Layer, First Edition. James Kempf.

Published 2025 by John Wiley & Sons, Inc.

powered by on-site DERs and local storage when the grid goes down, the DERs and storage form a building-integrated microgrid. In this chapter, Section 9.1 briefly discusses building-integrated microgrids.

For the BEMS to manage on-site DERs, protocols and APIs are required to communicate with the energy production and storage devices in the building. Unlike BAMS protocols, there are relatively few protocols that support direct communication between the BEMS and a DER. The most widely deployed application layer protocol for monitoring and control of DERs is Modbus [1], an older, industrial automation protocol widely used by industrial IoT devices. Section 9.2 discussed Modbus with focus on Modbus TCP, the Modbus application layer running on the IP stack over TCP/ TLS. The SunSpec Modbus profile [2], a Modbus specification targeted at DERs such as solar PV systems and BESSs, is the topic of Section 9.3. Section 9.4 briefly reviews another protocol used for DER monitoring and control, namely, controller area network bus (CANbus). While this chapter focuses on the use of Modbus for monitoring and controlling on-site DERs, Modbus is also used instead of BACnet in some large commercial BAMS deployments for HVAC control, lighting control, fire safety systems, and access control.

A more common approach deployed by DER manufacturers is to hide the actual communication protocol behind an API that provides convenient abstractions for BEMS developers and requires developers to access the API through a website maintained by the DER manufacturer. Manufacturer APIs are also very common on Wi-Fi-connected thermostats and controllers for higher load devices. These APIs can be particularly useful when dealing with proprietary systems or when advanced features are required. Section 9.5 examines a particular vendor API in detail, the Enphase Enlighten API [3], a battery and solar control API from Enphase, a vendor of solar microinverters, AC-coupled batteries, and building-integrated DER microgrid controllers.

For the BEMS to manage DERS in a way that reduces building operational carbon emissions requires protocols and APIs for communicating with the utility, so utilization of on-site DERs can be scheduled when the utility is expecting a grid peak event or marginal grid carbon emissions are high. This information is also required to schedule flexible loads so they run during times when the grid's marginal carbon intensity is low. In Section 9.6, the focus moves to the utility-to-BEMS connection. After a brief introduction to the topic of utility connection in Section 9.6, Section 9.7 discusses OpenADR [4]. OpenADR is a widely deployed protocol for utility-to-BEMS communication, which originated as a protocol for demand response (hence the "DR" in the name), where a utility could signal a grid event when loads should be reduced, among other functions. Section 9.8 presents an overview of IEEE 2030.5 [5], which was originally developed as a replacement for the Zigbee Smart Energy profile on the IP stack rather than a custom stack, but was repurposed in 2018 for utility-to-DER control and contains detailed models for controlling different kinds of DERs and loads. There are other protocols that are in use for utility-to-site communication; two are briefly reviewed in Section 9.9.

Section 9.10 discusses the topic of controlling building energy use based directly on information about marginal grid carbon emissions rather than price. Both OpenADR and IEEE 2030.5 have support for signaling the BEMS about periods when marginal grid carbon emissions are high. In addition, California provides the Market Informed Demand Automation Server (MIDAS) protocol [6] for information on grid greenhouse gas emissions, as well as access to time-varying rates for power on the California Independent System Operator (CAISO) wholesale market and CAISO grid event notifications called FlexAlerts. These technologies allow a BEMS to schedule flexible loads and increase usage of on-site renewable generation when carbon emissions are high. Finally, Section 9.11 summarizes the chapter.

9.1 DERs and Building-Integrated Microgrids

Building mounted solar PV systems, in the absence of energy storage, are inherently grid-tied solar systems. They require a stable grid connection to power building loads and feed any excess energy back to the grid. This dependence stems from the need to balance energy production with consumption and for the inverter to track or follow the grid frequency and other technical characteristics of the power system. Solar panels produce variable amounts of electricity based on sunlight availability, and without storage, this energy must be either immediately used or exported. Similarly, if the solar panels are not providing sufficient power to satisfy the building load, the remainder needs to be imported from the grid. In grid outage situations, grid-tied PV systems without storage are designed to shut down to prevent back-feeding electricity into the grid, which can be dangerous for grid maintenance personnel. This safety feature, known as anti-islanding, is crucial but renders solar PV systems nonfunctional during grid outages.

The integration of battery storage, microgrid controllers, and automated transfer switches can transform a building with solar PV into a self-sufficient microgrid capable of operating independently from the main grid. When the grid is functioning, the microgrid can operate in grid-tied mode by complying with the IEEE 1547 [7] standard, which governs the interconnection and interoperability of DERs with the grid. In the event of a grid outage, an automated transfer switch can isolate the building from the grid, a process called islanding, and a microgrid controller, which complies with IEEE 2030.7 [8], can maintain power to the building by managing the energy flow from storage and solar PV to loads. If the amount of battery storage is insufficient to power all the loads in the building, only essential loads such as refrigerators and lights can be powered through a separate circuit breaker subpanel. In addition, the microgrid controller can ensure that the frequency and other technical characteristics of the power supplied by the DERs and storage are maintained to the standards required if the grid were connected to protect sensitive building equipment from damage.

Expanding this concept to a larger scale, building-integrated microgrids can be interconnected to form a campus or community microgrid. Each building is equipped with its own solar PV and storage, but the individual building microgrids are connected to operate collectively as a unified community microgrid. Individual building storage provides resilience for the specific building in which it is located, while centralized storage or a generator enables the microgrid to continue functioning when individual building storage is exhausted.

By optimizing the use of renewable energy sources such as solar PV, buildings can significantly reduce their reliance on fossil-fuel-based grid electricity, thereby lowering their marginal carbon emissions. The integration of energy storage and smart management systems further enhances this capability, allowing buildings to store and use renewable energy efficiently. Additionally, connecting building-integrated microgrids into larger community or campus microgrids enhances the resilience of the individual building electricity supply in the face of climate-caused disruptions such as power outages.

9.2 Modbus

Modbus was developed in 1979 by Modicon, now a part of Schneider Electric, primarily for use with its programmable logic controllers (PLCs) [9]. PLCs are small, modular computers commonly used in industrial automation applications with customized instructions for performing a particular task. At that time, the industrial automation world was rapidly evolving, and there was a growing need for a standardized communication protocol that could connect PLCs and various other industrial devices. Before the advent of Modbus, each manufacturer had its own proprietary communication protocol. This lack of standardization meant that integrating devices from different manufacturers into a single system was difficult, often requiring custom interfaces or gateways. In 2002, the Modbus organization was formed by a group of independent users and suppliers to maintain and advance the protocol. The Modbus organization provides public access to the protocol, its specifications, and related information. The organization also ensures the Modbus protocol's continued availability, enhancement, and application in industrial automation markets.

Modbus is relatively simple compared to other protocols, making it easy to understand, implement, and debug. It is robust in industrial environments, which often have electrical noise and other harsh conditions. Modbus was one of the first truly open protocols for industrial applications and was released without royalties or restrictions, encouraging widespread adoption across manufacturers and industries. As an application layer protocol, it can also be used over various types of transport, link, and physical layers, from serial lines to Ethernet, which makes it highly adaptable to different industrial networks. Since it was designed for PLCs, Modbus naturally fits into many industrial control applications, implementing communication between PLCs and other devices such as sensors, actuators, control dashboards, and, more recently, even more complex systems such as the BEMS and DERs.

The protocol is based on a client–server architecture, where one device acts as the client and communicates with one or more server devices, such as sensors, meters, or other control devices. The client is typically a supervisory controller that initiates transactions, and the server devices respond by supplying the requested data or performing actions. Modbus can run on two different types of networks:

- Modbus Remote Terminal Unit (RTU) [10] – used for serial communication with a custom addressing scheme limited to a maximum of 256 addresses. Modbus RTU is commonly used over RS-232, RS-422, or RS-485 serial networks, such as BACnet MS/TP. Each message starts and ends with a silent interval, framing the message on each end with time gaps,
- Modbus TCP [11] – Modbus acts as an application layer protocol on top of the TCP/TLS transport layer, with the device address being an IPv4 or IPv6 address. Modbus TCP is more suitable for applications that require high data rates or need to connect to building local area networks and the Internet.

This section only discusses Modbus TCP., For more on Modbus RTU, see the Modbus specification [1].

9.2.1 Modbus TCP Message Format

Figure 9.1 contains an illustration of the Modbus TCP application layer message format. The application layer message consists of the basic Modbus application part, which is common to all transports, called the protocol data unit (PDU).

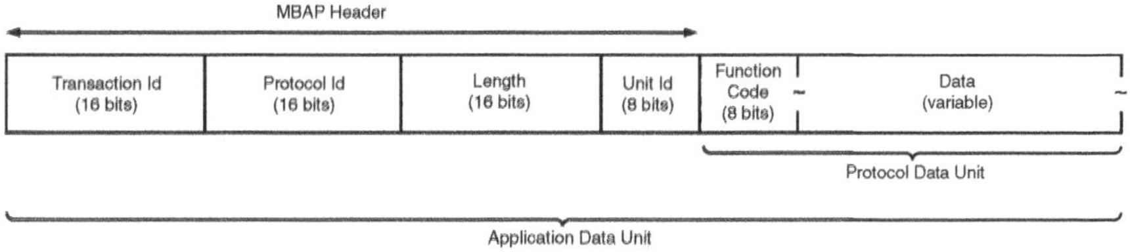

Figure 9.1 Modbus TCP Application Data Unit

The PDU, together with the additional fields required for adaptation to the transport layer, is called the application data unit (ADU). The PDU consist of the following fields:

- Function Code – a one-byte code indicating the type of the message. Valid codes are from 1 to 255, 0 is invalid. The range 128–255 is reserved for exception codes. The public function codes are listed in Table 9.1,
- Data – a variable-sized field containing data that is communicated between the client and the server. The Data field can be empty for some message types.

Table 9.1 Public Modbus Function Codes and Message Types

Name	Code	Modbus Table	Description
Read Coils	0x01	*Coil Status*	Reads the status of coils in a device.
Read Discrete Inputs	0x02	*Discrete Inputs*	Reads the status of discrete inputs in a device.
Read Holding Registers	0x03	*Holding Registers*	Reads the binary contents of holding registers in a device.
Read Input Registers	0x04	*Input Registers*	Reads the binary contents of input registers, typically used for analog inputs.
Write Single Coil	0x05	*Coil Status*	Writes a single coil in a device to either ON or OFF state.
Write Single Register	0x06	*Holding Registers*	Writes a single holding register in a device.
Write Multiple coils	0x0f	*Coil Status*	Writes multiple coils in a device.
Write Multiple Registers	0x10	*Holding Registers*	Writes multiple holding registers in a device.
Read File Record	0x14	—	Performs file record-oriented data transfer between a Modbus server and a client.
Write File Record	0x15	—	Performs file record-oriented data transfer from a Modbus client to a server.
Mask Write Register	0x16	*Holding Registers*	Modifies the contents of a specified holding register using a combination of an AND mask and an OR mask.
Read/Write Multiple Registers	0x17	*Holding Registers*	Performs a combination of one read operation and one write operation in a single Modbus transaction.
Read FIFO Queue	0x18	—	Reads the contents of a first-In-first-out (FIFO) queue of registers in a device.

The fields before the function code in Figure 9.1 are the adaption fields for Modbus TCP and constitute the Modbus Application Protocol (MBAP) header. The following is a list of these fields:

- Transaction ID – a 2-byte field containing a transaction identifier included in requests by the client and copied into responses by the server to enable matching of requests to responses,
- Protocol ID– a 2-byte field intended to identify protocols tunneled over Modbus but set to 0 in Modbus messages,
- Length – a 2-byte field giving the number of bytes in the following fields in both requests and reply messages,
- Unit ID – a one-byte address for devices on a Modbus RTU serial link connected to an IP network through a gateway.

Modbus TCP uses application port 502, and all messages are sent in big-endian (most significant byte first) format. The maximum size of a Modbus TCP ADU is 260 bytes, including the MBAP and PDU, and is constrained by the need to potentially route the PDU message over a serial line if the network contains a gateway between Modbus RTU and Modbus TCP.

9.2.2 Modbus Data Model

The Modbus data model is structured around four tables. Entries in the tables correspond to measurements and controls on the device. The following is a list of these tables:

- Discrete Input Table – contains read-only, single-bit registers representing the state of sensors or external switches,
- Coil Table – contains readable and writable, single-bit registers, used primarily for controlling outputs such as relays or actuators,
- Input Register Table – contains read-only, 16-bit registers representing measurements or sensor readings,
- Holding Register Table – contains readable and writable, 16-bit registers for storing and modifying values such as configuration parameters or control set points.

Memory locations in the tables are addressed by messages of different types. The Function Code field value specifies which table to access, while the starting address in that table, the data to be read or written, and the number of elements are specified in the Data field. The address can be any integer within the range of 0 to 65,535. The four tables are individually addressable, with each table having entries from 0 to 65,535.

Without any further context, the Modbus data model is abstract. The data model must be mapped or "bound" to a specific application or device type. This means that the abstract data model of Modbus (the four tables and their elements) must be related to the actual sensors and actuators of the device. This binding is vendor-specific and can vary greatly depending on the device's application and design. The way Modbus data tables are mapped to the functions and data of a specific device is determined by the device manufacturer. This mapping is unique to each device, meaning that the same Modbus address might correspond to different physical or logical entities in different devices.

As an example of how Modbus might be used in a DER application, consider a solar PV system with a Modbus-enabled controller. The controller monitors various parameters and controls different aspects of the system. For example, a *Coil Status* bit register could be used to turn the inverter on and off. Discrete input registers might represent the status of switches or breakers, such as whether a breaker is tripped. Input registers could store analog data such as the temperature, voltage, or power output of the solar panels. Holding registers could be used to set or change operational parameters, such as power thresholds or operational modes. These device specific functions are mapped to Modbus tables by the vendor. Other vendors may use different mappings, making it difficult to write high-level BEMS software that controls buildings with equipment from multiple vendors. Industry standards for particular device classes, such as the SunSpec Modbus profile [2] discussed in Section 9.3, can support services that are interoperable between different vendors' hardware.

9.2.3 Modbus Function Codes and Message Types

The Modbus protocol includes three categories of function codes:

- Public function codes range from 0x01 to 0x40 and 0x6f to 0x7f and cover common operations such as reading and writing coils and registers and diagnostic functions,
- User-defined function codes range from 0x41 to 0x48 and 0x64 to 0x6e and allow for customization and extension of the Modbus protocol to meet the specific needs of certain applications or devices,
- Exception function codes run from 0x80 to 0xff, are used for error handling, and are the mirror images of the standard function codes, with the most significant bit set to 1, indicating an error response. For instance, if a standard read function (code 0x03) encounters an error, the error response would use code 0x80 + 0x03 = 0x83. The exception codes inform the Modbus TCP client about problems encountered during the processing of a request.

Table 9.1 contains the public Modbus TCP function codes along with which table they access and a brief description of their function. These include function codes for reading coils, discrete inputs, and registers, and writing single and multiple coils and registers. Some function codes access other structured information in the device. For example, codes 0x14 and 0x15, *Read File Record* and *Write File Record*, are used to read-and-write file records, with data specifying for which file and records to access included in the Data field part of the Modbus PDU. Files are designated by a 2-byte number between 0x0000 to 0xffff. Modbus structures files as a sequence of 10,000 records, addressed from 0X0000 to 0X270F. The access specifications must contain a subrequest code 0x06 both in the request and response of the *Read File Record* and *Write File Record* messages, with the read subrequest containing the record numbers to be read and the write request containing the records to be written along with the new values. The read-and-write operations on files are modeled like the read-and-write operations on tables, with the locations in the file designated as registers. The code 0x18, *Read FIFO Queue*, accesses another data structure in the Modbus device, a FIFO queue. Each Modbus device contains a FIFO queue with 32x2-byte registers. The first register in the queue contains the count of the remaining queued registers. Other function codes not shown in the table, including codes for reporting diagnostics, are only used in Modbus RTU or have more specialized functions such as tunneling other protocols over Modbus.

9.2.4 Modbus Message Type Examples

In this subsection, a few examples of Modbus message types are discussed in detail, specifically the *Read Holding Registers* (0x03), *Write Multiple Registers* (0x10), and *Write Single Register* (0x06) messages. The *Read Holding Register* and *Write Multiple Registers* message types are required in the SunSpec Modbus profile, while the *Write Single Register* (0x06) message is optional [12]. No other message types are mentioned in the SunSpec Modbus profile.

9.2.4.1 *Read Holding Registers* (0x03)

The *Read Holding Registers* message (0x03) enables a client to read the contents of holding registers in a server. The format of the request is shown in Figure 9.2. The Function Code field is first and is set to 0x03. The 2-byte Starting Address field follows and indicates the address of the first holding register to be read. Finally, the 2-byte Number of Registers field specifies the number of holding registers to be read.

Function Code (8 bits)	Starting Address (16 bits)	Number of Registers (16 bits)

Figure 9.2 Modbus *Read Holding Registers* Request Message Format

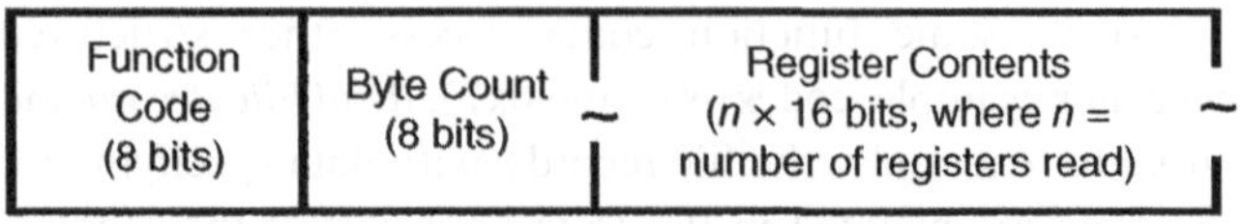

Figure 9.3 Modbus *Read Holding Registers* Response Message Format

Figure 9.3 illustrates the structure of the response message. The Function Code field is the first field and has the same value as in the request, 0x03. Next, the Byte Count field contains the number of bytes of register data to follow in the response. Since each register is 2 bytes, this count will be twice the number of registers read. The Register Values field contains the values of the registers, and contains the number of bytes indicated in the Byte Count field.

9.2.4.2 *Write Multiple Registers* (0x10)

The *Write Multiple Registers* function (0x10) allows a client to write values to a series of consecutive registers in the server (Figure 9.4). This function is particularly useful when multiple settings or parameters need to be updated simultaneously. The Function Code is in the first 1-byte field and has a value of 0x10. The next field is the 2-byte Starting Address field, which specifies the address of the first register to be written. The next field is the 2-byteNumber of Registers field, indicating the number of registers to be written, followed by a 1-byte Byte Count field with the number of remaining bytes in the message. The last field contains the New Register Values field and has length 2 multiplied by the value of the Number of Registers field since each register is 2 bytes wide.

The *Write Multiple Registers* response message starts with the 1-byte Function Code field set to 0x10, followed by the 2-byte Starting Address field containing the address of the first register written (Figure 9.5). The 2-byte Number of Registers field confirms the total number of registers that were written.

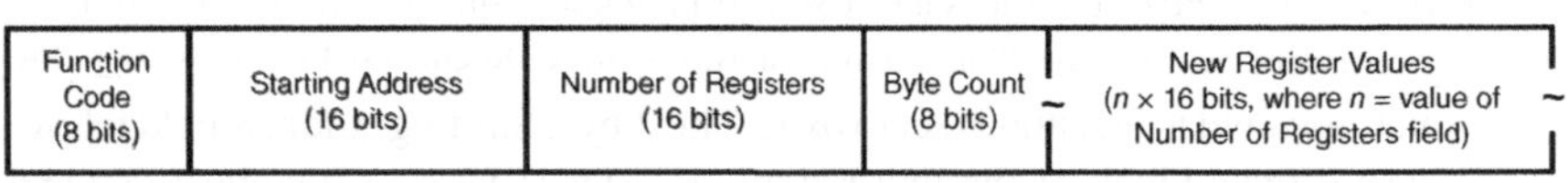

Figure 9.4 Modbus *Write Multiple Registers* Request Message Format

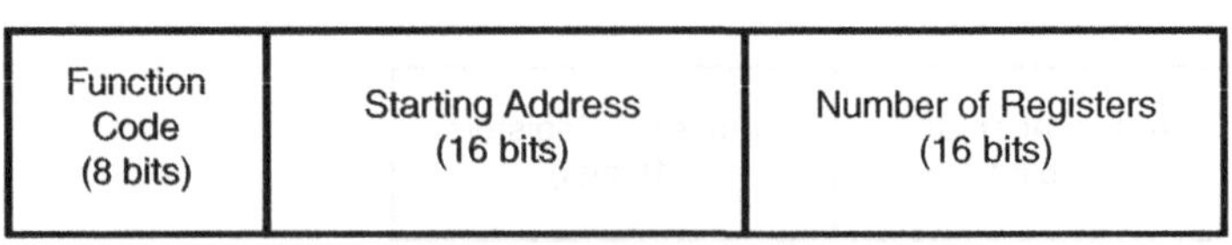

Figure 9.5 Modbus *Write Multiple Registers* Response Message Format

9.2.4.3 *Write Single Register* (0x06)

The *Write Single Register* request and response messages have the same format, as shown in Figure 9.6. The 1-byte Function Code field is set to the message type code, 0x06, and is followed by the 2-byte Register Address field with the address of the register to be written in the request or that was written in the response. The 2-byte New Register Values field contains the new value to write into the register in the request and is echoed back in the response.

9.2.4.4 Exception Responses

If an error occurs, the server responds with an exception message, as shown in Figure 9.7. The exception response has a modified function code in the Function Code field and an exception code in the 1-byte Exception Code field. The function code in the exception response is modified by adding 0x80 to the original function code (0x83 for function code 0x03, 0x90 for function code 0x10, and 0x86 for function code 0x06). The exception codes indicate the nature of the error. Common exception codes include the following:

- Illegal Function (Code 01) – Returned if the function code is not recognized or supported by the server,
- Illegal Data Address (Code 02) – Returned if the starting address or the range specified in the request is not available on the server,
- Illegal Data Value (Code 03) – Returned if a value in the request is not acceptable to the server,
- Server Device Failure (Code 04) – Indicates an unspecified failure in the server device.

There are a total of nine exception codes defined in Section 7 of the Modbus specification [13].

Function Code (8 bits)	Register Address (16 bits)	New Register Value (16 bits)

Figure 9.6 Modbus *Write Single Register* Request/Response Message Format

Figure 9.7 Modbus *Exception Response* Message Format

Function Code (8 bits)	Exception Code (8 bits

9.3 SunSpec Modbus Profile

The SunSpec Alliance [14] was formed by DER vendors to standardize device profiles for various DER equipment. SunSpec has approved device profiles for inverters, meters, solar PV modules, environmental sensors, string combiners, trackers, and BESSs, and is currently working on a device profile for EV chargers, including vehicle-to-grid inverters. SunSpec device models are defined using JSON as the canonical format, but a comma-separated value (CSV) format is also provided for spreadsheet use. From the standpoint of a SunSpec device model, Modbus is one of several "transport" protocols, with HTTP/XML being an additional protocol of note.

9.3.1 SunSpec Modbus Network Architecture

Figure 9.8 contains a high-level schematic diagram of the SunSpec Modbus network architecture. DER devices in the building are connected through a local area network or serial line to the building gateway. The devices implement SunSpec Modbus device models, which allows the building gateway to issue Modbus commands to the devices for monitoring and control. The building gateway is connected through the Internet to the BEMS, running in a public cloud. The BEMS receives information from the utility about the carbon emissions intensity, price of electricity on the grid, and grid events, and uses algorithms to decide when to schedule the DERs to supply the building load and when to release energy to the grid. The utility can also announce a grid event, causing the BEMS to schedule the DERs to service building load at the time of the event, or release energy to the grid if the utility program allows energy production to qualify as load reduction.

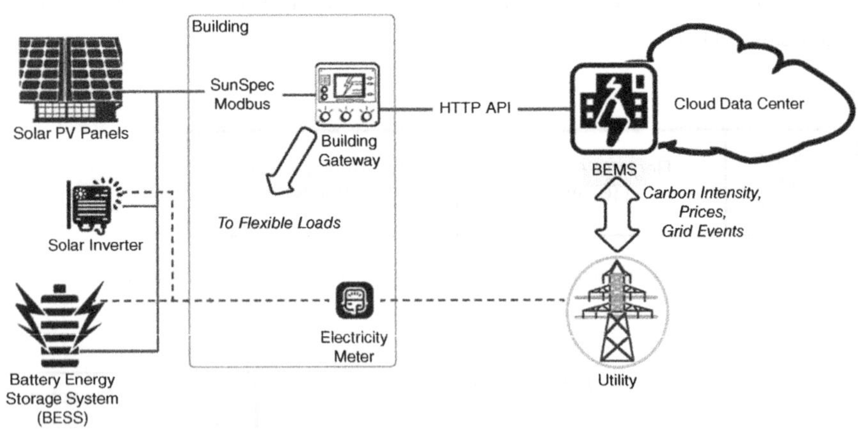

Figure 9.8 Schematic of SunSpec Modbus Network Architecture

9.3.2 SunSpec Device Information Model Structure

SunSpec distinguishes between the device information model definition, which is the specification of data and control points for a device in a particular class of devices, and the device information model instance, which is the mapping of the device information model definition to a set of Modbus registers in a device manufactured by a specific manufacturer. This distinction is similar to the distinction between a class and an instance in an object-oriented programming language, where the class acts as a template to create instances. The device information model definition is constructed in one of the two supported encoding languages, JSON or CSV, with JSON being the canonical language. The device information model instance is constructed by mapping the device information model definition onto Modbus registers as part of the device information model stack. The device information model stack is discussed in Section 9.3.4. The SunSpec Device Information Model Specification [2] contains instructions for both constructing the device information model definition and constructing the device information model instance by performing the mapping to Modbus for a particular device.

9.3.3 Device Model Elements

As shown in Figure 9.9, a device model is defined using a two-level hierarchical collection of grouping elements, along with point and symbol elements. The first level of the hierarchy is the model itself. A model is a high-level collection of points and point groups that represent a specific type of device or service. The second level of the hierarchy is the point group, used to form logical groupings of points within a model. A point group can be nested inside another point group. There are two reasons to form a point group:

- A repeating set of points can be grouped, creating multiple instances of the point group that can be accessed as an array of point groups.
- A set of points with synchronous operational requirements can be grouped, indicating that the points in the group must be read and written atomically.

A point constitutes the actual data reported by the device or by the control register available to exercise control when written. Each model defines the data and control points reported by devices of that type. A point can be associated with one or more symbols that provide human-readable representations of a point. Comments can be added to the device model to provide additional information about the point, such as the meaning of the value, the units of measurement, or the source of the data, but comments have no mapping into the Modbus device information model instance.

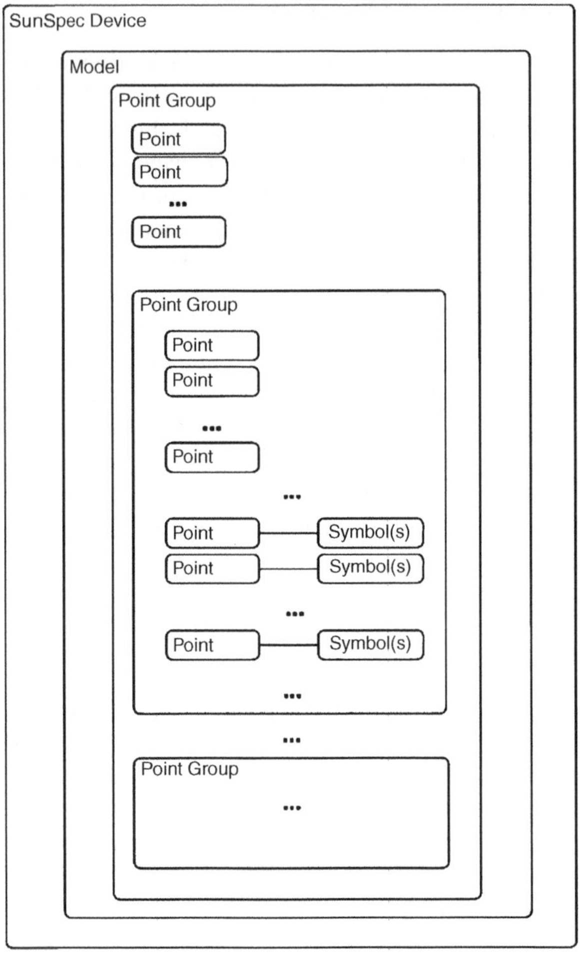

Figure 9.9 SunSpec Device Information Model Elements. *Source:* Adapted from SunSpec Alliance [15].

In addition to the hierarchical structure, attributes can be associated with the device information model elements. Attributes are used to refine an element specification. An attribute consists of an attribute key and one or more values. Table 9.2 contains a list of attribute keys with a short description of the allowed values and whether the attribute is required (R) or optional (O) for each of the device information model elements. If there is no indication that the attribute is either optional or required for a particular element, it cannot be included in the definition of the element. Note that the values of the *Count* and *Scale Factor* attributes can be another point defined in the model or a constant value. The point referred to needs to be defined in the top-level point group before the point or point group element that refers to it.

Table 9.2 SunSpec Modbus Data Model Elements and Associated Attributes

Attribute	Description	Model	Point Group	Point	Symbol
ID	Element identifier.	R	R	R	R
Points	An array of point definitions in a point group.		R		
Group	An array of point elements or other point group elements.	R			
Groups	An array of point group definitions in a point group.		O		
Value	A constant value associated with the element. Unnecessary if the element is not constant.			O	R
Type	Element type from Tables 3 and 4 in the SunSpec Device Information Model Specification [16].		R	R	
Count	Number of instances of the element.		O	O	
Size	The element size in units of 16-bit registers. It is mandatory for point elements when the type is a string.			O	
Scale Factor	The scaling factor point associated with the element.			O	
Units	The units associated with the element, for example, watts.			O	
Access	Access control on the element, one of read (`R`) or read/write (`RW`). If not present, defaults to `R`.			O	
Mandatory	One of mandatory (`M`) or optional (`O`). If not present, it defaults to `O`.			O	
Label	A short label associated with the element.	R	R	O	O
Description	A longer description associated with the element.	O	O	O	O
Symbols	A collection of symbol elements associated with the point.	O	O	O	O

Source: [17]/SunSpec Alliance.

The following provides more detail on some of the attributes described in Table 9.2:

- *ID* – holds the element name and is unique in the model top-level points or point group where it is defined. With the exception of the model element, the *ID* attribute value must be a string consisting of alphanumeric characters and the underscore (_) character. As discussed in Section 9.3.3.1, the *ID* attribute for a model element must be the numeric model identifier assigned by SunSpec,

- *Points, Groups, and Symbols* –represent collections of point elements, point group elements, and symbol elements as arrays. The *Points* attribute value contains an array of point elements, the *Groups* attribute value contains a list of point group elements, and the *Symbols* attribute value contains a list of symbol elements,
- *Count* – specifies the number of occurrences of the element in the model. The *Count* value is typically used to specify the number of occurrences of a point group element, but it may also be used to specify a single repeating point. The value must be static and not change,
- *Scale Factor* – used as an alternative to a `float32` or `float64` type. The point element value is represented by an integer with a signed value specified as the value of the *Scale Factor* attribute applied. A negative value explicitly shifts the decimal point to the left, and a positive value shifts the decimal point to the right by the number of places specified as the integer. If the *Scale Factor* value is another point defined in the model, the referenced point must be defined as the `sunssf` *Scale Factor* type. The `sunssf` type indicates a power of 10 multiplier if the value is positive or a divider if the value is negative. The *Scale Factor* value is constant.

In the following subsections, the device information model elements and the attributes that can be assigned to them are discussed in more detail.

9.3.3.1 Model Element

The model element is a container for the other elements in the model definition and acts as a top-level point group. The model element includes all points and point groups for the device definition. A model has a set of initial points defined outside of a point group, followed optionally by point group definitions. The initial set of points includes two required points as the first two elements in the model:

- A point with the identifier `ID` and value that is the overall model identifier,
- A point with the identifier `L` and value that is the overall model length, including all points and point groups defined within the model.

The value of the `ID` point must be a unique integer between 0x0001 and 0xffff, assigned by the SunSpec Alliance. The value of the `L` point is 0 in the model definition but is set to the length of the model in a model instantiation, in units of 2-byte registers. In the model instantiation, the ordering of the model members must be maintained since it will be mapped directly to Modbus register locations.

9.3.3.2 Point Group Element

The point group element includes point elements or other point group elements. The point group has a *Type* attribute whose value must be either `group` or `sync`. The `group` type is used to create a set of points and point groups. A point group can optionally have a *Count* attribute, and if the value is greater than 1, the point group repeats for the number of times specified by the *Count* attribute. The *Count* attribute value can be defined as a constant or specified as the value of another point in the model definition. If the point group *Count* value is specified by another point, the specifying point must be defined in the top-level point group of the model before the point group definition. The `sync` point group type designates points and point groups that must be read and written atomically as one unit. The device model instantiation must signal an error if all the members of a `sync` group cannot be read or written atomically. As with the model element, the ordering of the point group members must be maintained when mapped to Modbus register locations. In addition, all points at the top level in a point group must be defined before any embedded point groups are defined.

9.3.3.3 Point Element

The base element in the model definition schema is the point element. Point elements are required to have a *Type* attribute and optionally have a *Size* attribute. If the value of the *Type* attribute is `string`, the *Size* attribute is required and designates the number of characters in the string. A point element can also optionally have a *Count* attribute. If the value of the *Count* attribute is greater than 1, repeated instances of the point occur in the instantiation. If the point element has a *Value* attribute, the point is a constant. The value of the *Value* attribute can be specified as the value of another point in the definition. As with point groups, if the value is specified by another point, the specifying point must be defined in the top-level point group of the model before the point definition.

9.3.3.4 Symbol Element

The symbol element associates an identifier with a constant point value in a point definition. Symbol elements must be defined in association with a point definition, and a point definition may have multiple symbol elements associated with it. Each symbol identifier must be unique for the point definition. Symbols are used to define constants that serve as enumerated values or bitfields, such as an enumerated type, similar to the `enum` type in the OpenAPI specification from Section 8.4.2.

9.3.3.5 Comment Element

The comment element is a single string and can be associated with any model element other than another comment element. A model element can have multiple comments associated with it. Comments have no representation in the device information model instance in the actual device.

9.3.4 SunSpec Device Information Model Stack

A full SunSpec specification for a device consists of a model stack including at least three of the following four components:

- SunSpec Common Model – a mandatory data model that must be present at the start of every SunSpec device information model,
- SunSpec Standard Model – specifies common parameters and control points common to all devices in a particular class and is standardized by SunSpec,
- Vendor Model – defined by the device vendor and contains parameters and control points applicable only to the vendor's devices. Vendor Models do not need to follow the SunSpec Standard Model review process but must conform to the rules governing SunSpec model definition and must be assigned an identifier by SunSpec,
- End Model – marks the end of the device information model definition and is also mandatory.

A device information model definition must include either a SunSpec Standard Model or a Vendor Model and can include both. Each information model has a 2-byte identifier assigned to it and a length, permitting clients to browse the contents of a device and skip models that are not of interest.

9.3.4.1 Common Model Data Elements

The SunSpec Common Model is the first model in the model stack and provides important meta-data about the device and its structure. The SunSpec Common Model is described in the SunSpec Information Model Specification [15]. The Common Model includes the following elements:

- *ID* – the model identifier, having the value 0x0001, indicating that this is the Common Model,
- *L* – the model length, having the value 65 or 66, with models of length 66 containing a pad register as the final register,
- *Mn* – an alphanumeric string identifier with no embedded spaces indicating the manufacturer of the device. Manufacturers must register their ID with SunSpec to guarantee uniqueness,
- *Md* – an alphanumeric string identifier with no embedded spaces indicating the product model of the device,
- *Opt* – an alphanumeric string identifier with no embedded spaces indicating any product model options for this device,
- *Vr* – an alphanumeric string identifier with no embedded spaces identifying the firmware version of the device,
- *SN* – an alphanumeric string identifier containing the serial number having no embedded spaces, which uniquely identifies this device within the manufacturer's name space,

- *DA* – a protocol-specific value to address this device instance. For Modbus devices, this is the Modbus device ID,
- *Pad* – SunSpec pad16 register. Padding must be included for Common Models of length 66. The Pad is not included if the length is 65.

SunSpec requires that the result of concatenating the three strings *Mn*, *Md*, and *SN* returns a globally unique string across a manufacture's product line. This string can be used for logging and upgrade functions.

9.3.5 Model Encodings

SunSpec supports two encodings for model definitions, JSON and CSV, with JSON as the canonical encoding format.

9.3.5.1 JSON Encoding

SunSpec uses JSON to represent model elements in the machine-readable and human-readable form, with each SunSpec model element represented as a JSON object. The JSON model object represents the model schema directly in REST APIs. SunSpec has defined a prototype REST API for models [18], and also guides the mapping from the REST API to Modbus. The JSON key name for the device model definition elements corresponds to their model definition names. At the top level, the value of the `model` key is a JSON object containing the entire model element, while the values for `groups`, `points`, `symbols`, and `comments` keys are arrays holding JSON objects representing the respective elements. The `model` object contains a required `group` key, the value of which is the entire rest of the model. The `model` object also has a few other required keys. The value of `id` key, corresponding to the *ID* attribute, is a unique identifier assigned to the model by SunSpec. The model object also must have a `label` key corresponding to the *Label* attribute, the value of which is a short, human-readable string labeling the model. Two additional optional attributes are the `description` key, corresponding to the *Description* attribute, which can have a longer descriptive string for the model, and `comments` key, with an array of comment strings.

The elements of the model definition are similarly represented as JSON objects. A point group element is represented as a JSON object and must have a `name` key with the point group name, an `id` key with the model identifier representing the *ID* attribute, and a `type` key with the point group type representing the *Type* attribute. Other required and optional attributes contained within the point group are represented as JSON key-value pairs, including `groups` for embedded point group elements, `points` for point elements, and `comments` for comment strings, just as in the top-level `model` object, with the values being arrays holding data values for the representative element. Point and symbol elements are represented as JSON objects with the respective required and optional attributes as JSON key-value pairs. A point object is required to have a `name` key, an `id` key, and a `type` key, the latter two representing the respective required attributes, other keys matching attributes are optional.

Table 6 in the SunSpec Device Information Model Specification [19] contains the mapping between device information model type definitions and the JSON strings used to represent the elements and their attributes. A symbol object is required to have a `name` key, an `id` key, and a `value` key, with again the latter two representing the required attributes and other keys matching optional attributes. Comments are always strings in an array that is the value of a `comments` key since JSON has no native syntax for representing comments. Appendix A of the SunSpec Device Information Model Specification [20] contains a simple example of a device model in the JSON encoding, and a more complete list of SunSpec JSON models can be found in the SunSpec GitHub repository [21].

9.3.5.2 CSV Encoding

A CSV encoding is also supported for use in spreadsheets. The CSV encoding consists of a direct mapping between the JSON definition and the CSV model definition attributes. Each row in the CSV file defines a model definition element, such as a point element or point group element. Each column represents an attribute of the element. The sequence of element definitions from the model and from the JSON definition must be preserved in the CSV file definition. Table 7 in the SunSpec Device Information Model Specification [22] contains a list of the required column header names. A hierarchy of point group elements is represented by prepending the point group identifier to the element identifier, which is a member of the point group, separated by a period (.); for example, `power.output` for the output point in the power section of an inverter model definition. This convention is applied to all elements which are not directly defined in the top-level model element.

9.3.6 Modbus Mapping

In Modbus, the device information model definition stack is mapped into the 16-bit holding register address space table to form the device information model instance, as shown in Figure 9.10. The device information model instance must start at one of three registers having the following addresses: 0, 40,000 (0x9c40), or 50,000 (0xc350). The first two 2-byte registers must have the values 0x5375 and 0x6e53, forming the letters "`SunS`" or 0x53756e53 when concatenated. The Common Model is next, and it contains data points with basic information about the device, such as the manufacturer's name, the device model number, the device serial number, the firmware or software version number, etc., as described in Section 9.3.4.1. The device model stack continues with registers corresponding to the Standard Models and/or Vendor Models. Each device model instance section starts with an identifier register and a length register containing the number of 2-byte Modbus registers following the length register. No empty registers are allowed between points. Multiple device model sections are possible within the device model section. After the device model sections is the End Model. The End Model has an identifier of 0xffff and length 0 since it indicates the end of the device model instances.

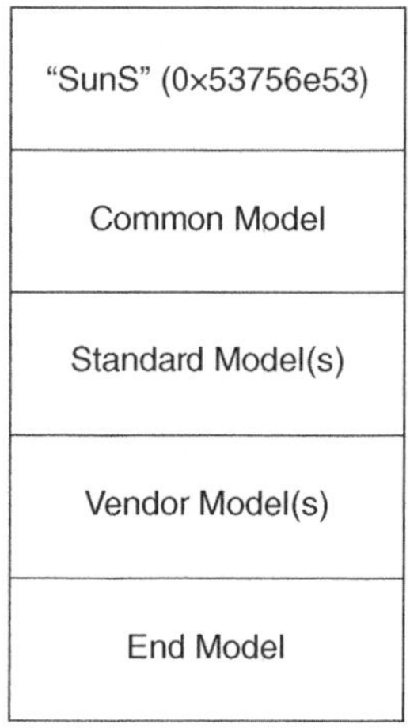

Figure 9.10 SunSpec ModBus Device Information Model Stack. *Source:* [23]/ SunSpec Alliance.

There is no required ordering for Standard and Vendor models, but the Common and End Models must be at the beginning and end, respectively, of the device information model instantiation. The point and point group elements must be mapped to registers in the order in which they appear in the depth-first traversal of the device information model definition represented as a JSON object tree without skipping any registers. Integers are stored in 1, 2, 4, or 8 16-bit registers depending if the value is 16, 32, 64, or 128 bits in length, and signed integers are stored in two's complement format. Strings are stored with two 8 bit characters (bytes) per 16-bit register, with a null (0x00) character used to terminate and pad the string out to the length specified in the device model definition. Floating point values are specified according to the 32-bit and 64-bit IEEE 754 standard [24]. The `sunsf` *Scale Factor* type is stored in two 16-bit registers, with a signed range of −10 to +10. All points in the model must be mapped into registers, even those specified as optional or unimplemented. Registers for unimplemented points set to the appropriate unimplemented value based on the point type. The unimplemented and pad values for different data types, along with more detail on register mappings, are given in Section 6.4 of the SunSpec Device Information Model Specification [25]. The SunSpec Information Model Specification [15] also provides more detail on mapping the device information model specification into Modbus.

9.4 Other DER Communication Protocols

There are a few other protocols used in monitoring and controlling utility-scale DER deployments, but the only other protocol which been used in some DER products for local communication in buildings is the CANbus protocol, or CANbus [26]. CANbus is documented in the ISO 11898 series of standards. It has been adopted for various industrial IoT use cases, including DERs, and is primarily a data link layer and physical layer protocol rather than an application layer protocol such as Modbus TCP. The CANbus protocol was developed by Bosch GmbH, a German electronics and engineering company that manufactures automotive components such as fuel systems, brakes, and electrical drives, in the early 1980s and standardized by ISO in 1993 as ISO 11898. By the mid-1990s, CANbus had become the preferred network protocol in the automotive industry because it enables a distributed architecture among multiple processors in an automobile without requiring a centralized controller. Some manufacturers of BESSs and solar inverters incorporate CANbus into their products. Because CANbus is universally used in automotive electronic systems, it is popular for manufacturers of EVs that also build BESSs or who provide components for original equipment manufacturers that manufacture BESSs [27].

9.5 DER Manufacturer APIs

Many manufacturers of solar PV equipment and BESSs have structured their device monitoring and control architecture around a manufacturer's website with an HTTP REST API that allows properly authenticated clients to access monitoring and control information on their DER devices. Figure 9.11 illustrates this network architecture. In contrast with the SunSpec Modbus architecture in Figure 9.8, the device manufacturer's website architecture does not allow the building gateway or the BEMS to access the devices directly. Access is always through the website intermediary. The actual control and monitoring protocol between the manufacturer's website and the devices at the building may be through a REST API, or the devices at the building may use Modbus or another protocol to communicate directly back to the manufacturer's website. In any case, the BEMS engineer and building manager do not have access to the protocol used by the devices to report to the website.

The advantage of this architecture over the Modbus architecture shown in Figure 9.8 is that the REST API can be tailored to provide the information required for energy monitoring and control in a form that is easily incorporated into the BEMS services. One disadvantage is that if the BEMS is designed as a to cover many buildings with different manufacturers' equipment, a separate plug-in must be implemented for each manufacturer's website since there is often little commonality between the URL structure and schema used by different manufacturers. Another disadvantage is that if a manufacturer goes out of business and shuts down the website, any building with that manufacturer's equipment installed is left without any control and monitoring capability. This has not happened yet for DERs, but it has happened for manufacturers of smart home devices, such as garage door openers and smart lighting [28].

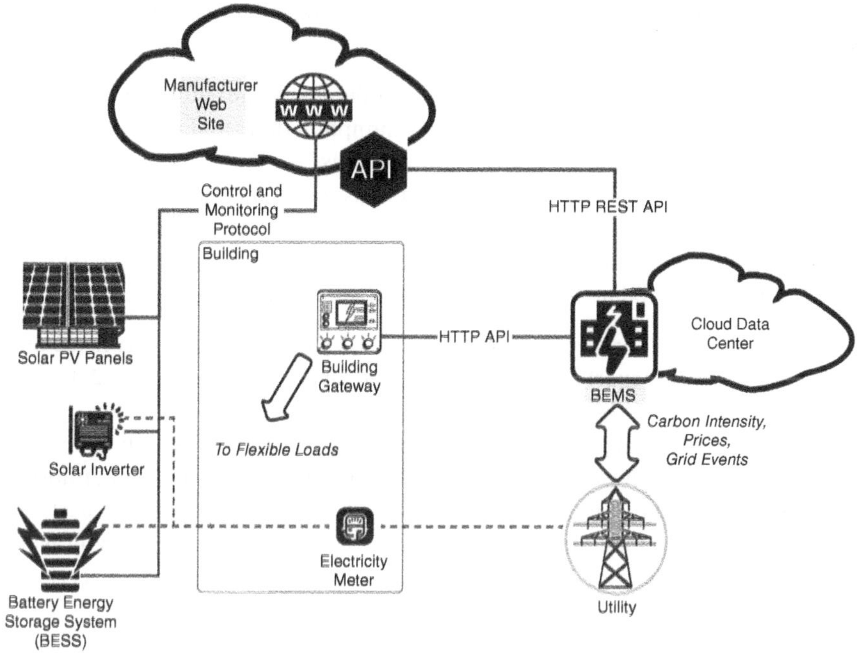

Figure 9.11 Schematic of Device Manufacturer Website Network Architecture

9.5.1 Enphase Enlighten API

An example of a manufacturer REST API is Enphase Enlighten API [29] Enphase manufacturers a range of hardware products useful for incorporating into building DER systems, including microinverters, AC-coupled batteries, microgrid controllers, and the IQ Combiner. The IQ Combiner contains the envoy energy production and storage monitoring system that communicates with the Enlighten cloud-based website, Enphase's manufacturer website for their API. The basic design of an Enphase solar PV system converts DC to AC at the solar panels through microinverters rather than using large string inverters to convert power from a string of several panels. The Envoy communicates with the microinverters and battery using a proprietary powerline communication protocol (PLC) similar to Ethernet over powerline discussed in Section 3.2.4. This protocol operates between the microinverters, battery, and Envoy communication hub over the existing AC electrical wiring in the building, eliminating the need for any additional signal wiring. The Envoy then communicates with the cloud-based website using a proprietary Internet stack protocol which Enphase does not make public. BEMS applications can access the data collected from the Envoy through the Enphase website using the public Enlighten REST API with JSON schema.

The latest version of the Enlighten API is Version 4 and is divided into two request groups:

- Monitoring API – allows the BEMS to access details on the equipment deployed at a site, system-level production and consumption data, and device-level production data and to modify device configurations,
- Commissioning API – allows registered Enphase partners to activate and update the devices in a system, manage which installers are allowed to use the commissioning API, create a building site user, which Enphase calls a "home owner," and retrieve and update the utility tariff in effect at the building site.

The Enlighten API and website require TLS 1.2 with server side certificate and HTTPS for basic transport security services, and an API key and OAuth 2.0 token for application-level IAM.

In the remainder of this section, the monitoring API is discussed in detail, but only an overview is presented on the commissioning API, since the monitoring API is public but the commissioning API is only available to Enphase partners and their subcontractors. Therefore, the commissioning API is not generally available to third-party service developers not involved with installing and servicing Enphase equipment.

9.5.2 Registering for a Developer Account and Obtaining IAM Credentials

The access credentials for both the monitoring and commissioning APIs are generated by an eight-step process through the Enphase Enlighten website [30] The developer or partner creates an account on the website with a username and password and signs up for one of three different account types. The account types include a free tier, and are distinguished by the allowed number of requests per minute and per month, with paid accounts having higher limits than the free account. Once the account is established, the developer or partner registers an application through a form by selecting the plan type for the application and the access control category for requests and filling in the application name, a short description of the application, and the name of the developer, partner, or company that developed the application. Access control is used to limit what types of data the application can access; for example, if the application is to monitor EV charging, then the access control category EV charging can be selected.

Once the application has been registered, Enphase issues credentials for the application to obtain OAuth 2.0 tokens and an API key for the application. The OAuth 2.0 credentials include a client ID and client secret for use in the OAuth 2.0 client credentials token generation grant flow, discussed in Section 8.6.4.2. The client id and client secret are then used to generate an access token and refresh token through an authorization URL that contains the client id as a query parameter. The building owner is required to approve the token generation request through a user interface display in the web application. The access token has a lifetime of 1 day, while the refresh token has a lifetime of 30 days.

9.5.3 Base URL and Request Header Format

The URLs for both the monitoring and commissioning APIs begin with the following host and initial path:

```
https://api.enphaseenergy.com/api/v4
```

An OAuth 2.0 access token must be included in an HTTP `Authorization` header using the `Bearer` scheme, and the API key must be included as the value of the `key` query parameter in the URL for every API request. An example of the HTTP header for a request to the monitoring API is

```
GET /api/v4/systems/{system_id}/summary?key=
<API key> HTTP/1.1
Host: http://api.enphaseenergy.com
Authorization: Bearer <access token>
```

9.5.4 Monitoring API

The monitoring API is split into six subgroups:

- System Details – retrieves detailed metadata about a specific Enphase solar PV deployment. These details include the system's size, location, the number and type of solar panels and microinverters at the site, the system's start date, and other relevant metadata,
- Site-Level Production Monitoring – provides data on the site-level energy production, such as total energy produced over various time frames (daily, monthly, and yearly), current production rates, and historical production data,
- Site-Level Consumption Monitoring – tracks the energy consumption of the site, including the amount of energy used and peak consumption times, and, in addition, telemetry information from any installed batteries,
- Device-Level Production Monitoring – provides more granular information on energy production or charge/discharge on individual devices within the solar PV deployment, including microinverters, solar panels, and batteries,
- System Configurations – retrieves the current configuration of batteries, essential loads, and connection to the grid. This group of APIs also contains requests to change the configuration of batteries and essential loads according to a few predefined criteria,

- Streaming – contains a real-time data streaming API request that provides data on system performance for a duration of 30 seconds by default. The data includes real-time production and consumption data, system status updates, and alerts. The duration can be configured between 30 and 300 seconds by including a query parameter in the URL with a key duration set to the desired duration in seconds.

9.5.4.1 Monitoring API Request Format

Most of the monitoring API requests in the system details, site-level production monitoring, site-level consumption monitoring, device-level production monitoring, and streaming API groups have the same initial path format:

/api/v4/systems/{system-id}/<remaining path>

where {system-id} is the integer identifier of the system for which the operation should be applied, and <remaining path> contains the rest of the path name that specifies the operation. The HTTP method type for all requests in the site-level production monitoring, site-level consumption monitoring, device-level production monitoring, and streaming groups is GET.

Four API requests in the system details group do not have this format:

- GET /api/v4/systems – returns all the integer system identifiers that the access token holder is allowed to access,
- POST /api/v4/systems/search – returns the system identifiers which the access token holder is allowed to access that match a list of criteria passed as a JSON object in the request body,
- GET /api/v4/systems/retrieve_system_id – returns the integer system identifier in a JSON object matching the Envoy controller serial number passed as the string value of the query parameter serial_num,
- GET /api/v4/systems/inverters_summary_by_envoy_or_site – returns a summary of the microinverters at a site in a JSON object if the integer site identifier is provided as the value of the site_id query parameter or which are connected to a particular Envoy controller if the Envoy serial number is provided as the string value of the envoy_serial_number query parameter.

The system configuration group also has a slightly different path format:

systems/config/{system_id}/<configuration type>

The system_id path parameter specifies which system to configure, the <configuration type> path component describes what type of configuration to retrieve or modify, and the HTTP method type for requests is GET or PUT, depending on whether the configuration is retrieved or modified. The system configuration group is discussed in more detail in Section 9.5.4.5.

In the following discussion, path names for API requests are referred to through their `<remaining-path>` components, except where the path name does not follow the general pattern. For example, the request for interval data on energy export from a site that is part of the site-level consumption monitoring group and has path name:

`/api/v4/systems/{system_id}/energy_export_telemetry`

is referred to as `energy_export_telemetry`.

9.5.4.2 Monitoring API Request Time Granularity

There are two basic types of monitoring requests:

- Calls that provide daily production or consumption time series for a site between two dates specified in `start_date` and `end_date` query parameters. A time series is returned with one data reading per day between the `start_date` and `end_date`, with the exception of the `rgm_stats` request, which returns data at 15-minute intervals.
- Calls that provide telemetry on production or consumption for either a site or device over an interval starting at a date specified in the `start_at` query parameter and running for the time period specified in the `granularity` query parameter, where the `granularity` can be set to `15mins`, `day`, or `week`, with the default being `day`. The number of data points returned depends on the request type and is discussed in the following.

The first type includes the `rgm_stats` and `energy_lifetime` requests in the site-level production monitoring group and the `consumption_lifetime`, `battery_lifetime`, `energy_import_lifetime`, and `energy_export_lifetime` requests in the site-level consumption monitoring group. The second type includes many calls in the site-level production monitoring, site-level consumption monitoring, and device-level production monitoring groups where the last component in the request path is `telemetry`. or where a path component includes `telemetry` as part of the name, for example, `latest_telemetry`. In the following discussion, the first type of request is referred to as producing time series data, while the second as producing interval data.

For the interval data-producing requests, the number of parameters and the actual granularity of the returned data depends on the specific API request. For example, the `telemetry/production_micro` request in the site-level production monitoring group returns three data points if the `granularity` parameter is set to `15mins` and 288 if the parameter is set to `day`, with each interval in both cases having a period of 5 minutes. In the site-level consumption monitoring group, the `energy_import_telemetry` request takes the same query parameters but only returns one data point if the granularity parameter is set to 15mins and 96 if the parameter is set to day, so the interval is 15 minutes. In neither case is the number for week specified in the documentation, but the assumption is that the number of returned data points corresponds to the interval.

9.5.4.3 Site Level Production Monitoring Group, and Site Level Consumption Monitoring Group Requests

In the site-level production monitoring group, the telemetry/production_meter request returns interval data for an entire site generated from production meters, while the telemetry/production_micro request returns interval data from the microinverters. To obtain time series data, the rgm_stats request returns time series data from revenue grade meters, while the energy_lifetime request returns time series data from either microinverters if the site is not metered or meters if it is. In addition, the production_meter_readings request returns the last known reading from the production meters. The distinction between production meters and revenue grade meters is due to the incorporation of current transformers in more recent models of the Envoy controller, which increases the accuracy.

The site-level consumption monitoring group provides electricity consumption monitoring data and import/export data for the site. The consumption_lifetime request returns time series data of energy consumption from a consumption meter, while the energy_import_lifetime, energy_export_lifetime, and battery_lifetime requests return energy import, energy export, and battery charge/discharge time series data, respectively. The corresponding interval data reporting requests are telemetry/battery for battery charge/discharge, telemetry/consumption_meter for consumption metering, energy_import_telemetry for energy import, and energy_export_telemetry for energy export.

9.5.4.4 Device Level Production Monitoring Group, and Streaming Group Requests

The device-level production monitoring group provides access to interval data from microinverters and two different types of Enphase battery products: the older AC Battery System (ACBS) and the Encharge battery, which is also AC-connected. All the requests return interval data. The <remaining path> template for these requests is as follows:

```
devices/<device-type>/{device_serial_number}/telemetry
```

where <device-type> is one of micros, acbs, or encharge, and {device_serial_number} is a path parameter specifying the serial number of the individual device; for example, the individual microinverter serial number for micros.

The streaming API has one request, which provides real-time data on all devices and grid power for a site. For the streaming API request, the <remaining path>

path component is `live_data`. The duration of the stream in seconds can be specified with a query parameter, `duration`, having a minimum value of 30, a maximum value of 300, and a default value of 30 if the query parameter is not specified. The API documentation does not specify the frequency of the event reporting. The streaming API uses server-sent events, discussed in Section 8.4.2.4, to return the stream to the client, with the content type of the response body being `text/event-stream`. The event data is formatted as a JSON object with key-value pairs reporting the important parameters for each device.

9.5.4.5 System Configuration Group Requests

In contrast with the other groups, the system configuration group requests are of two HTTP method types:

- HTTP `GET` – return a particular type of configuration,
- HTTP `PUT` – modify a particular type of configuration.

From Section 9.5.4.1, the `<configuration type>` path component for the system configuration group requests specifies the type of configuration:

- `battery_settings` – report or modify the battery charge/discharge mode and the state of charge held in reserve for backup,
- `storm_guard` – report or modify whether the battery backup settings should automatically change if a strong storm is forecast for the locality to enable sufficient battery capacity for backup power if the grid should go down,
- `load_control` – report or modify parameters for essential loads,
- `grid_settings` – report whether the system is connected to the grid or not. This request has a `GET` method only.

The controllable battery settings are available through the `battery_settings` request and include a `battery_mode` JSON object key-value pair that can be set to one of three battery operating modes:

- `"Self-consumption"` – when there is solar power available, building loads are prioritized then the batteries are charged. If there is still any excess solar-generated power, it is exported to the grid if the utility allows export. At night or when the panels are not producing power due to cloudy weather, building loads are powered by the batteries, and energy is imported from the grid if the building requires more power than the batteries and solar panels can provide.
- `"Savings Mode"` – used when a utility TOU tariff is in place. During the peak period, the building is powered by batteries, then solar panels and finally by the grid if insufficient power is available from the on-site DERs, reducing the cost of power bought from the utility. Any excess solar power is exported to

the utility if allowed. During the off-peak period, the batteries are charged from the solar panels, and if additional solar-generated power is available, it is used to power the building and then exported to the grid if the utility allows export. At night, power is imported from the utility.

- `"Full Backup"` – the solar panels charge the battery until it is full, and the battery is only used if there is a grid outage. After the battery is charged, solar PV-generated power is used to power the building, and, finally, power is exported when the building loads are satisfied.

In addition to the `battery_mode` key-value pair, the JSON object parameters in the request body for the `PUT` and reply bodies for the `GET` requests have the following keys:

- `system_id` – the system id from the path parameter,
- `reserve_soc` – an integer giving the percent state of charge that the battery should hold in reserve for grid outages,
- `energy_independence` – an `enum` string that can be either `"enabled"` or "`disabled"` which modifies the savings mode setting. If energy independence is enabled, the battery will discharge during peak hours to save on grid-imported energy. After peak hours, the battery will continue to discharge until the `reserve_soc` level is reached, then will charge again the next morning to prepare for peak hours. If energy independence is disabled, the battery will only discharge during peak hours,
- `charge_from_grid` – an `enum` string which can be either `"enabled"` or "`disabled"` depending on whether the utility allows grid charging. When enabled, the battery can charge from the grid when the battery state of charge is below the reserve capacity and it is in full backup mode or savings or self-consumption mode.
- `battery_shutdown_level` – an integer giving the percent state of charge at which the battery should shut down and wait to start charging again when the grid is down.

The controllable load settings are available through the `load_control` request, allowing control of essential loads if the grid goes down. Load control is only provided if the Envoy hardware components are in place and are connected to appliances with controllable loads or a generator [31]. These allow the Envoy to control load shedding during grid outages or solar PV shedding to maintain the proper solar-to-storage ratio if the Enphase system is off-grid.

Controllable loads are represented by the value of the `load_control_request` body JSON object `data` key, an array of JSON objects representing controllable loads, sent in the request body of the `PUT` and response body of the `GET` requests. One object provides configuration for each high-load appliance connected to the

IQ Controller auxiliary contacts. A load control `mode` JSON key in each object can be set to one of four load control modes:

- `"Basic"` – automatically turns off power to the load represented by the JSON object when the grid goes down,
- `"Scheduled"` – power for the load represented by the object can be scheduled to turn on and off at particular times if the grid goes down. The times are specified by the `essential_start_time` and `essential_end_time` keys in the load's configuration JSON object,
- `"Advanced_BatteryPowered"` and `"Advanced_GeneratorPowered"` – the source of power for the load if the grid goes down is either a generator or a battery, depending on which value is set. The JSON object representing the load also has a key `soc_low` that can be set to any value between 10 and 90, which determines the percent state of charge of the battery at which the load will be powered off,
- `"Manual_always_not_powered"` and `"Manual_always_powered"` – depending on which of these values the key is set to, the load will be either always powered or not powered. This allows the building owner or manager to control whether the load is powered, regardless of whether the grid is up or down.

The JSON load configuration object also has a `name` key whose value is a string with the auxiliary contact name in the Envoy hardware to which the load is connected and a `load_name` key, which is a human-readable mnemonic string for the load that is displayed in the UI.

9.5.4.6 Using the Enphase Monitoring API to Achieve Building Carbon and Energy Use Reduction

The monitoring group API requests support a variety of different ways to monitor the performance of on-site solar PV systems, batteries, and load, using the site-level production, site-level consumption, device-level production, and streaming groups. Time series, interval, and real-time tracking data and information on production can be displayed in a dashboard for the building owner or user to estimate carbon and energy savings due to on-site power use. When coupled with a utility tariff and information about compensation for any energy export programs the DERs may be enrolled in, the production data can be used to estimate compensation due to the building owner for exported power.

The HTTP `PUT` requests available in the system configuration group allow a limited degree of control over DERs and loads. If the utility has a TOU tariff or provides a feed of dynamic marginal grid carbon emissions intensity, the `load_control` `PUT` request can be used to regulate building carbon emissions. The `mode` key in the request body JSON objects representing the controllable loads can be set to `"Manual_always_powered"` to switch on high-load appliances such as pool pumps when the tariff is in the off-peak period or if the carbon emissions feed indicates low grid marginal carbon emissions. When the tariff switches to the peak period or the carbon emissions feed indicates high grid carbon emissions, the `mode` key can be set to `"Manual_always_not_powered"` to switch off the appliances. In addition, the controllers on the appliances need to be set properly to ensure any comfort constraints are maintained; for example, decreasing the HVAC setpoint temperature to achieve building precooling during the off-peak/low carbon emissions period. While the `"Scheduled"` setting looks as if it would be a better match for this purpose; unfortunately, the `"Scheduled"` mode is only available if the system is off-grid or the grid is down.

Production and battery charge information from the monitoring group time series, interval, and real-time tracking data requests can also be used to reduce building carbon emissions together with the `battery_settings PUT` requests. If there is no high TOU period but a feed of dynamic marginal grid carbon emissions intensity is available, the `battery_settings` request can be used to preferentially consume power generated from the solar PV or stored in the battery instead of using power from the grid when marginal grid carbon intensity is high. The BEMS can set the `battery_settings` request body parameter `battery_mode` key to `"Self-consumption"` when the monitoring API requests indicate that solar power or battery power generated from solar is available and the carbon intensity feed indicates that marginal grid carbon intensity is high, switch back to `"Savings Mode"` when the carbon intensity decreases below a pre-set level.

9.5.5 Commissioning API

The commissioning API is split into 11 subgroups:

- Activations – handles the commissioning of Enphase systems, which Enphase calls "activation." This group includes API requests for managing and tracking the activation status of Enphase devices, such as microinverters and batteries, within a site installation,
- Users –manages user accounts associated with the Enphase system both during and after commissioning. It includes requests for retrieving the accounts of users that are associated with a particular site activation, a site owner, or an employee of a partner company, and for updating user accounts for users that are associated with a site activation,
- Companies – manages partner company accounts in the Enphase system. It includes requests to register a new partner company user, retrieve information on a partner company or company user, or update a partner company user's account information,

- Home Owner – supports a request for creating a building owner account which Enphase calls a "home owner" user account, including contact information but without security credentials. These need to be registered by the building owner the first time they log into the Enphase website,
- Estimate – creates and returns estimates for monthly power production for a site,
- Grid Profiles – lists the grid profiles available to the user at a geographic location. A grid profile defines the utility-approved operating parameters for micro-inverters and batteries,
- Meters –contains requests for enabling and disabling meters, updating the operational date of meters, and obtaining meter configuration details,
- PvManufacturers –contains a request that provides information about solar PV panel manufacturers whose panels are compatible with or used in Enphase deployments,
- PvModels – similar to PvManufacturers, the PvModels subgroup contains a request that returns specific models of PV panels used in Enphase deployments,
- Tariff –containsrequests that retrieve and modify records of the current electricity tariff in effect at a site,
- Arrays – manages the records for the configuration of solar panel arrays in an Enphase deployment. It includes API requests for information about the configuration, orientation, and other technical details of the solar arrays installed, for modifying the configuration of an array, and for deleting an array.

The commissioning API is typically accessed by an Enphase partner installer through a mobile phone app. When the installer has completed installation of a system, they will update the information using the app, commission the system, and register an account for the building owner. The building owner then receives an email asking them to visit the login page and set up account credentials. Other parts of the commissioning API can be used to manage partner users and verify that the panels being installed are compatible with Enphase products.

9.5.6 Other Manufacturer Device Control APIs

Many manufacturers of DERs support device monitoring and control APIs but some charge a fee to access them. Other publically available device control REST APIs for DERs include the following:

- SMA [32] – SMA offers APIs that enable third-party applications to interact with SMA inverters and DER devices. These APIs provide access to real-time data, system configuration, and control,
- Fronius Solar [33] – Fronius provides a Solar API (JSON) for integrating with their solar products. This API allows for the retrieval of real-time and historical data directly from Fronius inverters rather than from a website intermediary,

enabling detailed monitoring and analysis of solar energy production and system performance,

- Tesla Energy [34] – The unofficial, open-source Tesla API, documented by Tim Dorr, offers a way to interact with Tesla products, including their solar panels and Powerwall battery systems, allowing users to access data, control charging, and manage settings.

Fronius' approach combines direct access to devices at the building site together with a REST API structured for convenient BEMS application development. This enables the DER devices themselves to act as REST API servers or designates one device at the site as a controller and allows applications to access the devices or the onsite controller directly, avoiding the need for a separate manufacturer website to support the API. The REST API code is embedded in the device firmware. In addition, this provides the building owner with recourse in the event the manufacturer shuts down their business due to bankruptcy or a pivot to a new business model.

EcoBee Smart Thermostat API

DER devices such as solar inverters and batteries, are not the only type of energy control device with a manufacturer API. Several Wi-Fi-connected thermostats also support APIs. In the video link above, a hobbyist explains how he built a controller for the EcoBee smart thermostat using a Raspberry Pi. After the first few minutes of showing the hardware, the video switches to how to register an application on the EcoBee website and delves into the Python code for accessing the EcoBee API.

The unofficial Tesla API is another alternative to remove dependence on a manufacturer's website. An API actively maintained by an open-source community avoids having the BEMS dependent on a manufacturer's website. However, developing such an API requires lots of time to decode how to access the devices and may not be supported by the manufacturer. In addition, maintaining the API to incorporate updates in the device firmware requires a solidly committed open-source community. Unless there is a thriving open-source community behind the project, the API code is likely to languish and get behind the manufacturer's firmware versions, which can leave the BEMS open to unpatched security vulnerabilities.

9.6 The Utility Connection

Reducing building operational carbon emissions requires information in advance about when the carbon emissions intensity of grid power will be highest so flexible loads can be prescheduled and then reduced during grid peaks.

Emissions typically peak when the grid load is highest, late afternoon in summer for cooling-dominated climate zones or early mornings in winter for heating-dominated climate zones, although they can peak at other times. During times of peak load, the utility or wholesale market operator needs to switch on gas-fired peaker plants increasing the grid's carbon emissions intensity. Grid peaks also tend to be when power is most expensive on the wholesale market, which is reflected in retail utility TOU tariffs through the peak tariff period. Utilities may need to institute rolling blackouts during grid peaks, since grid peaks are often when the load begins to exceed generation, , unless the grid load can be reduced.

Demand response is the traditional method for utilities to rebalance load and generation when load threatens to outrun generation. Utilities began prototyping demand response during the 1970s energy crisis, and in the 1980s, started offering financial incentives to commercial customers with large loads for reducing their electricity usage during peak times. These programs were initially manual, with utility employees contacting customers by phone to request load reductions. The commercialization of the Internet and cell phone technology in the period between the 1990s and 2000s enabled automated demand response in which the utility could notify customers using email or text messaging to reduce load, often a day ahead of an anticipated grid peak event. However, the customer still had to turn off appliances manually, and the lack of automated control meant that strategies such as preheating and precooling were unavailable to ensure that comfort and availability provided by essential loads were maintained during the grid event.

More recently, the introduction of protocols such as OpenADR 3.0, IEEE 2030.5, and ANSI/CTA 2045 for signaling grid peak events and controlling large loads has opened the potential for further automation, removing the need for manual intervention. In addition, the availability of APIs providing information on marginal grid carbon intensity increases the potential to further reduce building carbon emissions at night, when the overall grid load is reduced, but the utility may not be able to provide renewable power due to lack of energy storage for solar or lack of wind or hydropower alternatives. This subsection examines these technical solutions for automating the connection between the utility and the BEMS.

9.7 OpenADR

The OpenADR (Open Automated Demand Response) is a standardized API that enables utilities to send demand response signals to customers over an Internet connection and allows customers to automate their response to the signals rather than calling, sending emails, or texting. The signals can be either price-based, reflecting high electricity prices, or event-based, indicating specific grid stress situations. Residential, commercial, and industrial customers can voluntarily enroll in demand response programs offered by utilities. In addition, OpenADR can be integrated into the BEMS to control HVAC, EVSEs, hot water heaters, pool pumps, lighting, and other high-load appliances without human intervention. When the BEMS receives an OpenADR signal from the utility, it can execute a load reduction strategy to ensure that high-load appliances are turned off while maintaining user preferences for appliance service. The API supports real-time communication, allowing for timely responses to changing grid conditions. OpenADR provides a link between grid operators and the BEMS, enabling more effective demand-side management and contributing to the reduction of carbon emissions by optimizing energy use during peak demand periods.

OpenADR 2.0 and OpenLEDR

OpenADR 2.0 has been deployed widely by many utilities, while OpenADR 3.0 is relatively new, so many BEMS deployments may need to interface with it until OpenADR 3.0 becomes established. The video link above, from the Linux Foundation and OpenADR Alliance, provides a short introduction to using OpenADR 2.0 through the Linux Foundation OpenLEDR Python SDK. The Python SDK interface is easier to use than trying to program directly to the XML-based OpenADR 2.0 interface.

OpenADR originated from a collaboration between the California utilities and the Demand Response Research Center at Lawrence Berkeley National Laboratory in the early 2000s. The initial motivation was to address the state's early 2000s energy crisis. The aim was to create a standardized way to automate demand response, making it easier for utilities to communicate demand response signals to customers and for customers to automate their responses. The first version, OpenADR 1.0, was developed around 2002 as a prototype. It primarily focused on California's needs and was successful in enabling automated demand reduction. To further the development and ensure wider adoption of the API, the OpenADR Alliance was formed in 2010. The OpenADR Alliance is a consortium of industry stakeholders, government laboratories, and academic researchers working on developing the API. In 2011, the Alliance released OpenADR 2.0a [35] followed by the release of OpenADR 2.0 in 2012, and OpenADR 2.0b [36] in August 2013.

OpenADR 2.0a and OpenADR 2.0b are two separate profiles of the OpenADR 2.0 standard, designed to address different demand response use cases, with OpenADR 2.0a being a subset of the OpenADR 2.0b features. OpenADR 2.0a is simpler and is designed for basic demand response applications and is suitable for embedding into devices. OpenADR 2.0b is a more comprehensive and advanced profile, providing a full set of services and features. It supports a wider range of demand response events, more detailed event scheduling, and enhanced reporting capabilities. OpenADR 2.0b contains a subset of the features in the OASIS Energy Interoperation Specification Version 1.0 [37]. For example, the OpenADR architecture maps directly to the OASIS specification, but the transactive energy section, in which energy customers and providers can offer quotes and tenders for energy and bid on energy supplies in a real-time energy market, was omitted.

OpenADR 2.0a and 2b utilize web technologies, specifically XML [38] for representing object schema and the Simple Object Access Protocol (SOAP) [39] for communication over HTTP. In 2020, Linux Foundation Energy, an open-source group that works on software for energy applications, released OpenLEADR [40], a Python 3 implementation of OpenADR 2.0b. OpenLEADR was developed to enable wider adoption of OpenADR by providing a free, open-source framework that developers and companies can use to build OpenADR-compliant solutions. This implementation supports the core features of OpenADR 2.0b. In 2019, OpenADR was accepted by the International Electrotechnical Commission (IEC) as standard IEC 62746-10-1 ED1 [41].

In 2023, the OpenADR Alliance recognized that the XML and SOAP technologies were over 10 years old, and in the ensuing period, JSON for object schema and REST APIs for communication were developed and widely deployed for web APIs because they are much simpler to implement services with than XML. XML object descriptions tend to be long and wordy, while JSON schema are very compact. Because SOAP APIs are not stateless, such as REST, they often require complex negotiations between the client and server. Consequently, the Alliance decided to develop OpenADR 3.0 [42] based on the OpenAPI 3.0 API specification language and to use the OAuth client credentials flow for IAM. In addition to the written specification document, a reference implementation of the OpenAPI 3.0 API is also available [43]. The basic system architecture remains the same as OpenADR 2.0, except in the OpenADR 2.0b architecture, the business logic is deeply intertwined with the API server, while in OpenADR 3.0, the two are separated, and the specification only deals with the demand response communication. However, the Alliance decided not to declare OpenADR 2.0 at the end of life since there are so many deployments that depend on it. The intent is to have utilities and others gradually migrate to OpenADR 3.0. The rest of this section focuses primarily on OpenADR 3.0.

9.7.1 OpenADR 3.0 Architecture

Figure 9.12 illustrates the basic OpenADR architecture in the context of a utility running the distribution grid and/or an independent system operator (ISO) running the wholesale market or transmission grid. The OpenADR architecture consists of the following components:

- Virtual Top Node (VTN),
- Virtual End Node (VEN).

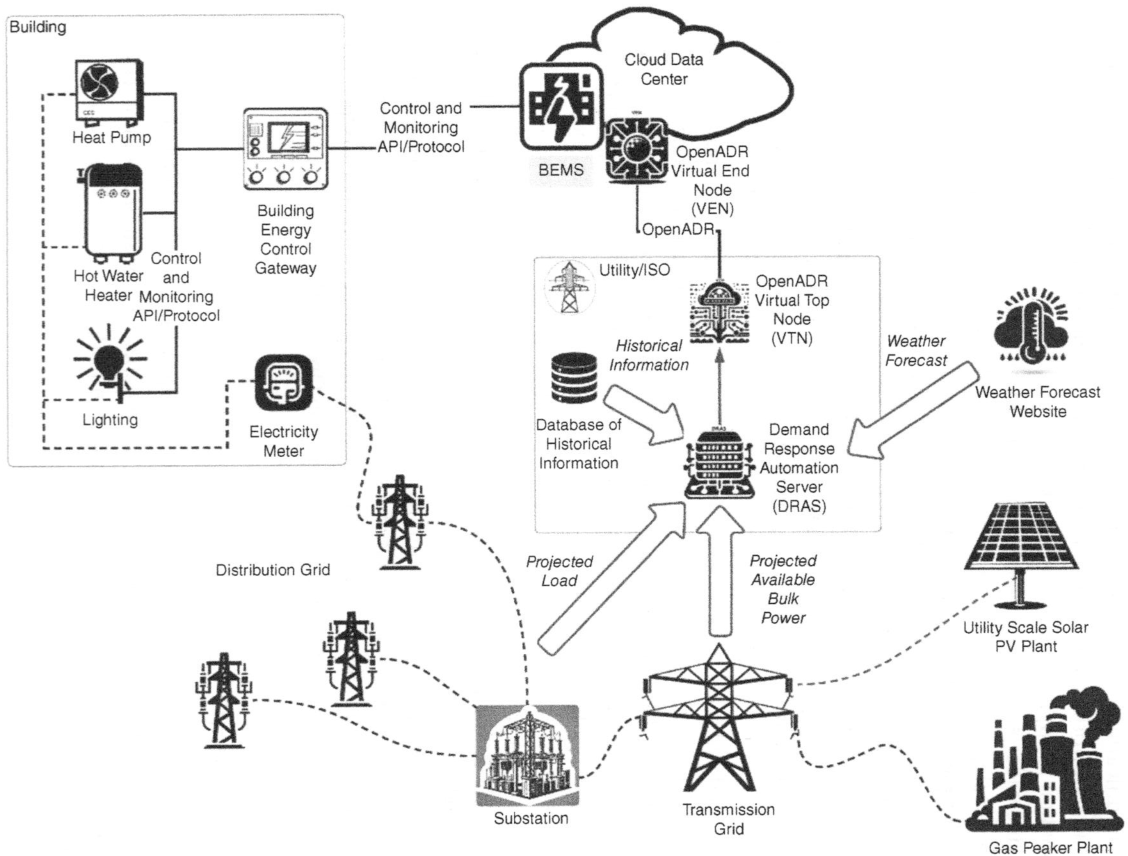

Figure 9.12 Basic OpenADR Architecture

These components are the same in both the OpenADR 2.0 and 3.0 specifications.

The power system operator (utility or ISO) monitors weather forecasts since many grid peak events are weather related, the projected available bulk power, including cases where a power plant will be offline due to maintenance, and the projected load on the distribution grid. It combines that information with historical information to generate a prediction of when a peak grid event is likely to occur, during which the available bulk power may not match the expected load, triggering a brownout or blackout. The Demand Response Automation Server (DRAS), while not a part of the OpenADR architecture, serves as the primary source of demand response signals and information. The DRAS is responsible for generating and dispatching demand response event information and instructions to the VTNs. These event dispatches may include notifications of expected energy demand peaks, requests for load reduction, and pricing information. The DRAS determines when and how demand response events are initiated based on grid conditions, energy prices, and other relevant factors. The DRAS also schedules demand response events, monitors real-time energy supply and usage, and analyzes the effectiveness of demand response strategies.

The VTN acts as the demand response signal dispatcher between the DRAS and the VENs. The VTN is responsible for receiving demand response event information, price signals, and other relevant data from the DRAS and then relaying these instructions to the VENs. The VENs are typically part of the BEMS, though they can also be incorporated directly into standalone high-load appliances, such as HVAC systems, hot water heaters, and lighting systems. The VEN's primary role is to transmit the demand response commands it receives to the BEMS. Upon receiving a demand response signal through the VEN, the BEMS translates this information into a strategy for actionable adjustments in the energy consumption of flexible loads or generation for a BESS and a schedule. This can involve reducing or shifting load, controlling the time of BESS charge or discharge, or adjusting operating schedules to align with demand response events, depending on the BEMS algorithms and the strategy necessary to satisfy the comfort and safety priorities of the building occupants. The VTN is also responsible for enrolling VENs into utility/ISO demand response and energy capacity management programs and for collecting reports from the VENs on their performance during peak grid events.

Note that if the devices in the building are configured to act directly as an OpenADR VEN, the BEMS can run a back-to-back VEN/VTR and control the devices using OpenADR. The alternative is to have the power provider's VTN talk directly with the device, but this architecture would forgo the opportunity to coordinate between devices and take occupant preferences into account. The back-to-back architecture is also used by utility aggregators that aggregate DERs such as BESSs and flexible loads and offer them to utilities as a controllable resource in capacity management programs. In the back-to-back VEN/VTN architecture, the power provider interacts with a VEN in the BEMS, the BEMS executes algorithms to determine which high-load appliances need to be prescheduled or when the BESS needs to charge or discharge and coordinates between devices to accommodate the comfort requirements of the building occupants. It then acts as a VTN and signals the VENs in the appliances and BESSs with OpenADR via the Building Energy Management Gateway. The appliances and BESS are then responsible for carrying out the schedule. In this use case, OpenADR acts as the control protocol between the BEMS, the Building Energy Management Gateway, and the high-load appliances and BESS. One of the goals of the OpenADR 3.0 design was to make this type of architecture easier to implement since REST APIs and JSON are much simpler to implement directly in devices than XML and SOAP APIs.

9.7.1.1 OpenADR 3.0 Architecture Refinements

In OpenADR 2.0, much of the business logic to obtain the credentials to sign up a VEN, create a utility or ISO demand response or energy capacity management program, and for the VEN to sign up for a program was not specified. In OpenADR 3.0, a separate User Guide [43], which is not part of the formal API specification, provides architectural refinements that expand on some actions peripheral to the actual sending of demand response events and reporting on the response. These refinements are not directly supported in OpenADR 2.0b, but in OpenAPR 3.0, they are all handled through API requests defined in the OpenAPR 3.0 specification. One refinement provides requests that enable a service provider to create and manage demand response, energy capacity management, and pricing programs, and update the VTN with those programs. Requests are also defined, enabling a VEN to subscribe to programs and manage their subscriptions.

Another refinement involves customer enrollment of their VEN with an energy service provider and the granting of credentials for the VEN to access the service provider's VTN. This process does not involve the use of the OpenADR protocol itself, so it must be accomplished out of the band using a website maintained by the service provider or through email, text message, or phone call. If the energy service provider handles enrollment through a website, the customer onboarding experience is similar to other web services.

The customer navigates to the website and establishes an account with a username and password. Once the account is established, the customer logs in and registers their generation devices and flexible load appliances available for inclusion in demand response programs. The enrollment website provides the customer with unique identifiers for each manageable energy resource. In addition, the website generates a unique client ID and client secret that the VEN can exchange with the VTN for an OAuth 2.0 token, assigns the customer's VEN a role that maps into an OAuth 2.0 scope, and determines which API requests the VEN has permission to make. The customer then provisions their VEN with the resource identifiers and initial credentials, and the VEN is ready to participate in the energy service provider demand response, capacity management, and pricing programs.

9.7.2 OpenADR 3.0 API

The OpenADR 3.0 API is divided into the following request groups:

- Programs – supports requests allowing the energy service provider to create and manage demand response, energy capacity, and pricing programs. This group also provides an operation to search for programs that match certain tags,
- Reports – enables the VEN to create and manage reports on its performance to the VTN. This group also provides an operation to search for reports that match a particular program identifier or client name,
- Events – supports operations allowing the creation and management of events. This group also provides an operation to search for reports that match certain tags,
- Subscriptions – allows VEN clients to sign up for notifications when HTTP operations are performed on various objects, including events. The notifications are delivered via a callback URL. This group also provides an operation to search for subscriptions that match certain tags or to search for subscriptions by object type or client name,
- VENs – supports two kinds of requests: requests to register a VEN and manage a VEN registration, and requests to manage the flexible load and generation resources provided by a registered VEN. Like the other groups, search operations are also available for searching based on certain tags,
- Authentication – contains one request allowing the VEN to obtain an OAuth 2.0 access token in an initial exchange for the client identifier and client secret it was provisioned with during commissioning. The access token contains permissions for the VEN, enabling the VTN to perform IAM on the VEN's requests.

9.7.3 OpenADR 3.0 URL Format

With the exception of the authentication group, the OpenADR request groups' URLs all have a similar format. Each of the groups has a base URL indicating the group object resource the requests operate upon:

- Programs – `/programs`
- Reports – `/reports`
- Events – `/events`
- Subscriptions – `/subscriptions`
- VENs – `/vens`

Table 9.3 OpenADR 3.0 `targetType` Query Parameter `enum` Values

`targetType` **Enumeration Value**	**Description**
POWER_SERVICE_LOCATION	Target names represent locations in the utility distribution system that are usually the point of service to a customer site.
SERVICE_AREA	Target names represent a geographical area that represents the utility's distribution system coverage.
GROUP	Target names are unspecified groups.
RESOURCE_NAME	Target names identify flexible load and generation devices.
VEN_NAME	Target names identify registered VENs.
EVENT_NAME	Target names identify events.
PROGRAM_NAME	Target names identify demand response, energy capacity, and pricing programs.

Source: Adapted from OpenADR Alliance [42].

The URL HTTP `GET` requests have no path parameters since they aren't specific to any object, but they do have query parameters allowing the client to filter for objects that match certain strings. The `/programs`, `/events`, `/subscriptions`, and `/vens` all have a `targetType` query parameter which can have one of the values in Table 9.3. This indicates the type of the actual tag values that are provided in the `targetValues` list query parameter. Most of the types are straightforward, but the `GROUP` target type can be used to group objects along any criteria for which a matching string can be formulated.

The HTTP `GET` requests for `/reports`, `/events`, and `/subscriptions` can have additional query parameters to narrow the search. The `GET /reports` request can have query parameters to limit the search to a particular program (`programID`) or a particular client (`clientName`). The values of these parameters are unique identifiers for the program and the client VEN, respectively. The `GET` request for `/events` can have a `programID` query parameter to limit the search to a specific program, while the `GET` request for `/subscriptions` supports a `clientName` query parameter and an `objects` parameter. The value of the `objects` parameter indicates what type of object should be included in the reply and can have one of the following values: `PROGRAM`, `EVENT`, `REPORT`, `SUBSCRIPTION`, `VEN`, or `RESOURCE`. The HTTP `GET` requests also support `limit` and `skip` query parameters which allow the client to limit the number of objects returned or indicate the number of objects to skip for pagination.

The `POST` requests in all the groups take no path parameters, but if the object is successfully created, one of the JSON keys in the object returned in the response body is `id`. This key's value contains a unique (to the VTN) string identifier for the created object, which is used in subsequent GET, PUT, and DELETE object management requests as a path parameter. The path parameter identifies the object to which the management operation can be applied. For example:

```
GET /subscriptions/der-755
```

returns the subscription object having identifier der-755. If the object needs to be modified or deleted and the client does not have the object identifier available, the identifier for the object can also be determined by using the search operation for the group to which the object belongs.

9.7.4 OpenADR 3.0 Transport Layer Security and IAM

OpenADR 3.0 requires that the VTN be provisioned with a certificate for server-side TLS and HTTP. However, mutual TLS is not required in order to avoid requiring the provisioning of client-side certificates, which could be difficult if the VEN is running in the firmware of an end device. TLS 1.2 is required as a baseline with TLS 1.3 or later optional.

The OpenADR 3.0 security schemes for IAM are defined in the securitySchemes/components section of the YAML definition. The securitySchemes section is shown in Figure 9.13. The securitySchemes/oauth2 key was discussed in Section 8.5.5.1. Two security schemes are defined as follows:

- oAuth2ClientCredentials – an OAuth 2.0 scheme that utilizes the client credentials flow and scopes to implement role-based access control on the API requests using a JSON token in the Authorization header.

```
securitySchemes:
   oAuth2ClientCredentials:
     type: oauth2
     description: Client credential flow.
     flows:
       clientCredentials:
         tokenUrl: auth/token
         scopes:
           read_all: VENs and BL can read all resources
           write_programs: Only BL can write to programs
           write_events: Only BL can write to events
           write_reports: only VENs can write to reports
           write_subscriptions: VENs and BL can write to subscriptions
           write_vens: VENS and BL can write to vens and resources
   bearerAuth:
     type: http
     scheme: bearer
     bearerFormat: JWT
```

Figure 9.13 OpenADR 3.0 Security Schemes

- `bearerAuth` – a generic HTTP scheme to be used with the HTTP `Authorization` header. The authorization scheme in the OpenAPI definition is `bearer`, and the format of the bearer token is listed as `JWT` for documentation purposes. This scheme was included primarily to support the UI in the reference implementation.

As required by the OpenAPI 3.0 specification, the OpenADR definition includes the authentication group request URL `auth/token` as the token endpoint where the client can exchange their client identity and client secret for a JWT. In the `scopes` section, a separate scope is listed for the operations that modify objects in each of the separate API groups, allowing fine-grained control over access. For example, the `write_programs` scope only allows the energy service provider (indicated by "BL" for "business logic") to create programs. The specification also includes the `read_all` scope, which allows any client to retrieve objects. The `read_all` scope provides access control for the `GET` operations in all groups.

An example of the write scopes is the definition of the `POST` request for `/programs` which only includes `write_programs` in the list of scopes:

```
/programs:
  …
  post:
    tags:
      - programs
    summary: create a program
    operationId: createProgram
    description: Create a new program in
    the server.
    security:
      - oAuth2ClientCredentials: [write_
        programs]
      - bearerAuth: []
  …
```

The security schemes for the `POST` and `PUT` operations in the other request groups reflect this pattern.

The OpenADR 3.0 specification implements the client credentials flow through the authentication group request `GET /auth/token`. The VEN uses this request to obtain an access token from the VTN. The VEN must include two custom headers in the request containing the credentials provisioned at commissioning, `clientID` with the client identifier and `clientSecret` with the client secret:

```
GET /auth/token HTTP/1.1
Host: building-ven.io
clientID: <client identifier>
clientSecret: <client secret>
```

If the VTN can validate the credentials provided by the VEN, the VTN returns an OAuth 2.0 access token encoding the VEN's identity and the scopes to which it has access. The VEN then uses the OAuth 2.0 JWT with the HTTP `Authentication` header and `Bearer` scheme to issue requests, for example:

```
GET / subscriptions/der-755 HTTP/1.1
Host: building-ven.io
Authorization: Bearer <access token>
...
```

Although the specification states that the `/auth/token` endpoint is not required but only a convenience for the reference implementation, many commercial implementations are likely to include it anyway.

9.7.5 Creating a Subscription to an Energy Service Provider's Program and Receiving Events

The heart of the OpenADR specification is the subscription group. The OpenADR 3.0 YAML definition [44] provides details on the data schema and operations. The subscription group allows a VEN to subscribe to a program from an energy service provider and receive callbacks on a URL provided in the initial subscription where a demand response, capacity management, or pricing event is posted by the energy service provider. Alternatively, if the VEN is unable to function in the HTTP server role and take incoming requests, it can poll for new events periodically by using the `GET /events` request and include the `programID` query parameter with the value set to the program identifier of the program to which it has subscribed.

Figure 9.14 illustrates the process. A registered VEN sends a `POST / subscriptions` request to the VTN, including a `subscription` JSON object in the request body having the following required keys:

- `clientName` – the unique client's name assigned to the VEN by the VTN when it is registered,
- `programID` – the unique program identifier assigned to the program by the VTN when the program was created,
- `objectOperations` – an array of objects and operations to which the subscription applies.

The `subscription` object also supports some optional key-value pairs; see the OpenADR 3.0 YAML definition for a list.

Since subscription creation can apply to objects in any of the API groups except for the authentication group, the VEN must include an indication that it wants to subscribe to events. The `objectOperations` array items are JSON

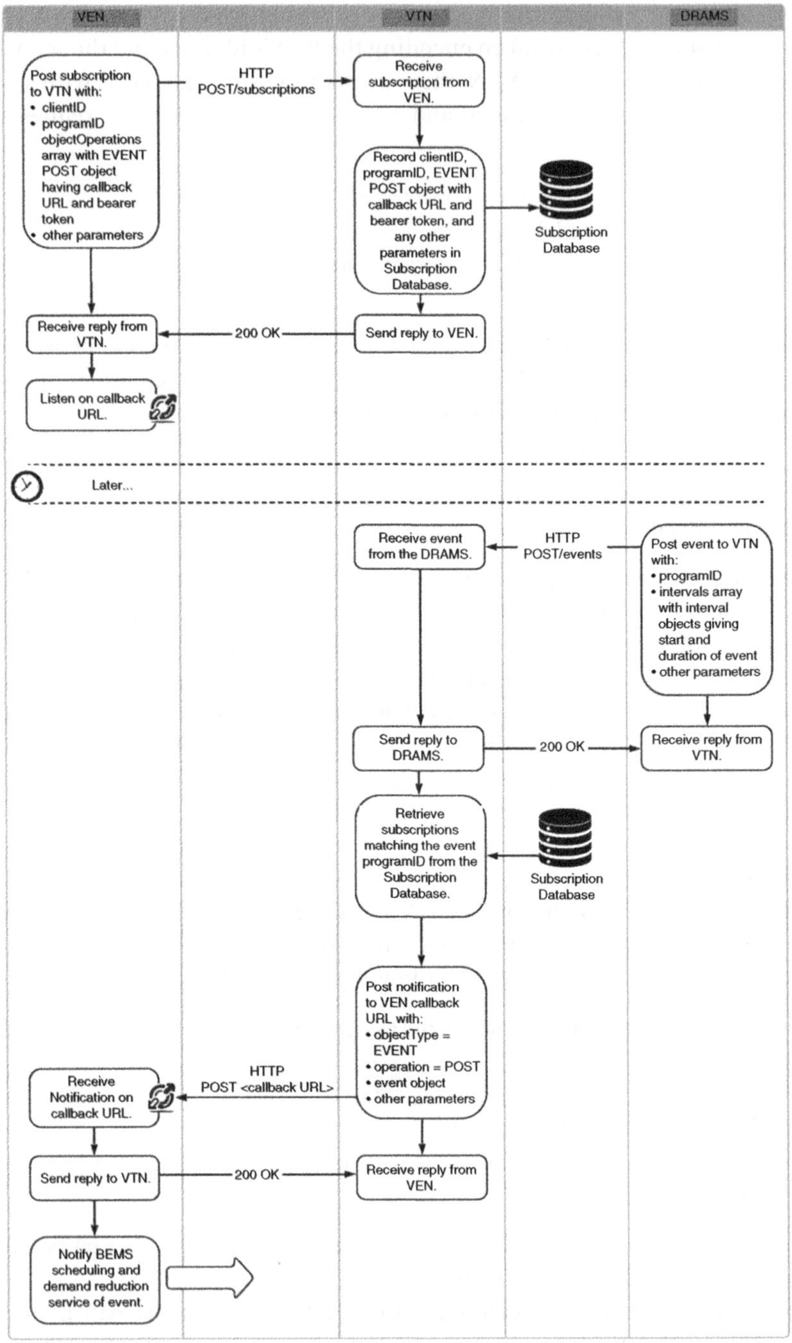

Figure 9.14 OpenADR 3.0 Subscription and Event Notification Process

objects that specify to which objects and operations the subscription applies. These objects have the following keys:

- `objects` – for subscribing to an event, this array must have a single element, the enumeration-defined string, "EVENT",
- `operations` – for subscribing to the creation of an event by the DRAMS, this array must have a single enumeration-defined string, "POST",
- `callbackUrl` – the callback URL hosted by the VEN, on which the VEN will receive notifications of events for the program,
- `bearerToken` – a bearer token for HTTP IAM on the VEN.

An example of a `subscription` for an event is the following:

```
{
    "clientName": "BEMS-65041",
    "programID": "DR-extra-477",
    "objectOperations": [
        {
            "objects": ["EVENT"],
            "operations": ["POST"],
            "callbackURL": "https://
            mybem.energyco.com/
            openAdrEvents",
            "bearerToken": 6853af09e386501
        }
    ]
}
```

The client named "BEMS-65041" is requesting a subscription to be notified when `POST` operations are performed with `EVENT` objects in the demand response program "DR-extra-477." The client has provided a URL for the VTN to send notifications and a bearer token for authentication.

The VTN receives the request and records the VEN's subscription in the Subscription Database, and replies to the VEN. The VEN then begins listening to the callback URL for notifications that an event has been scheduled.

Sometime later, a demand response event is called by the DRAMS. The DRAMS sends a `POST /events` to the VTN, including an `event` JSON object in the request body. The `event` object has the following required keys:

- `programID` – the unique program identifier matching the program for which the event is being posted,

- `intervals` – an array of `interval` objects that define the start time and duration of the event.

The `event` object has many more optional key-value pairs, including a priority in case multiple events are scheduled for the same time period and a report descriptor object if the energy service provider would like a report on the BEMS response to the event after it has concluded.

The `interval` object contains key-value pairs describing the event. The `intervalPeriod` key's value is an object in the `interval` object that bounds the event in time. The `intervalPeriod` object has the following keys:

- `start` – a string specifying the start of the interval in ISO 8601 date time format [45],[1] for example: `"2023-12-15T16:00:00.000-09:00"` for 4 PM Pacific Standard Time on 15 December 2023,
- `duration` – a string that may specify the duration of the interval, in ISO 8601 duration format, for example: `"PT2H30M"` for 2 hours and 30 minutes,
- `randomizeStart` – a string specifying a randomization interval in ISO 8601 duration format applies to the start time.

Only the `start` key is required. If the `duration` key is omitted, the event continues indefinitely until canceled.

The `payloads` key's value in the `interval` object is an array containing `valueMap` objects describing the interval. The OpenADR 3.0 specification uses the same `valueMap` type to define objects in different contexts that can have different types. The payload type in the `valueMap` object is specified by a `type` key which, in the case of an event interval, can have one of the enumeration values specified in Table 1 of the OpenADR 3.0 specification [46]. The `valueMap` key value is an array with elements that can match one of `number`, `string`, `boolean` or a `point` object, having two `number` keys, `x` and `y`. Depending on the program and event, the payload `valueMap` objects provide more information on the event.

---

1 ISO 8601 format is, for all practical purposes, the same as RFC 3339 format which was discussed in Section 8.4.2.1.

An example of an `event` object for the `DR-extra-447` program is the following:

```
{
    "programID": "DR-extra-477",
    "intervals": [
        {
            "id": 530,
            "intervalPeriod": {
                "start": "2024-08-24T16:
                00:00-8",
                "duration": "PT03H30M"
            }
            "payloads": [
                {
                    "type": "ALERT_
                    FLEX_ALERT",
                    "values": [ true ]
                }
            ]
        }
    ]
}
```

The DRAMS is posting a demand response (`ALERT_FLEX_ALERT`) event in the program "`DR-extra-477`" to start at 4:00 PM on August 24th, 2024, which will last for 3 hours and 30 minutes, or until 7:30 PM, in the Pacific Time Zone.

Upon receipt of the event request, the VTN searches its Subscription Database for VEN subscriptions matching the program identifier, retrieves the callback URL and bearer token, and posts a `notification` object to the VENs that have subscribed to events for the program. The notification object has the following keys:

- `objectType` – the enumeration-defined string, "`EVENT`",
- `operation` – the enumeration-defined string, "`POST`",
- `object` – the `event` object sent by the DRAMS defining the event.

An example of a notification object for the DR-extra-447 program is as follows:


```
{
    "objectType": "EVENT",
    "operation": "POST",
    "object": {
        "programID": "DR-extra-477",
        "intervals": [
            {
                "id": 530,
                "intervalPeriod": {
                    "start":
                    "2024-08-24T16:
                    00:00-8",
                    "duration": "PT07H30M"
                }
                "payloads": [
                    {
                        "type": "ALERT_
                        FLEX_ALERT",
                        "values": [ true ]
                    }
                ]
            }
        ]
    }
}
```


The VEN then notifies the BEMS scheduling and algorithmic load reduction service of the event, and the BEMS exercises the necessary supervisory control over its load and generation resources to respond to the event.

9.7.5.1 Callback Security

As with all resources exposed to the Internet, the callback URL requires security to ensure that the identity of the requester is verified and their permission to perform the operation is validated. The server instance hosting the callback URL should, therefore, be provisioned with a TLS certificate allowing the VTN to check that the URL has not been hijacked. Since the VTN is also provisioned with a server-side certificate, mutual TLS is possible. IAM can be performed by having the VEN check the bearer token provided along with the callback notification. Although the `bearerToken` is not a required parameter, it should always be included anyway, to identify the incoming request and ensure the sender is properly authorized to perform the callback.

9.7.5.2 Polling

For devices and other use cases where certificate provisioning and access token verification are impractical, as mentionedabove , polling can be used to check for events. The polling interval should be taken into account when a VTN is most likely to be updated with a new event. For example, if the VTN primarily posts day-ahead events at 4 PM, then polling at 9 AM is likely to miss the event for that day. However, the frequency of the polling should be limited to reduce the unwanted load on the VTN and the network.

Since the `GET /events` request has no way to limit the time interval over which to search, a naming convention can be used to limit which events should be returned. If the VTN names events with an embedded date, the `targetType` and `targetValues` query parameters can be used to limit the time interval, for example:

```
GET /events?
        programID="DR-235",
        targetType="EVENT_NAME",
        targetValues="DR2023-12-14"
        HTTP/ 1.1
```

returns the event with the event name "`DR2023-12-14`" in the "`DR-235`" program. In addition to the naming convention, the `event` objects need to include all intervals for the program to which they apply so that the VEN can retrieve all event intervals for that day.

9.8 IEEE 2030.5

IEEE 2030.5 [5], also called the Smart Energy Profile 2.0, originated from an initiative by the Zigbee Alliance that started in 2008 to define a protocol on the Internet stack for the DER management system (DERMS) of a utility. The Zigbee Smart Energy Profile 1.0 laid the groundwork for connecting devices within the home to communicate with each other and with the utility's smart meters, but the Zigbee protocol was not compatible with the IP stack and required a gateway to bridge to the Internet. With the increasing deployment of DERs and the need for more sophisticated grid management, the limitations of

IEEE 2030.5 Deep Dive

The video link above from Newcastle University in the UK provides a more detailed overview of 2030.5 in the context of managed ("smart") EV charging. The video includes an example of XML DER resource description, a discussion of the DER and flow reservation function sets, and an examination of cybersecurity for 2030.5. It also discusses a few specific managed charging use cases and presents how 2030.5 works in California Rule 21.

Smart Energy Profile 1.0 became apparent. The goal of the IEEE 2030.5 development effort was to transition the protocol from a home-centric focus to a more comprehensive grid-wide focus. IEEE 2030.5 emerged as a significant evolution from its predecessor, incorporating a wider range of technologies and broader industry input. The standard has undergone several important releases, the latest of which is the 2023 release. The 2018 release significantly expanded the standard's functionality beyond the original goal of matching the Zigbee Smart Energy Profile on top of an Internet stack, including improved security measures, enhanced interoperability, and broader support for different DER deployment use cases.

The IEEE 2030.5 standard is designed around the Common Information Model (CIM) from the IEC, which is specified in three different standards: IEC 61970 [47], IEC 61968 [48], and IEC 62325 [49]. The CIM defines abstract data schema and APIs for DER and their interaction with the electric grid. 2030.5 instantiates the CIM through technology from the Internet stack and several older web technologies from the W3C Foundation. DNS service discovery is used to enable devices to find 2030.5 servers. For the transport layer, TLS 1.2 is specified with mutual TLS, requiring that both devices and servers are provisioned with a certificate. A PKI architecture is defined with specific classes of certificates for different use cases.

At the application layer, 2030.5 uses HTTP for communication between the client devices and the utility server, XML [38] for defining data schema, and the web application development language (WADL) [50] for defining operations. Unlike SOAP, WADL is a specification language for defining XML based REST APIs that predates OpenAPI 3.0. However, as with XML and SOAP, WADL definitions tend to be very long and wordy because the language is based on the use of HTML-like tags to designate operations rather than the simple YAML or JSON key-value pairs. In addition, because one of the deployment environments originally targeted by IEEE 2020.5 was DER firmware, there are strict rules on the size of the uniform resource identifiers (URIs). URI path components may not be any longer than four characters, and one character path components occur in the standard. The overall length of the URI is between 80 and 255 bytes. Overall, this makes 2030.5 WADL definitions very difficult for developers to read.

IEEE 2030.5 supports the following aspects of DER management:

- Different deployment architectures – supports multiple different deployment architectures depending on the use case,
- Device discovery and registration – mechanisms for devices to be automatically discovered and registered on the network, facilitating easier integration and management,
- Support for utility programs – support for demand response, capacity management, and pricing programs, and load control initiatives, as well as for DERs to provide other utility services,
- Interoperability – a focus on ensuring that devices from different manufacturers can communicate effectively with each other,
- Real-time data exchange – Capabilities for exchanging real-time data such as energy usage, generation, and system status, including, as in OpenADR, the ability of devices to register a notification URI with the server to subscribe to updates,
- Support for specific DERs – schema allowing management of specific DERs such as solar panels, batteries, and EVs.

The IEEE 2030.5 specification contains no architectural guidance on how to deploy the protocol in a distribution grid context, however a reference implementation of 2030.5 written in the C programming language is available from the Electric Power Research Institute (EPRI) [51].

9.8.1 SunSpec Common Smart Inverter Profile (CSIP)

California Rule 21 is a regulatory framework introduced and maintained by the California Public Utility Commission, the regulatory body responsible for oversight of the electrical grid in California, that addresses the interconnection, operation, and safety of distributed generation and energy storage resources within California's electrical grid. Rule 21 was updated in 2016, mandating the use of one of three protocols, including the IEEE 2030.5 standard for smart inverters, to enable better control over DERs by utilities. SunSpec developed a profile for IEEE 2030.5 that addresses the Rule 21 requirements, called the Common Smart Inverter Profile (CSIP) [52]. Implementation of CSIP is required by the CPUC for inverter manufacturers that want to deploy their products in California. CSIP provides a detailed guideline for the implementation of IEEE 2030.5 in smart inverters, ensuring that inverters can effectively communicate with the electrical utility's DERMs platform. The CSIP allows inverters to provide ancillary services to the grid in addition to signaling demand response, energy capacity, and pricing events. Ancillary services are functions in addition to simply providing power, such as frequency control, that contribute to maintaining grid stability and help the grid recover after a power system event.

Figure 9.15 illustrates the overall CSIP deployment architectures. CSIP defines three main deployment architectures for smart inverter communication:

- Utility to Device – a client-to-server connection from the device to the utility's 2030.5 server allowing the utility to directly control the device, shown at the top in Figure 9.15,
- Utility to Aggregator – a connection from an aggregator, such as a BEMS, to the utility's 2030.5 server, shown in the middle in Figure 9.15,
- Utility to Generating Facility Management System – a connection between the utility's 2030.5 server and a utility-scale power plant with a generating facility management system, shown at the bottom in Figure 9.15.

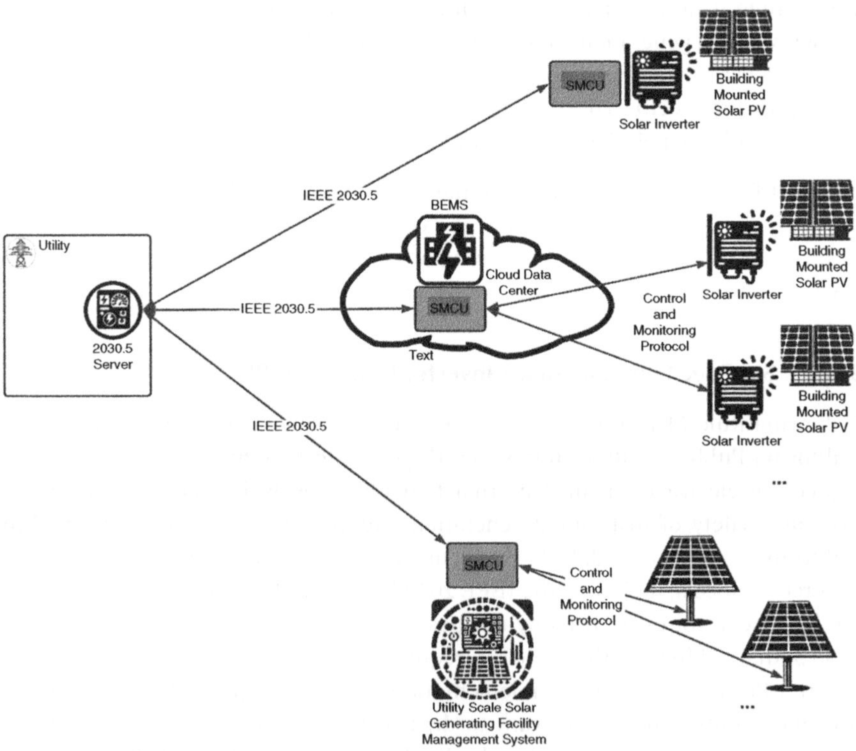

Figure 9.15 SunSpec Common Smart Inverter Profile Network Architecture. *Source:* Adapted from [53].

In the context of building energy management, only the second deployment architecture is of interest, since the BEMS functions in the role of an aggregator for one or more buildings.

The CSIP defines an architectural component called the smart inverter control unit (SMCU), which acts as an intermediary between the utility and the inverter-connected resources. The SMCU translates commands received from the utility over 2030.5 to controls for the inverters. If multiple inverters are downstream of the SMCU, the control protocol between the SMCU and the inverters is left unspecified. The SMCU is responsible for real-time monitoring and management of inverter functions, including adjusting power output, responding to grid disturbances, and participating in utility programs. The SMCU's ability to communicate bidirectionally with the utility via IEEE 2030.5 enables smart inverters to contribute to grid stability and reliability in addition to simply providing energy to the grid.

CSIP utilizes the 2030.5 grouping function to align DERs managed by aggregators with the utility system topology or a utility program. Some settings and commands can be sent to an aggregator's entire fleet of DERs, while others only message a subset of the DERs. The CSIP specification allows a DER to be assigned to a minimum of one and a maximum of 15 groups. DERs can reside in multiple groups; for example, in a topological group, such as a group of DERs on the same feeder, or a nontopological group, such as a demand response group.

9.8.2 Protocol Overview

In the IEEE 2030.5 protocol, resources are URI-addressable elements that can be accessed using a REST interface. Each resource defines a set of attributes and capabilities, representing physical devices such as solar inverters or electric meters, logical entities such as customer accounts, or pricing or demand response signals. For example, a metering resource encapsulates data related to energy usage and generation metrics. The resources are organized under different function sets. Each function set groups together resources that are related to a particular aspect of smart grid functionality. For instance, the Distributed Energy Resource function set includes resources specific to managing DERs. The resources and function sets are organized into three basic categories: Support, Common, and Smart Energy. These categories are similar to the high-level groups in OpenAPI defined using the `tags` YAML key.

9.8.2.1 Support Category

The Support category contains resources and function sets that provide support functions for the operation of the IEEE 2030.5 protocol, including basic infrastructure elements necessary for the establishment and maintenance of communication within the smart grid network. The Support category includes resources that provide device discovery, registration, and security services. The following function sets are defined:

- DeviceCapability – supports identifying the capabilities of devices within the smart grid network. It includes resources that describe what functions a device can perform, the types of measurements it can take, and any other specific features or limitations. Servers are required to support the DeviceCapability function set and must provide URIs for the additional services that the server supports,
- SelfDevice – provides the means for a device to effectively communicate its characteristics, status, and capabilities within the IEEE 2030.5 network, ensuring that the device can be integrated and managed. This includes resources for device identification, managing configuration, and reporting status, reporting health and diagnostics, interacting with other devices, and managing regulatory compliance-related settings and security parameters,
- EndDevice – deals with the representation and management of end devices such as appliances or smart meters. This function set includes resources for device identification, status reporting, and operational management,
- FunctionSetAssignments – used to assign specific function sets to devices or collections of devices, allowing a service provider such as an aggregator or utility to link specific function set instances to a particular program. For example, FunctionSetAssignments can be used to group inverters into operational areas based on the distribution grid topology, allowing them to be commanded through a single message,
- Subscription/Notification Mechanism – allows devices and systems to subscribe to updates or notifications rather than having to periodically poll for updates,
- Response – designed to handle acknowledgments and responses to requests made within the smart grid network. This function set allows a device to respond to requests indicating that the message was received, to opt-in/opt-out of messages, or to send other status information.

9.8.2.2 Common Category

The Common category contains resources and function sets that are related to device information, error reporting, and basic operational data, and acts as a bridge providing interoperable and standardized communication among different devices and systems. The following function sets are included in the Common category:

- Time – responsible for ensuring accurate time synchronization across all devices in the smart grid network. This function set allows devices to obtain time from either an authoritative source, such as the Network Time Protocol (NTP), or less accurate sources and mandates specific time-related behaviors,
- Device Information – provides static, manufacturer-specific information about a device. It is used to convey important details about the physical and operational characteristics of the device for identification, management, and interaction within

the smart grid network. The primary resource is the DeviceInformation resource, which contains details such as the manufacturer, model, and other specifics of the device. This function set is mandatory for any device implementing an HTTP server,

- Power Status – monitors and logs changes in the device's power source, such as switching between grid and battery, and battery charge levels,
- Network Status – provides management of the device's network interfaces and information on the device's network performance at both IP and link layers,
- Log Event – contains time-stamped LogEvents generated by the device, typically indicating out-of-bounds conditions or significant events. It supports a list of LogEvents with details such as event codes and descriptions,
- Configuration – centralizes read/write access to the device's operational configuration, facilitating configuration management and backup,
- File Download – supports secure, interoperable remote file downloads to devices, including software, logs, and configuration files.

9.8.2.3 Smart Energy Category

The Smart Energy category is focused on resources and function sets that directly pertain to energy management and smart grid applications, including advanced functions for managing DERs, demand response, metering, and customer energy management. The Smart Energy category is important for the active management and optimization of energy resources within the grid. The following function sets are defined:

- Demand Response and Load Control – facilitates the interface for demand response and load control. It enables server devices to expose load control events to client devices through EndDeviceControls resources, which can target specific device types and contain necessary attributes for devices to process events, including start times, durations, and randomization. It includes the DemandResponseProgram resource to model energy service provider demand response programs and uses a unique identifier for each program instance,
- Metering – defines operations to exchange measurement information on electricity,
- Pricing – provides information on the utility tariffs and is designed to support a variety of tariff types,
- Messaging – supports an interface for a text messaging service. Clients of this function set include wall-mounted displays in the building and other text messaging display devices,
- Billing – provides information from the utility on consumption and costs, estimates of future consumption, and historical consumption. It also provides a mechanism to allow the utility or other energy service provider to push down targets or challenges to help consumers with budget management,
- Prepayment – defines mechanisms for the conditional delivery of services based on credit or debt and provides an interface for clients to view, update, or act upon account balance information, measured either monetarily or by commodity,

- Flow Reservation – provides an interface for the exchange of flow reservation events, crucial for devices such as EVs and other large power-consuming devices. It enables the scheduling of high-demand periods and helps manage energy rates and demand tiers,
- Distributed Energy Resource – manages DERs, including generation and storage devices such as solar inverters and battery storage systems. Servers host DERPrograms resources to expose DERControl resource events to DER clients,
- Metering Mirror – provides a mechanism for constrained devices to efficiently post metering data to a metering server by minimizing the transactions required to create and maintain a mirroring relationship.

9.9 Other Protocols for Communicating Between the Utility and Flexible Loads/DERs

Two other protocols have seen some deployment for flexible load management: ANSI/CTA 2045 [54] and DNP3 [55]. ANSI/CTA 2045 was developed for controlling high-load residential devices, while DNP3 is primarily a Supervisory Control and Data Acquisition (SCADA) protocol for use within utilities but has been deployed in some cases for flexible load management on commercial campuses.

9.9.1 ANSI/CTA 2045

ANSI/CTA 2045 [54] was developed to standardize communication interfaces, both hardware and software, for flexible load devices, such as hot water heaters, heat pumps, EV chargers, and pool pumps, installed in residential and small commercial buildings. The first version was published in 2013, and the latest version, designated 2045-B, was published in 2021. The protocol was developed through a collaboration of industry experts, utilities, and technology providers under the Consumer Technology Association (CTA). The OpenADR Alliance runs a certification program called EcoPort that certifies 2045 implementations to ensure that they are interoperable [56], and regulatory agencies in different states are starting to examine whether 2045 should be mandated for appliances. For example, the states of Washington and Oregon require new hot water heaters to incorporate a 2045 interface, and similar regulatory actions are underway in California and several New England states.

ANSI/CTA 2045 defines a networking stack consisting of a physical layer, link layer, and application layer. At the physical layer, direct current (DC) powered devices are specified to use a Universal Communications Module (UCM) as the gateway to the utility and a Smart Grid Device (SGD) module on the appliance, with an RS-485 serial communication link between the two. The hardware for these devices is defined directly in the standard and is designed to allow plugging the UCM into the SGD, which is mounted on the appliance. For AC-powered appliances, an RS-485 link is specified. One drawback of the 2045 specification is that it leaves security as an exercise for the implementor, probably because the only physical layers included in the standard utilize the RS-485 serial bus.

The format of the link and application layer fields is binary, with opcodes and bitfields specifying messages of different types. In addition to transporting 2045 link layer and application layer messages, the link layer supports a 2-byte message type code that allows tunneling of 12 different protocols, including BACnet, Zigbee Smart Energy 1.0, 2030.5, OpenADR 1.0 and 2.0, and even IP. Tunneled packets are passed through to the appliance controller for decapsulation and processing. The 2045 protocol itself has four link layer message type codes: basic DR (code 0x0801) and intermediate DR (code 0x0802) at the application layer, and data link (code 0x0803) and commissioning and network support (code 0x0804) at the link layer.

The application layer messages are designated with two opcodes, the first giving the message type and the second used in some cases for additional information. The basic DR application message type specifies demand response and energy management messages that all devices are expected to support, the intermediate DR type is for advanced messages that only a subset of devices are expected to support. An example of a basic message is the Shed message, with opcode 1 set to 0x01 and opcode 2 set to the duration of the event. The Shed message is sent by the utility when a load shed event begins. An example of an intermediate DR application message is the Get/Set Temperature Setpoint, with opcode 1 set to 0x03 and opcode 2 set to 0x03. The Get/Set Temperature Setpoint gets or sets the current temperature setpoint on a device such as a thermostat.

9.9.2 DNP3

An example of another protocol that has achieved some deployment for connecting the utility's DERMS system to building DERs is Distributed Network Protocol 3 (DNP3) [55]. Development of DNP3 started in 1993 by GE-Harris Canada, initially as a North American standard to allow interoperability between various vendors' SCADA equipment for the electrical grid. SCADA systems are deployed by utilities to monitor the condition of equipment that is crucial to keeping the grid functioning, such as transformers and switches. Work on DNP3 occurred while IEC 60870-5 [57], the international standard for SCADA system applications, was still under development and not yet standardized. The IEEE standardized DNP3 as IEEE 1815 in 2010 and released several updated versions culminating in the latest release in 2015 [58]. In 2019, IEEE released IEEE 1815.2 [59], a DNP3 profile specifically for managing DERs.

DNP3 defines a full networking stack consisting of a link layer, transport layer, and application layer. The link layer offers services such as multiplexing, data fragmentation, error checking, link control, prioritization, and link layer addressing. The application layer defines functions and generic data types suitable for common SCADA applications. The DNP3 frame structure resembles that of IEC 60870-5 but is not identical to it. DNP3 supports transmission over both serial bus connections and TCP/IP networks. For communication over IP networks, the DNP3 layers are integrated into the application layer and tunneled over the UDP, TCP, or IP protocols. DNP3 includes robust security, including a secure authentication protocol, as part of the specification and recommendations for the use of TLS in IP networks.

Communication in DNP3 is between central master stations and remote units in a utility's SCADA system, with the master station serving as an interface between the human network manager and the monitoring system and the remote unit interfacing between the master station and the physical equipment being monitored and controlled at remote sites. The protocol uses 27 basic function codes for communication, enabling master stations to request and receive status information from remote devices, adjust their configurations, and control remote equipment. The protocol's object library includes schema for binary and analog inputs. In building energy management applications, DNP3 has been employed in a multi-building campus environment. For example, Endeavour Consult C.A., which is now part of Emerson's Test and Measurement Group, developed a solution for smart building applications using DNP3 [60]. This project involved monitoring and controlling multiple medium voltage disconnect switches in a smart building, utilizing several PCs as Human Machine Interfaces (HMIs) and a National Instruments CompactRIO controller configured as a DNP3 master. The DNP3 protocol was used for real-time interaction with these systems over an IP network.

9.10 Controlling Energy Use Based Directly on Reported Grid Marginal Carbon Emissions

Most of the discussion in this chapter has focused on reducing building energy consumption or switching to using on-site renewable generation during peak grid events as a strategy to reduce operational carbon emissions. Because utilities tend to bring up gas-fired peaker plants during grid events, shedding load and switching to on-site renewably generated power avoids importing power during periods when grid-supplied electricity has increased marginal carbon emissions intensity. These strategies have an added benefit in that they can save money on the utility bill, since utilities typically design their TOU retail tariff plans to align the peak tariff period with times when peak events tend to occur. However, there may be occasions when marginal carbon emissions intensity increases without a peak event occurring. For example, the utility may schedule maintenance on a utility-scale solar PV, wind, or energy storage facility and start up a gas-fired plant to cover the power lost from the renewable facility.

OpenADR 3.0 supports messages to inform the BEMS about grid marginal carbon intensity in the context of a utility program. An event `payloads` array having an item with the `valueMap type` keyset to the event payloads enumeration type `GHG` [4] can be used to signal that the item contains a marginal carbon emissions intensity value, with the `values` key array containing a marginal greenhouse gas emissions estimate in g/kWh as a `float`. For example, suppose a utility sponsored a program to inform customers when the marginal carbon intensity exceeds the average grid carbon intensity of 22.87 g/kWh by 2 standard deviations or 68.61 g/kWh. The following `event` object could be distributed as part of a `notification` update to the callback URLs of VENs that have subscribed to the program:


```
{
  "programID": "2sigma-ghg",
  "createdDateTime":
  "2023-11-17T16:30:00-08:00",
  "objectType": "EVENT",
  "intervalPeriod": {
    "start": "2023-11-17T16:30:00-09:00"
  },
  "intervals": [
    {
      "id": 0,
      "intervalPeriod": {
        "start": "2023-06-15T09:30:
        00-08:00",
      },
      "payloads": [
        {
          "type": "GHG",
          "values": [
            68.61
          ]
        }
      ]
    }
  ]
}
```

Since the `intervalPeriod` key's value object has no `duration` key, the event lasts until the VTN sends out a notification that the marginal grid carbon intensity has dropped below 2 standard deviations above the mean. Other event notifications can include `GHG valueMap` objects in their payload descriptions as well as payload descriptions related to the program's primary purpose.

IEEE 2030.5 also supports notifications on carbon dioxide emissions, in addition to emissions of sulfur dioxide and nitrogen oxide, through two XML objects:

- `CostKindType` – a flag indicating various environmental cost types associated with energy consumption, including carbon dioxide emissions, sulfur dioxide emissions, nitrogen oxide emissions, and the amount of renewable generation. The flag is set to the following values to indicate the content type:
  - 0 – carbon dioxide emissions,
  - 1 – sulfur dioxide emissions,
  - 3 – nitrogen oxide emissions,
  - 4 – renewable generation.

  The emissions are in grams per unit, while the renewable generation is in percent.
- `EnvironmentalCost` – the value of the cost for the indicated `CostKindType`.

These can be queried with the RateComponent resource in the Pricing function group to obtain information on the marginal grid carbon emissions intensity and for the utility server to notify the BEMS about the marginal carbon emissions intensity so that high-load devices can be scheduled when the amount of renewable generation is high if the utility supports providing this kind of information.

9.10.1 California MIDAS Server

With both OpenADR 3.0 and IEEE 2030.5, the BEMS is dependent on the utility for information about marginal grid carbon emissions. The information may or may not be available depending on the utility. In many jurisdictions, the utility buys electricity on a wholesale marketrun by an ISO, regional system operator (RSO), or regional transmission operator (RTO). If the electricity market in a regulatory jurisdiction has a wholesale market operator, information on the marginal grid carbon emissions intensity may be offered directly by the wholesale market operator. In late 2021, the California Energy Commission (CEC) launched the MIDAS, a service that offers access to data on grid electricity in the California wholesale market.

The data includes utility tariffs, greenhouse case emissions, and FlexAlerts from CAISO, which runs the wholesale market in California. CAISO provides FlexAlerts when a peak grid event is expected [61]. The service can be either directly connected to flexible loads or connected through a BEMs. The MIDAS database sources its information on utility tariffs from tariff information uploaded by load-serving entities such as utilities and community choice agencies, which are nonprofit electricity suppliers in California. Greenhouse gas emission data is sourced from WattTime's SGIP API [62], and CAISO FlexAlerts from the CAISO FlexAlert website [63].

9.10.1.1 Rate Identification Numbers (RINs)

MIDAS uses rate identification numbers (RINs) to identify tariffs, greenhouse gas emissions information, and CAISO FlexAlerts for different distribution operators and energy providers in California.

For utility tariffs, a RIN consists of a text string as follows:

```
<country><state>-<distributor><energy provider>-
<tariff rate>-<location>
```

The <location> portion may consist of from 1 to 10 characters, the others are two characters. The RIN for a tariff does not change over time. Rates with no specific location use the location code 0000.

As an example, in the following RIN, the <country> is the United States, the <state> is California, the <distributor> is Pacific Gas and Electric (PG&E), a utility in northern California, the <energy provider> is Marin Clean Energy, a community choice agency in Marin County, and the <tariff rate> is TOU-C, a time of use rate:

```
USCA-PGMC-CTOU-0000
```

For greenhouse gas emissions information, the MIDAS database provides 33 different RINs covering real-time, forecasted, and historic emissions in 11 regions of California. RINs having the format USCA-SGIP-SGRT-XXXX, USCA-SGIP-SGFC-XXXX, and USCA-SGIP-SGHT-XXXX provide access to the real-time, forecasted, and historical greenhouse gas emissions information, respectively, in the MIDAS database obtained from the WattTime SGIP API, with XXXX designating one of the 11 regions. The region codes are available from the WattTime SGIP API documentation. The SGRT RIN returns only one data point giving the current marginal carbon emissions for the specified region. The SGFC RIN returns the forecasted CO_2 levels at 5-minute intervals. The SGHT RINs return historical tables for the specified region. Once a real-time value is replaced, it is moved to the historical table.

For CAISO FlexAlerts, MIDAS provides real-time, forecasted, and historic values with the date and time of previous alerts. The format for these RINs is USCA-FLEX-FXRT-0000, USCA-FLEX-FXFC-0000, and USCA-FLEX-FXHT-0000 for the real-time, forecasted, and historical alerts, respectively. The RIN for historical alerts retrieves all previous alerts.

9.10.1.2 MIDAS REST API

MIDAS supports a simple REST API layered on top of the CAISO server [64]. Instructions for accessing the MIDAS API and links to open-source code in Python, R, and C# are available in GitLab [65]. The base URL for MIDAS is as follows:

```
https://midasapi.energy.ca.gov/api/<endpoint>
```

The API can accept or return data formatted as both JSON and XML, depending on the `Accept` header included with the request. The following `<endpoint>` elements correspond to the available request endpoint types:

- `Registration` – registers a new user or organization with the HTTP `POST` method. The request body contains a JSON or XML object with properties whose values are strings containing the organization's name (`organization` key), a username (`username` key), a password (`password` key), an email address (`emailaddress` key) for contacting the user, and the full name of the user (`fullname` key), all encoded using base64 encoding. If the registration process is successful, an email is sent to the email address, and the user must reply to the email before requesting an access token.
- `Token` – authenticate and retrieve an access token with the HTTP `GET` method. The request must have an `Authorization` header utilizing the `Basic` scheme, followed by the username and password separated by a colon and base 64 encoded as discussed in Section 8.5.3.1. The token is returned in the `Token` header. The token is only good for 10 minutes, after which it must be refreshed with another call to the `Token` endpoint.
- `ValueData` – retrieve tables, data types, and XML schema depending on the query parameters, discussed in more detail in Section 9.10.1.3.
- `HistoricalList` – return a historical list of all RINs and data type values with a short human-readable description for a distributor whose code is provided as the value of the `distributionCode` key and an energy provider whose code is provided as the value of the `energyCode` key.
- `HistoricalData` – return the historical data for a specific RIN indicated as the value of the `id` query parameter between a start date, which is the value of the `startDate` query parameter, and an end date, which is the value of the `endDate` query parameter. All query parameters are required.
- `Holiday` – return a table of registered holidays. Holiday records are returned as the code for the energy provider, a human-readable description for the energy provider, the local date and time of the holiday in RFC 3339 format, and a human-readable description of the holiday.

The API requires users to register and obtain a token before making requests for data. Individual users are allowed to issue HTTP `GET` methods to obtain data. Load-serving entities can additionally `POST` tariffs and holidays. Organizations that are load-serving entities must send an email to the CEC from their work account requesting verification before they can post. The `POST` method can only be used by individual users to register through the `Registration` endpoint. The HTTP method types `PUT` and `DELETE` are not used in the API and there is also no option for a client to subscribe to notifications, so a BEMS needs to poll periodically for real-time updates.

9.10.1.3 ValueData Endpoint

The `ValueData` endpoint enables querying for multiple types of data. If no query parameters are provided, a `ValueData` request returns the XML schema for MIDAS. Query parameters are used to specify whether real-time or historical data should be retrieved and to focus a query in other ways. The Midas API documentation provides more information on the format of the return values.

The following describes the information returned by the `ValueData` request in response to specific query parameters:

- all RIN lists – setting the `SignalType` query parameter to a code for the desired data type causes the `ValueData` request to return all RIN lists for the corresponding data type. The following codes are supported:
  - `0` – all data types,
  - `1` – tariff data only,
  - `2` – greenhouse gas emissions only,
  - `3` – FlexAlerts only.
- a particular RIN – setting the `id` query parameter to a particular RIN and the `QueryType` query parameter to one of `allData` or `realTime` causes the `ValueData` request to return either the full schedule or the real-time value, respectively,
- lookup table – setting the `LookupTable` query parameter retrieves the lookup table corresponding to the parameter value. The parameter can have value strings: `"Country,"` `"DayType,"` `"Distribution,"` `"EndUse,"` `"Energy,"` `"Location,"` `"RateType,"` `"Sector,"` `"State,"` `"TimeZone,"` and `"Unit"`.

Lookup tables contain a complete list of codes for different data types; for example, the distribution table contains the two-letter code for each distribution company in California, along with the name of the distribution company corresponding to the code. More information on lookup tables is available from Appendix C of the MIDAS API documentation [66].

9.11 Summary

This chapter examines network protocols and APIs for communicating between the BEMS and DER deployments such as solar PV systems and BESSs at the building site, and between the BEMS and the utility. DER deployments can play an important role in building decarbonization since the BEMS can switch over to using locally generated and stored renewable energy during periods when the grid is under stress or during peak tariff periods if the building is on a TOU tariff. These are peak load times when the utility will bring up gas-fired peaker plants to avoid the need for blackouts, causing the marginal grid carbon emissions intensity to rise. Energy from DER deployments can also be used together with flexible loads to ensure that occupant comfort requirements are met, including providing power if the grid goes down.

DERs are most useful when they are part of a building-integrated microgrid, the topic of Section 9.1. BESSs allow energy generated from renewable sources such as solar PV to be saved and released when it is most advantageous from the standpoint of mitigating building operational carbon emissions. The Modbus protocol is widely used to control on-site DER deployments and is discussed in Section 9.2. Modbus provides the basis for the SunSpec Modbus profile, a specification standardizing the Modbus interface for solar PV inverters and batteries, which was presented in Section 9.3. The SunSpec Modbus profile has been implemented by some inverter and battery manufacturers to allow direct control of DERs from a BEMS. Some manufacturers use other protocols, and one of these, CANbus, is briefly reviewed in Section 9.4. More popular, however, are manufacturer APIs, and Section 9.5 discusses an example API from the DER manufacturer Enphase. With a manufacturer API, the protocol between the DERs on-site and the manufacturer's website, where monitoring and control actions from the building owner or manager to the DERs originate, is not publicly disclosed. Public information is only provided about the API to the website.

The other side of managing building energy use is the connection with the utility. If the utility has a TOU tariff, the peak time can be used as a signal for when to increase DER use and turn down high-load appliances. However, many utilities have started offering demand response programs, where the BEMS can receive a signal when a peak event is occurring, and there is a risk of blackout. The building owner can then be compensated for reducing energy use or increasing energy export from on-site DERS during that time. OpenADR, discussed in Section 9.7, and IEEE 2030.5, discussed in Section 9.8, are two protocols for communicating demand response and other events from the utility to the BEMS. IEEE 2030.5 has additional support for allowing the building to provide ancillary services, such as frequency control, if the building is equipped with DERs. While OpenADR and IEEE 2030.5 are the primary protocols in use for communicating between the utility and customer-sited DERs and flexible loads, two others, namely, ANSI/CTA 2045 and DNP3, are briefly discussed in Section 9.9.

Finally, although most grid events are primarily concerned with reducing load or increasing supply to match a supply/demand imbalance, both OpenADR and IEEE 2030.5 provide support for signaling the marginal grid carbon emissions intensity directly. If the utility supports a program for it, this allows the BEMS to manage energy use based on the carbon content of grid electricity, a topic discussed in Section 9.10. Alternatively, the wholesale market operator may host a service, such as California's MIDAS service, that provides this information.

References

1 Modbus Organization. Modbus Application Protocol Specification, v1.1b3 [Internet]. [cited 2023 Nov 12]. Available from: https://www.modbus.org/docs/Modbus_Application_Protocol_V1_1b3.pdf.

2 SunSpec Alliance. SunSpec Device Information Model Specification, Version 1.1, May, 2022.

3 Enphase. The Enlighten Systems API [Internet]. [cited 2023 Nov 12]. Available from: https://developer-v4.enphase.com/docs.html.

4 Sandoval F., Ed. and Nordman B., Ed. OpenADR 3.0: OpenADR 3.0.1 Definitions. OpenADR Alliance, Document Number: 20231118-2, Revision Number 3.0.1, November, 2023.

5 IEEE. IEEE Standard for Smart Energy Profile Application Protocol. IEEE Std 2030.5-2018. New York: IEEE; December, 2018.

6 California Energy Commission. Market Informed Demand Automation Server (MIDAS) Documentation Version 1.2: Connecting to and Interacting with the MIDAS [Internet]. [cited 2023 Nov 13]. Available from: www.energy.ca.gov/publications/2021/market-informed-demand-automation-server-midas-documentation-version-12.

7 IEEE. IEEE 1547-2018: IEEE Standard for Interconnection and Interoperability of Distributed Energy Resources with Associated Electric Power Systems Interfaces [Internet]. [cited 2023 Dec 7]. Available from: https://standards.ieee.org/ieee/1547/5915.

8 IEEE. IEEE 2030.7: IEEE Standard for the Specification of Microgrid Controllers [Internet]. [cited 2023 Dec 7]. Available from: https://standards.ieee.org/ieee/2030.7/5941.

9 EMQX. Modbus Protocol: The Grandfather of IoT Communication [Internet]. [cited 2023 Nov 15]. Available from: https://www.emqx.com/en/blog/modbus-protocol-the-grandfather-of-iot-communication#history-and-origin-of-modbus-protocol.

10 Modbus Organization. MODBUS over Serial Line Specification and Implementation Guide V1.02 [Internet]. [cited 2023 Nov 16]. Available from: https://www.modbus.org/docs/Modbus_over_serial_line_V1_02.pdf.

11 Modbus Organization. MODBUS Messaging on TCP/IP Implementation Guide V1.0b [Internet]. [cited 2023 Nov 16]. Available from: https://www.modbus.org/docs/Modbus_Messaging_Implementation_Guide_V1_0b.pdf.

12 Section 6.3, Sunspec Modbus Profile, Version 1.1, p. 34.

13 Section 7, Modbus Application Protocol Specification, v1.1b3, pp. 47–49.

14 SunSpec Alliance. Open Standards for the Distributed Energy Industry [Internet]. [cited 2023 Nov 21]. Available from: https://sunspec.org.

15 SunSpec Alliance. SunSpec Information Model Specification, Version 1.9, Document #12041.

16 Table 3 and Table 4, SunSpec Device Information Model Specification, Version 1.1, p. 24 and 27.

17 Table 2, SunSpec Device Information Model Specification, Version 1.1, p. 22.

18 SunSpec Alliance. SunSpec Device Web Service API Specification [Internet]. [cited 2023 Nov 26]. Available from: https://sunspec.org/wp-content/uploads/2019/10/SunSpec-Device-Web-Service-API-Specification.pdf.

19 Table 6, SunSpec Device Information Model Specification, Version 1.1, p. 28.

20 Appendix A, JSON Model Definition Encoding Example, SunSpec Device Information Model Specification, Version 1.1, pp. 44–46.

21 SunSpec. sunspec/models [Internet]. [cited 2024 March 30]. Available from https://github.com/sunspec/models/tree/master/json.

22 Table 7, SunSpec Device Information Model Specification, Version 1.1, p. 31.

23 Figure 5, SunSpec Device Information Model Specification, Version 1.1, p. 33.

24 IEEE. 754-2008-IEEE Standard for Floating-Point Arithmetic [Internet]. [cited 2023 Nov 26]. Available from: https://ieeexplore.ieee.org/servlet/opac?punumber=4610933.

25 Section 6.4, SunSpec Device Information Model Specification, Version 1.1, pp. 35–38.

26 Wikipedia. CAN Bus [Internet]. [cited 2023 Nov 26]. Available from: https://en.wikipedia.org/wiki/CAN_bus.

27 safiery.com. CANbus Lithium Batteries [Internet]. [cited 2023 Dec 2]. Available from: https://safiery.com/canbus-lithium-batteries.

28 Higginbotham S. Insteon Is Down and May Not Be Coming Back [Internet]. [cited 2023 Nov 27]. Available from: https://staceyoniot.com/insteon-is-down-and-may-not-be-coming-back.

29 Enphase. Enlighten API – Documentation [Internet]. [cited 2023 Nov 27]. Available from: https://developer-v4.enphase.com/docs.html.

30 Enphase. Enlighten API – Quick Start [Internet]. [cited 2023 Nov 28]. Available from: https://developer-v4.enphase.com/docs/quickstart.html.

31 Enphase. Technical Brief – North America: Load and Solar Circuit Control Using IQ System Controller Auxiliary Contacts [Internet]. [cited 2023 Dec 5]. Available from:https://enphase.com/download/load-and-solar-circuit-control-using-iq-system-controller-auxiliary-contacts-tech-brief.

32 SMA. SMA APIs [Internet]. [cited 2023 Dec 2]. Available from: https://developer.sma.de/sma-apis.html.

33 Fronius. Fronius Solar API [Internet]. [cited 2023 Nov 27]. Available from: https://www.fronius.com/en/solar-energy/installers-partners/technical-data/all-products/system-monitoring/open-interfaces/fronius-solar-api-json-.

34 Dorr T. Energy Products: Tesla JSON API (Unofficial) [Internet]. [cited 2023 Dec 2]. Available from: https://tesla-api.timdorr.com/api-basics/products.

35 OpenADR Alliance. OpenADR 2.0 Profile Specification A Profile, Document Number: 20110712-1, 2012.

36 Herberg U., Ed., Zuber J., Ed., and Mashima D., Ed. OpenADR 2.0 Profile Specification B Profile. OpenADR Alliance, Document Number: 20120912-1, Revision Number: 1.1, November, 2015.

37 OASIS. Energy Interoperation Version 1.0 [Internet]. [cited 2023 Dec 9]. Available from: http://docs.oasis-open.org/energyinterop/ei/v1.0/cs02/energyinterop-v1.0-cs02.html#_Toc319920796.

38 W3C Foundation. Extensible Markup Language (XML) 1.0 (Fifth Edition) [Internet]. [cited 2023 Dec 7]. Available from: https://www.w3.org/TR/xml.

39 W3C Foundation. SOAP Version 1.2 Part 1: Messaging Framework (Second Edition) [Internet]. [cited 2023 Dec 7]. Available from: https://www.w3.org/TR/soap12.

40 LF Energy. OpenLEADR [Internet]. [cited 2023 Dec 7]. Available from: https://lfenergy.org/projects/openleadr.

41 IEC. IEC 62746-10-1:2018 Systems Interface Between Customer Energy Management System and The Power Management System – Part 10-1: Open Automated Demand Response [Internet]. [cited 2023 Dec 7]. Available from: https://webstore.iec.ch/publication/26267.

42 OpenADR Alliance. As of December, 2023, the Reference Implementation Repo has No Code Yet [Internet]. Available from: https://github.com/oadr3.

43 OpenADR Alliance. OpenADR 3.0: OpenADR 3.0 User Guide. Document Number: 20230706-2, Revision 3.0.0, September, 2023.

44 OpenADR Alliance. Specification Download [Internet]. [cited 2023 Dec 14]. Available from: https://www.openadr.org/specification-download.

45 ISO. ISO 8601: Date and Time Format [Internet]. [cited 2023 Dec 15]. Available from: https://www.iso.org/iso-8601-date-and-time-format.html.

46 Table 1, OpenADR 3.0: OpenADR 3.0.1 Definitions. OpenADR Alliance; pp. 20–22.

47 Wikipedia. IEC 61970 [Internet]. [cited 2023 Dec 17]. Available from: https://en.wikipedia.org/wiki/IEC_61970.

48 Wikipedia. IEC 61968 [Internet]. [cited 2023 Dec 17]. Available from: https://en.wikipedia.org/wiki/IEC_61968.

49 Wikipedia. IEC 62325 [Internet]. [cited 2023 Dec 17]. Available from: https://en.wikipedia.org/wiki/IEC_62325.

50 W3C Foundation. Web Application Description Language [Internet]. [cited 2023 Dec 15]. Available from: https://www.w3.org/submissions/wadl.

51 EPRI. IEEE-2030.5-Client [Internet]. [cited 2023 Dec 17]. Available from: https://github.com/epri-dev/IEEE-2030.5-Client.

52 SunSpec Alliance. Common Smart Inverter Profile. Version 2.1, March, 2018.

53 Figure 1, Common Smart Inverter Profile. Version 2.1; p. 4.

54 Consumer Technology Association. Modular Communications Interface for Energy Management. ANSI/CTA-2045-B, February, 2021.

55 Wikipedia. DNP3 [Internet]. [cited 2023 Dec 20]. Available from: https://en.wikipedia.org/wiki/DNP3.

56 OpenADR Alliance. Supplemental Information: EcoPort Relationship to CTA-2045 [Internet]. [cited 2023 Dec 18]. Available from: https://www.openadr.org/ecoport-supplemental.

57 Wikipedia. IEC 60870-5 [Internet]. [cited 2023 Dec 20]. Available from: https://en.wikipedia.org/wiki/IEC_60870-5.

58 IEEE. IEEE 1815.1-2015: IEEE Standard for Exchanging Information Between Networks Implementing IEC 61850 and IEEE Std 1815(TM) [Distributed Network Protocol (DNP3)] [Internet]. [cited 2023 Dec 20]. Available from: https://standards.ieee.org/ieee/1815.1/5699.

59 IEEE. IEEE 1815.2: Standard Profile for Communications with Distributed Energy Resources (DERs) using IEEE Std 1815 [Distributed Network Protocol (DNP3)] [Internet]. [cited 2023 Dec 20]. Available from: https://standards.ieee.org/ieee/1815.2/7731.

60 Hidalgo J. Controlling Multiple Outstations for Smart Buildings with DNP3 in Real-Time [Internet]. [cited 2023 Dec 20]. Available from: https://www.ni.com/en/innovations/case-studies/19/controlling-multiple-outstations-for-smart-buildings-with-dnp3-in-real-time.html.

61 CAISO. What Is a FlexAlert? [Internet]. [cited 2023 Dec 21]. Available from: https://www.flexalert.org/what-is-flex-alert.

62 WattTime. California Self-Generation Incentive Program: Greenhouse Gas Signal [Internet]. [cited 2023 Dec 21]. Available from: https://content.sgipsignal.com/api-documentation.

63 CAISO. FlexAlert [Internet]. [cited 2024 April 1] Available from https://www.flexalert.org.

64 Shepherd M, Cuffee D, and Wayland S. Market Informed Demand Automation Server (MIDAS) Documentation: Connecting to and Interacting with the MIDAS Database and Application Programming Interface. California Energy Commission; 2022. Publication Number: CEC-400-2021-009-ED2.

65 CEC. MIDAS-Documentation [Internet]. [cited 2023 Dec 24]. Available from: https://gitlab.com/CEC-MIDAS/midas-documentation.

66 Appendix C, MIDAS-Documentation [Internet]. [cited 2023 Dec 24]. Available from: https://gitlab.com/CEC-MIDAS/midas-documentation/-/blob/main/appendix-c.md.

Index

Symbols

Automating Building Energy Management for Accelerated Building Decarbonization: System Architecture and the Network Layer, First Edition. James Kempf.

Published 2025 by John Wiley & Sons, Inc.

C

D

J

K

L

M

Printed and bound by CPI Group (UK) Ltd, Croydon, CR0 4YY

17/09/2026

14964309-0008